BIG HISTORY

A JOURNEY FROM THE BIG BANG TO THE MODERN WORLD AND INTO THE FUTURE

GIUSEPPE FERRONE

ISBN: 978-0-646-82476-5

First published in Australia in 2021 by Giuseppe Ferrone

Copyright: Giuseppe Ferrone 2021

Big History – A Journey from the Big Bang to the Modern World
and into the Future.

Cover design by Ross MacLennan of www.bookcoversaustralia.com

Typesetting by www.allinonebookdesign.com.au

Author can be contacted at guiseppeferrone@outlook.com

A catalogue record for this book is available from the National
Library of Australia.

Ferrone, Giuseppe, 1959-,

ABOUT THE AUTHOR

GIUSEPPE FERRONE is a humanist and High School history teacher, with degrees in history, business and education. He lives in Perth, Western Australia, and is passionate about the important lessons we can learn from the past, and the critical thinking skills students of science and history can apply throughout their lives, for the benefit of humanity. The author welcomes your feedback by email at guiseppeferrone@ outlook.com

DEDICATION

This book is dedicated to my grandson Luca who, along with
all of the other young people born in the 21st century, will hopefully
play a significant role in the positive advancement of technology
for the benefit of all humanity, while protecting and
preserving the finely balanced ecosystems
of our beautiful fragile "pale-blue dot", planet Earth.

CONTENTS

BIG HISTORY AND CREATION STORIES

I FIRST BECAME AWARE OF BIG HISTORY WHEN I ATTENDED THE HISTORY Teachers Association of Australia annual conference at Perth College in Western Australia in 2012, where the keynote speaker was David Christian, a historian based at Macquarie University in Sydney; and the founder of the discipline and course. As I was sitting in the auditorium, surrounded by history teachers from around Australia, listening to him delivering his presentation, I had an epiphany. Here was a unique, thought provoking and entertaining multimedia course being taught to high school students around the world, which actually connected the multiple disciplines of human knowledge into one logical, continuous and seamless historical narrative, starting with the formation of the Universe and the Big Bang 13.8 billion years ago! It was our modern scientific and historical creation story. I was hooked! I wanted to learn it and teach it!

During the conference lunch, I introduced myself to David. I was impressed with his passion, enthusiasm and vision for this course. He explained to me the reason why he had conceived of the course. A specialist in Russian history who was born in Brooklyn New York, David grew up in Nigeria, was educated at Oxford University in England and then Ontario Canada, lived in Russia, and is now based in Sydney. He believes that history should not just be taught in schools in narrowly defined constructs, such as nations, ideologies, conflicts or personalities; but instead these time or idea specific historical studies should be complimented and enhanced by an overriding "big picture" view of history, where students can identify common themes, ideas and patterns from the past and apply these to the present and most importantly, into the future. When planning the course, David wanted students to go on a journey for the discovery of human knowledge, and to do this properly, he reasoned that the story should not start with the beginning of human civilizations and written sources, but instead should start with our earliest understanding of time and space; the Big Bang.

We have all shared the experience of going from one subject to another in classes at school, often drifting aimlessly and not being able to make meaningful connections with the various subjects in our timetables. What really impressed me on that day in the auditorium, while I was listening to David Christian, was the realisation that this course would give students a deep understanding of the logical connections between the different disciplines which are the product of centuries of human knowledge, but have been embedded into our education system in a disjointed manner. It wasn't always like this. Our current education system is primarily a legacy of the Industrial Revolution which started in Britain in the mid-eighteenth century and took hold worldwide during the nineteenth century. Public education became widespread during this time and its primary objective was to prepare young people with the necessary skills to participate in national narratives, ideologies, consumer economies and workforces. This was a major deviation from the original purpose of education in ancient and then late medieval-renaissance-enlightenment times, where young people, albeit those privileged children of the elite and mainly boys for that matter to begin with, were educated in a wide range of disciplines such as philosophy (which included the natural sciences), history, literature, music, oratory and drama; not so much for economic or political-ideological functional purposes, but instead to instil in them the joy of learning and the enhancement of their minds, their critical thinking skills and their world views. My inspiration for this joyful and curious approach to multi-disciplinary learning, for the sake of learning, and enhancing ourselves as thoughtful, creative, inspired and wondrous humans, is the great "Renaissance man" himself, Leonardo da Vinci.

Unfortunately, we seem to have largely lost this curiosity and passion for learning, in the early 21st century when many govern-

(1) A classroom during the later period of the Industrial Revolution.

ments are encouraging their youth to study "STEM" subjects (science, technology, engineering and mathematics). As important as these subjects are in preparing students for an increasingly complex technological world; they should not be a replacement for the humanities (history, philosophy, literature, languages, music, art and the social sciences). The STEM subjects and the humanities should be given equal prominence in an enlightened education system. It is the study of humanities that gives students the skills to become progressive critical thinkers; something that we need more and more of, in this society of fake news and misrepresentation running rampant; aided and facilitated by social media, extreme-nationalist manipulative politicians with their own self-serving agendas; and the demise of professional independent journalism and critical thinking. No wonder the seeking of truth based on reality, rational thinking, logic, common sense, non-bias, and empirical evidence, has suffered severely as a consequence.

"Count me in, I would love to teach this course!" I told David Christian at the conference. I knew that in order to introduce the Big History course into my high school, I would need to demonstrate to the school administration that it came with academic rigour, and a sufficient number of students needed to be enthusiastic about taking it as an elective. After discussing the course with my good friend and the very popular school physics teacher Bill Ellis, whom all of his students fondly address as "Bill", he too was impressed with its potential. In 2013 Bill and I organised weekly after-school classes, which would start at 3.15 and run for two hours. We provided hot chocolate and biscuits and promoted it to all of the cohorts from the thirteen years old Year 8s to the seventeen years old Year 12s. We also sent an email to all of the parents, introducing the course. The after-school sessions started with a group of around forty very enthusiastic students who had given up their own time to attend this course; not to pass a test, not to secure a particular career or job, but purely for the sheer love and joy of learning. Many of these students were not high academic achievers, but just kids who had an innate curiosity to learn more about the Universe and the place of humans within it. It grew from there as the word spread about how interesting the course was.

Microsoft founder Bill Gates was an early convert to the power of this course. After watching a series of video lectures which David Christian had recorded with the "Great Courses" US based self-learning organisation, Bill Gates approached David and expressed interest in funding the development of the Big History course into an online multi-media platform, accessible by teachers and students worldwide. This is what happened and as a result, the course is now taught to high school students in many countries around the world. An online public access course has also been developed. In 2014 I was successful in introducing the Big History course as an elective at my school in Perth, Western Australia, where I have been teaching it ever since. In collaborating, David Christian and Bill Gates both believed that a course like this would not only encourage young people to develop progressive critical thinking skills, but also it would encourage them to think globally about the challenges and opportunities facing humanity, rather than thinking parochially within the narrow constraints of their indoctrinated nationalities, ideologies, cultures and belief systems.

So what exactly is Big History? It is our modern scientific and historical creation story. Every civilization and society has a creation story. The difference between the Big History creation story and all of the others, is that Big History is based on the latest available scientific and historical evidence which is constantly being amended, challenged, debated and added to, as the boundaries of our knowledge progressively expand; whereas other creation stories are typically based on traditional mythology, passed down from one generation to another, and relatively static in nature. Nevertheless, all creation stories deserve respect and understanding because they are vitally important to the peoples living within their own unique cultures, who have faith and belief in them.

All mythology possesses the common theme of telling the story of ancestors and the origins of humans and the Universe or natural world, the gods, supernatural beings, and heroes and villains with super-human and usually god-given powers. To give readers a taste of the broad spectrum of creation stories that have existed throughout human history, I have included a selection below, just giving a brief outline of each one. Readers should note that in many instances, there is more than one version or interpretation of mythological creation stories for particular cultures and civilizations.

The ancient Egyptians had several creation myths. All begin with the swirling chaotic waters, called Nun. Amun-Ra, the supreme Sun god, willed himself into being, and then created a hill. Amun-Ra possessed an all-seeing eye. He spat out a son, Shu, god of the air, and then vomited up a daughter, Tefnut, goddess of moisture. These two gods were charged with the task of creating order out of chaos. Shu and Tefnut produced Geb, the Earth with all its natural landforms and life, excluding humans; and Nut, the sky and celestial bodies. First these two gods were entwined, but Geb lifted Nut above him. Gradually the world came into order, but Shu and Tefnut became lost in the remaining darkness. Amun-Ra removed his all-seeing eye and sent it to search for them. When Shu and Tefnut returned, thanks to the eye, Amun-Ra wept with joy. Where the tears struck the Earth, humans began to form. Some of the other prominent gods in Egyptian mythology were Anubis, who was the son of the underworld gods Osiris and Nebthet; Bastet, the cat-headed god of fertility; Horus, the falcon-headed sky god whose eyes were the Sun and the Moon; Isis, the major goddess of ancient Egypt, sister and wife of Osiris and mother of Horus, she typified the faithful wife and devoted mother; Osiris, the embodiment of goodness, who ruled the underworld after being killed by Set. The pharaohs were believed to be the reincarnation of Osiris. There were also Ptah, the chief god of the pantheon worshipped at Memphis; Sekhmet, the consort of Ptah, the craftsman god and daughter of Amun-Ra, the Sun-god; Set, the storm-god son of Geb and Nut; and Thoth, the god of reckoning, learning and writing. He was the inventor of writing, the creator of languages, the representative of the scribes, interpreters and advisers of the gods.

(2) Ancient Egyptian Mythology

In the classical Greek world, Chaos was the origin of everything and the very first phenomena that ever existed. It was a primordial void, the empty unfathomable space at the beginning of time, from which everything was created, including the Universe and all of the Greek gods, such as Gaia (mother earth), Tartarus (the underworld), Eros (love), Erebus (darkness) and Nyx (night). In the beginning, Chaos was a state of random disorder existing in primordial emptiness, and later a Cosmic Egg formed in its stomach, hatching and producing the first gods into the darkness. Chaos was a space that separated and divided the Earth, where mortal humans lived; from the sky, where the gods lived. Zeus was the Olympian god of the sky and the thunder, and was considered to be the king of all other gods and humans. He was the son of Cronus and Rhea and was unfaithful to his sister and wife, Hera. Zeus spawned many children including Athena (wisdom), Apollo (divinity), Artemis (hunting), Hermes (messenger), Dionysus (wine), Heracles (bravery), Helen of Troy (sailors), Hephaestus (fire), Hebe (youth) and Ares (war).

(3) Ancient Greek Mythology

The ancient Sumerian Mesopotamian creation myth involves a struggle against violent and destructive forces. In this story there is a struggle of the younger god Marduk, against the chaotic water gods; the male Apsu, representing fresh water and the female Tiamat, representing salt water. The other gods are the offspring of Tiamat, who is impregnated by Apsu, symbolising the deposition of silt in the river delta. In this creation story, the noisy and active younger gods upset the static tranquillity of the older Apsu and Tiamat. A cycle of violence then erupts, resulting in Marduk, the aggressive upstart, leading the gods to a final decisive battle against Tiamat. Marduk defeats Tiamat and splits her body, creating heaven and Earth. During this battle Marduk also slays Kingu, a supporter of Tiamat. Marduk then proclaims that human beings are to be created from the blood of Kingu. The moral of this Mesopotamian myth is that the human being is an insignificant part of a much larger struggle within the natural world; and that there is a monumental struggle between order and disorder. Although the Sumerian Mesopotamians worshipped hundreds of gods, there was a pantheon of seven prominent gods. These were, An, the most supreme god, believed to be the sky god and regarded as the lord of the heavens. The second major deity was Enlil, the god of the wind and storms, and the son of An and Ki. Enlil took over his father's role as king of the gods. Another important deity in Sumerian mythology was Enki, the god of wisdom, magic and incantations. Enki is also credited with the creation and protection of mankind. The most popular deity in the Sumerian civilization, based on literary texts which have survived, was Inanna, the goddess of sexuality, passion, love and war. In addition to An, Enlil, Enki and Inanna; there were three other deities that made up the seven most important gods and goddesses in the Sumerian Mesopotamian pantheon. One of them is Utu, the god of the Sun and justice, and believed to be the twin brother of Inanna. Another important deity was Ninhursag, who was worshipped as a mother goddess, associated with fertility, nature and life on Earth. Ninhursag was also the protector of pregnant women and children. The last of the pantheon of seven deities was Nanna, the god of the Moon and of wisdom. This god was believed to be the father of Inanna.

In the Zoroastrian religion of Ancient Persia, the world was created by the deity, Ahura Mazda. The great mountain Alburz, grew for 800 years until it touched the sky. From that point, rain fell, forming the Vourukasha sea and two great rivers. The first animal to be created, the white bull, lived on the bank of the river Veh Rod. However, the evil spirit Angra Mainyu, killed it. Its seed was carried to the Moon and purified, creating many animals and plants. Across the river lived the first

(4) Ancient Sumerian-Mesopotamian Mythology

man, Gayomard, he was as bright as the Sun. Angra Mainyu also killed him. The Sun purified his seed for forty years, which then sprouted a rhubarb plant. This plant grew into Mashya and Mashyanag, the first human mortals. Instead of killing them, Angra Mainyu deceived them into worshipping him. After fifty years, they bore twins, but they ate the twins, owing to their sin of false worship. After a very long time, two more twins were born, and from them came all humans.

(5) Persian Mythology   *(6) Kuba − Central African Mythology*

In the Kuba culture creation story of central Africa (now the modern day nation of the Democratic Republic of the Congo), before the Universe was created, there was only darkness and water, inhabited by the giant pale god, Mbombo, a white-coloured figure who had been ill for millions of years. The reason for his illness was his incurable loneliness. Mbombo vomited and produced the Sun, creating light and day. He then threw up a second time and created the Moon and the stars, which divided day and night. Again, he threw up and out came nine animals; a leopard, crested eagle, crocodile, fish, tortoise, heron, goat, beetle, and a scarab, a black and white cat-like animal that would eventually become lightning. The heron created more birds, the crocodile created more reptiles and serpents, and the goat more horned beasts. The fish created other fish, and the beetle created all the insects. An iguana, produced by the crocodile, made all of the animals without horns. Along with the animals, humans were finally also produced.

In the Maori culture of the Pacific Islands, creation was caused by the separation of the Earth and Sky deities, Rangi and Papa. Before they created the Universe, they were locked in an eternal embrace. They gave birth to only male children, who were trapped in the dark with nothingness between them. Eventually, their trapped children desired light, and forced themselves apart. This created space, stars, planets, humans, animals and plants for

(7)   Maori Polynesian Mythology

the Universe, which was formed as a result of a war between the god-children of Rangi and Papa that occurred after the separation.

In Viking Norse mythology, the three gods Odin, Vili and Ve, killed the primordial frost giant Ymir. His body was cut into pieces, and a

(8) Viking Norse Mythology

different aspect of the Universe emerged from each one. Odin, Vili and Ve created the Earth from Ymir's flesh, used his blood to fill the oceans, his bones and hair became the hills and trees, his brain became the clouds, his skull became the heavens, and his eyebrows formed Midgard, the realm of the humans.

In the ancient Japanese indigenous Ainu people's mythological creation story, Izanagi and Izanami glided down the rainbow-striped floating bridge of heaven. They stared into the oily, primordial ocean of chaos below. Izanagi dipped his jewelled spear into the ocean and stirred the swirling jellyfish-like mass; a glistening droplet fell from his spear point and was trans-

(9) Japanese Creation Mythology

formed into an island. Izanagi and Izanami descended to the island they created and built a tall, sacred column. Izanagi circled the column in one direction and Izanami in the other direction. When they met face to face, they married. Izanami then gave birth to the eight islands of Japan, the mountains, the seasons, the gods of land and water and all forms of nature. After also giving birth to the fire god Ho-musubi, Izanami died of a burning fever.

In the Cherokee North American Indian creation mythological story, before there was land, there was only a sky realm and a primordial ocean. All the animals lived in the sky realm and it was beginning to become crowded. They were all curious about what was beneath the water and one day Dayuni'si, the Water Beetle, volunteered to explore it. He went everywhere across the surface of the primordial ocean but could not find any solid ground. He then dived below the surface to the bottom, and all he found was mud. This mud began to enlarge in size and spread outwards until it became the land on Earth as we know it. After all this had happened, one of the animals attached this new land to the sky with four strings. Just after the land on Earth was formed, it was flat, water-logged and soft so the animals decided to send a bird down to see if it had dried. They eventually returned to the animals in the sky with a result. The land was still too wet, so they sent the Great Buzzard from Galun'lati to prepare it for them. The Buzzard flew down and by the time he reached the Cherokee land, he was so tired that his wings began to hit the ground. Wherever they hit the ground, a mountain or valley was formed. The Cherokee land still remains the same today with all the landforms that the Buzzard formed. The animals then decided it was too dark, so they made the Sun and put it on the path in which it still runs today. The animals could then admire the newly created Earth around them. Humans came after the animals and plants. At first there was only a brother and sister, until he struck her with a fish and told her to multiply; and so she did.

The Earth Mother of the South American Aztecs, Coatlicue ("skirt of snakes") is depicted in a fearsome way, wearing a necklace of human hearts and hands, and a skirt of snakes as her name suggests. The Aztec

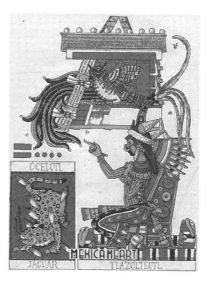

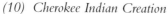

(10) Cherokee Indian Creation   *(11) Aztec Creation Mythology*

creation story starts when Coatlicue was impregnated by an obsidian knife and gave birth to Coyolxauhqui, the Goddess of the Moon, also to 400 sons, who became the stars of the southern sky. Later, a ball of feathers fell from the sky which, upon Coatlicue finding it and placing the ball in her waistband, caused her to become pregnant again. Coyolxauhqui and her brothers turned against their mother, whose unusual pregnancy shocked and outraged them. However, the child inside Coatlicue, Huitzilopochtli, the God of War and the Sun, sprang from his mother's womb, fully grown and armoured. He attacked Coyolxauhqui, killing her with the aid of a fire serpent. Cutting off her head, he flung it into the sky, where it became the Moon.

After the collapse of the Western Roman Empire in the fifth century CE, what is now the over 2,000 years old religion of Christianity gained prominence in Europe, and The Book of Genesis, first written around 3,400 years ago, took hold. It is the first book of the Hebrew Bible and of the Christian Old Testament and is an account of the creation of the heavens and the Earth by God in six days (the seventh being the Sabbath and day of rest). Genesis opens with God battling chaotic ocean waters in complete darkness, a stormy conflict without a beginning. Over the next six days, God creates everything by pushing the chaotic waters beyond the barriers of heaven and Earth. On the first day, light was created. On the second day, the sky was created. On the third day, dry land, the seas, plants and trees were created. On the fourth day, the Sun, Moon and stars were created. On the fifth day, creatures that live in the sea, and creatures that fly, were created. On the sixth day, animals that live on the land, and finally humans, made in the image of God, were created. By the seventh day, God had finished his work of creation, and rested, making the seventh day a special holy day (Sunday).

(12) *Christianity Creation in the Genesis*   (13) *Islam Creation*

In the Islamic religion which first originated in the Middle East in the seventh century CE, the Quran states that Allah created the Sun, the Moon and the planets including the Earth. Allah also created the night and the day. The Quran states that "the heavens and the earth were joined together as one unit, before We clove them asunder." Following a big explosion, Allah "turned to the sky, and it had been (as) smoke." He said to the sky and the earth: "Come together, willingly or unwillingly." They said: "We come together in willing obedience." As a result, the elements and what was to become the planets and stars began to cool, come together, and form into shape, following the natural laws that Allah established in the Universe. The first human beings, Adam and his wife Eve (in Islamic tradition called Hawwah), appear in the Quran; which states that they were created from clay, and were brought to life by the blowing of the soul into their bodies.

In Buddhism which originated in India in the sixth century BCE, space, time and nature was created naturally, has cycles, survives for a set time, then is destroyed and remade. Buddhists believe that everything depends on everything else, and everything is interconnected. Present events are caused by past events and become the cause of future events. Buddhism is human-centred and states that existence is endless because individuals are reincarnated over and over again, experiencing suffering throughout many lives. Only achieving liberation, or "nirvana", can free a human being from the cycle of life, death and rebirth. Buddhism has six realms into which a soul can be reborn. From the most to least pleasant, these are:

- Heaven, the home of the gods (devas); which is a realm of enjoyment inhabited by blissful, long-lived beings. It is subdivided into 26 levels of increasing happiness. In this realm of nirvana, life is a continual round of pleasure and enjoyment, with no suffering, anxiety or unfulfilled desires; and where human beings are rewarded for many past good deeds.

- The realm of Humanity. Although humans suffer, this is considered the most fortunate state, because humans have the greatest chance of enlightenment. In this realm, passionate and perceptive human beings experience many states of mind and have the most opportunity to free themselves from the cycle of death and rebirth, and hence progress to the heavenly realm.

(14) Buddhism Creation

- The realm of the Titans or angry gods (asuras); these are warlike human beings who are at the mercy of angry emotional impulses.
- The realm of the Hungry Ghosts (pretas); these unhappy human beings are bound to the fringes of human existence, unable to leave because of particularly strong attachments. They are unable to satisfy their craving, symbolised by their depiction with huge bellies and tiny mouths.
- The Animal realm. This is undesirable because animals are exploited by human beings, and do not have the necessary self-awareness to achieve liberation. This is a life of ignorant complacency and dullness, in which one does not look beyond avoiding pain and seeking comfort.
- The Hell realms. People here are horribly tortured in many creative ways, but not forever, only until their bad karma is worked off. This is a claustrophobic place of extreme hot or cold in which human beings cannot escape the torment of their own intense anger and hate.

In Hinduism, which also originated in India sometime between 2,300 and 1,500 BCE, it is believed that the world is created many times, over and over again, and not just once and for all. Hinduism also states that this Universe is one of many multiple universes in existence, with

other forms of life abounding and existing on different planes. In the Hindu religion, the creation story from the Vishnu Purana states that Vishnu, while laying on an ocean of milk on top of the serpent Sesha, sprung a lotus from his navel that contained the god Brahma. Having been sprung from Vishnu's navel, Brahma created all living things including humans; as well as the Sun, Moon, planets and a number of other gods and demigods.

Because of my proud and

(15) Hinduism Creation

strong connection with the country of my birth, Australia; and the up to 65,000 years indigenous Aboriginal culture being by far the longest surviving continuous culture on this planet; I am going to explore Australian Aboriginal creation mythology in somewhat more detail than the others I have mentioned above. The Australian Aborigines have their Dreamtime creation story which has existed for around 65,000 years. The Dreamtime is the beginning of knowledge, from which came the laws of existence. The Dreaming world was the original time of the Ancestor Beings who emerged from the Earth at the time of the creation. Time began in the world from the moment these supernatural beings were born out of their own eternity, from within the sacred land.

There was a time when everything was still. All the spirits of the Earth were asleep, with the exception of the Great Father of All Spirits, who was the only one awake. Gently, he awoke the Sun Mother. As she opened her eyes a warm ray of light spread outwards towards the sleeping Earth. The Father of All Spirits said to the Sun Mother: "Mother, I have work for you. Go down to the Earth and awaken the sleeping spirits. Give them forms." The Sun Mother glided down to Earth, which was bare at the time, and began to walk in all directions, and everywhere she walked, plants grew. After returning to the field where she had begun her work, the Mother rested, well pleased with herself. The Father of All Spirits came down and saw her work, but seeing that the job was unfinished, instructed her to go into the caves and wake the spirits. This time she ventured into the dark caves on the mountainsides. The bright light that radiated from her awoke the spirits, and after she left, insects of all kinds flew out of the caves. The Sun Mother sat down and watched the glorious sight of her insects mingling with her flowers. However, once again the Father urged her on.

The Mother ventured into a very deep cave, spreading her light around her. Her heat melted the ice, and the rivers and streams of the world were created. Then she created fish and small snakes, lizards and frogs. Next she awoke the spirits of the birds and animals, and they burst into the sunshine in a glorious array of colours. Seeing this, the

Father of All Spirits was pleased with the Sun Mother's work. She called all her newly-created creatures to her, and instructed them to enjoy the wealth of the Earth, and to live peacefully and in harmony with one another. Then she rose into the sky and became the Sun. The living creatures watched the Sun in awe as she crept across the sky, towards the west. However, when she finally sank beneath the horizon and darkness loomed, they were panic-stricken, thinking she had deserted them. All through the night they stood frozen in their places, thinking that the end of time had come. After what seemed to them like an eternity, the Sun Mother peeked her head above the horizon in the East. The Earth's children learned to expect her coming and going, and were no longer afraid.

At first the children lived together peacefully, but eventually envy crept into their hearts. They began to argue. The Sun Mother was forced to come down from her home in the sky to mediate the bickering. She gave each creature the power to change their form to whatever they chose. However, she was not pleased with the end result. The rats she had made, had changed into bats; there were giant lizards and fish with blue tongues and feet. The oddest of the newly formed

(16) Australian Aboriginal Dreamtime and the Creation

animals was one with a bill, like a duck, with teeth for chewing, a tail like a beaver, and the ability to lay eggs. It was called a platypus. The Sun Mother looked down upon the Earth and thought to herself that she must create new creatures, lest the Father of All Spirits be angered by what she now saw. She gave birth to two children. The son was the Morning Star and the daughter was the Moon. Two children were born to them, and these were sent down to Earth. They became the first Aboriginal ancestors. The Sun Mother made them superior to the animals, because they had part of her mind, and would never want to change their shape or form. To this day, the land is sacred to the indigenous Aboriginal peoples of Australia.

Creation stories are as old as humanity. All societies have had a need to create some meaning to the natural world around them and to their existence. Without the benefit of scientific evidence, they developed mythical creation stories which were traditionally passed down orally from one generation to another. An earthquake, storm, volcanic eruption, famine or pandemic pestilence, was often believed to be the sign of an angry god. These creation stories were very powerful and acted as the "glue" which embedded the peoples with a common purpose and within their respective cultures.

Big History is a story of increasing thresholds of complexity. This may seem at odds with the second law of thermodynamics in physics which states that the total entropy or disorder in the Universe will increase over space and time. If that's true, and the Universe is getting constantly more disordered, then why do we see ordered things like galaxies, stars, planets and complex life, like humans? In physics there is a difference between complexity and entropy. Entropy refers to the number of ways matter and energy can be re-arranged within a relatively stable system, in this case, the Universe. It is possible for something to grow in complexity and yet become less disordered at the same time. Typically, as entropy increases, disorder also increases, reaching a peak and then decreasing again. But within this large disorder are pockets of ordered complexity. This is where we will find galaxies, stars, planets and life in all of its complexity and diversity. This is how our Universe functions. In the beginning with the Big Bang, the Universe had little entropy and was very simple and homogenous. As the Universe expanded it became more complex in spite of the greater disordered entropy. At the end of its existence, the Universe will become simple again, as all of the galaxies and stars exhaust matter and energy, and disordered entropy becomes extremely high during this process of the Universe contracting and "dying."

A threshold of increasing complexity occurs when something appears for the first time, and is more complex than what existed previously. Big History identifies eight major thresholds of increasing complexity. These are the Big Bang and the formation of the Universe, 13.8 billion years ago; the first stars and galaxies, 200 million years later; the creation of new chemical elements from within the core of stars and also from dying exploding giant stars; the emergence of our solar system including the planets, 4.5 billion years ago, with a particular focus on our planet Earth; the evolution of life on Earth, originating 3.8 billion years ago; the evolution of hominines into humans starting around 7 million years ago and giving rise to the first humans, homo sapiens, around 300,000 years ago; the origins of agriculture and agrarian civilizations, around 12,000 years ago; and the modern revolution, which started approximately 250 years ago with the Industrial Revolution and continues unabated to this day.

A new threshold of increasing complexity can only arise if the previously disordered entropic combination of matter, energy, time and space is perfectly arranged in order to bring into existence, something new in the Universe for the first time. We call this perfection of order, the "goldilocks conditions." In order to understand goldilocks conditions better, I have outlined broadly, each of the major goldilocks conditions which gave rise to the eight thresholds of increasing complexity. These, of course, will be covered in greater detail in future chapters.

We are uncertain of the goldilocks conditions which created the Big Bang and gave rise to the Universe, 13.8 billion years ago. One theory states that it was quantum fluctuations within a pre-existing multiverse (multiple universe) which created the perfect conditions for the Big Bang. But the fact of the matter is; at this point in time, we just don't

(17) The Big Bang Singularity

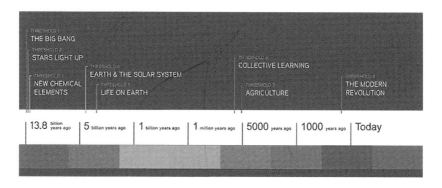

(18) Big History – Eight Thresholds of Increasing Complexity

know what created the Big Bang and what, if anything, existed before it happened. If it was "nothing," what exactly does "nothing" mean? The astrophysicist Lawrence Krauss, in his 2012 book, "A Universe from Nothing: Why There Is Something Rather Than Nothing", had this to say: "We have discovered that all signs suggest a universe that could and plausibly did arise from a deeper nothing – involving the absence of space itself – and which one day may return to nothing via processes that may not be comprehensible but also processes that do not require any external control or direction." In his book, Krauss goes on to say: "Why is there something rather than nothing? Ultimately, this question may be no more significant or profound than asking why some flowers are red and some are blue. 'Something' may always come from nothing. It may be required, independent of the under-lying nature of reality. Or perhaps 'something' may not be very special or even very common in the multiverse. Either way, what is really useful is not pondering this question, but rather participating in the exciting voyage of discovery that may reveal specifically how the universe in which we live evolved and is evolving and the processes that ultimately operationally govern our existence. That is why we have science. We may supplement this understanding with reflection and call that philosophy. But only via continuing to probe every nook and cranny of the universe that is accessible to us will we truly build a useful appreciation of our own place in the universe."

Following the Big Bang came the first stars and galaxies, approxi-mately 200 million years later. The goldilocks conditions giving rise to stars and galaxies were different variations of density and temperature in pockets of the Universe, combined with the force of gravity, a force created within a fraction of a second after the Big Bang, which gave rise to sufficiently high temperatures for fusion to occur and for stars to light up; and for the force of gravity, to hold hundreds of billions of stars together in the form of galaxies.

With the formation of stars and galaxies, came the third threshold of increasing complexity; heavier chemical elements. Out of the Big Bang came the lightest elements of hydrogen, helium and some traces of lithium. All of the other chemical elements in the periodic table originated either within the core of stars, for all elements up to and including iron; and the elements heavier than iron came from the death of giant and supergiant stars in explosions called, "supernovae." The goldilocks conditions required to create heavier elements were extremely high temperatures within the core of stars and even higher temperatures from dying supernovae stars, combined with the strong nuclear force, one of the four known forces of energy which originated within a fraction of a second of the Big Bang.

Now the pre-existing conditions existed for the fourth threshold of increasing complexity, the formation of planets. Planets are more complex than stars, because they require a greater combination of chemical elements. So the goldilocks conditions for the formation of planets were all of the chemical elements produced from stars, accumulating in clouds of gas called nebulae, out of which new star systems with planets were formed; combined with the force of gravity. Appearing for the first time were gaseous and rocky planets. In our Solar System, this process occurred approximately 4.5 billion years ago.

Next came the fifth threshold of increasing complexity, life. The goldilocks conditions for the earliest life to form on Earth around 3.8 billion years ago, were an abundance of complex chemicals including carbon, sufficient flows of energy, and the existence of water. The end result was prokaryotes, single-celled microorganisms with no nucleus, which we believe may have originated deep under the primordial oceans where volcanic vents emitted the necessary cocktail of chemicals allowing for the formation of the building block of life, DNA.

Now that life had formed on Earth, the goldilocks conditions giving rise to the process of evolution and natural selection over billions of years ultimately led to the emergence of hominines, around 7 million years ago. Hominines evolved from primates, once they left the trees and settled on the savannah of East Africa. Over the course of 7 million years, mainly due to climatic changes affecting food and other resource availability, hominines eventually evolved into Homo sapiens, us! Our ancestors first appeared in the Rift Valley of East Africa, approximately 300,000 years ago; and began a mass movement out of Africa, approximately 130,000 years ago.

The end of the last Ice Age and the warmer climate that followed, combined with an extended period of collective human learning, and a population increase which made foraging (hunting and gathering) for food increasingly difficult; created the perfect goldilocks conditions for the emergence of agriculture alongside great river systems, starting in the Fertile Crescent of Mesopotamia in the Middle East, approximately 12,000 years ago.

The eighth and final threshold of increasing complexity in Big History was the modern revolution, a time in history that began approximately 250 years ago when humans first began to exert sufficient control over the Earth's resources, to dramatically alter ecosystems and global climate. We call this period, the Anthropocene. The goldilocks conditions giving rise to this threshold were a rapid acceleration in human collective learning, particularly in science, and a need to increase our control and use of energy for technological advancement. The energy which we harnessed were fossil fuels; coal, oil and gas. This led to the rise of machines and consumer economics and ultimately to globalisation. We are still experiencing the modern revolution, albeit one that is driven largely by the internet, digital technology, genetic engineering and artificial intelligence; all which have accelerated, early in the 21st century. In fact, I will argue in a later chapter of this book that at around the middle of the 21st century, a ninth threshold of increasing complexity in the story of Big History will commence; the Technological Revolution which will culminate in the singularity; when artificial intelligence will exceed human intelligence; the foundation for which, was laid in the latter half of the twentieth century.

An interesting aspect of Big History is an attempt to predict what the future may look like, based on the "big" patterns and trends observed from the past. This future can be categorised into the future of humanity; the future of our planet Earth; and the ultimate future of our Milky Way galaxy and the Universe. We will look at these in some detail in a later chapter.

The single most significant difference between humans and all other living creatures on Earth is that we have developed sufficiently

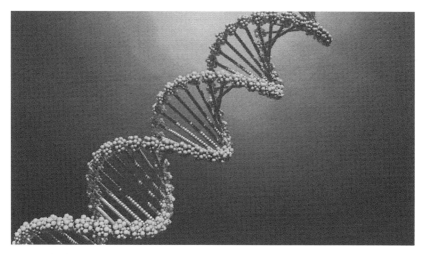

(19) From Organic DNA - originating 3.8 billion years ago ...

(20) ... to the Artificial Intelligence Singularity –
expected by the mid-21st century.

large complex brains to the point where we are self-conscious and self-aware sentient beings with the ability to engage in collective learning. As far as we are aware, no other life on Earth can learn collectively. Collective learning is the ability to pass on knowledge from one generation to another. This is the foundation of education, and the incredibly complex interconnected technology-driven global community we humans have created. Only we have the capacity to do this. It is this accumulation of collective learning over many thousands of years which has led to all human ideas, innovations, inventions and technology. You can teach your dog many tricks; but your dog cannot then teach other dogs the same tricks. This is why the human brain is the most complex thing in the Universe that we are aware of. We are about to embark on the ultimate journey; a journey of all human knowledge, accumulated over 300,000 years of collective learning. Big History is our modern scientific and historical creation story.

2

THE LONG AND WINDING ROAD TO THE BIG BANG THEORY

I T TOOK MANY CENTURIES FOR SCIENTIFIC KNOWLEDGE TO PROGRESS from a geocentric earth-centred view of the Universe, first developed in ancient times to our modern cosmological theory of the Big Bang, and an expanding Universe. In this chapter we will examine the lives and ideas of some of the greatest philosophers and scientists that ever lived, who advanced our knowledge of the Universe.

The story starts with the ancient Greeks, who began applying rational thinking and logic rather than faith-based mythology, in their attempts to understand the workings of the Universe. Thales, often called the father of Greek science and mathematics, applied philosophy to ask questions about the Universe rather than the actions of gods and demons. He provided a bridge between the worlds of mythology and the beginnings of scientific reason. Thales relied on the astronomical records of the Mesopotamian Babylonians and Egyptians before him, to accurately predict a solar eclipse in 585 BCE, for example. He also believed the Earth was flat and floated on water, like a tree-log in a vast ocean.

The Greek philosopher Aristotle, who lived in Athens from 384 to 322 BCE believed the Earth was a sphere. He studied under his mentor, the great philosopher Plato (who himself was inspired by Socrates) and later started his own school, the Lyceum. Aristotle had a geocentric view of the Universe, reasoning that the Earth was at the centre of the Universe and that the Sun, Moon, planets and all of the stars revolved in circular motions around the Earth. Aristotle's ideas were widely accepted by the Greeks during this time. The exception to Aristotle's way of thinking came with the Greek philosopher Aristarchus, who around a century after the death of Aristotle, was one of the earliest recorded believers in a heliocentric, or Sun-centred Universe. He was out of line with the thinking of the majority of other Greek philosophers at that time. In 129 BCE, Hipparchus, a prominent Greek astronomer, calculated the comparative brightness of up to 1,000 stars in the night sky. He also accurately calculated the distance between the Earth and the Moon.

Around four hundred and fifty years after Aristotle, between 127 and 141 CE, the first astronomer to attempt to construct scientific maps of the night sky was the Egyptian, Claudius Ptolemaeus, better known as Ptolemy of Alexandria. Under its Greek-Macedonian rulers at the time, Alexandria in Egypt had the largest library in the ancient world, attracting scholars from Greece and surrounding regions. At its peak

(21) Aristotle of Athens

this library is believed to have housed between 900,000 and 1,000,000 manuscripts. Tragically the original library of Alexandria was destroyed in a fire that occurred during Julius Caesar's occupation of the city in 48 BCE. Imagine how deeper our knowledge of the ancient world would have been if the manuscripts in the library of Alexandria had survived? My ancient history teacher in high school described it to me like this; he said: "Imagine looking through the key-hole of the entrance door to the school gymnasium. You would see a number of objects in the gym within a very narrow field of view, through that keyhole; that represents everything we know about the ancient world. All the objects within the large gymnasium you cannot see outside of your very narrow field of "keyhole" vision, represent everything we do not know about the ancient world, because the manuscripts and other primary sources of evidence have long been lost to us."

Ptolemy was able to combine his observations of the movements of stars with his knowledge of geometry, to predict the movements of the planets. In his manuscript, "The Almagetsi" he also claimed the Sun, Moon, planets and stars circled the Earth. He believed that each celestial body moved in epicycles (small circles) while orbiting around the Earth. Ptolemy reasoned that this explained why planets sometimes appeared to travel backwards in the

(22) Ptolemy of Alexandria

night sky. The concept of epicycles was first proposed by the Greek philosopher Hipparchus. This Ptolemic earth-centred or geocentric view of the Universe was largely accepted for around 1,500 years, until it was challenged by the Polish monk and astronomer, Nicolaus Copernicus.

In 1514, Copernicus reasoned that the Sun was at the centre of the Universe; a heliocentric view. He was not the first to have this theory. Some earlier astronomers had the same beliefs, but it was Copernicus who backed up this theory with his own observations of the movements of the planets in the sky, during the Renaissance, a time when science was gaining a greater respectability amongst educated Europeans. His heliocentric theory was highly controversial at the time because the Catholic Church, which was still very powerful and influential in medieval Europe, taught the Ptolemic view, that the Earth and mankind were at the centre of the Universe; so Copernicus' theory was considered to be heresy, against the teachings of the Church. This seems to be the reason why Copernicus did not publish his findings until two months before his death. When his theory was published in 1543, in a manuscript titled: "On the Revolution of the Celestial Spheres," it was met with great hostility and derision from the Church. There were to be three subsequent events which resulted in the eventual acceptance of Copernicus' theory; these were Tycho Brahe's precise naked-eye observations of the movements of the planets and other celestial bodies in the night sky; Johannes Kepler's three laws of planetary motion; and Galileo's use of the telescope for the first time, to directly observe planets, moons and their motions.

(23) Nicolaus Copernicus

(24) Johannes Kepler

One evening in 1572, the Danish astronomer Tycho Brahe observed what he thought was a brilliantly bright new star in the night sky, in the constellation Cassiopeia. We now know that what he had witnessed was a supernova, an exploding giant star. In 1604, a second supernova appeared. These "new visitors" to the night sky caused Brahe to seriously question Ptolemy's theory, that all of the stars had a static fixed position in the outermost sphere of the Universe. Tycho Brahe proposed a theory of the solar system, which contained elements of both the Earth-centred Ptolemaic system, and the Sun-centred Copernican system. In his theory, the other planets revolved around the Sun, but he believed the Sun revolved around the Earth. In 1609, in his capacity as an assistant to Tycho Brahe, and using Brahe's observational notes as evidence; it took the German astronomer Johannes Kepler to reinforce the Copernicus theory that all of the planets in the solar system, including Earth, revolved around the Sun. Kepler discovered three major laws of planetary motion, as follows; firstly, the planets move in elliptical orbits around the Sun; secondly, the time necessary to traverse any arc of a planetary orbit is proportional to the area of the sector between the central body (the Sun) and that arc (the "area law"); and thirdly, there is an exact relationship between the squares of the planets' periodic times and the cubes of the radii of their orbits (the "harmonic law"). Brahe and Kepler's discoveries turned Nicholas Copernicus's Sun-centred planetary system into a dynamic (not static) Universe, with the Sun actively pushing the planets around in noncircular, elliptical orbits.

In 1609, the Italian astronomer Galileo Galilei became aware of a new invention; a "spyglass" which used a glass lens that magnified the view of objects. He constructed one for himself and began to use it as a telescope to observe the heavens, magnifying the objects he observed, up to 20 times. When he turned his telescope on the planet Jupiter, he was astounded to discover four moons orbiting around the gas-giant. Galileo's observations confirmed Copernicus' and Kepler's views of the solar system; with the planets including Earth, orbiting around the Sun; and the moons orbiting around the planets. In a letter to his mentor and contemporary astronomer, Johannes Kepler, Galileo said: "I esteem myself happy to have as great an ally as you in my search for truth. I will read your work all the more willingly because I have for many years been a partisan of the Copernican view because it reveals to me the causes of many natural phenomena that are entirely incomprehensible in the light of the generally accepted hypothesis. To refute the latter, I have collected many proofs, but I do not publish them, because I am deterred by the fate of our teacher Copernicus who, although he had won immortal fame with a few, was ridiculed and condemned by countless people (for very great is the number of the stupid)." And in another extract of a letter to Kepler, he stated: "My dear Kepler, what would you say of the learned here, who, replete with the pertinacity of the asp, have steadfastly refused to cast a glance through the telescope? What shall we make of this? Shall we laugh, or shall we cry?"

The beginnings of modern science and the scientific method of gathering evidence to support theories can be largely attributed to the pioneering work of Galileo. He kept meticulous records of his observations to support his theories of planetary motion. In 1637 Galileo published a book which summarised his ideas and theories; it was titled: "Discourses and Mathematical Demonstrations Relating to Two New Sciences." The publication of his findings put him in direct conflict with the powerful Catholic Church Inquisition; and although imprisoned for his alleged heresy, he was fortunate not to have been burnt at the stake, as was the fate of his philosopher-scientist predecessor, Giordano Bruno in 1600, for his belief that the Sun was no different to the other stars, and the Universe was likely teeming with life. It wasn't until 1992 when Pope John Paul II acknowledged the Church had persecuted the great scientist, that Galileo was finally exonerated.

(25) Galileo Galilei and his telescope   *(26) Isaac Newton*

The birth of modern physics and the catalyst for the acceleration of the scientific revolution, can be attributed to Isaac Newton, who was born in 1642, the same year that Galileo died. Isaac Newton used mathematics, particularly calculus, to more accurately explain known scientific phenomena. He developed mathematical laws explaining how objects moved on Earth as well as in space. Newton attributed the movement of orbiting planets as being the result of motion along a straight line combined with the gravitational pull of the Sun. His laws were based on the idea that nothing is naturally at rest. He reasoned that all celestial bodies are constantly moving, with no limits on space and time. He was the first scientist to accurately identify gravity as a major force in the Universe. Newton published his findings in a manuscript titled "Philosophiae Naturalis Mathematica" ("Mathematical Principles of Natural Philosophy") in 1687. In this publication he outlined his three laws of motion. Firstly, that every object is in a state of uniform motion and will remain in that state of motion unless an external force acts upon it. Secondly, that force equals mass times acceleration; and thirdly that for every action there is an equal and opposite reaction. He called the "external force" gravity. His Law of Universal Gravity outlined how gravity causes planets to orbit the substantially larger-mass Sun and why the planets including Earth, have smaller moons in orbit around them. Newton also contributed to the development of reflecting telescopes which used mirrors to gather and magnify light, rather than refracting telescopes like Galileo's, which used a lens.

An astronomer who is often overlooked for her achievements was Henrietta Swan Leavitt. In 1895 she became a volunteer assistant at the Harvard Observatory, recruited as a "human computer" to catalogue the magnitude and brightness of stars. While cataloguing stellar magnitudes from photographic plates, she discovered 4 super-novas and around 2,400 variable stars. Her outstanding achievement in 1912, was her discovery that for a certain class of variable stars, called Cepheid variables, the period of the cycle of fluctuation in brightness is highly regular and is determined by the actual luminosity of the star. As a result of her period-luminosity curve for these stars; subsequent astronomers, Edwin Hubble and Harlow Shapley were able to calculate the distances of many Cepheid stars; and therefore also the distances of star clusters and galaxies in which they were observed. This paved the way for Hubble to use the data from a Cepheid variable star to determine the distance of the Andromeda galaxy; the first time the distance of a galaxy outside the Milky Way had been accurately measured. Although many of her colleagues recognised how brilliant Leavitt was, she did not receive any official honours or recognition from the scientific community other than a crater on the Moon and an asteroid being named after her. The Swedish mathematician Gosta Mittag-Leffler wanted to nominate her for the Nobel Prize in Physics in 1926, however this could not occur because she died five years earlier. I wonder how much the fact she was a woman, contributed to the lack of recognition given to her for her outstanding achievements?

(27) Edwin Hubble   (28) Albert Einstein

In 1915, Albert Einstein formulated a description of the Universe based on his Theory of General Relativity. He determined that the laws of physics are the same for all non-accelerating observers, and that the speed of light in a vacuum was independent of the motion of all observers. He also rationalised that massive objects cause a distortion in space and time, which is felt as gravity. Einstein concluded that space and time were interwoven into a single continuum known as space-time. Events that occur at the same time for one observer could occur at a different time for another. Einstein's theories and achievements paved the way for a new field of quantum physics and won him the Nobel Prize for Physics in 1921. Although scientific instruments can neither directly detect or measure space-time, several of the phenomena predicted by its warping have been confirmed since Einstein developed his theory. One phenomenon is gravitational lensing; where light around a massive object, such as a black hole, is bent, causing it to act as the lens for the matter that exists behind it. Astronomers routinely use this method to study stars and galaxies behind massive objects.

Another phenomenon which supported Einstein's theory of general relativity was the detection of gravitational waves. In 2014 astrophysicists discovered anomalies in photon light from the Big Bang that represented the first images of ripples in the Universe, called gravitational waves. This finding was direct proof of the theory of inflation, the idea that the Universe expanded rapidly in the fraction of a nanosecond after the Big Bang. But it was the collision of two black holes creating ripples in space-time, that resulted in gravitational waves actually being detected for the first time in 2015, by the Laser Interferometer Gravitational Wave Observatory (LIGO) in the United States. In 1917 Einstein applied his new theory of general relativity and the impact it had on gravity to the Universe as a whole. Initially he assumed that the Universe was static in time and possessed a uniform distribution of matter, which led him to the conclusion that rather than expanding, the Universe was static, finite and had a spherical spatial curvature. However, in an April 1931 report to the Prussian Academy of Sciences, Einstein finally adopted a model of an expanding Universe. In 1932 he teamed up with the Dutch theoretical physicist and astronomer, William de Sitter, and proposed an eternally expanding Universe which became the widely accepted cosmological model until the mid- 1990s.

A major breakthrough in our understanding of the structure and size of the Universe occurred in the 1920s with the work of Edwin Hubble. For centuries, astronomers believed that the Milky Way made up the entirety of the Universe. Hubble was the first astronomer to demonstrate that the fuzzy "cloudy" patches in the night sky when seen by telescopes, were in fact other galaxies, each containing hundreds of billions of stars, many light years away from our Milky Way galaxy. By looking at the red-shifting spectra of light emitted from these galaxies, Hubble was able to conclude they were moving away from us, evidence that the Universe was expanding. The person who is known as the "Father of the Big Bang" was Belgian astrophysicist and Catholic priest, Georges Lemaitre. In 1931 he proposed that the Universe began as a single primordial hot dense atom of energy that exploded, causing the Universe to rapidly expand. This was in effect, the foundation of the Big Bang Theory, although it was the astronomer Fred Hoyle who first flippantly coined the phrase "Big Bang" in 1949. At the same time, Georges Gamow, a Russian-American astrophysicist conceived of the Big Bang Theory as we know it today. He and his colleagues proposed that if a big bang had occurred, it would have left an "afterglow" in the form of cosmic microwave background radiation, which would still be present in the Universe today.

In 1965, astrophysicists Arno Penzias and Robert Wilson were searching for faint microwave signals from the outer regions of the Milky Way galaxy. While conducting this search they actually detected the "afterglow" of the Big Bang which had been predicted by Gamow. This cosmic microwave background radiation was important evidence that the Universe had begun with an exceptionally intensely hot big bang. In 1979, particle physicist Alan Guth completed mathematical calculations that led to the idea of "cosmic inflation," a period of rapid expansion in the early Universe. The concept of cosmic inflation addresses many questions first left unanswered with the original Big Bang Theory. It explains why the Universe is so big and why at least four different forces can be found in it today; gravity, electromagnetism, the strong and the weak nuclear forces. It also helps explain the origin of all matter in the Universe. Before the discovery of cosmic microwave background radiation provided strong evidence supporting the Big Bang as the accepted theory by most astrophysicists and cosmologists; the widely accepted theory was the Steady State Theory. This theory, developed in 1948 by British astronomers, Sir Hermann Bondi, Thomas Gold and Sir Fred Hoyle, concluded that the Universe had no beginning and no end. It describes an expanding Universe that stays in perfect balance. The balance is maintained by the continuous creation of matter from energy, according to the theory.

(29) Sir Fred Hoyle   *(30) Stephen Hawking*

In 1981, astrophysicist Stephen Hawking published a paper proposing a theory that the Universe is far less complex than what other "multiverse" (multiple universe) theories suggested. His theory was based around a concept called eternal inflation, which stated that immediately after the Big Bang, the Universe experienced a period of exponential expansion; and then it slowed down as energy converted into matter and radiation. However, according to his eternal inflation theory, some "bubbles" (or pockets) of space stopped inflating and in the process created small "dead-ends" of static (non-expanding) space. In the meantime, Hawking claimed, there were other "bubbles" of the Universe where inflation never stopped, because of quantum effects, which led to an infinite number of multiple universes. According to Hawking's theory, everything that we see in the observable Universe, is contained in just one "bubble" in which inflation has stopped, allowing for the formation of galaxies, stars and planets. Hawking explained: "The usual theory of eternal inflation predicts that globally our Universe is like an infinite fractal, with a mosaic of different pocket universes, separated by an inflating ocean. The local laws of physics and chemistry can differ from one pocket universe to another, which together would form a multiverse. But I have never been a fan of the multiverse. If the scale of different universes in the multiverse is large or infinite the theory can't be tested."

The astrophysicist Thomas Hertog, had this to say about Stephen Hawking's theory of eternal inflation: "The problem with the usual account of eternal inflation is that it assumes an existing background universe that evolves according to Einstein's theory of general relativity and treats the quantum effects as small fluctuations around this. However, the dynamics of eternal inflation wipes out the separation between classical and quantum physics. As a consequence, Einstein's theory breaks down in eternal inflation." Hawking's eternal inflation theory of the Universe was based on string theory, which attempts to reconcile Einstein's theory of general relativity with quantum theory, by replacing the point-like particles in particle physics with tiny, vibrating one-dimensional strings. String theory proposes that a volume of space can be described on a lower-dimensional boundary; so the Universe is like a hologram, in which physical reality existing in three-dimensional spaces can be reduced to two-dimensional surface projections.

In 1983, Stephen Hawking, in collaboration with physicist James Hartle, came up with the "no boundary theory" otherwise known as the "Hartle-Hawking state." They proposed that, before the Big Bang, there was no space and no time. Hawking said: "Asking what came before the Big Bang is meaningless, according to the no-boundary proposal, because there is no notion of time available to refer to. It would be like asking what lies south of the South Pole." So when the Universe began, it expanded from a single point, with no boundary. According to this theory, the early Universe did not have a boundary; allowing Hawking to come up with more reliable predictions about the structure of the Universe. His prediction was that the Universe, on the largest scales, is reasonably smooth and ultimately finite in structure. Hawking and Hartle's no-boundary proposal has fascinated astrophysicists for four decades. The proposal represented a first attempt at a quantum description of the Universe. Soon, an entire new field, quantum cosmology, emerged as researchers came up with alternative ideas and theories about how the Universe could have come from "nothing."

Today, many cosmologists are focused on the ultimate future of the Universe and understanding mysterious dark matter and dark energy. Data suggesting that the Universe is expanding at an accelerating rate was published in 1998. For more than ten years, astronomers studied the expansion of the Universe by measuring the redshift and brightness of distant supernovae and galaxies. By 1998, enough information had been gathered to lead astronomers to the surprising conclusion that the expansion of the Universe is not slowing, but instead accelerating. This increasing acceleration of the expansion of the Universe seems to be caused by an unidentified form of energy, currently called "dark energy." In addition to dark energy, approximately 85% of the mass of the Universe is made up of material that astronomers cannot directly observe, known as "dark matter." This material does not emit light or energy. The familiar material in the Universe, known as baryonic matter, is composed of protons, neutrons and electrons and makes up all of the known chemical elements in the periodic table. Dark matter may be made of baryonic or non-baryonic matter. Astrophysicists and cosmologists have deduced, in order to hold the identifiable elements of the Universe together, this mysterious dark matter must make up approximately 85% of the matter in the Universe. Potential candidates for dark matter include dim brown dwarf stars, white dwarf stars, neutron stars, supermassive black holes (found in the centre of all galaxies), weakly interacting massive particles (WIMPS) and neutralinos, which are massive hypothetical particles heavier and slower than neutrinos. So at this point in time, we don't know the form of energy that is driving the accelerated expansion of the Universe; and we don't know what 85% of the matter in the Universe, actually consists of.

It will be an exciting time when cosmologists finally discover what dark energy and dark matter are comprised of. This is one of the things that makes cosmology such an exciting career path for young scientists; it represents the cutting edge of human knowledge and relies on advanced mathematical computer modelling and cutting-edge astrophysics to formulate theories connected with the formation and current state of the Universe.

3

THE BIG BANG AND BIRTH OF OUR UNIVERSE

THE BIG BANG IS THE FIRST THRESHOLD OF INCREASING COMPLEXITY in our Big History story. It is the current widely accepted model for the formation of the Universe in which space, time, matter and energy, were created from a cosmic singularity. The Big Bang Theory states that in the 13.8 billion years since the Universe began, it has expanded from an extremely small but incredibly dense and hot primordial starting point, to the unimaginably enormous Universe that exists today. At this starting point, the estimated temperature of the infant Universe was 10^{32} degrees Celsius. The fundamental forces of gravity, strong and weak nuclear forces and electromagnetism were also formed within the first fractional second of the Big Bang; as well as the fundamental particles of quarks, electrons, photons and neutrinos.

According to the Big Bang Theory, there were a number of stages from the primordial starting point, to the Universe as it exists today. These stages, or epochs, are mainly linked to changes in the temperature of the Universe. The earliest identifiable time period was the Planck Epoch, which is the closest that astrophysics can get to the absolute beginning of time. At this starting point, the Universe had an estimated temperature of 10^{32} degrees Celsius. The second stage, also within minute fractions of the first second, was the Grand Unification Epoch, when the force of gravity separated from the three other fundamental forces, and the earliest particles (and anti-particles) were created. The third stage, also within minute fractions of the first second was the Inflationary Epoch which was triggered by the separation of the strong nuclear force. In this epoch, the Universe underwent an extremely rapid exponential expansion, known as cosmic inflation. The size of the Universe during this period of a tiny fraction of a second increased to around 10 centimetres, or the size of a plum. The particles remaining from the Grand Unification Epoch, comprising hot, dense quarks, became distributed very thinly across the Universe.

What followed next was the Electroweak Epoch, also within a fraction of the first second. As the strong nuclear force separated, particle interactions created large numbers of exotic particles and Higgs bosons. The Higgs bosons slowed down these particles and attributed mass to them, allowing a Universe that was made entirely out of cosmic radiation to support particles with mass. The next stage also within fractions of the first second was the Quark Epoch, when quarks, electrons and neutrinos began to form in large numbers as the temperature of the Universe cooled off to below 10 quadrillion degrees Celsius, and the four fundamental forces assumed their present forms. Quarks and anti-quarks annihilated each other with contact, but in a process known as "baryogenesis," a surplus of quarks (around one for every billion pairs) survived and ultimately combined to form the first chemical matter. The Hadron Epoch followed, also within

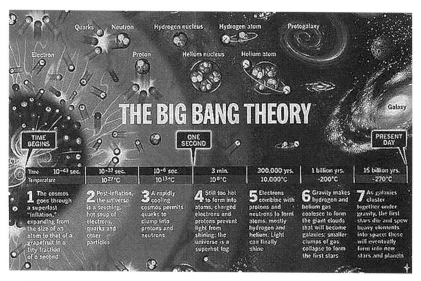

(31) The Big Bang Theory illustrated from the beginning of time to the present day

the first second of the Big Bang. During this epoch the temperature of the infant primordial Universe cooled to around one trillion degrees Celsius. This was cool enough to allow quarks to combine to form hadrons, such as protons and neutrons. Electrons colliding with protons fused to form neutrons and give off massless neutrinos, which continue to travel freely in space to this day, at or very close to the speed of light.

What followed next was the Lepton Epoch, which occurred from one second to three minutes after the Big Bang. After the majority, but not all, of the hadrons and anti-hadrons annihilated each other at the end of the Hadron Epoch, leptons, such as electrons, and anti-leptons, such as positrons, dominated the mass of the Universe. As electrons and positrons collided and annihilated each other, energy in the form of photons was freed up; and colliding photons in turn created more electron-positron pairs. What followed was the Nucleosynthesis Epoch, which lasted from three minutes to twenty minutes after the Big Bang. At this stage, the temperature of the Universe fell to around one billion degrees Celsius, a point at which atomic nuclei could begin to form as protons and neutrons; and combine through nuclear fusion to form the nuclei of simple chemical elements such as hydrogen, helium and lithium. After around twenty minutes, the temperature and density of the Universe had fallen to the point where this nuclear fusion could not continue.

The longer Photon Epoch came next, starting from three minutes and ending around 240,000 years after the Big Bang. During this long period of gradual cooling in temperature, the Universe filled with plasma, a hot, opaque "soup" of atomic nuclei and electrons. After most of the leptons and anti-leptons had annihilated each other at the end of the Lepton Epoch, the energy of the Universe was dominated by photons, which continued to interact frequently with charged protons, electrons and nuclei. The Recombination and Recoupling Epoch followed, from 240,000 to 300,000 years after the Big Bang; during which the temperature of the Universe fell to around 6,000 degrees Celsius, which is around the same as the surface of the Sun. The density of the Universe continued to fall as ionized hydrogen and helium atoms captured electrons (in a process known as recombination), therefore neutralizing their electric charge. With the electrons now bound to atoms, the Universe finally became transparent to light, making this the earliest epoch observable today. Photons were also released throughout the Universe which until the current time, had been interacting with electrons and protons in an opaque photon-baryon fluid, in a process known as "decoupling." These photons travelled freely and formed part of the cosmic microwave background radiation, which is visible to this day. By the end of this epoch, the known matter in the Universe consisted of about 75% hydrogen and 25% helium, and minute traces of helium. *lithium*

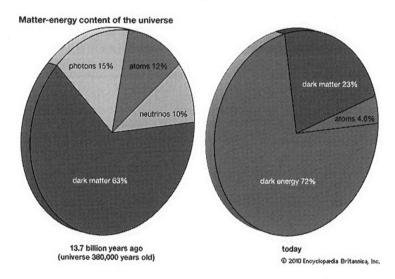

(32) *The Matter and Energy content of the Universe*

The Dark Epoch (or Dark Age) occurred next, from 300,000 to 200 million years after the Big Bang. This was a period after the formation of the first atoms but before the first quasars, stars and galaxies were formed. Although, as we have seen, photons existed, the Universe was literally dark at this time, with no stars having formed yet, to emit light. With only very thin scatterings of matter and low energy levels remaining, activity in the Universe dropped off dramatically. At this stage, the Universe was dominated by the mysterious "dark matter" of which as we have seen, we currently know very little about. Following the Dark Age was the Reionization Epoch, which lasted from around 200 million years to 1 billion years after the Big Bang. During this time, the first quasars (incredibly bright celestial objects powered by black holes) formed as a result of gravitational collapse, and the intense radiation they emitted re-ionized the surrounding Universe. From this point onwards, most of the Universe was composed of ionized plasma.

The Star and Galaxy Formation Epoch in the evolution of the Universe started 200 million years after the Big Bang. The force of gravity, amplified slight irregularities in the density of the primordial gases. These gases became increasingly dense, even as the Universe was continuing its rapid expansion. These relatively small dense clouds of gas (nebulae) started to collapse under their own gravity and became hot enough, at 15 million degrees Celsius, to trigger the process of nuclear fusion reactions between hydrogen atoms, creating the first stars. These first stars were very short-lived supermassive stars, over a hundred times the mass of our Sun. These were known as Population III (or metal-free) stars. Eventually less-massive Population II and then Population I stars also began to form, from the supernova remnants of the original supermassive Population III stars. The bigger the mass of stars the quicker they burn-out and explode in massive supernova events, when their chemical element remnants go towards the formation of subsequent generations of stars. Large volumes of matter collapses to form galaxies and gravitational attraction pulls galaxies towards each other to form galactic groups, clusters and superclusters. Today, around 13.8 billion years after the Big Bang, the accelerated expansion of the Universe and the recycling of older generation stars into new generation stars, continues. This is how our Sun was formed, from recycled earlier generation supermassive stars that exhausted their energy and exploded into supernovas.

The Big Bang Theory is supported by three major forms of evidence. Firstly, it is supported by the expansion of the Universe as deduced from the distant redshift relationship of galaxies and stars as first outlined by Edwin Hubble. If the observed redshift expansion is extrapolated in reverse, the conclusion can be reached that at some

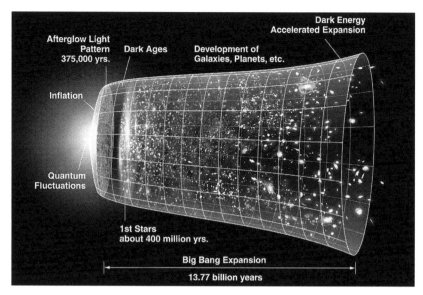

(33) Big Bang expansion over a 13.77 billion year time period

point in the distant past, all matter in the Universe must have been contained within one minutely-small starting point in space and time. Secondly, the abundance of the lightest chemical elements of hydrogen, helium and lithium is consistent with their creation in the Big Bang and not subsequently within the core of stars or exploding supernovae. Thirdly, as a result of the expansion of the Universe, astrophysicists predicted that radiation from the Big Bang would have cooled to a temperature of around 3 degrees Celsius at the present epoch in time. The detected cosmic microwave background radiation, currently permeates the Universe at a temperature of 2.725 degrees, close to its predicted temperature in the Big Bang Theory.

Although the Big Bang Theory remains the generally accepted theory for the formation and subsequent expansion of the Universe, there are a number of other theories, not necessarily dispelling the Big Bang Theory, that have circulated within the community of cosmologists, astrophysicists and philosophers in recent years. Some are more outlandish than others but I have included a selection of them here to give readers a taste of just how much the study of theoretical cosmology and astrophysics is at the cutting edge of human thinking, knowledge and sophisticated mathematical computer modelling.

The "Clashing Branes" multiverse theory contemplates our Universe being a membrane which is floating in higher dimensional space, repeatedly smashing into neighbouring universes. According to a branch of string theory called "braneworld" there are large additional dimensions of space, and although gravity can reach out to these other dimensions, we are confined to our own "brane" universe, which has only three dimensions. Neil Turok of Cambridge University and Paul Steinhardt of Princeton University have applied this theory to explain how the Big Bang could have been triggered when our Universe violently clashed with a universe in another dimension. They claim that these clashes of universes happen repeatedly, producing new "big bangs" resulting in a never-ending stream of universes in the cosmos.

In the "Evolving Universes" multiverse theory, when matter at the centre of a black hole is compressed to extreme densities, it bounces back and creates a new "baby universe." The laws of physics in this new offspring universe might be different to those that exist in the parent universe; resulting in constantly evolving universes according to cosmologist, Lee Smolin. The universes producing the most number of black holes will spawn more new universes, and so will dominate the population of the multiverse.

In the "Superfluid Space-Time" theory, space and time is actually a superfluid substance, flowing in the Universe with zero friction. If the Universe is rotating, superfluid space-time would be scattered in violent whirling masses called vortices, according to astrophysicists Paul Mazur and George Chapline. These vortices might have seeded the formation of galaxies. Mazur suggests our Universe might have been born within a collapsing star, where the combination of stellar matter and superfluid space-time could have spawned dark energy, the mysterious force that is accelerating the expansion of the Universe.

In the "Goldilocks Universe" theory, the Universe has properties that are "just right" to permit more complexity such as the formation of life; compared to a multitude of other universes that may not have achieved these "goldilocks conditions." This theory is guided by the

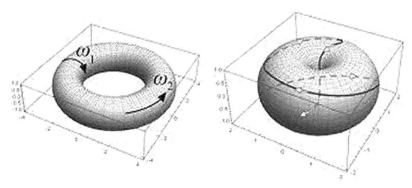

(34) The Superfluid Space-Time Theory of the Universe

Braneworld Theory
Randall-Sundrum Formulation

- **All particles and forces in our universe, except gravity, are stuck on our brane**

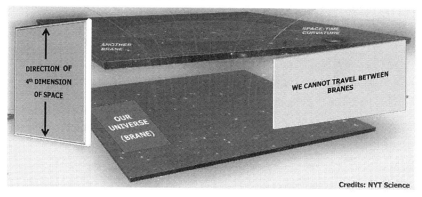

(35) The Braneworld Theory of the Universe

"anthropic principle" which states that the Universe we can see, must be hospitable, otherwise we would not be in existence as self-conscious sentient creatures, and be able to observe it. This idea has gained some strength with cosmologists because the theory of inflation does suggest that there may be an infinite number of universes in existence; and string theory suggests that each of these universes may have an infinitely diverse range of properties and laws of physics.

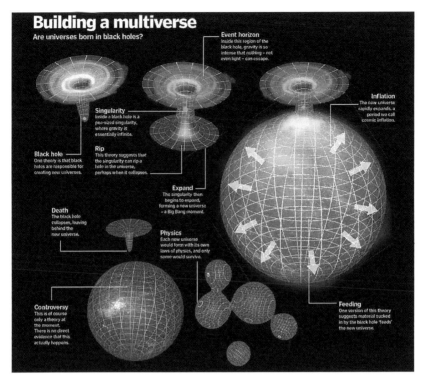

(36) The Multiverse Theory of the Universe

The "Modified Newtonian Dynamics" or "MOND" theory suggests that dark matter may actually not consist of any particles, but instead could be attributed to the "odd" behaviour of gravity. The theory states that gravity does not fade away as quickly as current theories claim it does. This "stronger gravity" may actually be the dark matter holding together galaxies which might otherwise fall apart.

The "Cosmic Ghost" theory is an attempt to resolve three mysteries of modern cosmology, in one "ghostly" presence. After making some adjustments to Einstein's General Theory of Relativity, a team of astrophysicists identified a strange substance coming from their new and revised computer mathematical modelling, which they called the "ghost condensate." It can produce repulsive gravity to drive cosmic inflation in the Big Bang; and it may also account for the current rapid acceleration of the Universe which has been attributed to unknown dark energy. It is also suggested that if this ghost condensate combines into particles of mass, it could account for dark matter.

The "Sterile Neutrinos" theory suggests that dark matter could be made of very elusive particles known as "sterile neutrinos." They are hypothetically heavier than known neutrinos and may interact with other matter only through the force of gravity, making them impossible to detect. Sterile neutrinos may also be essential in the formation of stars and galaxies.

Some theorise that our Universe may not be real at all. In the "Matrix" theory, the philosopher Nick Bostrom claims that we may be living inside a computer virtual simulation created by an extremely advanced civilisation that developed the technology to simulate consciousness. This may involve multiple universes, all simulated with differing properties and laws of physics. This theory was brought into mainstream popular culture with "The Matrix" movies; about a virtual reality universe created by advanced artificially intelligent beings first developed by humans in the 21st century. This virtual reality universe called "The Matrix" is a large extremely sophisticated computer system powered by the collective brain downloads of humans. How can we be certain that we are not all creations that exist within a sophisticated computer simulation, created by advanced intelligent beings; either developed by humans, or originating from another star system? Was the philosopher Rene Descartes right when he tried to justify the reality of our existence and independent self-consciousness by saying: "I think, therefore I am?" In the Matrix movies, Morpheus tells

(37) Renee Descartes   (38) The Matrix

Neo that human existence is just a façade, that it's not real. In the reality of the movie, humans are being "farmed" as a source of energy by a race of self-conscious advanced machines. People actually live their entire lives in pods, with their brains fed sensory stimuli which gives them the illusion of leading "ordinary lives." Morpheus explains that, up until then, the "reality" perceived by Neo is actually a computer-generated dream world; a neurological interactive simulation, called the Matrix.

We can juxtapose the "advanced computer simulation" plot of the Matrix movies, with this quotation from the 17th century French philosopher, Rene Descartes:

"Everything I have accepted up to now as being absolutely true and assured, I have learned from or through the senses. But I have sometimes found that these senses played me false; it is prudent never to trust entirely, those who have once deceived us. Thus, what I thought I have seen with my eyes, I actually grasped solely with the faculty of judgment, which is in my mind."

4

STARS, GALAXIES AND OTHER PHENOMENA IN THE UNIVERSE

We have seen that the first stars appeared in the early Universe approximately 200 million years after the Big Bang. Before this second threshold of increasing complexity in the journey of Big History, the Universe was a dark, foreboding place, devoid of any photon light emissions. It was a plasma Universe.

Stars are formed in large clouds of gas and dust called nebulae, which are the remnants of exploding earlier generations of stars. The clouds slowly contract and then begin to collapse onto a number of cores (or points) within the nebula cloud, as a result of the immense force of gravity. In the middle of these cores, it is extremely hot and dense. When a certain temperature level is reached within this core, nuclear fusion starts, and as a result, a star is born. This is called stellar ignition.

The sudden burst of energy and light created from a newly formed star blows away much of the remaining gas, but within the remaining nebula cloud, enough chemical elements exist for the formation of planets within a star (solar) system. Shortly after its formation, a star becomes relatively stable when the outward pressure from nuclear fusion is balanced with the inward pulling force of gravity; the Universe's own version of an amazingly finely tuned tight-rope.

A typical star, like our Sun, will have a life cycle of around 10 billion years, until it eventually runs out of chemical fuel. All stars in the Universe go through a life cycle in the same way that we humans do, only in the case of stars of course, the lifespan is much longer. When stars eventually do exhaust all fuel, their life will end in a number of differing spectacular ways, depending on the mass and type of star. We will examine this in more detail, later in the chapter.

What are the goldilocks conditions required for the formation of stars? For stars to be created, hydrogen and helium atoms are required. As we have seen, these elements were formed in the very early stages of the Big Bang. Sufficient variations in density and temperature in the early Universe was also a critical ingredient for the formation of stars, as well as enough gravitational force to create high enough temperature heat for nuclear fusion to occur.

The stages in the formation of a star of similar mass to our Sun, can be summarised as follows. Firstly, there will be an initial collapse of a nebula cloud which will cause it to heat up and in which protostars will form. Protostars are intense concentrations of hydrogen, helium and other gases, not yet fully formed into stars. Although cool, the protostar is very large, initially up to twenty times the diameter of the Sun. It's surface area is so great that it's overall luminosity (brightness based on energy emitted) is very high. Secondly, as the protostar radiates away its energy, gravitational collapse pulls it inwards rapidly. Its temperature rises during this process, but this is offset by a decrease in the size of the protostar, resulting in a significant decrease in its luminosity. Thirdly, once the core temperature of the protostar reaches around 10 million degrees Celsius, the process of nuclear fusion commences. At this point, the protostar becomes a star. Hydrogen fuses to form helium nuclei, releasing energy in the process. The star's surface temperature will increase significantly, compensating for its reduction in size. Finally, as the rate of nuclear fusion increases due to higher core temperature, the outward gas and radiation pressures eventually will match the inward gravitational force. As a result, the star will stabilise and achieve a state of what is called, hydrostatic equilibrium; and will then settle into the main sequence. Getting to this stage may take a few tens of millions of years.

The longest stage in the life of a star is called the main sequence. Most stars of similar mass to our Sun, are in the main sequence for around 10 billion years. Stars that are more massive than our Sun, may be in the main sequence for only around 10 million years; whereas stars

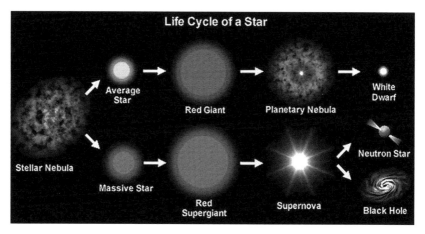

(39) The Life Cycle of a Star – Starting from a Stellar Nebula

that are less massive than our Sun may remain in the main sequence for hundreds of billions of years. Generally speaking, the larger the mass of a star, the shorter its life span is, and vice versa.

During its life span, a star will mainly burn hydrogen into energy. In the later stages of its existence the star will fuse from hydrogen to helium, then carbon, oxygen, neon, magnesium, silicon and iron; losing layers like the peeling of an onion. Iron is the heaviest chemical element created from within the core of a star. All elements heavier than iron can only be created from exploding supergiant massive stars, called supernovas. It is incredible to imagine that all of the chemical elements which gave rise to the evolution of life on Earth, including Homo sapiens, us, came from stars. We are literally made from star matter! I can empathise with the Ancient Egyptians, who worshipped the Sun-God Atum-Ra as their most supreme creator god. The astrophysicist Lawrence Krauss had this to say about the connection between stars and humans, in his 2012 book, "A Universe from Nothing: Why There Is Something Rather Than Nothing": "The amazing thing is that every atom in your body came from a star that exploded. And, the atoms in your left hand probably came from a different star than your right hand. It really is the most poetic thing I know about in physics; you are all stardust. You couldn't be here if stars hadn't exploded, because the elements – the carbon, nitrogen, oxygen, iron, all the things that matter for evolution – weren't created at the

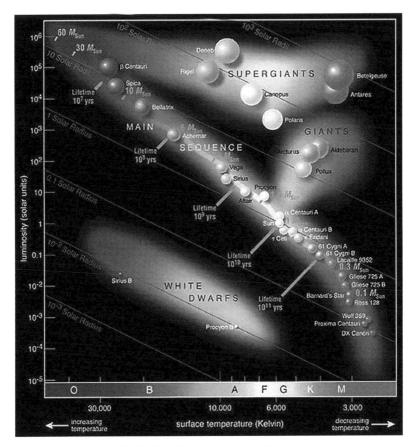

(40) The Main Sequence of Stars

beginning of time. They were created in the nuclear furnaces of stars, and the only way they could get into your body is if those stars were kind enough to explode. So forget Jesus. The stars died so that you could be here today."

Giant stars are around 10 to 100 times larger than our Sun and as they cool and expand, they glow red in colour. As the lifespan of these stars ends, the outer layers escape into space, resulting in their collapse into a white dwarf; an incredibly dense, hot dim star, smaller in size than the Earth.

Supergiant massive stars are typically over 100 to 1,000 times bigger than the Sun. With short lives of around 10 million years, when their life cycle ends, they explode into a supernova. The incredible heat and energy created from this explosion results in the formation of all chemical elements heavier than iron, including gold. It fascinates me when I think that my gold necklace, which I always wear, given to me by my mother for my 21st birthday, came from chemical elements created in an exploding supernova, around 4.5 to 5 billion years ago. Stars are element-creating factories! After exploding, a supernova will then eventually become either a neutron star, which is a small dense ball of neutrons that spins at an incredible speed; or a black hole, which contracts even more than a neutron star and is so dense that light cannot escape from it.

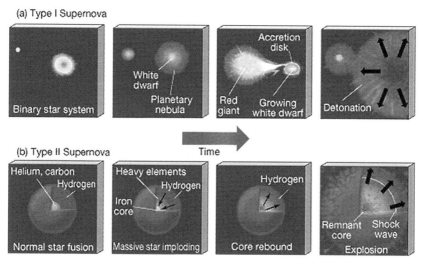

(41) Types of Supernovas

So the creation of chemical elements from within the core of stars and from exploding supernova stars, is the third threshold of increasing complexity in our Big History journey. The goldilocks conditions to create all of the chemical elements heavier than hydrogen and helium were extremely high temperatures created within the core of dying stars for all chemical elements up to and including iron; or even greater extremes of temperature created from exploding supernovas for all chemical elements heavier than iron. All of these chemical elements laid the foundation for the creation of more complexity in the Universe, such as gas and rocky planets, and all forms of life including humans.

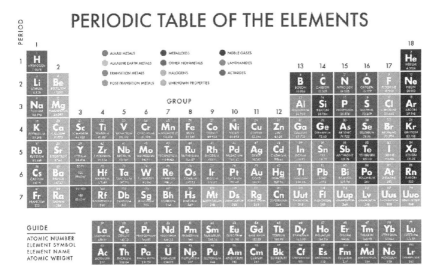

(42) The Periodic Table of the Chemical Elements

How were chemical elements so logically and brilliantly arranged into the Periodic Table of the Elements? The most significant contribution made for the orderly classification of chemical elements came from the Russian chemist Dimitri Mendeleev, who came up with the Periodic Table of the Elements. Mendeleev grew up in Siberia, Russia. He was the youngest of over a dozen siblings, and soon after his birth in 1834, ill health caused his father Ivan, a high school teacher, into retirement. Dimitri went on to study in St Petersburg, graduating in 1855 with a degree in chemistry. In 1856 Mendeleev completed a master's thesis on the relationships between the specific volumes of substances and their crystallographic and chemical properties. Shortly afterwards, the University of St Petersburg appointed Dimitri as a chemistry tutor, allowing him access to its laboratory. In 1859 he received state funding for two years of study in advanced chemistry at Heidelberg University in Germany, where he did research on several topics including surface tension, capillarity and evaporation, and intermolecular forces. In 1860 he attended a conference in Germany where the Italian chemist Stanislau Cannizzaro delivered a ground-breaking paper on atomic weights (now called relative atomic masses). This was a crucial step towards the development of the periodic table, because previously there had been a lot of disagreement amongst chemists over the assigning of atomic weights to the chemical elements. After returning to St Petersburg in 1861 Mendeleev resumed teaching at the university, while also lecturing at the city's Technological Institute.

(43) Dimitri Mendeleev came up with the Periodic Table of the Elements

In 1865 Mendeleev completed his doctoral thesis on chemical solution theory and two years later the university appointed him as a professor of general chemistry. He was required to lecture on inorganic chemistry, and since there was no satisfactory Russian textbook, he began writing one. This focused his mind on the challenge of arranging the chemical elements into some kind of orderly and logical pattern. Several other chemists, including Leopold Gmelin in Germany, Jean Baptiste Dumas in France, and John Newlands in England, had attempted to do this, all with limited success. The breakthrough for Mendeleev came in 1869 when he was preparing for an industrial tour to investigate cheese-making techniques. While on the tour, he started thinking seriously about an orderly table of the chemical elements, using a card system; recalling the process as quoted from his book, "Principles of Chemistry" published in 1905, he said: "I began to look about and write down the elements with their atomic weights and typical properties, analogous elements, and like atomic weights on separate cards, and this soon convinced me that the properties of the elements are in periodic dependence upon their atomic weights."

Mendeleev laid out his cards in columns and rows, as if in a game of solitaire, a favourite pastime of his during long railway journeys across the vast expansive landscapes of Russia, and while visiting various European countries, for lectures and conferences. The vertical columns listed the known chemical elements in order of increasing atomic weight, with a new column being started whenever this enabled him to fit elements with similar characteristics into the same horizontal row. As other chemists had previously noted, a few groups of chemical elements, in particular the alkali metals and the halogens, clearly belonged together. But many others, especially the rare earth elements, presented problems, no matter how they were arranged. Unlike his predecessors though, Mendeleev refused to give up, and he persevered. If an element in his table seemed to be placed incorrectly, he was willing to adjust its atomic weight to give it more compatible companions. For example, he proposed that the formula for beryllium oxide was BeO, rather than the accepted Be_2O_3. This lowered the atomic weight of beryllium, enabling him to locate it within magnesium rather than aluminium.

On the 6th of March 1869, the first rough sketch of his Periodic Table was presented to the Russian Chemical Society. Later that year, the society's journal published a more refined, detailed and accurate version. Even though it attracted little attention outside of Russia, Mendeleev continued to persevere with the laying out of more "element" cards on his table. The revised diagram Mendeleev published in 1871 was much closer to the final version. To compile it, he made further logical assumptions. For example, he lowered the atomic weight of tellurium, making its neighbour iodine the heavier of the two. This allowed him to place iodine within the category of halogens, and tellurium with sulphur and selenium. At that time, Mendeleev could not have foreseen that atomic number rather than atomic weight would later become the table's ordering principle, or that the identification of isotopes by mass spectrometry would eventually explain these and other anomalies. With great self-belief and confidence, Mendeleev enhanced the coherence of his table by leaving gaps for as-yet undiscovered chemical elements to complete the pattern he envisaged. In addition to predicting their chemical character, he also assigned these "gaps" notional values for physical properties such as specific gravity and melting-point. Incredibly, many of these "gaps" were filled by chemical elements discovered in later years; such as gallium, discovered by the French chemist, Paul Lecoq de Boisbaudran in 1875; and the discoveries of scandium by the Swedish chemist, Lars Fredrik Nilson in 1879; and germanium by the German chemist, Clemens Alexander Winkler in 1886.

In his private life, Mendeleev was defiantly unconventional. He had his hair cut and beard trimmed only once a year, declining to change his habits even when he met with the Czar. His family arrangements were also unorthodox. In 1862 he married Feosva Lescheva, having been introduced to her by a well-meaning elder sister of his who thought it was time he settled down. The couple had two children, but following a period of increased mutual unhappiness, they agreed to separate. Several years later, Dimitri fell in love with Anna Popov, a 17 years old art student. When Anna's parents sent her away to continue her studies in Rome, Dimitri followed her, and in 1881 the 47 years old proposed marriage. Anna accepted, but even after Dimitri and Feosva were divorced, a further obstacle remained. The Russian Orthodox Church recognised civil divorces, but demanded a seven-year interval before a subsequent marriage could occur. Nevertheless, in 1882 Dimitri found a priest willing to perform the marriage ceremony prematurely for a substantial fee. Despite their ambiguous and techni-cally bigamous situation, the couple lived happily together and raised four children. In politics, Mendeleev was also a rebellious maverick. He was an outspoken liberal who resigned his professorship in 1890 to disassociate himself from the Czarist government's harsh suppression of student protests. This controversial gesture was applauded by his students, but provoked hostility from university officials and the government.

In 1905, the Royal Society of London honoured Mendeleev with its Copley Medal, already having received its Davy Medal in 1882. In 1906, he was nominated for the Nobel Prize, but even though the chemistry panel supported his candidature, the awards committee ruled that his discovery was not recent enough to justify him for consideration. It is believed this decision was heavily influenced by the Swedish chemist Svante Arrhenius, who was a member of the Nobel committee and had clashed with Mendeleev in the past. Arrhenius apparently blocked Mendeleev's selection because he was unhappy about the Russian's longstanding and open criticism of Arrhenius's ionic dissociation theory, the idea that electrolytes dissociate in water to form ions. Almost fifty years after his death in 1907, Mendeleev was to receive the posthumous honour of having a chemical element named after him, when in 1955, physicists at the University of California bombarded element 99 (einsteinium) with alpha particles, and produced traces of a new element 101. Officially named mende-levium, this new element embedded his name as an icon, and one of the founders of modern chemistry, Of course today, everybody has recollections of studying and memorizing the Periodic Table of the Elements, in their high school science classes.

Stars are classified by their size (mass), colour, and surface temperature. Not all stars are the same colour because different chemical elements emit different spectra of colours when they burn, such as red and blue. The colour of light emitted from a star will give us some idea of its surface temperature. For example, hot stars are blue and white in colour whereas cooler stars are red and orange in colour. Astronomers call this variation in the temperature of stars, the spectral class. All stars can be classified into the spectral classes of O, B, A, F, G, K and M. The surface temperature of stars decreases as we move from O stars to M stars. The system for classifying stars is called the "Harvard spectral classification scheme" which was developed at Harvard University observatory in 1872 by Henry Draper who took the first photograph of the spectrum of the star Vega, and refined to its present form by the astronomer Annie Jump Cannon in 1924.

Let us briefly examine each of these star spectral classes. "O" class stars are the hottest and are typically dark blue/violet in colour

Spectral Classification

class	T_{eff} (K)	colour	mass ($M_\odot$)	H lines	other lines	Fraction MS stars
O	>33,000	blue	>16	weak	multiple ionised atoms	~0.0003 %
B	10,000-33,000	blue/ white	2.1-16	medium	no ionised He, neutral He instead	~0.1%
A	7500-10,000	white/ blue	1.4-2.1	strong	ionised Ca lines visible, no He, neutral metals	~0.6%
F	6000-7500	white	1.0-1.4	medium	ionised Ca stronger, metals	~3%
G	5200-6000	yellow	0.8-1.0	weak	strong metal lines	~7.5%
K	3700-5200	yellow/ orange	0.45-0.8	very weak	neutral Ca, TiO	~12%
M	<3700	orange/ red	<0.45	very weak	strong Ca, TiO, neutral metals	~76%

Oh Be A Fine Girl Kiss Me

(44) The Spectral Classification System for Stars

and have a surface temperature range of 33,000 to 100,000 degrees Celsius. These stars have an abundance of ionized helium atoms, emit high amounts of ultraviolet radiation and will typically burn out in a few million years. Examples are the stars, Zeta Orionis, Idran and Lacertra.

"B" class stars are blue in colour and have a temperature range of 10,000 to 33,000 degrees Celsius. They are also composed of an abundance of helium, more so than hydrogen. Examples are the stars, Regulus, Rigel and Spica.

"A" class stars are light blue in colour with hydrogen lines at up to maximum strength and many spectral lines caused by the presence of metallic elements. They have a temperature range of 7,500 to 10,000 degrees Celsius. Examples are the stars, Sirius, Vega and Altair.

"F" class stars are slightly hotter than the Sun and are white in colour, with high levels of ultraviolet radiation and a temperature range of 6,000 to 7,500 degrees Celsius. Examples are the stars, Procyon, Polaris and Canopus.

"G" class stars are yellow in colour. They have absorption lines of neutral metallic elements such as iron, calcium, sodium, magnesium and titanium. They have a temperature range of 5,200 to 6,000 degrees Celsius. Our Sun is a "G" class star and so are Alpha Centauri, Delta Cephei and Capella.

"K" class stars are orange in colour and are the second most common main sequence stars. They have a temperature range of 3,700 to 5,200 degrees Celsius. Examples are the stars Arcturus, Epsilon Eridani and Alderbaran.

Finally, there are the "M" class stars which are red in colour and have a temperature range of 2,000 to 3,700 degrees Celsius. They are the most common main sequence stars in the Universe and are so cool that molecules including water, carbon monoxide and titanium oxide are detectable. They typically have less than 50% of the mass of our Sun but some are actually giants and supergiants, like the red supergiant Betelgeuse, which is nearing the end of its life and will eventually explode into a fiery supernova. Located in the constellation Orion, Betelgeuse is around 1,000 times bigger than our Sun and around 640 light years away from Earth. Its brightness has been dipping to the lowest point in the past 100 years. It is expected to explode within the next million years. Even though the Earth will be safe from this explosion because of its distance; when it does explode into a supernova, Betelgeuse will outshine the full moon in the night sky for over a month. Other examples of M class stars are Antares, Mira and VY Canis Majoris.

Within each of these seven broad categories are subclasses numbered from 0 to 9. A star midway in the range F0 and G0 for example, would be classified as a F5 star. Our Sun is classified as a G2 type star.

Stars are ranked on the Hertzsprung-Russell Diagram (the H-R Diagram) according to their temperature and absolute magnitude (brightness). Most stars, including our Sun, will fall into the middle of the diagram, which is called the main sequence. Stars that are outside of the main sequence are relatively fewer in number and will mainly comprise supergiants, giants and white dwarf stars.

What happens when a main sequence star exhausts the hydrogen within its core? When this happens, nuclear fusion will stop, shutting off the outward radiation pressure. Inward gravitational attraction will cause the helium core of the star to contract, converting gravitational potential energy into thermal energy. Although fusion is no longer taking place in the core, the rise in temperature will heat up the shell of hydrogen surrounding the core until it is hot enough to start hydrogen nuclear fusion. The end result will be the production of significantly more energy than when it was a main sequence star. This "shell-burning" will cause the outer layers of the star to expand in order to maintain its required level of pressure. As the gas expands, it cools. This expansion and cooling will cause the overall temperature to drop and as a result the star will appear red in colour. Convection forces will transport the energy from the shell-burning region to the outer layers of the star. As a result, the star's luminosity (brightness) will eventually increase by a factor of up to 1,000 times. During this stage of expansion, the star becomes a red giant and will no longer be on the main sequence. This will be the fate of our Sun in around 5 billion

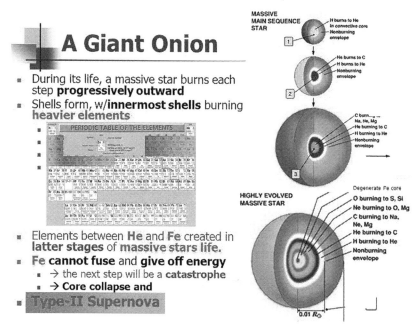

(45) The Layered Shell-Burning of Chemical Elements in Stars

years from now. It will become a red giant and will grow to have a radius of around 100 times its current size, engulfing the inner planets of Mercury and Venus. By then, due to variations in gravity caused by the fate of the Sun, the Earth will be expected to migrate to the current orbit of Mars, but may also be engulfed by the red giant Sun.

Supergiant stars such as Betelgeuse, Rigel and Antares are very prominent in our night sky and visible over vast distances in space, because of their extreme luminosities. These high-mass stars are rare and have very short lifespans compared to other lower-mass stars. Supergiant stars will consume their hydrogen at extremely rapid rates so may only survive on the main sequence for millions rather than billions of years. Once a supergiant uses up all of its fuel, the core will contract and heat up, due to the force of gravity. This will trigger helium-burning in its core. When moving off the main sequence, the overall temperature of the supergiant will drop, as its outer layers expand. This decrease in temperature will balance the increasing radius of the star, so as a result, its luminosity will remain constant. The energy released by the fusion of helium in the core will raise the temperature of the surrounding hydrogen shell, to the extent that the hydrogen will also begin the process of nuclear fusion.

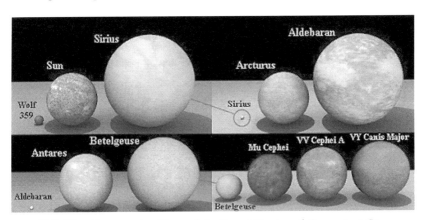

(46) The Comparative Sizes of the Sun, Giant, and Supergiant Stars

Over a period of time, the helium in the core will also be used-up, resulting in further core collapse and gravitational heating as described above. This will then trigger carbon fusion, producing sodium, neon and magnesium. Depending on the mass of the star, as each of the core fuels is extinguished, further collapses will lead to even higher temperatures, which will trigger the fusion of heavier chemical elements, right up to and including iron, for the most massive supergiant stars. The core region of a supergiant will resemble the layers of an onion, with a dense iron core surrounded by shells of silicon, sulphur, oxygen, carbon, helium and an outer shell of hydrogen. Eventually, as the temperature within the core continues to increase exponentially to around 600,000,000 degrees Celsius, the red supergiant will lose its outer layers and ultimately will be ripped apart, exploding into a supernova.

During the final destruction of the star in a supernova explosion, a significant amount of neutrons will be released as the iron nuclei in the core is ripped apart. These neutrons can be captured by many of the heavy nuclei. In this way, nuclei of chemical elements such as lead, gold and other heavier elements all the way up to uranium can be produced (synthesised) and dispersed into space, during the supernova explosion. Out of these dispersed chemical elements, a new generation of stars with orbiting planets and moons, will be formed. It is believed that our Solar System was formed from the remnants of a nearby supernova, 4.5 to 5 billion years ago.

Our Solar System is unusual in that it contains only one star, the Sun. Most star systems are binary; in which two stars orbit around a common centre of mass and the planets in the system orbiting around the two stars. Around 85% of the stars in the Universe are in binary systems. There are also star systems with three (triple) or even more stars orbiting around a common centre of mass. So if our Solar System was binary, there would be two Suns rising in the morning and setting in the evening, but not at the same time!

Stars with masses similar to our Sun will ultimately end up as white dwarfs. These stellar remnants have unusual properties. Firstly, they are very small in size. Their fuel is fully depleted so no nuclear fusion takes place and there is no outward radiation pressure to prevent gravitational collapse. More massive white dwarf stellar cores will experience stronger gravitational force so as a result, will compress even more into a smaller size. For example, a 0.5 solar-mass white dwarf star (half the mass of our Sun) has a radius of around 1.9 times that of the Earth. A white dwarf is composed mainly of carbon and oxygen ions mixed in with a myriad of degenerating electrons. It is this degeneration pressure coming from the electrons which will prevent the further collapse of the white dwarf.

It is incredible to imagine that a white dwarf with a mass equal to that of the Sun, will be compressed into a size not much greater than our Earth, giving it an extraordinarily high density. Although its surface temperature will be around 10,000 degrees Celsius, the core temperature of such a white dwarf will be as high as 100,000,000 degrees Celsius. This heat trapped within a white dwarf will gradually be radiated into space, but with its small radius, the heat will not escape quickly. It can take hundreds of billions of years for a white dwarf to radiate away all of its heat and then cool down to a black inert clump of carbon and degenerate electrons. Because the Universe is not old enough for this to have happened yet, all of the white dwarfs that have ever formed in the 13.8 billion-year history of the Universe, are still undergoing this process. White dwarf stars are believed to comprise about 10% of all the stars in the Universe and because they are so faint, they are difficult to detect from Earth. Nearby examples include Sirius B and Procyon B, both of which are found in binary star systems. They are stars that have come to the end of their lives.

As we have seen, when a massive or super massive star undergoes the collapse of its core, a supernova is the end result. After the supernova explosion, if the remaining mass of the star is less than around 3 solar masses, the ongoing collapse of its core is halted by the degeneracy pressure of its neutrons. The result is one of the most mysterious and interesting objects in the Universe, a neutron star. Neutron stars consist of degenerate neutron matter with extreme density. A bottle-cap full of this material would have a mass of around 10,000,000,000 tonnes! Neutron stars typically range in mass from 1.4 to 3 solar masses and are around 10 kilometres in diameter. So here is a really exotic object with say two to three times the mass of the Sun, jam-packed into a sphere which is the size of a small city! What is also unique about neutron stars is that they spin at an incredibly rapid rate. To give the reader an idea of just how rapid they spin; the Sun will rotate on its axis approximately once every month, whereas a neutron star can rotate on its axis many hundreds of times, every second! A particularly interesting type of neutron star is a pulsar. Pulsars are highly magnetized rotating neutron stars that emit incredibly intense beams of electromagnetic radiation from their magnetic poles.

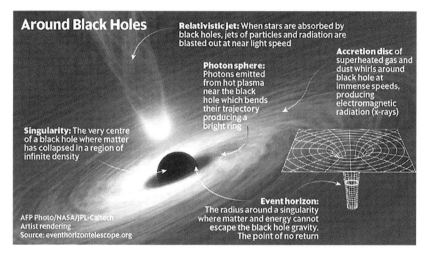

(47) The Characteristics and Structure of a Black Hole

When a star is so massive that the mass of the material left-over exceeds the limit that even neutron degeneracy pressure can withstand, then the remaining material keeps on collapsing inwards in a continuously repetitive process, until the entire mass becomes concentrated at a single point, called a singularity. The end result is a black hole. Black holes are even more mysterious than neutron stars. With all of the mass concentrated at a single point, they have extremely high gravitational fields. The reason they are referred to as "black holes" is because not even photon light can escape from them, once that light has passed a region or boundary which is known as the event horizon. At the event horizon, the escape velocity of the light equals the speed of light. With no light escaping from them at any waveband, black holes are understandably difficult to observe. Rather than directly observing them, astronomers were able to detect the presence of black holes because of the effect they had on surrounding matter in space. There is much we still don't know about black holes, but much of what we do know is thanks to the cosmologist Stephen Hawking. He helped give more solid mathematical equational backing to the concept of black holes, the existence of which was predicted by Albert Einstein's 1916 Theory of General Relativity. "Hawking actually proved some rigorous mathematical theorems about Einstein's equations for gravity that showed that, under quite general circumstances, there were places where the equations broke down – what are called singularities," said Tom Banks, a professor of physics and astronomy at Rutgers University in the United States. "And in particular, the region inside of a black hole is such a singularity", he said. But it was Hawking's investigation of the nature of black holes which would prove revolutionary. Initially, his work suggested that a black hole could never get smaller; specifically, that the surface area of its spherical event horizon, the point beyond which nothing can escape, could never decrease. Similarly, the second law of thermodynamics holds that the "entropy" or disorder, of a closed system can never decrease. In the early 1970s, physicist Jacob Bekenstein explicitly connected the concepts, proposing that a black hole's entropy is linked to the area of its event horizon. Hawking was originally sceptical of this idea; after all, entropy and black holes didn't seem to go together; black holes were supposed to radiate no energy of any sort – hence the name – and you can't have entropy without radiation. But then Hawking crunched the numbers in a way that nobody had done before; and showed that if you added quantum mechanics to the concept, you could show that in fact black holes were not really black at all; they actually emitted radiation. Hawking stated the radiation came from virtual particles which were constantly popping into and out of existence within the bizarre quantum realm. They did so in matter and antimatter pairs, one of which had positive energy, and the other negative energy. Ordinarily, these pairs immediately annihilate each other, but if this pair-popping occurred at the boundary of a black hole's event horizon, one particle could theoretically get gobbled up, while the other rocketed off into space. If the negative energy particle were consumed, the black hole's mass would shrink by a tiny amount, and the object would emit a miniscule amount of radiation. Hawking came up with this theory in 1974, which is why this hypothesized black hole light is known as Hawking radiation. On the 10th April 2019, the world was treated to something unprecedented – the first ever image of a black hole. The image captured the supermassive black hole at the centre of M87, a supergiant galaxy in the Virgo constellation; leaving no doubt about the existence of black holes.

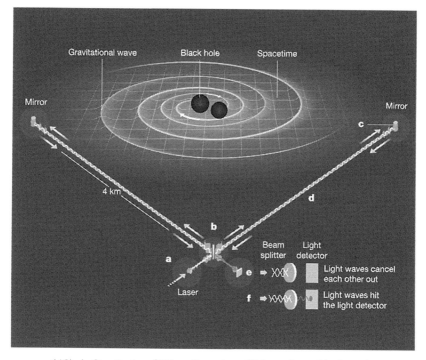

(48) A Gravitational Wave from the collision of two Black Holes

Gravitational waves are "ripples" in space-time caused by some of the most violent processes in the Universe. Albert Einstein predicted the existence of gravitational waves in 1916 in his General Theory of Relativity. The strongest gravitational waves are produced by the most cataclysmic events in the Universe, such as the collision of black holes, supernovas and colliding neutron stars. Gravitational waves are also believed to be caused by the extremely rapid rotation of neutron stars; and the earliest remnants of gravitational radiation created from the creation of the Universe in the Big Bang. The existence of gravitational waves was first detected on 14th September 2015, coming from the collision of two black holes, 1.3 billion light years away.

Let us now turn to the formation and types of galaxies. Galaxies are vast groupings of billions of stars and other material bound together by the significant force of gravity. They are often found in clusters which can include dozens to hundreds of galaxies, within each cluster. Cosmologists and astronomers do not yet have a complete understanding of how galaxies actually formed, but we do know that they were formed early in the history of the Universe, and that they have evolved over time, even colliding with each other! Most galaxies are between 10 billion and 13.4 billion years old.

The most widespread theory for the formation of galaxies is the "hierarchical" or "bottom-up" model. In this model, the Universe contains an abundance of cold dark matter. In this case, the "cold" refers to the average speed of the dark matter particles, being relatively slow compared to the speed of light. The existence of cold dark matter created slight differences in the density of the early primordial Universe. These irregularities have been detected in the cosmic microwave background radiation originating from the Big Bang. As the Universe expanded, these slight irregularities and the resulting gravitational instability caused gas clouds of nebulae to collapse, forming extremely high-mass stars. These stars merged into clusters of stars on a continuous basis, eventually giving rise to the first galaxies.

The galaxy formation process has not stopped in our constantly evolving Universe. Small galaxies are frequently gobbled up by larger ones. Our Milky Way galaxy may contain the remains of several smaller galaxies it has swallowed-up during its 13.5 billion-year lifetime. Even now, the Milky Way is in the process of digesting at least two smaller galaxies. The merger of galaxies is a common event because the Universe is crowded with galaxies which interact gravitationally. For example, the Milky Way spans about 100,000 light years from one end to the other; and the nearest major galaxy to us, the larger Andromeda galaxy, is around 2.5 million light years away. This means

(49) The Andromeda Galaxy – the closest galaxy to our Milky Way

the distance between these two galaxies is only about 25 times greater than the sizes of the respective galaxies, which doesn't leave a lot of vacant space in between them. These two galaxies are predicted to collide in around 4.5 billion years from now! There is so much space in between individual stars however, that even when galaxies collide, it is very rare for actual stars to collide with each other, but instead, the massive force of gravity will result in the two galaxies merging into one larger galaxy.

Galaxies are classified into three major categories; elliptical, spiral and irregular. These galaxies span a wide range of sizes, from dwarf galaxies containing as few as 100 million stars to giant galaxies with billions and trillions of stars. Elliptical galaxies comprise around one third of all galaxies and vary in shape from nearly circular to very elongated "cigar" shapes. They possess comparatively little gas and dust, instead they are comprised mainly of older stars and relatively few stellar formation nebulae of gas clouds. As a result, there are few new stars being formed in elliptical galaxies. The largest and rarest of these galaxies, are called giant ellipticals and are up to 300,000 light years across. Much more common are dwarf ellipticals, which are only a few thousand light years wide.

Spiral galaxies appear as flat, blue-white disks of stars, gas and dust with yellowish bulges in their centres. These galaxies are divided into two groups; normal spirals and barred spirals. In normal spirals, the spirals of stars originate from the central bulge, like tentacles of an octopus; whereas in the barred spirals, a bar of stars runs through the central bulge. Spiral galaxies are abundant with actively forming stars and comprise the largest proportion of all the galaxies in the Universe. Our Milky Way and the nearby Andromeda galaxies are both spirals.

Irregular galaxies have very little clouds of gas or dust and are neither elliptical or spiral in shape; instead, they comprise a variety of irregular shapes. These galaxies were abundant in the early Universe, before elliptical and spiral galaxies developed.

(50) The Three Types of Galaxies in the Universe

We have seen that an essential element in the formation of galaxies is dark matter. We currently don't know much about dark matter. In the late 1970s, astronomer Vera Rubin made the surprising discovery of dark matter. She was studying how galaxies spin when she realized the vast spiral Andromeda galaxy seemed to be rotating in a strange manner. In an apparent violation of the Newtonian laws of physics, the material at the edges of the galaxy was moving just as fast as the material near the centre; even though most of the mass in the galaxy was concentrated at its centre. It appeared that some extra "non-visible mass" given the name of "dark matter," appeared to be holding the galaxy together. She soon discovered that this dark matter was present in all galaxies.

Even though to this day, we don't actually know what dark matter is; we do know that it comprises around 85% of all matter in the Universe. The remaining 15% consists of the known chemical elements of the periodic table. We know it's there because its mysterious yet invisible presence affects how stars move within galaxies, how galaxies attract each other and how matter originally clumped together to form the first galaxies in the early Universe.

We humans have great difficulty comprehending the sheer unimaginable vastness of the Universe because of the understandable limitations of our brains. Let me give some context. If we could travel in a spacecraft at the speed of light, which equates to around 300,000 kilometres per second, it would still take us just over 4 years to get to the closest star, other than our Sun, Alpha Centauri! This is the closest star! In our proverbial back yard! I have often looked up at the night sky on a cloudless summer night, away from the city lights, and seen the centre of our majestic Milky Way spanning across the southern hemisphere sky, with myriad stars twinkling like diamonds. Each one of those stars have planets orbiting around them, some gaseous and others rocky, just like our Earth.

There are estimated to be 200-250 billion stars in our Milky Way galaxy alone. Our galaxy is just an ordinary spiral in a Universe which is estimated to have over two trillion galaxies! Each one of these galaxies have an average of hundreds of billions of stars, like our Sun. Some of the larger galaxies have trillions of stars! The nearest galaxy to our Milky Way, the spiral Andromeda is estimated to have one trillion stars. So although we don't have proof of the existence of intelligent life elsewhere in the Universe, I find it very hard to believe that with over two trillion galaxies, teeming with stars, there aren't other rocky planets just the right distance from their stars, to harbour life.

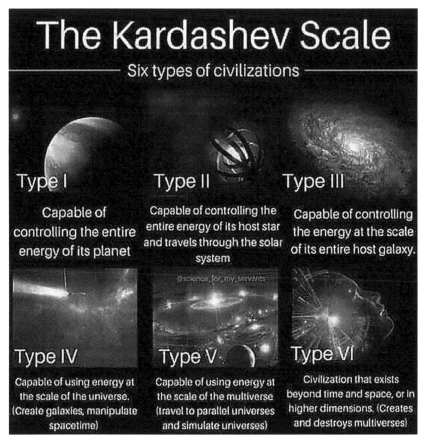

(51) The Kardashev Ranking for Advanced Civilizations

Why haven't we made contact, if intelligent life exists elsewhere in the Universe? Firstly, as we have seen, the Universe is incredibly vast in distance and scale. Even travelling at the speed of light, it would take many years for electromagnetic transmissions or even alien visitors, to make contact with us. Secondly, it may be that intelligent beings are aware of us and for whatever reason they have chosen either not to be detected even though they may be amongst us, or more likely, they have not visited Earth. Thirdly, when it comes to humans making contact, we just don't have the level of technology yet. We are still a techno-logically infant planet. Take our form of space travel, for example. We are using conventional rocket power for space exploration, where the maximum speed our spacecraft travel is around 50,000 kilometres per hour. At this speed, it would take around 78,000 years to get to the nearest star outside of our solar system, Alpha Centauri. Finally, it may be the case that when an alien civilization gets to a certain relatively advanced level of technology it destroys itself either with sophisticated weapons of mass destruction (think nuclear weapons), or destroys its planetary ecosystems (thing global warming), or consumes all of its resources (think overpopulation and food/water shortages). Is this the trajectory of our human civilization? I hope not. One of the purposes of the Big History course taught to high school students around the world is to inspire future leaders to think and resolve problematic issues globally, not nationally.

It is my deep hope that we will make contact with intelligent civilizations from another planet during my lifetime. If not, this amazing milestone in the history of humanity may be enjoyed by my children or grandchildren. From a pure mathematical probability point of view, is it logical that amongst the over two trillion galaxies in this vast Universe, each with hundreds of billions of stars with planets orbiting around them, that we are the only intelligent life form in existence? I think not. I believe the Universe is teeming with all forms of life, including intelligent life within a myriad of diverse civilizations at differing stages of technological development; some behind us and others far ahead of us.

One of my favourite movies which depicts first contact with an intelligent alien civilization is the movie "Contact" made in 1997 and inspired by the Carl Sagan book by the same name. Before I talk about the movie let me briefly explain who Carl Sagan was. He was one of the most famous astronomers of the 1970s and 1980s who fascinated audiences around the world with his television series "Cosmos" in which he laid out the wonder and vastness of the Universe in all of its chaotic violence and glory; and speculated on the purpose of life and humanity living on this beautiful "pale blue dot" and whether we are alone. He was a brilliant science communicator and inspired many people, like me, to take an active lifelong interest in science in general and the Universe in particular.

In the movie "Contact" astrophysicist Ellie Arroway, played by Jodie Foster, races against time to interpret a message originating from the

Vega star system. When the alien message is eventually deciphered, she is astounded to find it contains the engineering blueprint to construct an incredibly sophisticated advanced machine capable of transporting a human through a wormhole to the outer edges of our Milky Way galaxy. After dealing with military agencies, religious fanatics, and terrorists along the way, the machine is eventually constructed from funds provided by a mysterious billionaire who is dying of cancer and lives in a space station orbiting the Earth, because the zero-gravity environment reduces the rapid spread of his cancer. Ellie travels in this machine and makes first contact with intelligent beings who have been using this process over a long period of time, when the civilizations have progressed to a sufficient stage where they are ready for first contact. When Ellie returns to Earth, the whole experience only lasted a "few earth seconds" so nobody believes her incredible story of first contact, other than her close friend and spiritual advisor.

In spite of the enormous mathematical odds, if it turns out we are the only intelligent beings in the Universe, that in itself would be an absolutely amazing and incredible thought to ponder. All of the magnificent majesty of the Universe just for us and us alone, to enjoy and explore?

(52) A scene from the movie: "Contact" – based on a Carl Sagan book

5

THE SOLAR SYSTEM AND THE FORMATION OF PLANETS

Around 4.6 to 5 billion years ago, a nearby supergiant star exploded into a supernova and released an abundance of rich chemical elements into space in the form of gas and dust. Over time, these elements led to the formation of our Solar System, 4.5 billion years ago. Solar systems could not be established without the fourth threshold of increasing complexity in our Big History story, the formation of planets. For planets to form, a rich diversity of chemical elements is required; elements that can only be produced within the core of stars and from exploding supernovas. This chemical material needed to accrete (accumulate) over long periods of time, and combined with the force of gravity, spherical planets were formed.

As a result of the process of accretion, cosmic dust and rocks were lumped together to form particles; then small spheres, then bigger spheres, then tiny planets called planetesimals. As the planetesimals became larger, their number decreased; and the number of collisions between remaining planetesimals and asteroids also decreased. What emerged from this fourth threshold were rocky and gaseous planets, new astronomical bodies with more physical and chemical complexity and the potential to generate even more chemical complexity, such as life, in all of its diverse forms and characteristics.

The formation of our Solar System with our Sun, planets and moons gave rise to the fourth threshold of increasing complexity in our journey of Big History. As we have seen, the formation of planets is a more complicated process than the formation of stars, like our Sun. In order for planets to form, an abundance of diverse chemical elements is required. These can only come from exploding supernova stars. This is why we know our Sun is at least a second-generation star, meaning one which has been formed from the remnants of a previously existing star. Because the Universe is 13.8 billion years old and our Sun is 4.5 billion years old, it is likely the Sun is a third or even earlier generation star.

The perfect goldilocks conditions which led to the formation of our Solar System included the diversity of chemical elements, as previously mentioned, which were bound together by gravity into balls of matter orbiting the newly forming Sun. Around 99.8 percent of all the chemical elements originating from the exploding supernova and within the solar nebula, in the form of gas and debris went toward the formation of our Sun. The remaining 0.2 precent went towards the formation of the planets, moons and other celestial bodies in our Solar System.

The emerging Solar System initially formed a solar nebula, a spinning, swirling disk of gas and debris. At the centre of this solar nebula, gravity pulled more and more of this material inward. Eventually the temperature and pressure in the core was so great that hydrogen atoms

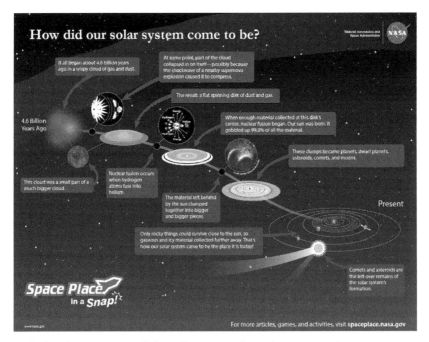

(53) The Formation of the Solar System from the remains of a Supernova

began to combine and form helium, and in the process released a significant amount of energy. As a result, nuclear fusion occurred and our Sun was born, eventually amassing more than 99% of all available matter.

Chemical matter and elements further away within the disk were also clumping together. These clumps constantly and violently smashed into one another, forming larger and larger objects through a gravitational process known as accretion. Some of these clumps grew big enough in order for gravity to shape them into spheres, which eventually became planets, dwarf planets and moons. In other cases, planets did not form and instead this matter remained as asteroids. The asteroid belt in between Mars and Jupiter is an example of this; a planet which never formed. Other smaller leftover pieces developed into comets and meteoroids.

The heavier rocky chemical elements and matter within the solar disk could withstand the heat emitting from the young Sun and as a result accreted into the four inner planets; Mercury, Venus, Earth and Mars. These are the terrestrial planets with solid rocky surfaces. At the same time, the lighter gaseous chemical elements and ice debris settled in the outer reaches of the newly formed Solar System. Ultimately the force of gravity pulled these lighter chemical elements into the two gaseous giants; Jupiter and Saturn; and the two ice giants Uranus and Neptune.

A Rocky Body Forms and Differentiates

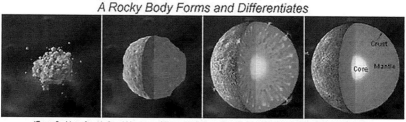

(From Smithsonian National Museum of Natural History - http://www.mnh.si.edu/earth/text/5_1_4_0.html)

(54) The Formation of Rocky Planets by the Process of Accretion

In addition to the eight planets in our Solar System, there are over 150 known moons. Mercury and Venus are the only planets that do not have moons orbiting around them as far as we are aware. Jupiter and Saturn have by far the most moons in orbit around them. In addition to the moons, there are five identified dwarf planets in our Solar System; these are Pluto, Ceres, Haumea, Eris and Ceres. There are believed to be many more dwarf planets in the outer Solar System that are yet to be identified. A dwarf planet is a smaller planet that although also in orbit around the Sun, has not cleared the gravitational attraction of other objects within the proximity of its orbit, unlike the eight larger planets.

Let us now examine further the major celestial objects in our Solar System, starting with our Sun. As we have seen, the Sun is a yellow category G star located in the centre of our Solar System. The Sun has a magnetic field that spreads throughout the Solar System in the form of solar wind. It is the Sun's gravity that holds the Solar System together, keeping all of the planets and other objects like asteroids and comets in its orbit. It is the interconnection between the Earth and the Sun that is responsible for the climatic seasons, ocean currents, weather, radiation belts, aurorae and of course, all life on Earth. As we have seen, the Sun is just an average star, similar to the trillions and trillions of other like-stars in the Universe. Nevertheless, it would take 1.3 million Earth's to fill up the Sun. The Earth is 150 million kilometres away from the Sun. Within the Solar System, distance is measured in Astronomical Units ("AU"). One Astronomical Unit is the equivalent average distance between the Earth and the Sun. It takes approximately 8 minutes for the light emitted from the Sun to

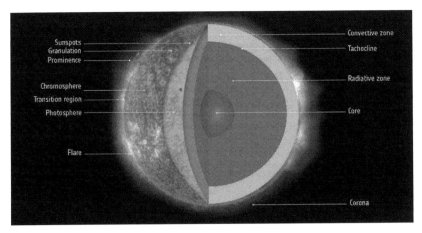

(55) The Composition of the Sun – Its Interior and Exterior

reach the Earth; so in theory, if the Sun was to explode, we would not know about it until 8 minutes later! In fact, every time we look at the stars in the night sky, we are looking back in time. So, for example, when we observe the closest star, Alpha Centauri, which is around 4 light years away, we are observing the star as it existed 4 years ago, because this is the amount of time in light years, for the light from this star to reach the Earth. Over 98% of all the chemical elements making up the Sun consist of hydrogen (71%) and helium (27%) with traces of mainly oxygen, carbon, nitrogen, silicon, magnesium, neon, iron and sulphur, making up the remaining 2%.

Let us now examine each of the eight planets, starting with the planet closest to the Sun, Mercury. It is the smallest planet, only slightly larger than the Earth's moon. Mercury has no significant atmosphere so

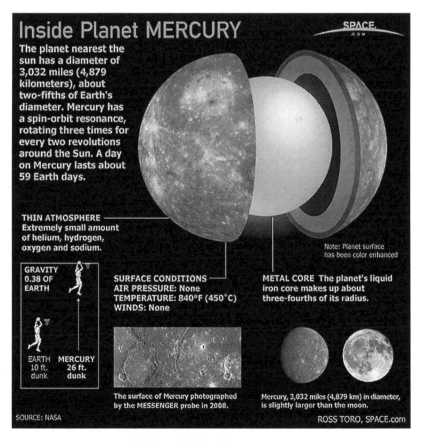

(56) Inside the Planet Mercury

as a result, it is littered with craters from asteroid and meteorite impacts over its lifespan. Being so close to the Sun, the surface temperature of Mercury can reach 450 degrees Celsius during the day and then with no atmosphere to trap the heat, plummets to minus 170 degrees Celsius at night! Mercury completes an orbit around the Sun every 88 Earth days, during which it can be as close as 47 million kilometres and as far as 70 million kilometres away from the Sun. Even though it is the closest planet to the Sun, in 2012, the NASA Messenger space-craft discovered traces of water ice in craters close to the planet's north pole, where regions may be permanently shielded from the heat of the Sun. Astronomers are unsure if it was comets or meteorites that delivered the ice to the planet, or whether water vapour may have emitted from the planet's interior and then frozen at the poles.

The second most distant planet from the Sun is Venus, of similar size, mass, density, gravity and composition, to the Earth. It is approxi-mately 80% the size of the Earth. Venus has a metallic iron core, and a molten rocky mantle with a thickness of 3,000 kilometres. The planet's crust is made up mostly of basalt and is estimated to have a thickness which varies between 10 and 20 kilometres. Although further away from the Sun than Mercury, Venus is in fact the hottest planet in the Solar System. This is because the planet's dense atmosphere effectively traps much of the heat bouncing off the surface of the planet in an extreme version of the greenhouse effect that is warming the Earth. As a result of this process, the temperatures on Venus can be as high as 471 degrees Celsius, more than hot enough to melt lead! The atmosphere of Venus consists mainly of carbon dioxide and clouds of sulphuric acid with small traces of nitrogen and water. It is an atmosphere heavier than any other planet, resulting in a surface pressure on Venus which is over 90 times greater than the Earth. This is the equivalent pressure that exists 1,000 metres underneath the oceans of the Earth.

Astronomers believe that in the distant past, Venus may have been habitable with a wide variety of lifeforms, which rapidly became extinct due to the extreme greenhouse effect on the planet as described above. If this is the case, we do not know what triggered the extreme changes in the climate of Venus. Could Earth's fate in the distant future, become

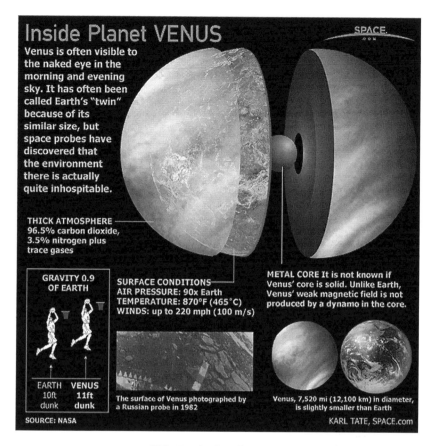

(57) Inside the Planet Venus

similar to that of Venus, if humans do not take drastic action to reverse the greenhouse effect giving rise to alarming levels of mainly carbon di oxide and methane induced global warning on our planet? Is there a tipping point, where not only does the greenhouse effect become irreversible but also rapidly accelerates?

Approximately two-thirds of the surface of Venus consists of flat, smooth plains covered with thousands of volcanoes, many still active, ranging in size from 1 to 240 kilometres in width, with lava and magma flows carving long winding canals that are up to 5,000 kilometres in length. Six mountainous regions make up the remaining third of the planet. The largest of these mountain ranges, the Maxwell Range, is around 870 kilometres long and reaches up to a height of over 11 kilometres. It takes Venus 243 Earth days to rotate on its axis, compared to 24 hours, on Earth. This is by far the slowest of any of the planets in the Solar System. Because it takes Venus 225 Earth days to complete an orbit around the Sun, we would expect one of its Earth days to be longer than its Earth year! However, because of Venus' unique retrograde rotation on its axis, the actual time from one sunrise to the next is around 117 Earth days. The average distance Venus is from the Sun is just over 108 million kilometres.

(58) The Venus Express Spacecraft – launched by the European Space Agency

The Venus Express spacecraft operated by the European Space Agency (ESA) orbited Venus between 2005 and 2014, finding evidence of continuous lightning on the planet, forming within clouds of sulphuric acid, unlike the lightning on Earth which forms within clouds of water vapour. This lightning on Venus is unique within our Solar system. Astronomers are particularly fascinated by the existence of lightning on Venus because it is possible that electrical charges created by lightning could contribute towards the formation of molecules needed to trigger life; a process some scientists believe helped to initiate life on Earth. A massive continuously fluctuating cyclone system has also been observed on Venus, with chemical elements constantly breaking apart and reforming. In August 2014, when the ESA's Venus Express completed its mission, the Earth-based controllers deliberately plunged the unmanned spacecraft into the outer-layers of Venus' atmosphere for a month. Venus Express survived this dangerous manoeuvre and was then moved into a higher orbit of the planet where it spent a further few months, before running out of fuel and finally burning up in the atmosphere of Venus.

Of course, the third rocky planet from the Sun is our beautiful home, the Earth. It is the only planet known to have an atmosphere which contains an abundance of oxygen, oceans of water on its surface, and life. Earth is the fifth largest planet in our Solar System. Our planet has a diameter of around 13,000 kilometres and has an average distance of around 150 million kilometres from the Sun. The core of our planet is comprised mainly of iron and nickel, which is responsible for the Earth's magnetic field that helps deflect harmful radiation coming from the Sun. Above the core is the mantle, which is not

(59) The Earth Rising from the Moon

completely solid but instead flows slowly. Earth's crust floats on the mantle. It is convectional energy originating from the mantle which causes the movement of continental and oceanic plates, creating earthquakes, volcanoes and the formation of mountain ranges, particularly at the plate boundaries. The Earth's atmosphere is approximately 78% nitrogen and 21% oxygen with traces of carbon di-oxide and other gases. Our planet has one moon orbiting around it. In the next chapter we will examine in detail the formation and history of our unique planet Earth and how the Moon was formed.

As we move further away from the Sun, the next planet is Mars. The reddish colour of this planet is due to the presence of an abundance of iron-rich minerals embedded within the rock and dust covering its surface. Mars is home to both the highest mountain and the deepest, longest valley in the Solar System. The highest mountain Olympus, is around 27 kilometres high, three times the height of Mount Everest. The Valles Marineris system of valleys, named after the Mariner 9 space probe that discovered it in 1971, has a depth of 10 kilometres and extends over 4,000 kilometres, close to the equivalent width of Australia. Mars also has the largest volcanoes in the Solar System, with Olympus itself, around 600 kilometres in diameter, being one of them. Channels, valleys and gullies exist all over the surface of Mars which seems to suggest that at some point in its history, liquid water may have flowed across its surface. Astronomers believe that water may still lie in cracks and pores within underground rocks. A 2018 expedition by the European Space Agency's "Mars Express" space craft suggested that salty water below the Martian surface could contain a considerable amount of oxygen, which may indicate the existence of microbial life on the planet. During the same mission, what looked like a body of underground water around 20 kilometres wide, was detected. Later in that same year, Mars Express also identified a huge icy zone in the planet's large Korolev crater. Large deposits of what appear to be finely layered stacks of water ice and dust extend from both of the planet's poles. All of these discoveries hold hope for the existence of primordial life forms underneath the surface of the planet.

Being further away from the Sun, Mars is much colder than the Earth. The average temperature on the surface of the planet is minus 60 degrees Celsius, although it can fluctuate from minus 125 degrees near the poles to a mild plus 20 degrees at midday near the equator. The carbon di-oxide rich atmosphere of Mars is around 100 times less dense than that of the Earth, but is still thick enough to support a weather system of clouds and winds. The dust storms on Mars are the largest in the Solar System, capable of engulfing the entire planet, and can last for many months. The average distance from Mars to the Sun is around 230 million kilometres. The two irregular shaped moons of Mars, Phobos and Deimos are much smaller than our Moon and appear to be made up of carbon-rich rock mixed with ice, dust and loose debris.

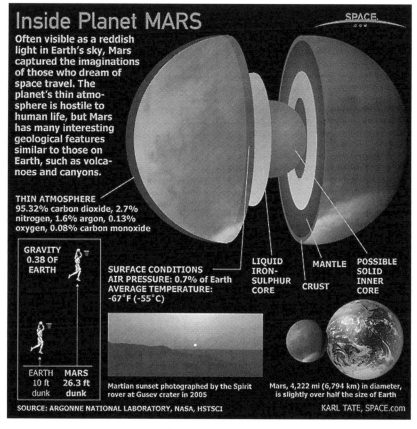

(60) Inside the Planet Mars

Efforts are underway to send human missions to Mars at some point in the not too distant future. NASA has determined that a manned mission to the red planet should be feasible by the 2030s. Elon Musk, the founder of Space X has also expressed an ambition to send a manned mission to Mars. The single biggest challenge of a manned mission will be the length of time it takes to reach Mars and then return to Earth. It would be a round trip of up to 21 months; 9 months to get there, at least 3 months on Mars and then a further 9 months to return to Earth. This means that astronauts would be exposed to harmful solar radiation and microgravity for this extended period of time, with devastating effects on the human body. Research is being conducted on the International Space Station in permanent orbit around the Earth, to study the medical impacts of these harmful longer term radiation exposures to humans.

(61) The NASA 2021 Mars Perseverance Mission Rover

The 2015 film "The Martian" starring Matt Damon is a realistic depiction of the challenges humans would face on a manned mission to Mars. Set in the not too distant future of 2035, the crew of the NASA spacecraft Ares III expedition lands safely on the eerie arid red landscape of Mars. Whilst exploring the red planet the team of astronauts is hit by a large unpredicted dust storm and are forced to abort their mission and return to Earth. During the evacuation, one of the crew members, Mark Watney, is hit by a projectile from the storm and stranded on the planet. Thinking he has died, the remaining crew leave the planet without him. Watney is in fact very much alive and recovers from his injuries. He must now come up with a plan to survive what he believes will be the next four years on the planet before the next scheduled NASA Mars mission arrives. Being a botanist, he manages to survive by cultivating potatoes using rehydrated human solid waste as fertiliser; creating water by extracting hydrogen from the hydrazine in the rocket fuel; and scavenging parts from the debris of previous missions. When NASA learns he is alive they devise a rescue plan which involves Watney driving the Mars exploratory roving land vehicle on a seven-month journey across the surface of the red planet from his home base habitat to the launch pad built for the next arriving mission in four years. When he arrives, sitting at the launch pad is an already-assembled Mars Ascent Vehicle (MAV), which is the rocket that will be used to transport the future crew from the surface of the planet to their orbiting spacecraft. NASA's ambitious plan is to have Watney fly this MAV into space and dock with the Hermes spacecraft which includes some of his earlier crew members, and has been sent by NASA to rescue him.

Located between Mars and Jupiter is the Main Asteroid Belt. Scattered in orbits around the Sun, are pieces of rock debris left over

(62) A Scene from the 2015 Movie: The Martian

from the formation of the Solar System, 4.5 billion years ago. This belt contains millions of asteroids, most of which are relatively small, ranging from the size of boulders to significantly larger ones. One of the questions astronomers have asked, is whether the Main Asteroid Belt is the remains of a planet that never formed. However, according to NASA scientists, the total mass of this belt is less than that of our Moon, not enough for accretion to occur and a planet to be formed. Instead, scientists believe it was the powerful gravitational attraction of nearby Jupiter that prevented this rock debris from accreting into and as part of the formation of the relatively nearby rocky planets of Mars and Earth.

Most of the asteroids in the Main Belt are made of rock and stone, but a small proportion contain iron, nickel and carbon-rich materials. Some of the more distant asteroids contain ice, with evidence suggesting some of this ice may contain water. There are more than 16 asteroids in the Main Belt that we are aware of, with a diameter greater than 240 kilometres. The largest asteroids, Vesta, Pallas and Hygiea are each over 400 kilometres in diameter. At over 950 kilometres in diameter, Ceres is around a quarter the size of our Moon and is classified as a dwarf planet, like Pluto.

In 2007, NASA launched a mission called "Dawn" to visit Ceres and Vesta. Dawn arrived at Vesta in 2011 and remained there for over a year, taking extensive photographic images of the asteroid, before moving on to reach Ceres in 2015. Dawn discovered that the inner Solar System's only dwarf planet is an ocean world where water and ammonia reacts with silicate rocks. The unmanned space probe will remain in orbit around Ceres until the end of its mission. Meanwhile the Dawn mission has delivered valuable data to NASA about the structure and composition of asteroids and dwarf planets; including hints of organic material identified on the icy Ceres.

The first of the gas planets in the Solar System is Jupiter, the biggest planet of all, being twice the mass of all the other planets combined. It is the equivalent of over 1,300 planet Earths. Another way of contextualising this is that if Jupiter were the size of a basketball, Earth would be the size of a grape. Jupiter has a dense core, the composition of

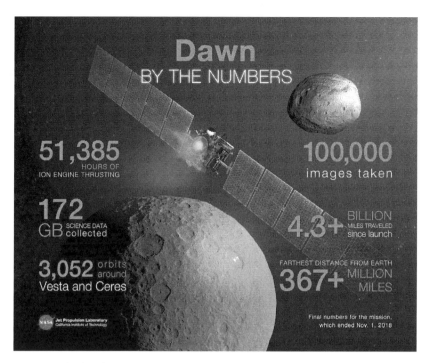

(63) The NASA Dawn Mission to the Main Asteroid Belt

which remains uncertain; but this core is surrounded by a helium-rich layer of fluid metallic hydrogen which extends out to up to 90% of the diameter of the gas giant planet.

The atmosphere of Jupiter resembles that of the Sun, being made up mostly of hydrogen and helium. In fact, many astronomers believe that Jupiter is a failed star, not achieving sufficient mass and core temperature for fusion to occur, when the Solar System was being formed over 4.5 billion years ago. The colourful light and dark bands that surround Jupiter are created by strong east–west winds in the planet's upper atmosphere travelling at speeds in excess of 540 kilometres per hour. The white clouds in the light zones are made up of crystals of frozen ammonia. The most extraordinary feature of Jupiter is its Great Red Spot, a giant hurricane-like storm that has lasted for over 300 years. At its widest, the Great Red Spot is around twice the size of the Earth, and its edge spins counter-clockwise at its centre at speeds of 680 kilometres per hour. The colour of this storm, which varies from brick red to slightly brown, appears to come from

(64) The Planet Jupiter - captured by the Hubble Space Telescope

traces of sulphur and phosphorous in the ammonia crystals in Jupiter's clouds. The temperature in the clouds is around minus 145 degrees Celsius; although near the planet's centre, the temperature of around 24,000 degrees Celsius, is hotter than the surface of the Sun.

Jupiter has a massive magnetic field, the strongest of all the planets and around 20,000 times stronger than the Earth's. The giant planet also rotates on its axis at a faster rate than any of the other planets, taking only a little under 10 hours to complete a turn on its axis, compared to 24 hours for Earth. This rapid spin has created a bulge at the planet's equator and flattened the poles. Jupiter's average distance from the Sun is over 778 million kilometres, over 5 times the distance from the Sun compared to the Earth. It takes Jupiter 11.8618 Earth years to complete an orbit around the Sun. Jupiter and Saturn are sometimes referred to as the "vacuum cleaners" of the Solar System because their

(65) Jupiter's four largest moons – compared to our Moon

massive gravitational attraction plays a major role in steering wayward asteroids away from the inner rocky planets, including our Earth.

This planet has more moons than any of the others orbiting around it, with 79 known moons; the four largest, all with rocky surfaces, being Ganymede, Io, Europa and Callisto; first observed by Galileo Galilei in the 17th century. Ganymede is the largest moon in our Solar System and is larger than Mercury and Pluto. It is also the only moon known to have its own magnetic field. It has at least one ocean between layers of ice and may contain several layers of ice and water. Ganymede will be the main target of the European Space Agency's "Jupiter Icy Moons Explorer" (JUICE) unmanned spacecraft which is scheduled to launch in 2022.

Io is the most volcanically active celestial body in the Solar System. The sulphur its volcanoes emit gives Io a yellow-orange appearance. As this moon orbits Jupiter, the planet's enormous gravitational attraction causes tides on Io's solid surface that rise up to 100 metres in height, generating enough heat to create volcanic activity.

The frozen crust of Europa is made up mostly of water ice and may conceal an underground liquid ocean that contains twice as much water as our planet Earth does. Some of this water ice spouts from the surface in sporadic plumes at Europa's southern pole. NASA is planning a mission to explore this icy moon, scheduled to launch later this decade; the unmanned space probe will be expected to perform up to 45 flybys around this moon to detect any evidence of possible life.

Callisto has an icy surface covered by craters of various shapes and sizes. Data gathered by the Galileo spacecraft indicates Callisto may have an ocean up to 250 kilometres below the surface. If there is an ocean of water, there may be a possibility of life.

When NASA's Voyager 2 spacecraft explored Jupiter in 1979 it detected three bands of rings around the planet's equator; each being much fainter than the rings of Saturn. Seven unmanned space exploration missions have flown past Jupiter; these are Pioneer 10 (1973) and 11 (1974), Voyager 1 (1977) and 2 (1977), Ulysses (1992), Cassini (2000) and New Horizons (2007). Only two missions to date have orbited Jupiter; these are NASA's Galileo (1995-2003) and Juno (2016) missions.

The twin Voyager 1 and 2 spacecraft are continuing on their more than 40-year journey since being launched in 1977. In August 2012, Voyager 1 left the solar system and made the historic entry into interstellar space, the region between stars, filled with material ejected by the death of nearby stars, millions of years ago. Voyager 2 entered interstellar space in November 2018. Both spacecraft are still sending scientific information about their surroundings back to Earth through the Deep Space Network (DSN). Both of the Voyager spacecraft are carrying a Golden Record which consists of 115 analogue-encoded phonographs, greetings in 55 languages, a 12-minute montage of sounds on Earth, and 90 minutes of music. This information about humanity serves as a time capsule in case the spacecraft are examined by an intelligent civilization during the journey through interstellar space. The following is a selection of the items on the record:

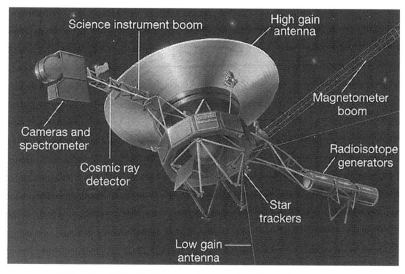

(66) The NASA Voyager Spacecraft – Launched in 1977

- A silhouette drawing of a human male and a pregnant female.
- A series of illustrations mapping out the complex structure of DNA.
- Images demonstrating human eating, drinking and licking.
- Images of Olympic sprinters.
- The Taj Mahal, chosen as an example of impressive architecture.
- The Golden Gate Bridge as an example of structural engineering.
- An excerpt from a book to give a glimpse of our written language. It is a page from Sir Isaac Newton's book, "System of the World", describing the means of launching an object into orbit.
- Recorded greetings in 55 languages; including the final greeting in English, from then 6-year old Nick Sagan, the son of astronomer Carl Sagan, where he said; "Hello from the children of planet Earth."
- A recorded greeting from a whale.
- A recording of a kiss planted by a man on the cheek of a woman.
- A recording of human brain-waves that sounded like "a string of exploding firecrackers" according to Carl Sagan's wife, Ann Druyan.
- The Georgian Chanting Chorus, "Tchakrulo."
- Chuck Berry's song, "Johnny B. Goode."

The next planet in our Solar System is the majestically ringed Saturn, the sixth planet in distance from the Sun and the second largest. Saturn is a gas giant made up mostly of hydrogen and helium. The ringed planet has the least density of all the planets and the only one less dense than water; which means if there was a bathtub full of water big enough to hold it, Saturn would float! The yellow and gold bands seen in the atmosphere of Saturn are the result of superfast winds in the upper atmosphere which can reach speeds of up to 1,800 kilometres per hour around its equator. Saturn completes a rotation on its axis every 10.5 hours; a high speed spin that causes it to bulge at its equator and flatten at the poles. It takes 29 Earth years for Saturn to complete an orbit around the Sun, and it has an average distance of 1.4 billion kilometres from the Sun.

(67) The planet Saturn captured by the Hubble Space Telescope

The medieval astronomer Galileo Galilei was the first person to observe the rings of Saturn in 1610. The rings of Saturn in fact consist of billions of particles of ice and rock debris, ranging in size from a grain of sugar to the size of a house. These particles are believed to be the left-over debris from comets, asteroids, shattered moons and dwarf planets. The main rings are typically only around 9 metres in thickness, although the NASA Cassini-Huygens spacecraft revealed vertical formations of ice up to 3 kilometres in height.

Saturn has 62 known moons. The largest, Titan, is slightly larger than Mercury and is the second largest moon in the Solar System after Jupiter's Ganymede. A number of Saturn's moons have unusual features such as Pan and Atlas, which are shaped like flying saucers; or Lapetus which has one side as bright as snow and the other as dark as coal. Enceladus appears to have "ice" volcanoes and a hidden ocean that is gushing out water and other chemicals from over 100 geysers spotted at the moon's southern pole.

Titan is the largest of Saturn's moons and the only moon in the Solar System known to have a significant atmosphere. Nitrogen and methane extend around the moon 10 times as far into space as Earth's atmosphere, sometimes falling back onto the surface as methane rain. Its atmosphere makes Titan one of the best potential candidates for the existence of life. This moon hosts many hydrocarbon-filled lakes as well as extremely tall mountains. Another of Saturn's moons, Dione, is believed to have a dense rocky core surrounded by water-ice. This moon hosts a thin oxygen atmosphere and may have a liquid ocean beneath its surface. Rhea is a heavily cratered moon and lacks a core at its centre. Instead, the entire moon is composed of ice, with traces of rock, causing it to resemble a dirty snowball. Even though Rhea is the second largest moon of Saturn, it is only around half the size of the Earth's moon.

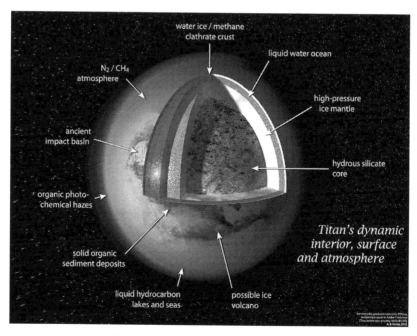

(68) The Structure of Titan — Saturn's largest moon

The first spacecraft to reach Saturn was Pioneer 11 in 1979 which flew within 22,000 kilometres of the ringed planet. Images from this mission confirmed to astronomers the presence of the two outer rings and a strong magnetic field. The Cassini spacecraft, launched in 1997, was the largest interplanetary spacecraft ever built and was specifically launched to orbit around Saturn. As well as collecting much data about Saturn, Cassini also carried the Huygens probe, which penetrated into Titan's atmosphere and successfully landed on the surface of this moon. The Cassini mission concluded in 2017, and when low on fuel, the spacecraft was deliberately crashed into Saturn in order to avoid the slight possibility of the craft crashing into and contaminating one of Saturn's moons.

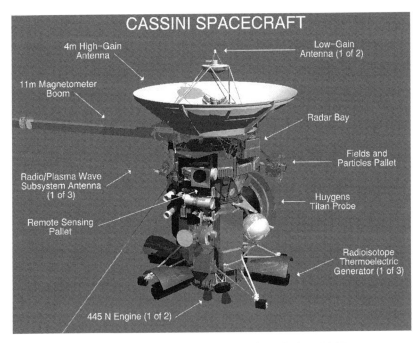

(69) NASA's Cassini spacecraft – launched in 1997

If we continued further afield from Saturn, the next planet we would encounter is Uranus. This planet is blue-green in colour as a result of the existence of methane mixed into its mainly hydrogen and helium atmosphere. Uranus is often referred to as an ice giant, because at least 80% of its mass comprises a fluid mix of water, methane and ammonia ice. Unlike the other planets in the Solar System, Uranus is tilted so far on its axis that it effectively orbits the Sun on its side. This extreme orientation may have been caused by Uranus colliding with a planet-size object, possibly twice the size of the Earth, soon after it was formed in the early stages of the Solar System. Uranus has the coldest atmosphere of any of the planets, even though it is not the most distant from the Sun. The reason for this is because Uranus has virtually no internally generated heat to supplement the heat from the Sun.

The atmosphere of Uranus comprises 82.5% hydrogen, 15.2% helium and 2.3% methane. Its internal structure is made up of a mantle of water, ammonia and methane ices; and a core of iron and magnesium silicate. Uranus' average distance from the Sun is 2.9 billion kilometres and it takes 84 Earth years for this planet to complete an orbit around the Sun. Uranus has 27 known moons, the largest of which are Oberon and Titania. In 1986, Voyager 2 visited Uranus and discovered 10 additional moons, all with a diameter of between 26 and 154 kilometres.

Interior Structure

• Both Uranus and Neptune may have
   – A rocky core
   – A mantle of liquid water and ammonia
   – An outer layer of liquid hydrogen and helium
   – A thin layer of atmosphere

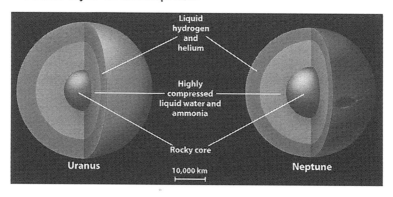

(70) The Interior Structure of Uranus and Neptune

The most distant of the planets in our Solar System is Neptune. Its cloud cover has a bright blue colour that is mainly due to the absorption of red light by the presence of methane in its mostly hydrogen and helium rich atmosphere; although the bluish colour also appears to be caused by an as yet unidentified compound also in the atmosphere. Neptune has a rocky core which is roughly equal to the Earth's entire mass. The wind system in the planet's atmosphere can reach speeds of up to 2,400 kilometres per hour, the fastest detected in the Solar System. By studying the cloud formations on this gaseous-ice giant planet, astronomers have been able to calculate that a day on Neptune lasts for just under 16 Earth hours. Neptune has an average distance from the Sun of 4.5 billion kilometres and completes an orbit around the Sun every 165 Earth years.

The atmosphere of Neptune comprises 80% hydrogen, 19% helium and 1% percent mainly methane and traces of other elements. This planet has a magnetic field which is 27 times more powerful than the Earth's. By mass, the overall composition of the surface of Neptune is 25% rock, up to 70% ice and around 5% hydrogen and helium. Internally Neptune possesses a mantle of water, ammonia and methane ices and a core of iron and magnesium silicate. Neptune has an unusual ring system that is not uniform and possesses bright thick clumps of dust called arcs. The rings are believed to be relatively young and short-lived. NASA's Voyager 2 was the first and as yet only spacecraft to visit Neptune in 1989. It was Voyager 2 that discovered Neptune's rings and 6 of its moons.

Neptune has 14 known moons of which only Triton is spherical; and the other 13, irregularly shaped. Triton is unique because it is the only large moon in the Solar System to circle its planet in a direction which is opposite to its planet's rotation. This "retrograde orbit" suggests that Triton may once have been a dwarf planet which Neptune "captured" gravitationally. Triton is extremely cold, with temperatures on its surface reaching around minus 235 degrees Celsius; making it one of the coldest places in the entire Solar System. Nevertheless, when Voyager 2 flew by, it detected geysers emitting icy matter up to 8 kilometres from the surface, which appears to indicate it has a warm interior. Astronomers believe there may be a subsurface ocean on this icy moon.

In addition to the planets, moons and asteroids, there are also comets to be found in our Solar System. Comets are icy bodies in space that release gas or dust. They consist primarily of dust, ice, carbon dioxide, ammonia and methane. Astronomers believe that comets are leftovers from the material which initially formed in our Solar System, 4.5 billion years ago. Some scientists think comets may have originally brought the water and organic molecules to Earth that led to the formation of life on this planet. Although comets orbit around the Sun, most inhabit an area far beyond the orbit of Pluto, known as the Oort Cloud. Occasionally a comet will make a spectacular entry into the inner Solar System; some do so regularly and others only once every few centuries.

The solid core or nucleus of a comet consists mainly of ice and dust coated with dark organic material; with the ice composed mainly of frozen water, but also other frozen substances such as ammonia, carbon dioxide, carbon monoxide and methane. The organic compounds found

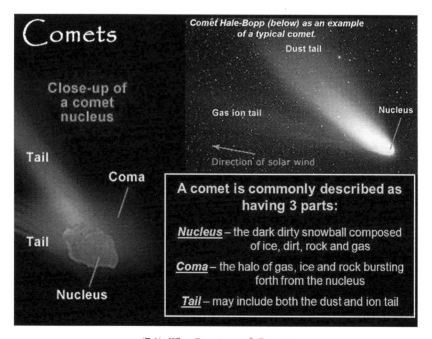

(71) The Structure of Comets

on comets, including combinations of carbon, hydrogen, nitrogen and oxygen, are the building blocks of life; giving rise to a theory that life on Earth may have originated from these organic compounds delivered by comets. The nucleus will often contain a small rocky core. As a comet moves closer to the Sun, the ice on the surface of the nucleus is vaporised by the Sun's heat, turning into gas and forming a cloud known as the coma. Radiation from the Sun pushes dust particles away from the coma, forming a long dust tail. Since the tails of comets are shaped by sunlight and the solar wind, they always point away from the Sun. Although comets and asteroids appear similar in their make-up, the difference is mainly the presence of a coma and tail in comets. The nuclei of most comets have a diameter of less than 16 kilometres; but their comas can reach a width of 1.6 million kilometres and the tails can be up to 160 million kilometres long.

Astronomers classify comets based on the duration of their orbits around the Sun. Short-period comets are those that take less than 200 years to complete an orbit around the Sun; whereas long-period comets are those that take longer than 200 years. There are also single-apparition, or hyperbolic comets that are not bound to an orbit around the Sun and therefore often originate from outside of our Solar System. Hyperbolic comets fly through our Solar System at high speed before heading out to interstellar space, never to return. In recent years, astronomers have also discovered comets located within the Main Asteroid Belt, between Mars and Jupiter. It is possible that these comets may have been a source of water for the inner rocky planets, including our Earth. Short-period comets mainly originate from a disk-shaped band of icy objects known as the Kuiper Belt, located beyond the orbit of the planet Neptune. It is believed that gravitational interactions with the four outer gaseous planets drags these bodies into the inner Solar System, when they then become active comets. On the other hand, long-period comets are believed to originate from the nearly spherical Oort Cloud, at the outer edge of our Solar System, which get slung inward by the gravitational attraction of nearby passing stars. Some comets, called sun-grazers, will get so close to the Sun that they will either collide with it, or break up and evaporate.

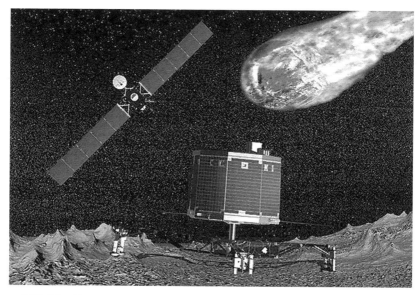

(72) The European Space Agency's Rosetta Spacecraft – launched in 2004

Most comets are named after the people who first discover them. For example, Comet Shoemaker-Levy 9 was given this name because it was the ninth short-periodic comet discovered by Eugene and Carolyn Shoemaker and David Levy. Those comets first discovered by spacecraft are given the name of the mission attached to the spacecraft. A number of recent missions have explored comets. The most prominent missions included NASA's Deep Impact mission, where an unmanned space probe collided into Comet Tempel 1 in 2005 and recorded the dramatic explosion, which revealed the interior structure and composition of the nucleus. In 2009, NASA announced that samples the Stardust mission collected from the surface of Comet Wild 2, revealed an organic substance which is one of the building blocks of life. In 2014, the European Space Agency's Rosetta spacecraft entered orbit around Comet Churyumov-Gerasimenko. The Philae lander was released by Rosetta and touched down onto the surface of the moving comet on 12th November 2014. Among the Rosetta mission's many discoveries was the first detection of organic molecules on the surface of a comet; the possibilities that the comet's odd shape may be due to it either spinning apart, or as a result of two comets fusing together; and the finding that comets may possess hard, crispy exteriors and cold but soft interiors, similar to ice-cream! On 30th September 2016, Rosetta ended its mission and was intentionally crashed into the comet.

The Shoemaker-Levy 9 Comet collided spectacularly with Jupiter in 1994. The giant gas planet's gravitational pull ripped the comet apart and created at least 21 impacts which were observed by astronomers. The largest collision created a fireball that rose around 3,000 kilometres above Jupiter's cloud tops as well as a giant, massively explosive dark spot more than 12,000 kilometres across, about the size of Earth.

The most famous comet in human history is Halley's Comet, which appears every 76 years and so therefore is a short-period comet, originating from within our Solar System. One of the earliest known sightings of Halley's Comet may have occurred in 466 BCE in ancient Greece. Accounts of the incident talk about a "wagon-sized" meteorite that landed in the Hellespont region of Greece, but the sources note the impact was accompanied by a "huge fiery body" that was visible in the night sky for 75 days. This timetable matches perfectly with the comet's projected appearance in the fifth century BCE. The next reference to the comet can be found in Han Dynasty China's "Records of the Grand Historian," which describe a "broom star" that appeared in the sky in 240 BCE. Other early sightings come from the Mesopotamian Babylonians, who recorded the comet's 164 BCE and 88 BCE visits on clay tablets. The Romans make reference to the sighting of the comet in 12 BCE.

Halley's comet inspired both fascination and horror in ancient and medieval observers. The comet was often considered to be a bad omen and was linked to all kinds of negative events including the death of kings and natural disasters. The ancient historian Flavius Josephus described the comet's visit in 64 CE as a "star resembling a sword" and considered it a warning for the destruction of Jerusalem by the Romans. A number of centuries later, the appearance of the comet in 451 CE was believed to signal Attila the Hun's defeat at the Battle of the Catalaunian Plains. In 837 CE, the Holy Roman Emperor Louis the Pious feared that the comet was a signal of his downfall; so he tried to counteract its influence by fasting, praying and helping the poor. The most famous appearance of Halley's Comet in human history occurred in 1066, when it coincided with William the Conqueror's Norman invasion and conquest of England. According to the Anglo-Saxon Chronicle, in the months before William set sail from Normandy to England, "a portent such as men had never seen before was seen in the heavens." Royal observers at the time considered the "long-haired star" to be a bad omen for the English King Harold II. This "prophecy" was fulfilled when William defeated and killed King Harold at the Battle of Hastings in that year.

(73) Halley's Comet depicted at the Battle of Hastings in 1066

The 1222 appearance of Halley's Comet is credited with inspiring the conqueror Genghis Khan to launch his Mongol army into a massive invasion of Europe. After viewing the comet in the night sky in 1298, the Italian artist Giotto is said to have depicted Halley's Comet as the star of Bethlehem in his famous painting, "Adoration of Magi." The 1456 return of the comet overlapped with the Ottoman Empire's invasion of the Balkans. In the 16th and 17th centuries, with the advent of the Renaissance and Enlightenment, people began to view the comet more from a scientific perspective rather than a bad omen. Nevertheless, the comet was still causing anxiety as recently as 1910. As the comet was nearing the Earth in that year, the New York

Times reported that a French astronomer named Camille Flammarion had warned that poisonous cyanogen gas in its tail might "impregnate the atmosphere and snuff out all life on the planet." Although most scientists dismissed it as nonsense, the prediction sparked a minor panic particularly amongst more superstitious, uneducated people. Before the comet passed without any incident, many people sealed up their homes to keep out the "fumes," stocked up on gas masks, and went to churches to pray for salvation.

Halley's Comet most recent return in 1986 marked the first time that astronomers were able to study it with more sophisticated scientific technology. High-powered telescopes were focused on the comet and five unmanned space probes conducted flybys as it hurtled within the inner Solar System. One of the probes, the European Space Agency's "Giotto" was able to get to within 600 kilometres of the comet's nucleus and take high quality images, providing a fascinating insight into Halley's Comet, including proving once and for all that its core is a solid mass primarily composed of dust and ice. The next appearance of this famous comet will be in July 2062.

(74) There are a number of asteroid-impact craters on Earth

6

EARLY EARTH

Geologists have divided the history of the Earth into a series of time intervals. These time intervals are not equal in length, like the hours in a day are. Instead, the time intervals vary in length. The reason for this variation is that geological time is divided up based on significant events occurring in the history of our planet Earth. We call these intervals, boundary events. For example, the boundary between the Permian and Triassic periods is marked by a global extinction in which a large percentage of the Earth's plant and animal species were eliminated. This is the largest mass extinction in the history of the Earth, in which nearly 96% of marine species and 70% of land species were wiped out around 250 million years ago.

Geologic Time Scale

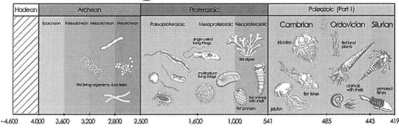

millions of years ago

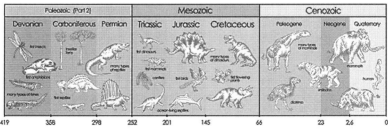

millions of years ago

(75) The Geological Earth Time Scale

The longest period of time in Earth history is called an Eon. There are four Eons in the history of our planet. They are hundreds of millions of years in duration. The four Eons from oldest to recent are the Hadean, 4.5 billion to 3.8 billion years ago; the Archean, 3.8 billion to 2.5 billion years ago; the Proterozoic, 2.5 billion to 540 million years ago; and the Phanerozoic, 540 million years ago to the present. Eons are divided into smaller time periods known as Eras. For example, the most recent Phanerozoic Eon is divided into three Eras; Cenozoic, Mesozoic and Paleozoic. Very significant events in the history of the Earth are used to determine the boundary of the Eras. Eras are further subdivided into Periods. The boundary events separating each Period, although widespread in their extent and influence, are not as significant as the boundary events separating the Eras. The Paleozoic Era for example, is subdivided into the Permian, Pennsylvanian, Mississippian, Devonian, Silurian, Ordovician and Cambrian Periods. Whereas, the Triassic was the last Period of the Mesozoic Era.

We have seen in the previous chapter that Earth and the other planets were created from the remains of the nebula cloud that largely went towards the formation of our Sun, 4.5 billion years ago. We saw that 4.6 to 5 billion years ago, our Solar system was nothing but a cloud of cold dust and gas particles swirling through empty space. This cloud of gas and dust was disturbed, perhaps by the explosion of a nearby supernova; and the cloud of gas and dust started to collapse as the force of gravity pulled everything together, forming a solar nebula, a huge spinning disk. As it spun, the disk separated into rings and

(76) The Early Earth during the Hadean Eon

the furious motion made the particles white-hot. The centre of the disk largely accreted to form the Sun, whereas the particles in the outer rings turned into large fiery balls of gas and molten liquid that cooled and condensed to take on a solid form. Around 4.5 billion years ago, these large fiery balls began to form into the planets of the Solar System, including our Earth.

As we have seen, the first Eon in which the Earth existed was known as the Hadean. This name comes from the ancient Greek word "Hades" which means "underworld", referring to the condition of the planet at the time. During this time, the Earth initially had no atmosphere and was under a continuous bombardment by meteorites, asteroids and comets, with the surface of the hellish-red molten planet covered with super volcanoes. This massive volcanic activity originated from substantial geothermal convectional heat flows coming from within the newly forming mantle of the planet. The gases emitted by these super volcanoes produced the earliest primordial atmosphere, which consisted mainly of methane, ammonia, neon and water vapour. Condensing water vapour became more prominent in the

(77) The Formation of the Moon – after the collision of Theia with the Earth

atmosphere as a result of ice delivered by comets which slammed into the early Earth. This water vapour accumulated in the atmosphere and contributed significantly towards the cooling of the molten exterior of our planet, forming a solid crust and ultimately, producing literally millions of years of continuous rain, which led to the oceans.

It was also during the Hadean Eon, roughly 4.48 billion years ago, that the Earth's only satellite, our Moon, was formed. Astronomers believe that the Moon was formed after a planetoid (planet in the process of being formed) about the size of Mars (referred to as Theia), slammed into the early Earth at an angle which wasn't sufficient to destroy the Earth. Nevertheless, the collision was powerful enough to vaporize some of the Earth's outer layers and melt both bodies into one, ejecting a portion of the mantle into orbit around the Earth, which through the process of accretion and the force of gravity, eventually formed into the Moon.

The Hadean Eon ended around 3.8 billion years ago with the onset of the Archean Eon. Just like the Hadean, this Eon also takes its name from an ancient Greek word, which in this case means "beginning" or "origin" referring to the first life on Earth. Most life forms today could not have survived in the Archean atmosphere, which lacked oxygen and an ozone layer. In spite of the apparently inhospitable atmosphere, it is widely understood that it was during this time that

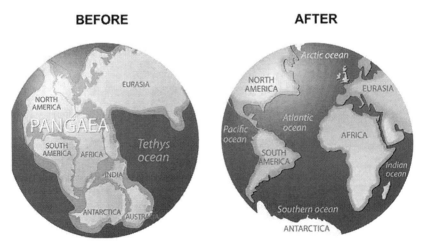

(78) From the Super-Continent of Pangaea to the current continents

the first primordial life began to take form, initially as single-celled microbial prokaryotes, deep under the ocean. We will examine the evolution of life on Earth in greater detail in the next chapter.

At the beginning of the Archean Eon, the mantle was much hotter than it is today, possibly as high as 1,600 degrees Celsius. As a result, the planet was much more geologically active, with the processes of heat convection and plate tectonics occurring much faster, and subduction zones at the boundaries between the crust and mantle being more common. The first formation of continental crust can be traced back to the late Hadean and early Archean Eons. What was left of these first small continents was called cratons. It was these cratons from which today's continents grew. As the surface of the Earth continually reshaped itself over the course of the subsequent Eons, continents formed and broke up in cycles. The continents migrated across the surface of the early Earth, occasionally combining to form a supercontinent. Approximately 750 million years ago during the Proterozoic Eon, the earliest known supercontinent, called Rodinia, began to break apart, then recombined around 600 - 540 million years to firstly form into Pannotia, and then ultimately into the supercontinent Pangaea, which was formed at the beginning of the Phanerozoic Eon. It was this latest supercontinent that broke apart 180 million years ago, ultimately forming into the configuration of continents which exist today.

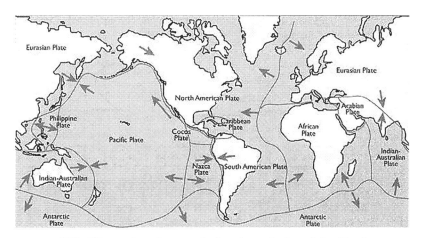

(79) The major plates and plate boundaries of the Earth

At this stage in our story, let us examine the concept of continental drift and plate tectonics. This story cannot be told without first examining the life of the man who is considered to be the heart and soul of this fascinating branch of geology. In 1915, a German explorer and scientist published a work that would revolutionise our understanding of planet Earth. In "The Origin of Continents and Oceans," Alfred Wegener proposed the then radical idea that the Earth's continents were originally a vast single land mass, hundreds of millions of years ago, and then broke apart, with those broken pieces of crust eventually drifting to their current positions. The mountains, forests, deserts, great plains and civilizations along with everything else, rest on a bedrock that is not static, but instead shifts very slowly, Wegener proposed.

(80) Alfred Wegener – the first to come up with the Theory of Continental Drift

There were other scientists before Wegener who had argued the possibility of continental drift over a long period of time; but it was Wegener who first outlined his theory at a geological conference in 1912. He took his theory seriously, curating a collection of supporting evidence, which found its way into his book that outlined the concept of continental drift to the public for the first time. Amongst other things, Wegener noted the close resemblance of fossils of animals and plants on continents located on opposite sides of the oceans; and of the geological similarities of rocks aligning the coasts of continents, such as the west coast of Africa and the east coast of South America. He claimed that these similarities existed because around 300 million years ago, the continents had once been joined as one supercontinent, which he named Pangea (all lands). He claimed that Pangea began to break up around 180 million years ago, starting the process of continental drift which continues to this day. "At the time, scientists believed continents had been linked by land bridges over which species had spread but which sank and disappeared," says palaeontologist Richard Cifelli, of Oklahoma University. "Wegener dismissed the idea of land bridges and argued instead that the continents had once been united," states Cifelli. The trouble with Wegener's theory at the time was that he could not provide a scientific explanation for what actually caused the continents to drift over time; and as a result he was heavily criticized and even ridiculed for over fifty years, within the scientific community. The highly respected British documentary-maker, David Attenborough, recalls asking his geology lecturer at Cambridge University in the late 1940s, why he was not giving lectures about continental drift. The lecturer's response was, "the idea is moonshine!"

During the 1920s, 1930s and 1940s, there was fierce debate between the relatively few geologists who supported Wegener's theory, the "drifters" and the majority who opposed his theory, the "fixists." It was not until the 1950s and 1960s that concrete evidence began to emerge which supported Alfred Wegener's theory of continental drift. In the 1960s a worldwide array of seismometers was installed to monitor nuclear bomb explosions at the height of the Cold War. These instruments revealed a surprising geological phenomenon, by showing that the majority of earthquakes and volcanoes occurred along the boundaries of the Earth's tectonic plates. Also, with the widespread use of radar after the Second World War, there was the discovery of sea-floor spreading in the 1950s, which causes oceans like the Atlantic and Pacific to expand. We now know that the Earth's outer shell, its crust, consists of continental and oceanic plates that glide over the mantle, the rocky/viscous inner layer that lies above our planet's core. Some plates slowly move apart, and when they do, hot magma originating from the mantle seeps up through the gaps, driven by convectional energy. When tectonic plates pull apart, this is called divergence. Islands are formed by diverging plates. Another example of divergence is when the diverging Eurasian and North American tectonic plates forced Europe and North America apart, and in the process created a mid-Atlantic ridge, forming the island of Iceland near its apex.

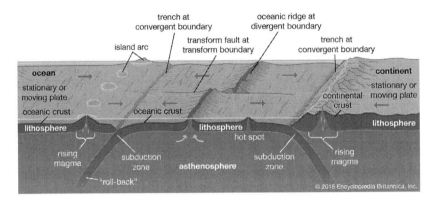

(81) The different plate movements occurring at plate boundaries

However, when tectonic plates push or collide together, the heavier plate is forced down underneath the lighter plate, or subducted, creating regions of intense volcanic activity along plate boundaries. This collision of plates is called convergence. Mountain ranges can be created from converging plates. For example, the Himalayan mountain range was formed around 50 million years ago when the Indian plate collided with the Eurasian plate. The third type of interaction between tectonic plates occurs when they slide against each other at the boundary. When this happens, the plates are transformative, and huge amounts of energy is released, resulting in earthquakes. An example of transformative plates occurs at the San Andreas fault in California, where the Pacific and North American plates intersect, causing periodic earthquakes along it's fault line, impacting the big cities of Los Angeles and San Francisco. The San Francisco earthquake of April 1906 destroyed over 80% of the city and ranks as one of the most significant ever recorded in human history. The earthquake and resulting fires killed over 3,000 people and left half of the city's 400,000 residents, homeless. Many seismologists predict there is another major earthquake due in or around San Francisco before 2032.

(82) The San Francisco earthquake and fires in April 1906

As for Wegener the person, his biographer, Mott Greene, describes him as "friendly, informal, somewhat introverted and deeply committed to his work." Unfortunately, that commitment ultimately cost him his life. In 1930, he and a colleague, Rasmus Villumsen, were on an exploratory expedition in Greenland when Wegener suddenly died of what is believed to have been a heart attack, shortly after celebrating his 50th birthday. Villumsen buried his body and marked the location of the body with skis, while he continued his trek across the Greenland icefields trying to reach a nearby camp. Villumsen didn't arrive at the camp and he was never seen again. A later expedition, led by Wegener's brother Kurt, built a pyramid-shaped mausoleum around Alfred's body, which over the years has become buried under layers of ice and snow. Incredibly and fittingly, his ice entombed body is drifting westward at a rate of around two centimetres each year, as it is carried by the North American tectonic plate. As Cifelli stated in an article in the scientific journal "Nature," which he co-wrote with Marco Romano of Sapienza University of Rome: "Wegener would have been glad to know that his body will have travelled some 20 kilometres in a million years' time, in accordance with his visionary theory." Wegener joins a long list of scientists like Aristarchus, Mendel, Copernicus, Kepler, Bruno, Galileo, Darwin and Harvey who throughout human history, have been ostracized and ridiculed for their radical ideas and theories, only to be proven correct by future generations.

In his 1915 book, "The Origin of Continents and Oceans", Alfred Wegener offered this timely reminder to some of his doubting and highly critical fellow geologists: "Scientists still do not appear to understand sufficiently that all earth sciences must contribute evidence toward unveiling the state of our planet in earlier times, and that the truth of the matter can only be reached by combining all this evidence. It is only by combining the information furnished by all the earth sciences that we can hope to determine 'truth' here, that is to say, to find the picture that sets out all the known facts in the best arrangement and that therefore has the highest degree of probability. Further, we have to be prepared always for the possibility that each new discovery, no matter what science furnishes it, may modify the conclusions we draw."

More recent scientific research has raised the intriguing question: "Does a planet need plate tectonics for life to develop?" Research published in the journal, "Physics of the Earth and Planetary Interiors" in 2016, argues that plate tectonics may be a phase in the evolution of rocky planets. If this is the case, this has major implications for the habitability of exoplanets, which are planets that have been identified from other star systems in our Milky Way galaxy. So far, over 4000 exoplanets have been detected, some rocky, others gaseous, and a few even of similar composition to our Earth. For example, Kepler-452b is an exoplanet orbiting the Sun-like star Kepler-452, 1402 light years from Earth, in the constellation Cygnus. It was identified by the Kepler space telescope and its discovery was announced by NASA on 23rd

July 2015. This planet is the first near-Earth-size planet discovered in the "habitable zone" around a Sun-like star. The habitable zone is the ideal distance a rocky planet with an atmosphere is away from its star; not too close where it is so hot that liquid water would vaporize; and not too far where liquid water would ice up. Kepler-452b is 60 percent larger in diameter than Earth and is considered a super-Earth-size planet. Even though it is larger than Earth, its 385-day orbit year is only 5 percent longer. The planet is 5 percent further from its parent star Kepler-452 than Earth is from the Sun. Kepler-452 is 6 billion years old, making it 1.5 billion years older than our Sun; although the star is also a G classification type, has a similar temperature and has a diameter 10 percent greater than our Sun. This exoplanet offers us the incredibly intriguing possibility of harbouring life.

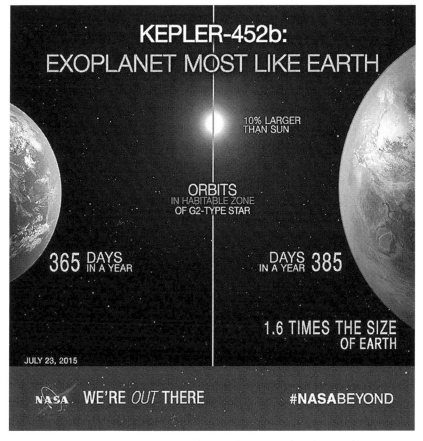

(83) The exoplanet Kepler-452b compared to the Earth

The two things that make Earth unique in our Solar System is that
it has plate tectonics and complex life. According to this research, these
two characteristics may be interrelated. The theory proposes that it
required a stable planet for complex life on Earth to evolve over a 3.8
billion-years period. Evolution of life was possible because the Earth's
surface had a sufficient temperature range to allow for liquid water. This
is a remarkable level of stability for such an incredibly long period of
time, especially when we consider the Sun has become 30% brighter
over that same time period, meaning that the Earth's atmosphere has
evolved, becoming less of a greenhouse than it was over 3 billion
years ago. These researchers believe that plate tectonics provides the
mechanism for this global temperature-controlled thermostat. Most
volcanic activity on the Earth's surface occurs at plate boundaries in
response to the movement of tectonic plates. The two most important
by-products of volcanoes, by sheer mass, are two greenhouse gases;
carbon dioxide and water vapour. This is the thermostat effect. If the
Earth gets too hot, high levels of rainfall and erosion start bringing
down the levels of carbon dioxide in the atmosphere, thus cooling
the planet. If the Earth gets too cold and freezes over ("the snowball
effect") the erosion mechanism stops; but active volcanism triggered by
ongoing plate movements continues pumping carbon dioxide into the
atmosphere, eventually creating a sufficient greenhouse level and heat,
to melt the icecaps. These researchers claim it was this mechanism that
allowed Earth to recover from a global "snowball" Ice Age, around 600
million years ago; and for complex life to rapidly evolve and flourish
thereafter.

The association between the presence of complex life and plate
tectonics has become so entrenched, that the search for habitable
exoplanets has focused on what are called "super earths." These are
rocky planets larger than Earth where the odds for plate tectonics
to occur are thought to be higher. As part of their research, scientists
have posed the question: "How do Earth-like planets evolve from their
primordial hot, violent beginnings to their eventual moderately cooler
climatic conditions, radiating their excess heat into space?" They found
that the evolutionary track an Earth-like planet takes depends not only

on its size, but also on how it is formed. For example, two planets identical in every other way, but with different starting temperatures, may evolve down very different evolutionary paths. These scientists also found that plate tectonics may simply be a phase in the evolution of Earth-like rocky planets, and that planets may begin and also end, in a static form, with no drifting plates. Planetary scientists have long accepted that as the Earth loses its internal convectional heat, the process of plate tectonics will cease, resulting in a future static planet which will resemble the Mars of today.

We believe that the chemistry which enabled the earliest single-celled micro-biotic life forms to appear on Earth around 3.8 billion years ago, originated deep under the ocean from chemical elements emitted by super-volcanoes which ultimately gave rise to DNA; thus forging a direct connection between plate tectonics, volcanic activity and primordial life.

7

LIFE ON EARTH

AROUND 3.8 BILLION YEARS AGO, DEEP WITHIN THE PRIMORDIAL oceans, something mysterious and astounding began to happen around the vents of suboceanic volcanoes; a mixture of complex chemicals in conjunction with energy flows radiating from the volcanoes within the oceanic environment of an abundance of water, gave rise to complex molecules that were bound together chemically and physically into single cells which were capable of reproducing. This gave rise to the fifth threshold of increasing complexity in our journey of Big History, the formation of life on Earth.

The earliest life forms that we know of were microbes called prokaryotes, singular cells without a nucleus, but enclosed by a membrane, which were fully developed by 3.5 billion years ago. Let us

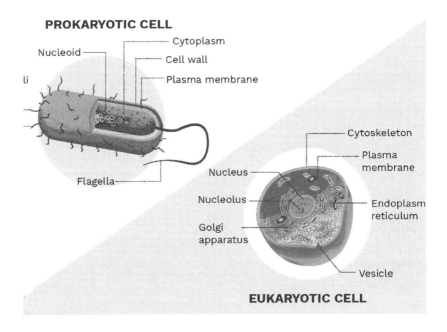

PROKARYOTIC CELL
- Nucleoid
- Cytoplasm
- Cell wall
- Plasma membrane
- li
- Flagella

- Nucleus
- Nucleolus
- Golgi apparatus
- Cytoskeleton
- Plasma membrane
- Endoplasm reticulum
- Vesicle

EUKARYOTIC CELL

(84) Prokaryotic and Eukaryotic cells

survey the major developments in the evolution of life on our Earth. The first microbial life was based on ribonucleic acid (RNA), organic chemicals which then later evolved into the more complex deoxyribonucleic acid (DNA). At this very early stage, a common ancestor gave rise to two major groups of microbial life; bacteria and archaea. We are uncertain of how, when and in what order, this split happened.

By 2.4 billion years ago, the "Great Oxidation Event" had occurred. Prokaryotes had evolved into cyanobacteria and produced an abundance of oxygen as a by-product of the process of photosynthesis. They produced food using water and the Sun's energy; releasing oxygen as a result. This oxygen bubbled up to the surface of the oceans and was gradually released into the atmosphere resulting in an increase of oxygen in the atmosphere from around 1 percent 3 billion years ago to just over 20 percent 2 billion years ago. As a result, the environment became less hospitable for other species of microbes that could not tolerate oxygen, and therefore became extinct in the process. Not long after the Great Oxidation event, around 2.3 billion years ago, the entire planet became frozen in what was the first "Snowball Earth." We believe this was due to a significant reduction in super-volcanic activity

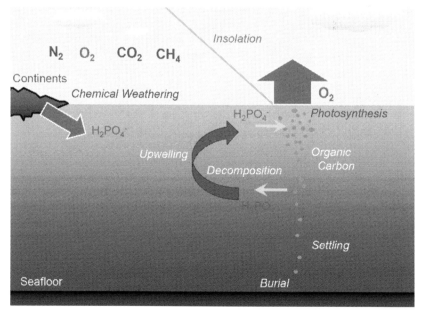

(85) The Great Oxidation Event – the first known mass extinction event

around the planet. When the situation subsided and the ice eventually melted, even more oxygen was released into the atmosphere.

Around 2 billion years ago, cells had evolved to the point of developing internal organs, known as organelles. The key organelle which evolved was the nucleus, or control centre of the cell, in which the genes are stored in the form of DNA. These cells with a nucleus are known as eukaryotes. Eventually, eukaryote cells absorbed photosynthetic bacteria, forming a symbiotic relationship with the bacteria. Mitochondria, the organelles that process food into energy, evolved from these mutually beneficial relationships. The absorbed bacteria also evolved into chloroplasts, the organelles that give green plants their colour and allow them to extract energy from sunlight. Different lineages of eukaryotic cells acquired chloroplasts in this way on at least three separate occasions. One of these resulting cell lines went on to evolve into all green algae and green plants. By 1.5 billion years ago, the eukaryotes had branched out into three groups which ultimately became the ancestors of modern plants, fungi and animals; each group evolved separately from this point in time. At this stage, all life on Earth was still single-celled.

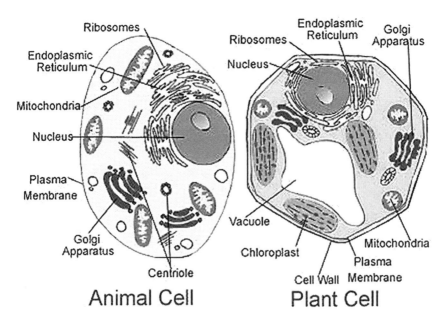

(86) Multicellular cells in animal and plant organisms

The first multi-cellular life developed around 900 million years ago. Biologists are unclear exactly how or why this happened. By 800 million years ago the early multicellular animals underwent their first split with one branch comprising the sponges and the other branch all other multicellular life, known as eumetazoa. Around 20 million years later, a small group called the placozoa broke away from the rest of the eumetazoa. Placozoa were thin plate-like creatures about 1 millimetre in length and consisted of only three layers of cells. Some biologists believe the placozoa may have been the last common ancestor of all animals. Then 770 million years ago the planet underwent another Snowball Earth, once again completely engulfed in ice.

Around 630 million years ago, some animals evolved into bilateral symmetrical shapes for the first time, meaning that they had a defined top, bottom, front and back. We know very little about what caused this to happen. It is believed that small worms known as acoela may be the first closest surviving relatives of the first ever bilateral animal. Almost certainly, the first bilateral animal was a type of worm. Bilateria, or animals with bilateral symmetry, underwent a major evolutionary split when they divided into the protostomes and deuterostomes, 590 million years ago. The deuterostomes went on to include all the vertebrates with backbones; and the protostomes became all arthropods such as insects, spiders, crabs and worms. Interestingly, the two groups can be distinguished by the way their embryos develop. The first hole that the embryo acquires, called the blastopore, becomes the anus in deuterostomes and the mouth in protostomes.

By 580 million years ago there was a proliferation of seafloor creatures, with a variety of body shapes, that lived alongside sponges for 80 million years. Their fossil evidence can be found in sedimentary rocks around the world. During this time, oxygen levels continued to rise, approaching levels sufficient to support oxygen-based life. The early sponges may actually have helped boost oxygen levels by eating bacteria, therefore removing them from the decomposition process. Tracks of a species of sponge organism named dickinsonia costata suggest that it may have been moving along the bottom of the ocean, possibly feasting on mats of microbes.

(87) The Cambrian Explosion of life started around 535 million years ago

The next big step in the long journey of the evolution of life on Earth was the Cambrian Explosion that began 535 million years ago, which resulted in a significant explosion of new life forms. This was when a myriad of new body formations and designs evolved, including more active animals with clearly defined heads and tails for directional movement to chase prey. It was at this time that the first fully developed vertebrates, animals with a backbone, appeared. They probably evolved from a jawless fish which had a notochord, a stiff rod of cartilage, instead of a true backbone. The first fully formed vertebrate was a fish. By 500 million years ago, animals had well and truly left the oceans and were exploring the land. It is believed the first animals on the land were more than likely euthycarcinoids, thought to be the missing link between insects and crustaceans. Not long after, around 465 million years ago, a diversity of plant species began to flourish on the land. Then 460 million years ago, fish species split into two major groups: the bony fish and cartilaginous fish, with skeletons of cartilage rather than hard bone. By 440 million years ago the bony fish had split into two major groups; the lobe-finned fish with bones in their fleshy fins, and the ray-finned fish. This was a landmark event in the evolutionary journey because the lobe-finned fish eventually gave rise to amphibians, reptiles, birds and mammals. The ray-finned fish also thrived and gave rise to most fish species that are in existence today.

By 417 million years ago lungs were fully developed for the first time. They first appeared in lungfish. Although they were definitely fish complete with gills, lungfish also had a pair of quite sophisticated lungs, which were divided into a number of smaller air sacs to increase their surface area. This allowed them to breathe outside of a water-based environment and therefore to survive when the ponds they lived in dried out. The oldest known clearly identifiable insects appeared for the first time around 400 million years ago and around this same time some plants evolved woody stems. Then 397 million years ago the first four-legged animals, or tetrapods appeared. These tetrapods went on to conquer the land and ultimately gave rise to the reign of the dinosaurs followed by mammalian primates, hominines and humans. Splitting from tetrapods around 310 million years ago were two distinct groups, the sauropsids and the synapsids. The sauropsids went on to include all dinosaurs, birds and modern reptiles; whereas even though the first synapsids were also reptiles, they had distinctive "mammal-like" jaws and eventually went on to evolve into mammals.

The pelycosaurs, the first major group of synapsid animals, dominated the land between 320 and 250 million years ago. The most well-known of these was the dimetrodon, a large predatory reptile with a sail on its back. In spite of its appearances, it was not a dinosaur. Between 275 and 100 million years ago, the therapsids, close cousins of the pelycosaurs, evolved alongside them and eventually replaced them. The therapsids survived until the early Cretaceous Period, 100 million years ago. During this Period, a group of them called cynodonts developed dog-like teeth and eventually were to evolve into the first mammals. Around 250 million years ago, the Permian Period ended with the greatest mass extinction in the history of the Earth, leading to the wiping out of a wide range of species. The most likely cause of the Permian Extinction was the warming of the Earth's climate and associated changes to the temperature of the oceans which appears to have been caused by a substantial increase in volcanic activity all over the planet, leading to a massive outpouring of magma and lava.

As the ecosystem of the planet recovered, it underwent a signif-icant shift. At this time the sauropsids, mainly in the form of dinosaurs,

(88) The dinosaurs dominated the Earth for over 160 million years

had developed. The over 160 million reign of the dinosaurs had begun! Our mammalian ancestors were mainly small nocturnal creatures that were eaten by the dinosaurs. In the oceans, several groups of reptiles developed into the great marine creatures of the dinosaur era. As the Triassic Period came to an end 200 million years ago, another mass extinction occurred, paving the way for the dinosaurs to become more prominent than their synapsid cousins, and dominate the planet. We are not certain what caused this extinction event, but the global climate changed significantly, sea levels rose and there was a sudden release of carbon di oxide into the atmosphere. An asteroid impact may have been the triggering cause. The earliest dinosaur fossils date to 233 million years ago. The reptilian ancestors of dinosaurs had set the stage nicely for the dinosaurs to dominate the planet, since they were already distributed around the planet.

Once dinosaurs first evolved, they rapidly spread to every continent. As ecosystems stabilized, dinosaurs settled into their niches. This included evolving many traits suited for particular environments. For example, Tyrannosaurus Rex evolved strong legs that allowed it to run through forested river valleys in search of animal prey, becoming a ruthless predator in the process. Another dinosaur species with unique features was Dreadnoughtus Schrani, the largest known terrestrial animal ever to have walked on the planet. This massive dinosaur flourished during the Late Cretaceous Period, 100 million to 66 million years ago. It was about 26 metres long and may have weighed more than a dozen African elephants. Its name literally means "fear nothing," and the immense size of this herbivore put it at the top of the food chain, with virtually no predators. It evolved a very long neck in order to reach plant food high up in tall trees. Another well-adapted species of dinosaur was Ankylosaurus, well designed with spikes on its head and side, a back covered with armoured plates, and a tail that ended in a massive club of bone; significantly protecting it from predators.

During this time, proto-mammals evolved into warm blooded creatures, with the ability to metabolise by regulating their internal temperature, regardless of the external conditions. The first split also began in the early mammalian population. The monotremes, a group of mammals that laid eggs, broke away from the other mammals that internally gave birth to offspring. Only a few monotremes survive to this day, including the duck-billed platypus and the echidnas. By 140 million years ago, placental mammals had split from their marsupial cousins. Marsupials such as kangaroos, give birth when their young are still tiny, and nourish them in a pouch for the first months of their lives. It is interesting to note that the majority of marsupials today live in Australia. They first originated in South-East Asia and then spread into North America, which was attached to Asia by a land bridge at this time. Then they moved on to South America and Antarctica before finally settling in Australia around 50 million years ago. Between 105 and 85 million years ago the placental mammals had split into four major groups; the laurasiatheres, a very diverse group that included all the hoofed mammals such as horses and also whales, bats and dogs; the euarchontoglires, which included all primates and rodents; the xenarthral, which included anteaters and armadillos; and finally the afrotheres, including elephants and aardvarks. It is unclear what the cause of this four-way split was.

Around 100 million years ago during the Cretaceous Period, the dinosaurs reached the peak of their size. A little earlier in the Earth's history about 93 million years ago, the oceans became starved of oxygen, possibly as a result of massive under-ocean volcanic eruptions. During this time 27 percent of all marine invertebrates became extinct. Also at this time, the ancestors of the modern primates split from the ancestors of modern rodents and lagomorphs such as rabbits. The rodents went on to become incredibly successful, making up about 40 percent of all modern-day mammals. A ground-breaking event unfolded around 70 million years ago when grasslands began to evolve around the planet. This was to have a significant impact upon human civilization as we shall see in later chapters.

(89) The asteroid impact 66 million years ago that wiped-out all dinosaurs

Another major event which massively impacted the future course of life on Earth and particularly the fate of mammals ultimately leading to us was the extinction of dinosaurs 66 million years ago when an asteroid slammed into the Earth at the Yucatan Peninsula in proximity to modern day Mexico. This extinction event wiped out all species of dinosaurs and cleared the way for mammals to dominate the Earth because they were no longer hunted for food by the dinosaurs.

Not long after this extinction event, around 63 million years ago, the primates split into two groups, known as the haplorrhines, or dry-nosed primates and the strepsirrhines, or wet-nosed primates. The strepsirrhines eventually evolved into modern lemurs, while the haplorrhines developed into monkeys, apes and ultimately humans.

The Palaeocene/Eocine Extinction Event occurred 55 million years ago when there was a sudden rise in greenhouse gases in the atmosphere which sent temperatures soaring, transforming the ecosystems and climate of the planet, and wiping out many species in the ocean depths, but sparing shallow-sea and land-based species. Gorillas branched off from the other great-ape species around 7 million years ago and began the process of evolution from primates, to a number of species of hominines and ultimately leading to humans, around 300,000 years ago.

We owe so much to the great British naturalist Charles Darwin for our knowledge of evolution and natural selection. It is amazing to think that before the mid-nineteenth century, we had very little understanding of the natural processes of a shifting and ever-changing Earth which directly impacted upon the evolution of life on this planet. Let us examine the story of Charles Darwin and the controversy he endured particularly from a group of sceptical devout Christians who questioned his findings and beliefs.

Charles Darwin was born on 12th February 1809 into a progressive family which included two enlightened grandfathers; Josiah Wedgewood, an industrialist and anti-slavery campaigner; and Erasmus Darwin, a medical doctor whose book "Zoonomia" set out the radical and highly controversial idea that one species could "transmute" into another. Both of Darwin's grandfathers also belonged to the "Lunar Society," a group of natural scientists and philosophers who would meet regularly to discuss the latest scientific and philosophical ideas. Following in the footsteps of his father and grandfather, Darwin studied medicine at the University of Edinburgh. He did not last long however because in a period long before the invention of anaesthetics, he found the brutal butchery techniques of surgery too violent to cope with. Instead, after abandoning his plans to become a doctor, he

considered a career in the Church, going on to study Theology and
Divinity at Cambridge University in 1827. He found that studying
to be a clergyman didn't particularly appeal to him either. His real
passion was biology and the natural world, so he spent much of his
time collecting beetles and observing the natural environment. Shortly
after he graduated in 1831, the opportunity of a lifetime presented
itself to him.

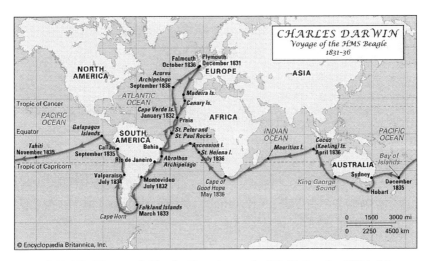

(90) The Voyage of Charles Darwin on the HMS Beagle: 1831-36

Darwin's tutor at Cambridge recommended him as a "gentleman
naturalist" on an around-the-world voyage that was being planned on
the HMS Beagle. In 1831 he took this exciting opportunity with open
arms. Over the following five years, with Darwin on board, the Beagle
visited four continents. During this time, he collected many plant
and animal specimens and investigated the local geology and natural
environments of the places he visited. He also read widely, including
Charles Lyell's recently published "Principles of Geology" which had a
major impact on his thinking during the voyage; particularly regarding
the long history of the Earth and how natural processes on the planet
can change over a very long period of time. In 1835, after leaving
South America, the HMS Beagle stopped at the Galapagos Islands,
965 kilometres off the coast of Ecuador, for five weeks. The Galapagos
archipelago is a group of small volcanic islands, each with a distinctive

landscape. While there, Darwin began to closely study and sketch the natural features of finches, tortoises and mockingbirds; collecting many specimens in the process. Darwin's finches are a classic example of adaptation and natural selection arising from differing environments. The common ancestor of finches arrived on the Galapagos islands around two million years ago. Over time, Darwin's finches had evolved into 15 recognizable species, differing in body size, beak shape, song sound and feeding behaviour. Darwin observed these different species of finch and came to the conclusion that changes in the size and form of finch beaks had enabled different species to utilize different food resources on the respective islands; such as insects, seeds, nectar from cactus flowers, as well as blood from iguanas; all driven by what Darwin ultimately called natural selection.

After arriving back in England in 1838, Darwin showed his specimens to a number of biologists and began to write diary records of his travels. While doing this, a powerful and profound idea began to circulate within his mind. Charles Darwin realised that on his travels, he had witnessed how the theory of transmutation proposed by his grandfather, actually happened. He had observed that animals more suited to their environment survived longer and had more offspring. He realised that transmutation, which he began to call "evolution" occurred as a result of a process that he called "natural selection." This was a profound conclusion which directly clashed with the teachings of the Christian Church. He was aware that his grandfather

(91) Charles Darwin   *(92) Thomas Huxley*

had been severely humiliated when he wrote about transmutation, and Darwin feared the same fate awaited him. So instead of immediately announcing his findings, he decided to gather more evidence to support his theory of evolution and natural selection.

In the meantime, tragedy struck his family when in 1851, his dear daughter Anne, was struck with a sudden sickness and died. This devastated him. During this time Darwin also suffered from severe bouts of anxiety which seemed to have been related to his dread of the prospect of having to publish his scientific findings and go against the teachings of the Church, to which he still remained faithful. By June of 1858, Darwin had written over two hundred and fifty thousand words on his theory of evolution and natural selection, but had still not published anything. But then, he received a letter from the English scientist, Alfred Russel Wallace, who was an admirer of Darwin. When reading this letter, it was clear to Darwin that Wallace had independently arrived at the same theory of natural selection as Darwin had. In writing the letter, Wallace was seeking Darwin's advice on whether he should publish his findings. This prompted Darwin to take drastic action because if he didn't go public and publish his findings, Wallace would get all of the credit for the ground-breaking theory of how life developed on Earth. Nevertheless, Darwin was an honourable gentleman who wanted to ensure Wallace also received due credit for his findings. So in July 1858, Charles Darwin finally went public with his highly controversial but scientifically brilliant theory of evolution by natural selection; making sure that Wallace also received deserved credit for the findings.

Darwin's ideas were presented to Britain's leading natural history and biology group at the time, the Linnean Society, founded in 1788. After consulting with colleagues, Darwin agreed that extracts from his and from Wallace's papers should be jointly presented at the same meeting. Wallace was appreciative of Darwin for agreeing to this. Tragically for Darwin, he was unable to attend the Linnean presentation because his son had died of scarlet fever at age 18 months, at around the same time. Still riddled by doubt and concern about the Church's reaction, Darwin finally mustered up enough courage to publish his new theory of evolution in November 1859; the book was titled "On the Origin of Species by Means of Natural Selection." This book was destined to become one of the most profound scientific treatises ever written. Darwin described writing this book as akin to "living in Hell." He feared losing his reputation, just like his grandfather Erasmus had. When the book was released it was fiercely criticised by the Church. Many people were deeply disturbed by the insinuation from his book that human beings were directly descended from apes.

Darwin was very reluctant to defend his ideas in the public forum so it was left to some of his closest supporters to do this for him; the most notable being a young biologist and anthropologist named Thomas Huxley, who took the fight right up to dissenting scientists, elements of the media and the Church establishment. During the 19th century in Britain, scientific lectures were a form of popular entertainment, so any debate about this controversial theory of evolution was certain to attract big crowds. Huxley relished this. His most famous clash came at

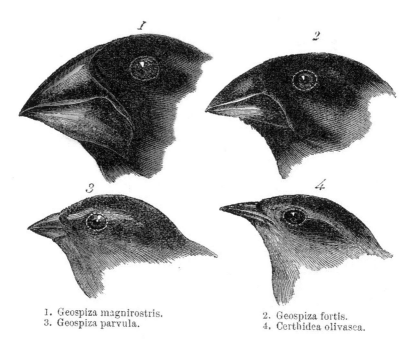

1. Geospiza magnirostris.
3. Geospiza parvula.
2. Geospiza fortis.
4. Certhidea olivasea.

(93)   Darwin's study of finch beak adaptations at the Galapagos Islands

a meeting of the British Association for the Advancement of Science, in June 1860. In what many commentators saw as a head-to-head clash between science and God, Huxley aggressively debated Darwin's theory of evolution against the Anglican Bishop Samuel Wilberforce and his biblical account of creation. Both sides claimed victory; but undoubtedly the debate shook Victorian society to its core and elevated the profile of Charles Darwin at the same time. In his strong defence of Darwin's theory of evolution, Huxley was quoted as saying: "A man has no reason to be ashamed of having an ape for his grand-father. If there were an ancestor whom I should feel shame in recalling, it would rather be a man — a man of restless and versatile intellect — who plunges into scientific questions with which he has no real acquaintance, only to obscure them by an aimless rhetoric, and distract the attention of his hearers from the real point at issue by eloquent digressions and skilled appeals to religious prejudice."

His book, On the Origin of the Species became a worldwide bestseller and was printed in multiple editions. With each subsequent edition, Darwin strengthened his arguments even further with new evidence. As part of the fifth edition, he introduced the phrase "survival of the fittest," a phrase first used by the philosopher Herbert Spencer; even though it has been closely associated with Darwin ever since. More than a decade after he published his first book, Darwin found the further courage and determination to apply his theory to humans by publishing "The Descent of Man" in February 1871. This book unambiguously focused on the evolution of humans and outlined how humans had evolved from primates, creating even greater controversy and heavy criticism from the Church establishment. In the decades following the publication of this book, Darwin's ideas gained wider acceptance to the point that there were two camps of thinking; the "Darwinists" who believed in evolution and the "Creationists" who continued to believe the Christian faith and an almighty creator.

Although he was afflicted by ongoing poor health, Charles Darwin continued his research and work until the end, dying a virtual recluse on 19th April 1882, surrounded only by his loyal wife Emma and a few devoted friends. In the final months of his life Darwin had been looked after by Emma, who stood by him even though she remained devoutly Christian. Knowing his death was imminent, Darwin described his family graveyard as "the sweetest place on Earth." His scientific followers had bigger plans for Darwin. Led by Thomas Huxley, arrangements had been made for Charles Darwin to be fittingly buried at Westminster Abbey in the centre of London.

Here are some of my favourite direct quotes from Charles Darwin. In expressing exasperation for the disrespect shown by some quarters of society for the scientific method, he said: "Ignorance more frequently begets confidence than does knowledge; it is those who know little, and not those who know much, who so positively assert that this or that problem will never be solved by science." In this quote Darwin directly attacks creationists, by stating: "I cannot persuade myself that a beneficent and omnipotent God would have designedly created parasitic wasps with the express intention of their feeding within the living bodies of caterpillars." In praise of the scientific method, Darwin said: "I have steadily endeavoured to keep my mind free so as to give up any hypothesis, however much beloved (and I cannot resist forming one on every subject), as soon as facts are shown to be opposed to it."

What is the theory of evolution by natural selection and why did it radically and profoundly impact and upturn the scientific community, religious beliefs and views on humanity and the natural world? Evolution by natural selection is the process by which living organisms change over time in response to changes in the natural environment of the planet, which trigger changes in physical and behavioural traits. The theory has two overriding principles; firstly, that all life on Earth is diverse and interconnected in a complex ecological web; and secondly, this diversity of life is the result of modifications of species by a process Darwin called natural selection, whereby some traits are favoured over others, in order to enhance the survival of species in ever-changing environments and ecosystems. The theory is sometimes referred to as the "survival of the fittest," but this can be easily misunderstood because in this context "fittest" does not refer to an organisms physical or athletic strengths, but instead to its ability to survive and reproduce.

Examples of natural selection in the animal world include the following. In a particular habitat there may be red bugs and green bugs. If the birds prefer the taste of the red bugs, there may eventually be a larger population of green bugs and only a few remaining red bugs. So the green bugs will reproduce in greater numbers and the red bugs may eventually become extinct. Another example is an environment in which there are long-necked and short-necked giraffes. If a shift in climate causes low-lying shrubs to die out, the giraffes with short necks will not get enough food, leading to their ultimate extinction; and the long-necked giraffes surviving and flourishing because they will be able to access food in the higher trees. Finally, there is the example of insects. Insects become resistant to pesticides very quickly, sometimes within one generation. If an insect is resistant to a particular chemical pesticide, most of the offspring will also be resistant. When we consider that insect generations last a matter of weeks, insects within a particular environment can become immune to a chemical within a matter of months.

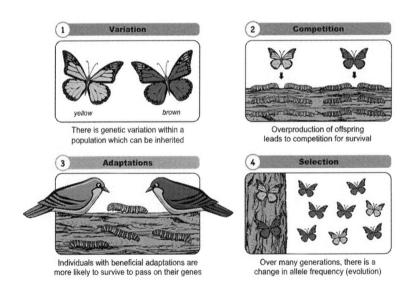

(94) The process of Adaptation and Natural Selection leading to Evolution

An example of natural selection as it applies to changes in society and its impact on human social psychology and behavioural evolution, otherwise known as "Social Darwinism" was a study of 1,900 students, published in the journal "Personality and Individual Differences" in October 2017. This study found that in our complex modern society, many younger people have trouble finding a mate or partner because of rapidly changing social technological advances that are evolving faster than humans. Nearly 50% of the students in this study faced considerable difficulty in securing a mate when going through the dating process. The researchers found that in most cases, these difficulties were not due to anything the young people were doing wrong, but instead they experienced difficulties when dating due to people living in an environment which was radically different from the environment in which they were brought up in by their parents, only a decade or two earlier; resulting in rapidly changing dating rituals and customs.

The most notorious and extreme example of the hijacking of Social Darwinism is the manner in which it was perversely applied by Adolf Hitler and his Nazis to justify the superiority of Aryan Germans to all other ethnic races, and the mass murder of over 6 million Jews in the Holocaust which came out of this obscene abuse of Darwin's ideas. Hitler adopted a very warped interpretation of the Social Darwinist concept of the survival of the fittest, and attempted to implement it in Germany. He believed the German master race had grown weak due to the influence of non-Aryans in Germany. As far as Hitler was concerned, the survival of the German Aryan race was dependent on its ability to maintain the purity of its gene pool. The Nazis targeted for extermination, particular groups or races that they considered to be biologically inferior. These groups included Jews, Roma (gypsies), Poles, Russians, homosexuals and people with disabilities.

In the first edition of his book, "On the Origin of the Species" published in 1859, Charles Darwin speculated about how natural selection could cause a land mammal to evolve into a whale. As a hypothetical example, Darwin cited North American black bears, which were known to catch insects by swimming in the water with their mouths wide open. In his book, he said, "I can see no difficulty in a race of bears being rendered, by natural selection, more aquatic in their structure and habits, with larger and larger mouths, till a creature was produced as monstrous as a whale." This idea was not received very well by the general public and some elements of the scientific community, at that time. Darwin was so humiliated by the ridicule and criticism he received that the "swimming-bear" passage was removed from later editions of his book. With the benefit of their knowledge of DNA and genetics, biologists now understand that Darwin had the right idea, but the wrong animal! Instead of looking at bears, he should have instead looked at cows and hippopotamuses, which do have a genealogical connection to whales, in the evolutionary tree of life.

The story of the origin of whales is one of evolution's most fascinating stories and one of the best examples that biologists have, of natural selection; so let's examine this evolutionary journey in some detail. To understand the origin of whales, we need to have a basic understanding of how natural selection actually works, changing a species in continuously small ways, and causing a population of species to change in features, function, colour or size over several generations. Biologists call this process, "microevolution." In addition to these small progressive changes, natural selection is also capable of much bigger changes. If there is enough time (remember, the Earth is 4.5 billion years old) and enough accumulated smaller changes, natural selection can eventually create entirely new species, known as "macroevolution." This longer term process of evolution can turn amphibians into reptiles; dinosaurs into birds; land mammals into whales; and the ancestors of apes into humans. So if we turn back to whales, biologists know that the transition of early whales from land to the ocean occurred in a series of predictable steps. The body parts of the immediate land ancestors of whales would have changed; such as the nose moving further back on the head; front legs becoming flippers; back legs disappearing altogether; bodies becoming more streamlined; and the development of tail flukes to more efficiently propel the whale through the water.

In his book, Charles Darwin also described a form of natural selection that depends on an organism's success at attracting a mate, a process known as "sexual selection." The colourful feathered plumage of peacocks, male skunks releasing bursts of skunk scent, and the antlers

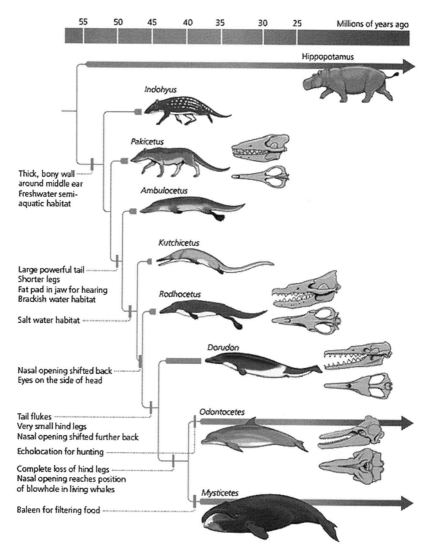

(95) The Evolution of the whale from land-based mammals

of male deer are three examples of sexual traits that have evolved under this type of selection. What makes Darwin's theory of natural selection by evolution even more astounding is that he did not know anything about DNA and genetics at the time; which came later, in 1953 with the discovery of the double helix, the "twisted ladder' structure of deoxyribonucleic acid (DNA), by James Watson and Francis Crick, marking a significant milestone in the history of science and the birth of genetics and molecular biology. This discovery led to an understanding of how genes encode different biological and behavioural traits; and also how genes are passed down from parents to offspring. The integration of Darwin's theory and the science of genetics is known as "modern evolutionary synthesis." We now know from genetics that the physical and behavioural changes that make natural selection possible happen at the level of cellular DNA and genes. These changes are called "mutations," the raw materials that literally give rise to evolution. Mutations can be caused by random errors in the replication or repair of DNA, and by chemical or radiation damage to cells. In most instances, mutations are either harmful (like cancer) or neutral; but in rare instances, a mutation can be beneficial to the survival of the organism and the evolutionary future of its offspring. If this is the case, that mutation will become more dominant in the offspring and future generations, and will spread throughout the population of that species. This is how natural selection guides the evolutionary process over a longer period of time and creates new species of life. Importantly, mutations are random in nature, but selection for them is not random. Natural selection is not the only mechanism by which organisms can evolve. Genes can also be transferred from one species population to another when organisms migrate or immigrate, in a process known as "gene flow." Also, the frequency of genes being transferred can also change at random; this is called "genetic drift."

Even though biologists could predict what early whales should look like, they initially did not possess the fossil evidence to support their claim. Religious creationists took this absence of evidence as proof that evolution did not occur. They ridiculed the idea that there ever could have been a "walking whale." By the mid-1990s things changed with a critical piece of evidence being discovered in 1994, when palaeontologists found the fossilized remains of "Ambulocetus Natans," an animal whose name literally translates to "swimming-walking whale." Its forelimbs had fingers and small hooves, but its hind feet were very large. It was clearly adapted for swimming, but it was also capable of moving on land, albeit in a clumsy manner, like a seal. When it swam, this ancient creature moved like an otter, pushing back with its hind feet and undulating its spine and tail. Modern whales propel themselves through the water with powerful beats of their horizontal tail flukes, but Ambulocetus still had a whip-like tail and had to use its legs to provide most of the propulsive force it needed to move through the water. Excitingly, in recent years, more and more of the fossilized remains of these transitional species, or "missing links," have been discovered, further supporting Darwin's theory as it applies to the evolution not only of whales, but all of the species on Earth.

Of course, fossil links have also been discovered to support the evolution of humans. More recently, in early 2018, a fossilized human jaw with accompanying teeth was found in Israel, estimated to be up to 194,000 years old, making this discovery at least 50,000 years older than other modern human fossils found outside of Africa. In spite of the overwhelming abundance of evidence from fossils and modern genetics, some people still question the validity of evolution and natural selection. This situation is not helped when some politicians and religious leaders denounce the theory of evolution; and instead claim that a higher-being is the creator and designer of all living organisms that have ever existed on this planet; and especially humans. It seems that this struggle between "science" and "creation" will never end, even though there are many people, including some scientists, who have deep religious or spiritual beliefs and also accept evolution.

(96) and (97) The Scopes Trial of 1925 questioned the legitimacy of Evolution

The controversy between believers in evolution and creationists came to a crescendo in 1925 during the "Scopes Trial" in the United States. Formally known as The State of Tennessee v John Thomas Scopes, this was an American legal case in which a high school biology teacher, John Scopes, was accused of violating Tennessee's Butler Act, which made it unlawful to teach human evolution in any state funded public school. The trial was deliberately staged in order to attract the maximum publicity to the small rural town of Dayton in Tennessee. Scopes passionately believed in freedom of speech and the freedom to teach the theory of evolution as it applied to humans, in his biology classes. So even though he knew the consequences of his actions, he taught evolution as a matter of principle. Scopes was found guilty and was fined $100, the equivalent of around $1,500 in today's dollars; but the verdict was appealed and overturned on a matter of legal technicality. Nevertheless, the trial had served its purpose. It brought international attention to the issue of religious-based censorship and the curtailing of freedom of speech and scientific expression. This trial brought to a head, the polarizing views of "modernists" who believed evolution was not inconsistent with religious beliefs, and "fundamentalists" who believed the word of God as literally revealed in the Bible, took priority over all human scientific knowledge.

Let us now turn to the fundamental question of "what is life?" We all instinctively feel what is alive and what isn't, but how can we make the distinction rationally and scientifically? What is it that defines life? How can we tell if one thing is alive and another isn't? Outlined here, are a number of characteristics that are unique to all living organisms. All of these characteristics, when combined, are only found in living organisms. Firstly, living things are highly organized, with structured and coordinated parts. For example, all living things are made up of one or more cells containing DNA, which are the essential building blocks of life. In the case of multicellular organisms, the cells are specialized to perform different functions and are organized into tissues, such as muscle and nervous tissues. Tissues in turn make up organs such as the heart, brain or lungs.

Characteristics of Life

1. **GROWTH & DEVELOPMENT**
   – *get bigger, more complex, or develops in some way*

 2. **ENERGY METABOLISM**
   – *eat, breathe, excrete waste; energy usage*

 3. **HOMEOSTASIS**
   – *maintain a relatively controlled internal environment*

 4. **ADAPTATION**
   – *changes over time due to natural selection and mutation*

5. **RESPONSE TO STIMULI**
   – *respond to things in their external environment (often as movement and adaptation over a period of time)*

 6. **CELLS**
   – *made of at least one cell containing highly complex structures organized chemical processes*

7. **REPRODUCTION**
   – *generate offspring with new combinations of parent DNA*

(98) The Unique Characteristics of Life

The second characteristic of all living organisms is the ability to metabolise. Living things must use energy and consume nutrients to carry out the chemical reactions that will sustain life. The aggregate total of all the biochemical reactions occurring in an organism is called its metabolism. There are two types of metabolism; anabolism and catabolism. In anabolism, organisms produce complex chemicals from simpler ones; whilst in the case of catabolism, they do the opposite. Anabolic processes will typically consume energy whereas catabolic processes will make stored energy available for the organism to consume.

The third characteristic of all living organisms is the ability to regulate their internal environment and temperature to maintain the relatively narrow range of conditions for their cells to function. This is called homeostasis. For example, the body temperature of a human needs to be kept relatively close to 37 degrees Celsius for the body to function in a healthy and stable manner.

The fourth characteristic common to all living organisms is regulated growth. Individual cells will become larger in size, and in the process, multicellular organisms will accumulate many cells as a result of the process of cell division. All humans for example, start life as a single solitary cell, yet when fully developed, the typical human possesses tens of trillions of cells in his or her body! Cellular growth will depend on anabolic pathways that build large and complex molecules such as DNA, the raw material of all genes.

The fifth essential characteristic found in all living organisms is the ability to reproduce, creating new organisms. Reproduction can be either asexual, which involves a single parent, or sexual, which requires two parents. Single-celled bacterium for example, can reproduce asexually by simply splitting into two. In sexual reproduction, two parent organisms produce sperm and egg cells containing half of their genetic information respectively. When these cells fuse together a new individual organism will be formed with a new set of genes. This process is called fertilization.

The sixth essential characteristic displayed by all living organisms is an ability to react to any changes in their environment. This is called response. For example, any human will instinctively pull their hand away from an open, intense flame; many plants will turn to the Sun as it moves across the daylight sky (like sunflowers); and unicellular organisms may migrate towards a source of nutrient chemicals or away from toxic chemicals. All of these are responses to a change or stimuli, in the external environment.

Finally, all living organisms will undergo a process of evolution, which means that the genetic makeup of the population of a species may change over an extended period of time. In some cases, evolution will occur as a result of natural selection, in which a particular inherited trait, such as a darker fur colour, or a narrower beak shape, will allow organisms to survive and reproduce better in a particular type of environment. Over a number of generations, these inherited preferred traits will become more and more common in the population, making that species better suited to its environment. This process is called natural adaptation.

It should be pointed out that some non-living things may possess some of the seven characteristics of life, outlined above, but not all of them. For example, crystals of ice are organized (even though they don't have cells) and can grow under the right cold temperature conditions; but do not meet any of the other essential characteristics of life. Similarly, a fire can grow and reproduce by creating new fires; as well as respond to external stimuli. But fire is not organized, does not maintain homeostasis and cannot evolve.

What constitutes life and what doesn't, is not always clear cut. For example, are viruses living organisms? Viruses are tiny structures of protein and nucleic acid that can only produce inside the cells of hosts (other organisms, including humans). Viruses possess some of the properties of life, but they cannot reproduce of their own accord and outside of a host. They also do not carry out their own metabolism and do not appear to maintain homeostasis. Should viruses be classified as living organisms? This is debateable, even amongst biologists. They certainly have the ability to create major disruption to human health and society as was witnessed during the 2020 Covid19 Pandemic which struck the planet in a big way. Many researchers believe the Covid19 virus started in the Chinese city of Wuhan, and then expanded at first across China and then internationally. It is believed the virus originated in bats, but exactly how it jumped to people is unknown. Many of the earliest people diagnosed with the virus had visited a meat and animal wet market in Wuhan. It is possible the virus found its way into animals sold at the market, such as foxes, racoons and sika deer, who may have been infected by bats. As of early May 2021, nearly 18 months after the virus first appeared, over 155 million people worldwide had contracted the virus, and 3.2 million people had died. The virus also triggered a world-wide recession, causing havoc to the lives of people. The good news is that as I write this book in early May 2021, at least three effective vaccines are in global circulation. In spite of this, the virus is still rampant in many parts of the world. Also, the threat of new viral epidemics emerging in the future remains ominously real. This is the reality of the Anthropocene we humans have created. We don't control the planet and the natural world as much as we think we do.

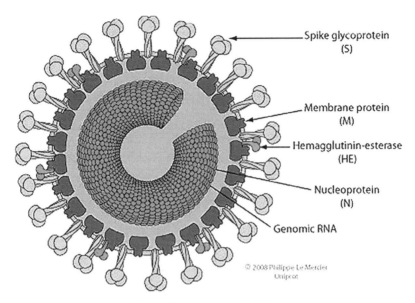

Spike glycoprotein
(S)

Membrane protein
(M)

Hemagglutinin-esterase
(HE)

Nucleoprotein
(N)

Genomic RNA

© 2008 Philippe Le Mercier
Uniprot

(99) The Structure of a Virus

Here are what a selection of world leaders had to say about the Covid19 pandemic during 2020. Australian Prime Minister Scott Morrison said: "In the absence of a vaccine, we may have to live this way for years." Brazilian President Jair Bolsonaro said: "This is the reality; the virus is there. We have to face it, but face it like a man, damn it, not like a kid. We'll confront the virus with reality. That's life. We're all going to die one day." After falling seriously ill and then recovering from Covid19, British Prime Minister Boris Johnson said: "It was a tough old moment, I won't deny it. They had a strategy to deal with a 'death of Stalin'-type scenario. I was not in particularly brilliant shape and I was aware there were contingency plans in place. The doctors had all sorts of arrangements for what to do if things went badly wrong." Canadian Prime Minister Justin Trudeau said: "We're on the brink of a fall that could be much worse than spring." President Xi Jinping of China said: "We should enhance solidarity and get through this together. We should follow the guidance of science, give full play to the leading role of the World Health Organization and launch a joint international response. Any attempt of politicizing the issue, or stigmatization, must be rejected." French President Emmanuel Macron said: "We must as long as possible allow our fellow citizens in nursing homes to continue having interactions. We want to avoid the radical and massive response we made in March-April (2020) when we had less understanding of the virus and had fewer tests available." Russian President Vladimir Putin said: "We are ready to share experience and continue cooperating with all states and international entities, including in supplying the Russian vaccine which has proved reliable, safe, and effective, to other countries." Indian Prime Minister Narendra Modi said: "Not one, not two, but as many as three coronavirus vaccines are being tested in India. Along with mass production, the roadmap for distribution of the vaccine to every single Indian in the least possible time is also ready." US President Donald Trump said: "We must hold accountable the nation which unleashed this plague onto the world, China. The Chinese government, and the World Health Organization, which is virtually controlled by China, falsely declared that there was no evidence of human-to-human transmission. Later, they falsely said people without symptoms would not spread the disease. The United Nations must hold China accountable for their actions." After her 78 years old father died from Covid19, Silvia Bertuletti from Bergamo in northern Italy said: "My father was left to die alone, at home, without help. We were simply abandoned. No one deserves an end like that." Finally, the Catholic Pope Francis had this to say: "If the pandemic has called into question the scale of values that sets money and power over all else; it has toppled the shaky pillars that supported a certain model of development."

It is amazing to think that every human being who ever lived on this planet, won the ultimate race. They won the race amongst the average of 180 million sperm cells ejaculated by their father to fertilize the female egg, at the time they were conceived as an embryo within the wombs of their mother. So regardless of how a human chooses to live their life, and whatever good fortune or misfortune may afflict them; they all start their journey of life with the massive good fortune of having fertilized that egg! But how and why did humans evolve on this planet. This fascinating story is recounted in the next chapter.

8

EVOLUTION
OF HUMANS

The sixth threshold of increasing complexity in the journey of Big History was the appearance of Homo sapiens, humans, in the Rift Valley of East Africa around 300,000 years ago. The goldilocks conditions for the emergence of Homo sapiens was a long preceding period of evolution which led first to mammalian primates, then to numerous species of hominines (not fully apes and not fully human), each developing more complex manipulative, perceptive and neurological capacity over an extended period of time, which ultimately led to us. The history of evolution from primates to Homo sapiens is by no means complete; there are numerous missing links yet to be connected, so as a result, paleoanthropologists differ in their opinions on timelines and the journey evolution took, which ultimately led to humans. Here is my attempt to piece together this journey of human evolution.

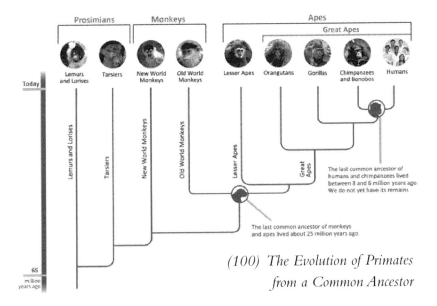

(100) The Evolution of Primates from a Common Ancestor

The path for the emergence of primates was cleared with the asteroid impact in the Gulf of Mexico 66 million years ago which led to the sudden extinction of dinosaurs along with 75 percent of all other species living on the planet at the time, after the impact created massive earthquakes and tsunamis and caused a dramatic shift in the climate of the planet. This was significant for the evolution of humans because our small mammalian ancestors were severely restricted from further evolving due to the fact they were primarily a food source of the carnivorous dinosaurs. After the dinosaurs became extinct small mammals were able to adapt and evolve to the point where around 55 million years ago, the earliest species of primates emerged. These animals would move quickly by swinging from one tree to the next using only their long arms. By 25 million years ago the first hominids, or great apes had emerged, with enlarged brains and no tails.

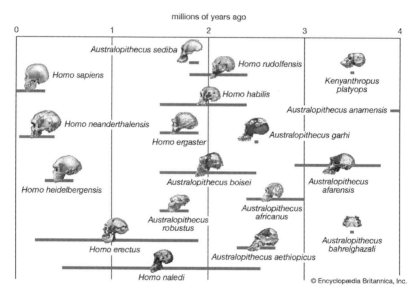

(101) *The Evolution of Hominines*

Then something significant began to happen around 7 million years ago when the earliest hominines appeared in Africa. The hominines were groups that had evolved from ape-like hominids but developed a series of features that distinguished them from their hominid ancestors, such as apes, chimpanzees, gorillas and orang-utans. The first of these hominines to emerge were Sahelanthropus, Orrorin and Ardipithecus; all three of which differed from their hominid ancestors because they had reduced canine teeth and a partial-limited bipedal capability, which meant they could walk on two legs for relatively short periods of time. Sahelanthropus tchadenis is one of the oldest known species in the ancestral human (hominine) family tree. This species lived between 7 and 6 million years ago in west-central Africa. Walking upright intermittently may have helped our ancestors survive in diverse habitats including forests and grasslands. They had a combination of ape-like and human-like features. Their ape-like features included a brain similar in size to that of a chimpanzee, a sloping face, very prominent brow-ridges and an elongated skull. The human-like features included small canine teeth, a short middle part of the face, and a spinal cord opening underneath the skull instead of towards the back as seen in non-bipedal apes. Some of the oldest evidence of a human-like species moving about in an upright position comes from Sahelanthropus.

By 4.2 million years ago the earliest hominines were replaced by Australopithecus, the first fully bipedal species, which also originated in Africa. Australopithecus afarensis is one of the longest lived and best known early human ancestral species. Paleoanthropologists have uncovered the remains of more than 300 individuals from this species. Existing between 3.85 and 2.95 million years ago in eastern Africa, this species survived for over 900,000 years. Similar to chimpanzees, their children grew rapidly after birth and reached adulthood earlier than modern humans. Australopithecus afarensis had both ape and human characteristics. Members of this species had ape-like facial proportions which included a flat nose and a strongly projecting jaw, as well as a brain slightly larger than chimpanzees, but only one-third the size of a modern human brain. They had long strong arms with curved fingers which were adapted for climbing trees; small canine teeth like all other early human ancestors; and a body that stood on two legs and regularly walked upright. Their adaptations for living both in the trees and on the ground helped them survive for over 900,000 years in spite of changes to the climate and environment over that period of time.

The most famous Australopithecus discovery was that of "Lucy" in 1974. Lucy was one of the first hominine fossils to gain worldwide recognition. At the time of Lucy's discovery, her skeleton was around 40% complete, by far the most complete early hominine known. The discovery occurred on 24th November 1974 when paleoanthropologist Donald Johanson was exploring the ravines and valleys of the Hadar river in the Afar region of north eastern Ethiopia in Africa when he spotted an arm bone fragment protruding out of a slope. Johanson later recounted that his heartbeat quickened with excitement when he realised it belonged not to an ape or monkey, but a hominine. As his team found more and more bone fragments, they began to realise they were uncovering an exceptionally unique skeleton. The full excavation of Lucy took three weeks to complete. Lucy's skeleton consisted of 47 out of 207 bones, including parts of her arms, legs, spine, ribs and pelvis; as well as the lower jaw and several skull fragments. Unfortunately, most of the hand and foot bones were missing. Nevertheless, they were able to determine from the shape of the pelvic bone, that the skeleton was a female. Lucy measured 1.05 metres in length and weighed around 28 kilograms, yet an erupted wisdom tooth and the fusing of certain bones indicated she was a young adult. She was given the name "Lucy" because the excavation team would often play the Beatles classic hit song "Lucy in the Sky with Diamonds" on their tape recorder back at the camp, while relaxing at night under the majestic stars of the Milky Way. The "Lucy" skeleton was formally dated at 3.18 million years old. Initially, Johanson believed Lucy was either a member of the genus Homo or a small Australopithecine. Only after analysing other fossils subsequently uncovered nearby and at Laetoli in Kenya did scientists establish these ancestral remains represented a new species, Australopithecus afarensis, four years after Lucy's discovery. After studying the injuries to Lucy's bones, a group of paleoanthropologists determined in 2016 that she died from injuries sustained from falling out of a tree. However, another group of researchers who have also examined the injuries to Lucy's bones have disputed this claim, concluding instead that Lucy was trampled to death by a stampede of wild animals.

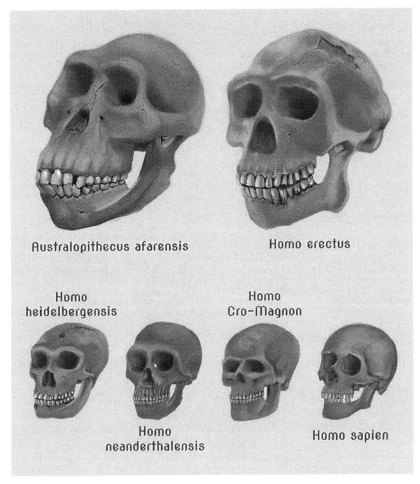

(102) Comparison of Hominine Skulls

In 2015 the discovery of a new fossil pushed evidence for the origin of the human genus, Homo, back hundreds of thousands of years earlier than previously thought. Discovered on a hilltop in Ethiopia, the 2.8 million-year old jaw containing five teeth gave paleoanthropologists a valuable insight into a period that we have little evidence about; the changes to our human ancestors between 2 and 3 million years ago. This fossil discovery did not provide one simple answer, but it did provide some clarification on the transition from Australopithecus fossils with smaller brains and ape-like jaws, such as the famous "Lucy" discovery; and members of the Homo genus with progressively larger brains that eventually led to modern humans. The fossil discovery was also intriguing because it displayed traits from both Australopithecus and Homo, suggesting a rapid transition from one species to the other. The whole topic of the earliest species of Homo is hotly debated amongst paleoanthropologists, but it is becoming increasingly clear there were multiple early species in East Africa and that the earliest changes (or adaptations) included a decrease in tooth and jaw size, possibly related to changes in the type of foods that were being eaten. The discovery of this fossil and others adds weight to the likelihood that the earliest species of Homo emerged from Australopithecus afarensis. The discovery of a skull in Ethiopia in 1985 helped define another early distinct ancestor, the species Paranthropus aethiopicus which lived from 2.7 to 2.3 million years ago. This species had a strongly protruding face, large teeth, a powerful jaw and a well-developed crest on top of its skull, indicating it had large chewing muscles, with a strong emphasis on the muscles that connected towards the back of the crest, creating a powerful chewing force on the front teeth. There is evidence that Paranthropus aethiopicus was one of the first species to produce rudimentary stone tools to chop through animal flesh, bone and tree-bark.

Between 2.4 and 1.4 million years ago an early species of Homo roamed the plains of eastern and southern Africa, Homo habilis. This species had a slightly larger brain and a smaller face and teeth than its predecessors, the Australopithecus and other older hominines; but it still retained long ape-like arms. Its name, which means "handy man" was given in 1964 because this species is believed to be one of the first makers of stone tools for cleaving meat from bone. Homo habilis was discovered between 1960 and 1963 by a team of scientists led by Louis and Mary Leakey; two of the most prominent fossil hunters of the twentieth century, known for their many discoveries relating to early human evolution. Mary Leakey originated many of the methods paleoanthropologists use today. She was the most systematic and scientifically trained of the pair. Her intuitive husband Louis, would select a site based on a hunch, and Mary then followed through with an exhaustive and meticulous search, often while Louis was abroad raising money for their future expeditions. Although Louis was usually credited in the popular media for their discoveries, and was often pictured with the skulls they excavated, he never actually found a skull himself. It was Mary who made the discoveries, but she later said that both of them complimented each other when conducting their research expeditions and excavations. As part of his research into early humans, Louis wanted to collect more information on the behaviour and social interactions of living apes. With this in mind, he recruited three young women; Jane Goodall, Dian Fossey and Birute Galdikas, to conduct their research respectively on chimpanzees, gorillas and orang-utans in the wild.

(103) Jane Goodall   (104) Dian Fossey

Let us briefly examine the achievements of these three remarkable women. Englishwoman Jane Goodall, is considered to be the world's most renowned expert on chimpanzees and is best known for her 60-year study of the social and family interactions of wild chimpanzees in Tanzania Africa, starting from 1960. Instead of allocating a number to each chimpanzee she studied, they were given their own names such as "Fifi" and "David Greybeard." She discovered that chimpanzees have unique and individual personalities, stating: "It isn't only human beings who have personality, who are capable of rational thought and emotions like joy and sorrow." Goodall observed human-like behaviours in chimpanzees like hugs, kisses, pats on the back and tickling. Her findings concluded that humans and chimpanzees not only share 98% identical DNA, but also similarities in emotion, intelligence, family and social relationships. Goodall had this to say about chimpanzees: "In what terms should we think of these beings, nonhuman yet possessing so very many human-like characteristics? How should we treat them? Surely we should treat them with the same consideration and kindness as we show to other humans; and as we recognize human rights, so too should we recognize the rights of the great apes? Yes."

Dian Fossey was an American zoologist who was best known for her extensive 18-year study of mountain gorillas in their natural rainforest habitat in Rwanda Africa. Fossey began her studies by imitating the behaviour of these gorillas, such as feeding, munching and scratching. She soon gained the trust of the mountain gorillas and began to observe their social behaviour at close range. Fossey became very attached to one male gorilla called "Digit." She first encountered him in 1967 when he was around 5 years old. Tragically, Digit was stabbed to death by poachers in 1977 while trying to defend his group. Fossey used his death to increase awareness world-wide about the mountain gorillas and to raise funds for anti-poaching patrols. Tragically, Fossey was hacked to death in her mountain cabin in December 1985. Although the murderers were never brought to justice, it is believed she was killed by poachers. Her 1983 autobiography "Gorillas in the Mist" became a best seller and was made into a movie of the same name in 1988. Here are two quotes from her book: "There was no way that I could explain to dogs, friends, or parents my compelling need to return to Africa to launch a long-term study of the gorillas. Some may call it destiny and others may call it dismaying. I call the sudden turn of events in my life, fortuitous." She also said: "The more you learn about the dignity of the gorilla, the more you want to avoid people."

German born Birute Mary Galdikas is an anthropologist, conservationist and educator who has studied and worked closely with the orang-utans of Indonesian Borneo in their natural habitat; becoming the world's foremost expert on orang-utans in the process. She has extensively observed and documented the ecology and behaviour of orang-utans in the wild. After 40 years of observational research, Galdikas has conducted one of the longest continuous studies of any wild mammal in the world. Amongst her findings was determining that orang-utans have an average birth interval of 7.7 years and that they consume over 400 types of food. She has also extensively documented their societal structure and mating habits. Galdikas is quoted as saying: "We must ensure a future for great apes in the wild or we will be left alone on this planet with no family to call our own except for ourselves."

(105) Meave Leakey

For over 50 years, British-born paleoanthropologist Meave Leakey has been unearthing fossils of our early ancestors in Kenya's Turkana Basin. Her discoveries have changed how we think about our origins. Instead of an orderly ape-to-human progression, her work suggests different pre-human species lived simultaneously. Leakey has co-written a book in 2020, "The Sediments of Time: My Lifelong Search for the Past," with her daughter Samira. The book reflects on her life as a paleoanthropologist and pieces together the story of the climate-driven evolution of our species. Leakey is part of a famous family of paleoanthropologists. Her husband, Richard Leakey, and his parents, Louis and Mary (referred to earlier in this chapter), are known for their discoveries of early hominines. In 1999, Meave Leakey was part of a team that found the skull of an early hominine that was approximately the same age as the famous Lucy (Australopithecus afarensis), the 3.18 million years-old fossil skeleton discovered

(106) Richard Leakey

in Ethiopia in 1974. Leakey named the discovered hominine Kenyanthropus platyops, also known as the flat-faced man from Kenya.

In an interview she gave about this discovery, Meave Leakey said: "Lucy got a huge amount of publicity. She was always projected as the common ancestor of humans. I always felt that it didn't make sense, because if you looked at any other animal lineage there were always so many species. I thought; there has to be diversity in the early hominines. When we found this specimen, it was crushed and broken, so it took a long time to make sense out of it. But you could tell it was something completely new, and different from Lucy. It was living contemporaneously with Lucy, but had this really flat face. The significance was far reaching; it showed Lucy was not necessarily the ancestor of all later hominines."

With regard to periodic attempts to dispute Africa as the birthplace of our human ancestors, Meave Leakey had this to say: "Early paleoanthropologists didn't believe that humans could have come from Africa. There was a prejudiced insistence that humans must have originated in Europe. The work to convince the scientific community and the world otherwise was started by my parents-in-law and continued by my husband, myself and my daughter Louise. As I have gone through my career it has become more and more accepted. Definitely Africa is where it all began. The climate and the vegetation were right. And, for me, east Africa is most likely, because if you look at where non-human primates are distributed today, they concentrate around the tropics and the equator."

Finally, in relation to the question of how our early human ancestors evolved their tremendous brain power and ability to walk on two legs, Meave Leakey had this to say: "Evolution happens because of changing habitats driven by climate change. Driven by a drying trend, towards more open savannah, I suspect our ancestors started coming down out of the trees to the ground. They found if they stood on two legs they could reach food, like berries and fruits on bushes, better; and they could travel further. Big brains came later, after bipedalism and increased dexterity. Brains are expensive in terms of calories. To develop a big brain, you have to have a good source of food. When our ancestors started a way to hunt and catch a lot of meat, they were able to evolve bigger brains."

The earliest known evidence of our human ancestors outside of Africa was found in the Shangchen region of southern China when in 2018 paleoanthropologists excavated stone tools that have been dated to around 2.12 million years ago. Prior to this discovery the earliest human ancestral skeletal remains and artefacts discovered out of Africa were found in Dmanisi, in the Republic of Georgia in 2000. These were believed to be linked to Homo erectus and were 1.85 million years old. At this stage, we don't know if the discovery in China in 2018 is connected to Homo erectus or another species. We have evidence to substantiate that the Homo erectus species originated in Africa around 1.89 million years ago and were in existence as recently as 110,000 years ago, when they became extinct. They are the earliest human ancestors to have possessed modern human-like body proportions including relatively elongated legs and shorter arms compared to the size of their torso. These features are believed to have evolved as adaptations to a life lived entirely at ground level on the savannah plains, completely severing the tree-climbing links of their ancestors. Homo erectus had larger brains and the ability to walk upright and run long distances, a handy trait when escaping predators. There is fossil evidence to indicate that this species cared for their elderly and weaker individuals. They are also responsible for the first major innovation in stone-tool technology, hand-axes. We know that Homo erectus began to migrate out of Africa 1.8 million years ago first to Eurasia and then arriving at the Indonesian island of Java, around 1.5 million years ago. Homo erectus appear to have been

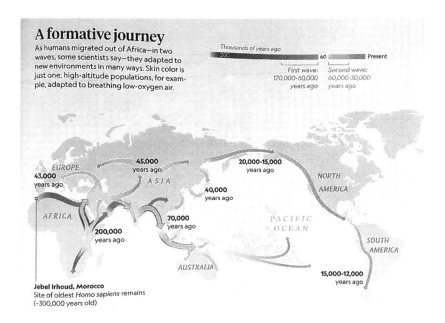

A formative journey

As humans migrated out of Africa—in two waves, some scientists say—they adapted to new environments in many ways. Skin color is just one; high-altitude populations, for example, adapted to breathing low-oxygen air.

Thousands of years ago

200   60   Present

First wave: Second wave:
120,000-60,000   60,000-30,000
years ago   years ago

EUROPE
43,000 years ago

ASIA

45,000 years ago

20,000-15,000 years ago

NORTH AMERICA

40,000 years ago

AFRICA

70,000 years ago

200,000 years ago

PACIFIC OCEAN

SOUTH AMERICA

AUSTRALIA

15,000-12,000 years ago

Jebel Irhoud, Morocco
Site of oldest *Homo sapiens* remains (~300,000 years old)

(107) The movement of Humans out of Africa

the first hominine species to use and control fire for cooking meat and for defensive purposes against predators. Then 900,000 years ago there is evidence they were using flint scrapers for preparing animal hides for clothing.

Approximately 800,000 years ago there appeared another early human ancestral species called Homo heidelbergensis with a very large brow-bridge and an even larger brain than earlier species. They were believed to be the first human ancestors to live in colder climates such as in northern Europe, and possessed short, wide bodies which were more than likely an adaptation to the colder climate, helping them to preserve body heat. There is evidence that they were very capable of using and controlling fire and were the first human ancestors to actively hunt large animals for food and clothing hides. Homo heidelbergensis also achieved a major milestone, being the first species to build shelters in the form of simple dwellings made out of wood and rock. Homo heidelbergensis is an important link in our evolution story because it appears to be the common ancestor of Homo sapiens, Neanderthals and Denisovans. It is believed Homo heidelbergensis left Africa 600,000 to 500,000 years ago.

Our closest extinct relatives were Homo neanderthalensis, better known as Neanderthals. They first emerged around 500,000 years ago and became extinct approximately 40,000 years ago. They had distinctive skulls, including a large middle part of the face, angled cheek bones, and a very large nose for humidifying and warming cold, dry air. Their bodies were shorter and stockier than humans, another adaptation to living in cold environments; and they had brains equal to, or a little larger than the human brain. Neanderthals made and used a diverse range of quite sophisticated tools, effectively controlled fire, lived in shelters, made and wore clothing, were skilled hunters of large animals, ate a variety of plants, and occasionally made symbolic or ornamental objects. They developed symbolic art in the form of marine shells painted with mineral pigments, 115,000 years ago, apparently before humans did. There is also evidence to suggest that Neanderthals deliberately buried their dead and occasionally even marked the graves with offerings, such as flowers. No other primates, or earlier species of hominines, had ever practised this kind of sophisticated and symbolic behaviour.

There is also compelling evidence that Neanderthals interbred with Homo sapiens. Neanderthals have contributed between 1-4% of the genomes of non-African modern humans; although a modern human who lived 40,000 years ago has been found to have between 6-9% Neanderthal DNA. The evidence we have of interbreeding between Neanderthals and Homo sapiens helps us to shed light on

(108)   Comparison of a Neanderthal on the left with a Human on the right

the migration patterns of modern humans out of Africa. Neanderthals could not have contributed to the genomes of modern African peoples because by the time Homo sapiens emerged in Africa 300,000 years ago, Neanderthals had already moved out of Africa. We now have evidence of Neanderthal-human interbreeding going as far back as 100,000 years ago. This of course means that the first humans must have left Africa at least 100,000 years ago. This is confirmed because there is evidence demonstrating humans began their exodus out of Africa between 200,000 and 130,000 years ago.

One big mystery which confounds paleoanthropologists is what caused the Neanderthals to become extinct 40,000 years ago? Neanderthals were widespread across Europe and western Asia for a long period of time, starting from at least 400,000 years ago. Things then began to dramatically change when populations of Homo sapiens, our direct human ancestors, migrated from Africa to Europe, right up until 45,000 years ago. Literally 5,000 years later, not a single Neanderthal remained. What happened? The most commonly accepted theory amongst paleoanthropologists is that competition for resources and changes to their habitat resulting from climate change, were the two main factors which led to their demise. Neanderthals had adapted to hunt large, Ice Age animals. When the last Ice Age began to end 40,000 years ago, the climate became warmer and the large animals they hunted, such as mammoths and large bears, became extinct. This may have resulted in the Neanderthals literally starving to death. We need to understand that unlike Homo sapiens by 40,000 years ago, the Neanderthals had not migrated around the world, but instead had settled in what today is northern Europe and Siberia where the climate was colder, which suited them. So because they were not widely dispersed around the planet, when the Ice Age ended and their source of food vanished, they were unable to survive. On the other hand, our direct human Homo sapiens ancestors had established long-distance trade networks which buffered them against times of climate change when their preferred foods were not available. We had an extinction insurance policy of being scattered around the world and being able to adapt to different climates, whereas the Neanderthals did not.

(109) The seven million year Evolution from Primates to Hominines and to Humans

Another extinct Homo species who shared the common ancestor of Homo heidelbergensis with Neanderthals and Homo sapiens, were the Denisovans, the remains of whom were first discovered in the Denisova cave in Siberia in 2010. It appears the Denisovans settled in Siberia and South East Asia. The dating of the few Denisovan fossils that have been discovered indicate this species first appeared 500,000 years ago and became extinct around 30,000 years ago. So, it seems that between 765,000 to 550,000 years ago, one group of Homo heidelbergensis left Africa and expanded into Eurasia, when they then split. Those that moved west into the Middle East and Europe evolved into Neanderthals; and those that moved east into Siberia and South East Asia evolved into the Denisovans. The other group of Homo heidelbergensis which remained in Africa, evolved into us, Homo sapiens. There is evidence that Homo sapiens and Denisovans interbred when our human ancestors arrived in South East Asia between 120,000 and 65,000 years ago. Research conducted on Denisovan DNA in Papua New Guinea reveals that the Neanderthals and Denisovans also interbred around 46,000 and 30,000 years ago. As of 2020, the only Denisovan skeletal remains recovered from a different site to Siberia were discovered in May 2019 in the Baishiya Karst cave in Gansu China. These fossil remains were found to be at least 160,000 years old. Interestingly, modern Tibetans, Melanesians and Australian Aborigines carry around 3-5% of Denisovan DNA. It appears that the Neanderthals became extinct 40,000 years ago, 10,000 years before the Denisovans did, 30,000 years ago; making the Denisovans the most recently lived Homo species apart from our own.

Sometimes it appears that evolution goes backwards rather than forwards. When a hominine fossil is examined by paleoanthropologists, it can sometimes be difficult to determine whether the species being examined retained a primitive trait from an earlier ancestor, such as a smaller brain size, or whether in fact it lost the characteristic and then re-evolved it. The strange case of Homo floresiensis may be an example of the latter. This human ancestral species lived on the island of Flores in Indonesia as recently as 50,000 years ago, yet in many ways resembled some of the founding members of the hominine genus who lived more than two million years earlier. Not only did Homo flore-siensis have a small body, but it also possessed a remarkably tiny brain for a Homo species, about the size of a chimpanzee's brain. The best estimate of paleoanthropologists is that this species descended from a "larger-brained" ancestral Homo species that became marooned on the island of Flores, and as a result, over a period of time evolved to a much smaller size in order to adapt to the limited food resources available on the island. As part of this adaptation process, Homo flore-siensis appears to have reversed a defining long term trend in the evolution of Homo species; the continuous expansion in the size of their brains. Nevertheless, despite its small brain, evidence has been uncovered to show that Homo floresiensis was still capable of making stone tools, hunt animals for food and cook, using open fires.

The next big event in the evolution of the Homo species was the emergence of Homo sapiens, humans, around 300,000 years ago in the Rift Valley of East Africa. Prior to 2017, our species was believed to be 250,000-200,000 years old; but in 2017 paleoanthropologists accurately re-dated a long overlooked Homo sapiens skull found in a Moroccan cave, to 300,000 years ago. It was during a period of dramatic climate change that our species evolved. Homo sapiens were nomadic hunters and gatherers; hunting animals and gathering plants, for food. They also evolved behaviours which enabled them to respond to the challenges of survival in what were unstable environments. Physiologically, Homo sapiens can be characterized by the lighter frame of their skeletons compared to earlier Homo species. Modern humans have very large brains which vary in size from population to population and also between males and females; although this variation in brain size within the genders of modern human species does not impact upon intelligence. Accommodating this bigger brain involved the reorganization of the skull into what is considered to be a "modern" thin walled, high vaulted skull with a flat and near vertical forehead. The faces of modern Homo sapiens also show very little if any of the heavy brow ridges prevalent in earlier Homo species. Our jaws are also less heavily developed, and have smaller teeth.

Let us now examine the lifestyles of our early Homo sapiens ancestors and how they survived. Prehistoric Homo sapiens not only made and used stone tools, they also specialized them for different purposes, making a variety of smaller, more complex and refined tools, including composite stone tools, fishhooks, harpoons, bows, arrows, spears and sewing needles. Our direct early ancestors had to find their own food. They spent a large part of each day hunting animals, a role mainly carried out by the males; and gathering plants, which was mainly carried out by the females and children. By 164,000 years ago

(110) Early Human foraging hunters and gatherers

humans were collecting and cooking shellfish, and by 90,000 years ago had begun making special fishing tools.

Homo sapiens have proven to be especially adaptable to changes in the environment. This trait may have developed because our species mostly stayed in Africa for the first one hundred thousand years of their existence. Whilst in Africa, some experts have argued, Homo sapiens evolved as a population of interconnected subgroups that spread across the continent, splitting up and reuniting again and again over tens of thousands of years, which allowed for periods of evolution in isolation, followed by opportunities for interbreeding and the cultural exchange of ideas, in other words, collective learning. This evolutionary path in Africa may have equipped Homo sapiens to become especially adaptable hominines, who were able to later migrate out of Africa and survive in a wide variety of climatic conditions and environments around the world.

Homo sapiens also evolved a unique combination of physical and behavioural characteristics, mainly attributed to their more complex brains. They were able to control and harness fire in a sophisticated way, build increasingly elaborate shelters, and create social networks with other tribal groups. Early Homo sapiens were nomadic in nature. They constantly moved around from one place to another, following seasonal patterns to secure their sources of food, be they land animals,

(111) Early Human village communities

fish or plants. They typically moved in groups of 50 to 100 men, women and children, all related to each other. As these groups encountered other groups during their nomadic wanderings, they became increasingly sophisticated with their social connections; including developing various ceremonial customs involving vocal chants and bodily movements; the first forms of symbolic communication, tribal "drum" music and dance.

Homo sapiens also began to develop the earliest forms of language. Their larger brains facilitated a level of self-consciousness and self-awareness never seen in any earlier Homo species. They were communicating with each other. We now see the early development of what was to become the most powerful attribute of Homo sapiens which distinguished them not only from all earlier Homo species but also all other living organisms on the planet; this was the capacity to engage in collective learning. Collective learning is the ability to acquire knowledge over a period of time and then pass it on to future generations. It is a very powerful force which humans were able to harness and exponentially develop. Collective learning is the single most influential reason why humans have developed so rapidly, technologically, socially and culturally over the last 300,000 years to the point of dominating the planet, as we do, for better and for worse.

Some of the other landmark achievements of Homo sapiens was their movement out of Africa into Eurasia initially in relatively small numbers, beginning around 200,000 years ago, and then as we have seen, in much greater numbers starting 130,000 years ago. They started using clothing made from animal hides on a widespread basis, around 170,000 years ago; and began to create art in the form of etchings on rock decorated with symbols, 100,000 years ago; and then with red ochre directly onto cave walls 73,000 years ago. At around the same time 75,000 years ago, they began to make jewellery from the threaded beads of shells.

Although humans had left Africa earlier, as described above, the larger exodus of human migration out of Africa started 130,000 years ago, more than likely as a result of a major change in climate which depleted food resources in east and central Africa, forcing waves of

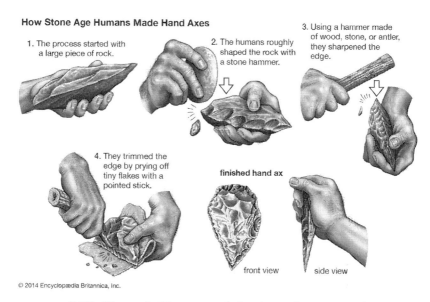

How Stone Age Humans Made Hand Axes

1. The process started with a large piece of rock.

2. The humans roughly shaped the rock with a stone hammer.

3. Using a hammer made of wood, stone, or antler, they sharpened the edge.

4. They trimmed the edge by prying off tiny flakes with a pointed stick.

finished hand ax

front view   side view

© 2014 Encyclopædia Britannica, Inc.

(112) How early Humans made hand axes from stone tools

humans to either venture into southern Africa or move north and then leave the continent entirely. This major change in climate triggered severe droughts that resulted in our human ancestors being driven close to starvation. After leaving Africa, Homo sapiens gradually spread around the entire planet in larger numbers; starting with the Middle East 130,000 years ago; central Asia 120,000 years ago; Europe 115,000 years ago; South East Asia 100,000 to 90,000 years ago; and ultimately into Australia 65,000 years ago.

The Australian Aborigines are the oldest continuous human culture on the planet. They first came to Australia during the last Ice Age, when sea levels were lower, allowing them to "island-hop" and sail from South East Asia in dug-out wooden canoes made from tree trunks. After they settled in Australia, the sea levels had risen after the end of the last Ice Age 12,000 years ago, resulting not only in the Aborigines remaining stranded in the Australian continent, but also a wide range of native animals and plants. By 20,000 years ago our human ancestors had crossed a land bridge connecting eastern Siberia and Western Alaska and moved into North America. That land bridge became submerged by the Pacific Ocean when the last Ice Age ended 12,000 years ago, and now forms the Bering Strait.

One of the most puzzling questions paleoanthropologists are seeking to answer is how our Homo sapiens ancestors interacted with the Neanderthals. It is important to understand that the Neanderthals were not our ancestors, but instead a species which evolved alongside and separate to us, originating from a common ancestor, Homo heidelbergensis. There is evidence to suggest both species, Homo sapiens and Neanderthals, were skilled fighters and aggressive warriors. This is an evolutionary trait. Predatory land mammals are territorial pack hunters. Like lions and bears, Homo sapiens and Neanderthals were cooperative big-animal hunters who sat atop the food chain and therefore had few predators of their own. So as both species increased in population, conflict would have inevitably developed in Africa over hunting territory not only between themselves, but also with other Homo species that were in Africa at the same time. We know that this territoriality has deep evolutionary roots in humans. It is possible that territorial disputes and competition for dwindling resources between the differing Homo species may have forced the Neanderthals to leave Africa in the first place. Territorial conflicts are also intense in our closest relatives, the chimpanzees. The brilliant 60-year longitudinal research of Jane Goodall demonstrated that male chimpanzees routinely ganged up to attack and kill males from rival groups, a behaviour eerily similar to human warfare. This suggests that an aggressive territorial warfare trait evolved in the earliest primate common ancestor of chimpanzees, Neanderthals, Homo sapiens and other Homo species, going back around 7 million years ago.

Homo sapiens and Neanderthals were very similar in their skull and skeletal anatomy, and also shared 99.7% identical DNA. Behaviourally, Neanderthals were very much like us. As we have seen, they made fire, buried their dead, created jewellery from seashells and animal teeth, made artwork and stone shrines. It is logical to assume if Neanderthals shared so many of our creative instincts, they more than likely shared many of our destructive aggressive instincts as well. The archaeological record certainly confirms Neanderthals were skilled big animal hunters, using spears to attack deer, bison and mammoths amongst other large mammals. It stands to reason Neanderthals would have

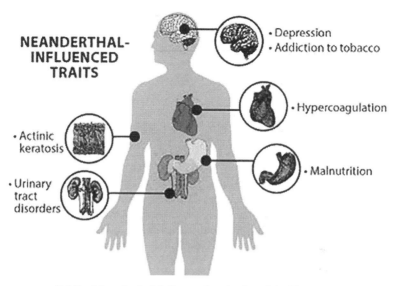

NEANDERTHAL-INFLUENCED TRAITS

- Depression
- Addiction to tobacco
- Hypercoagulation
- Malnutrition
- Actinic keratosis
- Urinary tract disorders

(113) Neanderthal-influenced traits found in Humans

used the same aggression to protect their tribal groups and territory if they were threatened. The skeletal remains of Homo sapiens and Neanderthals give us an intriguing insight into conflicts within, and quite possibly also between, these two groups, after Homo sapiens left Africa. Skulls excavated from both species often show trauma to the skull; almost certainly inflicted by clubs, which were fast, brutal, efficient and precise weapons, that had the capacity to shatter skulls. Another sign of possible warfare was evidence of fractures and breaks to the lower arms of both species, almost certainly caused by attempts to block blows inflicted by clubs. One Neanderthal skeleton excavated from a cave in Iraq was impaled by a spear to the chest. There is evidence of widespread trauma in the skeletal remains of young male Neanderthals. Of course, some of these injuries could have been sustained while hunting; but many of the injuries are what would be expected from warfare between tribal groups, albeit, small scale intense conflict, dominated by guerrilla-style raids and ambushes, rather than prolonged battles.

Although we do not have direct evidence of large-scale conflict between Homo sapiens and Neanderthals, the timing of both groups leaving Africa in large numbers and where they settled for extensive

(114) Neanderthals and Humans hunted mammoths to their ultimate extinction

periods of time, gives us logical clues and some insight. The best indirect evidence that Neanderthals not only fought Homo sapiens, but also prevailed at least in the earlier stages, is that they encountered us and were not immediately subjugated and controlled. Instead, for around 100,000 years the Neanderthals put up a resistance to human efforts to move into their territory in Europe the Middle East and western Asia. Why else did it take Homo sapiens so long to leave Africa, compared to the Neanderthals? The reason does not appear to entirely relate to a hostile environment giving rise to resource shortages unable to sustain a growing population of humans in Africa; but also seems to relate to the fact the Neanderthals were already thriving in Europe and western Asia and were able to successfully defend their territory from any human migratory incursion. The Neanderthals must have had the strategic advantage of intimately knowing the Middle Eastern/ European region north of Africa; such as its topography, landscape, the seasons, water sources and how to live off the native plants and animals; because they had occupied these lands for hundreds of thousands of years since leaving Africa. Also, when they encountered our human ancestors, their stronger muscular builds must have made them into very effective fighters in close-range in fierce "face-to-face" combat. Their much larger eyes would have given them superior low-light vision, allowing them to outmanoeuvre humans in night-time ambushes and sun-set or dawn raids.

So for a long period of time, there was a territorial stalemate with most humans still in Africa and Neanderthals well and truly settled in the Middle East, western Asia and Europe. But around 130,000 years ago this stalemate was broken when Homo sapiens finally were able to successfully migrate first into the Middle East-western Asia, and then into Europe on a larger scale. We don't exactly know what broke this stalemate. Was it because by this time Homo sapiens had invented superior weapons such as spear-throwers, which allowed the light-ly-built humans to attack the bulkier Neanderthals from a distance, using hit-and-run tactics? Or was the reason due to Homo sapiens developing superior hunting and gathering techniques which enabled them to feed bigger tribes, resulting in the humans outnumbering the Neanderthals in battle? Or was there another unknown reason? We just don't know.

What we do know is that even after the primitive Homo sapiens first broke out of Africa initially in small numbers, around 200,000 years ago, it took them over 150,000 years to conquer the Neanderthal lands. In Israel and Greece for example, there is archaeological evidence that shows the archaic Homo sapiens successfully occupied these lands, only to be pushed out again by a Neanderthal counteroffensive; before the humans regrouped in a final offensive starting 125,000 years ago, resulting in the final elimination of the Neanderthals. If this theory is correct, the Homo sapiens ultimately prevailed because the Neanderthals became extinct around 40,000 years ago. More than likely, I believe it was a combination of severe climatic change which eliminated their food sources; and direct conflict with our early human ancestors when they migrated out of Africa; that led to the ultimate extinction of the Neanderthals.

There are a number of theories of human evolution which have been developed by paleoanthropologists. I will highlight some of them very briefly. I think the reality of human evolution may involve a combination of events referred to in these theories. The facts are that at this stage, we are uncertain of all the conditions that led to the emergence of humans on this planet. The savannah hypothesis states that hominines were forced out of the trees they lived in and onto the expanding savannahs. As they emerged from the trees, they developed the ability of bipedalism, to walk upright on two feet; and superior vision. This idea was further expanded into the aridity hypothesis, which states that the savannah was expanding as a direct result of climate changes, creating increasingly arid conditions in east Africa, reducing the number of trees and food sources, and forcing early hominines onto the land. So as a result of intense aridity, hominines had to adapt and evolve in order to survive those conditions. The turnover pulse hypothesis states that extinctions due to extreme changes in environmental conditions adversely impacted specialist species with less ability to adapt, more than generalist species that could adapt better and therefore survive. In this hypothesis, our direct ancestors were generalist species and hence adapted, evolved and survived. A similar idea called the red queen hypothesis, states that species must constantly evolve in response to changes in the environment, in order to successfully compete for limited resources with co-evolving animals around them. The social brain hypothesis states that continuously improving cognitive abilities as a direct result of increasing brain sizes influenced by dietary changes, allowed our hominine ancestors to influence other co-evolving species and better control resources within their environment.

Around 74,000 years ago a massive super volcano erupted on the Indonesian island of Sumatra. The eruption was so substantial that it is believed volcanic ash was deposited as far as the Indian ocean, the South China Sea and perhaps into the African continent. It is also believed that gases ejected into the atmosphere may have caused average global temperatures to drop by as much as 18 degrees Celsius for a number of years after the eruption. This event, known as the Toba Catastrophe was over one hundred times more damaging than the mount Tambora eruption of 1815, the worst in recorded history. In his 2019 book, "End Times: A Brief Guide to the End of the World," the author Brian Walsh summarized a 2009 research paper on Toba's potential climate effects, as follows:

"Precipitation would have fallen by 45 percent, and vegetation cover would have shrunk dramatically, with broadleaf evergreen trees and tropical deciduous trees dying out. Imagine a winter that lasted for years, like something out of Game of Thrones, shrivelling life on land."

For our Homo sapiens ancestors, concentrated mainly in Africa, the Middle East, and western-southern Asia at the time of the Toba super eruption, life would have become very precarious and challenging. Genetic evidence suggests that between 50,000 and 100,000 years ago, our species experienced an extreme population reduction, dropping to as few as 2,000 individuals on the entire planet, from an earlier population that would have numbered in the hundreds of thousands. After existing for up to 230,000 years to that point in time, Homo sapiens almost became extinct; and the Toba super volcanic eruption may have been the cause of it.

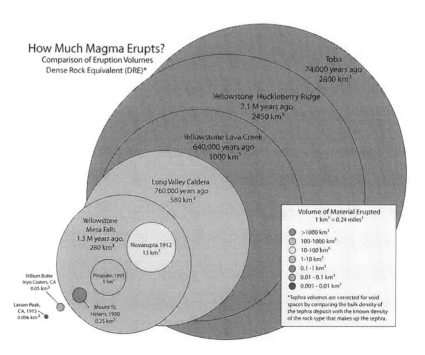

(115) The incredible magnitude of the Toba super-volcano eruption 74,000 years ago

How did our ancestors survive? Stanley Ambrose, an anthropologist at the University of Illinois and one of the main proponents of the Toba Catastrophe theory, hypothesizes that cooperation may have been the difference between survival and extinction. "You might think that in an apocalypse event that people would be stealing from each other," Ambrose has said. "It's not true. In a stable environment, when population density is high, you don't need to rely on your neighbours to get the next meal, and you may even need to defend yourself from them. But in a situation where it's the equivalent of a small lifeboat and everyone needs to cooperate, selfish people will be weeded out. You end up with a population that is more sharing and caring."

The Toba Catastrophe theory is controversial amongst anthropologists; many of whom have legitimate doubts about its impact on humans at the time. A number of them for example, question whether a worldwide ecological disaster actually did occur after the volcanic eruption. They cite evidence suggesting that humans in Africa were not affected. Other studies seem to show that vegetation in Africa did not suffer catastrophic die-offs. They also state that the Neanderthals living mainly in Europe at the time, did not appear to suffer any negative impacts from the eruption and subsequent adverse climatic effects.

Supporters of the Toba Catastrophe theory disagree with the conclusions of the above studies, and offer additional evidence which they claim shows that other large mammals also experienced severe drops in population around the time of the super volcanic eruption 74,000 years ago. We may never know conclusively whether the Toba eruption nearly brought about the extinction of our species, but as Walsh makes very clear in his book, the idea that our ancestors were almost wiped off the face of the Earth is very provocative. In his book, he states:

"You and I and everyone we know – everyone who came before us and everyone who might come after us – are here because the human beings who lived through Toba found a way to survive the eruption and its long, cold aftermath. Without their resource-

fulness, the human story could have ended in its earliest chapters.
Extinction was a possibility."

As the Homo species evolved, particular adaptations became more
prominent, because they allowed our hominine ancestors to survive in
changing environments mainly brought about by naturally occurring
Earth processes such as climatic change caused by atmospheric
disturbances and volcanic activity originating from convection forces
in the mantle. By the time our species emerged in Africa around
300,000 years ago, the following progressive adaptations allowed our
Homo sapiens ancestors to survive and flourish. Firstly, there was
bipedalism, the ability to walk upright on two legs. Advantages to
be found in bipedalism included the freedom to free up the hands
for labour and less physically exerting movement. Walking upright
allowed for longer distance travel and hunting; it also allowed for a
wider field of three-dimensional vision, important in the hunting
and gathering of food and for defence against looming predators.
Bipedalism also exposed less skin to the burning hot African sun; and
directly resulted in skeletal changes to the legs, knee and ankle joints,
as well as the spinal vertebrae, toes and arms. Even more significantly,

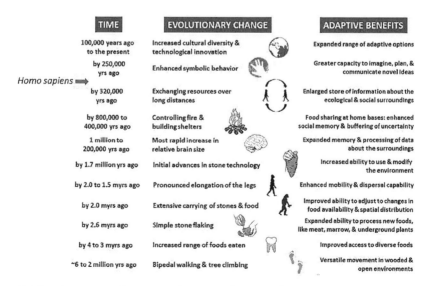

(116) Hominine evolutionary adaptations from 6 million years ago to the present

the human pelvis became shorter and more rounded, with a smaller birth canal in females, making the birth of offspring more difficult in humans compared to other primates. This resulted in shorter gestation periods of pregnancy, because babies needed to be born before their heads became too large; and more helpless infants who were not fully developed at birth and were totally dependent on their mothers.

Another significant adaptation which began with our earliest hominine ancestors was a progressively larger brain size. The ability of the human brain to grow after birth meant that language, creative thinking, self-consciousness, self-awareness and collective learning were all possible. It is believed that one of the biggest catalysts of brain growth was the cooking and eating of meat, after our human ancestors developed more sophisticated stone tools and the ability to harness fire. Over the course of human evolution, the size of the brain has tripled. Humans developed reduced sexual dimorphism or differences between males and females, as well as a hidden oestrus, which meant the female was fertile all year-round and showed no obvious external signs of fertility. The human sexes still do have some differences between them, with males generally being slightly larger and having more body hair and less body fat, compared to females. These changes may be related to the bonding of the sexes for the long-term raising of offspring. Other notable human adaptations included an overall reduction of body hair to better cope with a warmer environment, compared to earlier Homo species; the development of a chin; a descended larynx ("voice box") to enable more effective use of language; and an emphasis on vision for survival instead of smell.

Although hunter-gatherer societies still exist in small pockets within today's modern world, this lifestyle was the main way in which our early human ancestors lived right up to the transition to an agrarian-farming based lifestyle, starting around 12,000 years ago when the last Ice Age ended. In hunter-gatherer cultures, human beings obtain their food by hunting, fishing, scavenging, and gathering wild plants. Our early human ancestors often lived in groups of around fifty to one hundred people, consisting of several family units. They developed tools to help them survive and were dependent on a plentiful supply of food within their immediate area. If the available amount of food ran low, they would move on to greener forests or grasslands. Generally, as we have seen, the men would hunt wild animals and the women and children would forage for plants. There were a variety of lifestyles experienced by hunter-gatherer societies, so there was no single set of characteristics that applied to all of them. As time progressed, these societies became more complex. For example, stone tools became more sophisticated and specialised for particular purposes such as hunting and building basic shelters. Their knowledge of the environment, seasonal conditions and sources of food also became more sophisticated over time. The age of hunter-gatherer societies is referred to as the Palaeolithic Age; which can be further divided into the Lower Palaeolithic, lasting from 2.6 million years ago to 250,000 years ago; the Middle Palaeolithic from 250,000 to 50,000 years ago; and the Late Palaeolithic from 50,000 to 12,000 years ago, when the Ice Age ended and agriculture took hold, initially alongside the great fertile river systems of the world.

We have seen that our genus of Homo first developed within the abundant space of Africa, and it is here where the first hunter-gatherers appeared. There were particular locations in central, eastern and southern Africa where the land contained lush forests and ample grasslands which provided the ideal conditions for our early human ancestors to flourish. In southern Africa for example, sites such as Swartkrans cave and Sterkfontein provide evidence of early human occupation, although these are dated more recent than sites in eastern Africa, such as in Ethiopia, where stone tools dated from 2.6 million years ago have been uncovered. One of the oldest sites is Lake Turkana in Kenya, which was home to our early Australopithecine ancestors, including the famous Lucy.

As we have seen, early groups of Homo erectus were most likely amongst the first early human ancestors to venture out of Africa in large numbers, around 2 million years ago, migrating all the way to Eurasia, China and Indonesia 1.7 million years ago. Having said that, paleoanthropologists have also uncovered the remains of earlier unknown species dated as long back as 2.6 million years ago, in these regions as well. Europe was more than likely not explored by our early human ancestors until Homo heidelbergensis arrived there out of Africa between 800,000 and 700,000 years ago. Once in Europe, this species flourished and eventually evolved into Neanderthals, who then migrated as far north as Siberia, where Denisovans have also been recovered. By the end of the Middle Palaeolithic, almost the entire Eurasian landmass had been reached by our human ancestors. By the end of the Late Palaeolithic 12,000 years ago, humans had spread to South East Asia, Australia, the Pacific Islands and into North and South America. There was no environment which humans did not eventually learn to adapt to.

(117) Early Humans using fire – one of the greatest innovations of our ancestors

When it comes to shelters, most prehistoric hunters and gatherers would have used natural shelters as living space, such as overhanging cliffs and caves, which would have provided a place to escape wind, rain and extreme heat. The living spaces of the earliest hunters and gatherers were basic and not clearly structured. They would have included huts or tents with wooden supports, or even made from mammoth and other large mammalian bones. They also would have been illuminated at night by open fires in hearths. During the Late Palaeolithic, elaborate cave paintings appeared, indicating a level of development in our ancestors that enabled them to express themselves artistically and symbolically. The Chauvet and Lascaux caves in France both provide exceptional examples of hunter-gatherer art. By this stage, it was clear that self-awareness and collective learning was becoming a powerful force within the societies of our early ancestors. As their level of stone-age technology continued to develop, our ancestors were able to adapt to and control a variety of challenging environments around the world, ranging from scorching deserts, dense tropical forests and to frigid icy tundras.

The exact types of food hunter-gatherers consumed definitely varied depending on the environment in which they lived and the available flora and fauna. Some groups specialized in the hunting of prehistoric mega-animals such as giant elks, bears, woolly mammoths and large rhinoceros; whereas others might have focused more on trapping smaller wild animals, or on fishing. Scavenging was just as important to our early ancestors as hunting and gathering was. The earliest humans did not have sophisticated hunting tools or strategies capable of bringing down large wild animals, but they were meat eaters. Of course, once they obtained their food, they had to process it; either by using their powerful teeth to grind down tough plants with strong molars, and biting into non-butchered animal flesh; or rudimentary stone tools that would do the grinding and meat butchering for them. As our early ancestors evolved, their teeth size reduced and brain sizes increased. They made up for their smaller teeth by developing a more sophisticated stone tool culture and using fire to cook meat.

A paleoanthropological field and research study conducted in 2016 gives us a rare glimpse into the plant diet of hominine ancestors living at Gesher Benot Ya'aqov in Israel, around 790,000 years ago. An incredible 55 types of food plants were found there that included seeds, fruits, nuts, vegetables, and tuber roots. This diverse diet indicates that these early ancestors had a good knowledge and understanding of which edible plants could be found within their environment, and in which season they were available. In addition to the diverse variety of plants, this particular hunter-gatherer society also ate a variety of meat and fish. There is evidence to show this group also used fire to cook their meat. "The bigger the animal, the better" was definitely the philosophy of our early ancestors; especially when feeding a group of around 50 to 100 of them, all leading active lives. The pre-historic period which appears to have been the most favourable for a plentiful source of wild animal food for our early human ancestors was the Middle-Late Palaeolithic ranging between 120,000 and 12,000 years ago, and particularly in Eurasia stretching all the way into eastern Siberia. During this time in those regions, our early ancestors would have had access to an incredibly diverse range of mega-fauna including mammoths, rhinoceros, Lena horse bison, deer, bears and boars. Also during this time and place plants such as legumes, grasses, fruits, seeds and nuts were in plentiful supply and would have formed an integral part of their diet.

The tools used by hunter-gatherers can trace their humble beginnings to around 2.6 million years ago. Simple stones were used as choppers, hammers and scrapers, in order to cut the meat off animal bones and access the nutritious marrow inside bones, as well as processing plants and seeds. This simple stone-tool technology was brought out of Africa into the Middle East and Asia by Homo erectus. Around 1.7 million years ago, first Homo erectus and later Homo heidelbergensis, developed more sophisticated stone-wooden tools like hand axes, picks and cleavers, which enabled these early ancestors to use their hand grips for greater force when killing their prey. By the time of the Late Palaeolithic, blade tools were created along with bone and ivory artefacts, as well as more technologically advanced spear throwers and bows and arrows.

In addition to the development of tools, the control of fire had a significant impact on the further progression of early ancestral societies. The use of fire meant that for the first time our ancestors could keep warm, huddle around it for protection from wild animals, see at night, and of course, cook their meat, as well as clearing of forests for smaller nomadic settlements. The earliest evidence we have for the use of fire by hominines dates back to around 1.8 million years ago, at sites that show reddened patches and stones altered by heat. But these earliest African sites show no evidence of hearths, used to cook meat. So it is possible that at this very early stage, hominines did not possess the knowledge to start and control fire but instead were reacting to natural fires arising in forests from lightning strikes. It is clear from excavations, that by around 400,000 years ago, our hominine ancestors wandering around Africa, Middle East and Europe in and around caves, knew how to create and control fire, with clear evidence of hearths being used during this period. Over the next 100,000 years the use of fire became more widespread, becoming a key part of the hunter-gatherer lifestyle.

There was clearly a well-developed social side to the prehistoric hunter-gatherer lifestyle, with small groups sharing and organizing themselves within a self-contained living space, and actively working towards keeping each other alive. Paleoanthropological research indicates that a social network existed quite early in the history of hominines. There were connections which not only included family members, but also to wider groups, helping to foster a spirit of cooperation and mutual support. This social connectivity would have enhanced their chances of survival because they would have shared food and protected each other from predators and natural-environmental threats. Hominine tribal groups would have undoubtedly encountered other groups whilst nomadically moving from one place to another, sharing and exchanging food, stone-wooden tools and ornaments with these other groups; as well as mating.

Without a doubt, the most profoundly unique characteristic of modern humans is our ability to engage in self-conscious collective

(118) The earliest Human cave art found is over 50,000 years old

learning, passing our knowledge onto future generations. Without our
early ancestors developing the ability to use language for communication, we could not have developed intellectually the way we have, as
a species. Let us now examine the evolutionary path that led to the use
of language. In this section I will outline the most accepted ideas and
theories on how language developed; because we do not have a concise
knowledge of how and why this happened with our early ancestors.
How did the transition from basic, crude ape-like communication to
fully-fledged sophisticated human language actually occur? Most paleoanthropologists believe language emerged in stages, as our ancestors
evolved the necessary adaptations for speech. Let us begin by defining
what language actually is, and how it differs from the communication
methods used by our primate ancestors, such as the great apes. In human
language, arbitrary sounds and signs represent specific words, which
can be learnt, added to and infinitely combined within grammatical
structures, syntax and conventions. These qualities make language an
extraordinarily sophisticated communication system found exclusively
in humans and which can be applied across large populations.

There are three elements of language only present in hominines.
The first, is a fine control over our vocal chords. Primates such as apes
are born with a more limited range of vocalizations. The difference
is mainly due to how our brains are configured. Humans have
direct connections between the neurons controlling our voice box
and the motor cortex region of our brain, responsible for voluntary
movements. Brain scans have shown these connections are lacking in
other primates. Secondly, is our tendency to communicate for the
sake of communicating, and not just for survival, referred to by the
American biologist David Fitch as, "the drive to share thoughts."
Whereas chimpanzees use a limited set of vocal sounds and gestures
to convey the essential survival messages such as food, sex and danger
from predators; humans will talk in order to bond and exchange ideas,
and ensure they are understood by other humans. Scientists refer to
this difference as "theory of the mind," the understanding that others
have thoughts. Chimpanzees demonstrate a limited theory of mind,
whereas humans are self-aware that other humans are thinking beings,

and constantly use language to uncover and influence those thoughts. The last difference is hierarchical syntax. Phrases and sentences have a clearly established structure and these provide meaning beyond the simple sequence of words. One of the world's leading linguists on the theory of hierarchical syntax as the key to language, is Noam Chomsky, who has been researching this field for over 60 years.

Noam Chomsky is an American theoretical linguist whose research since the 1950s has revolutionized the field of linguistics by treating language as a uniquely human, biologically based cognitive ability. Through his contributions to linguistics, cognitive psychology and the philosophy of mind and language, Chomsky played a big role in what came to be known as the "cognitive revolution." According to Chomsky, human conceptual and linguistic creativity involves several brain/mind functions and relies on the existence of some kind of pre-existing hierarchical mental syntax organization. During his academic career as a professor of linguistics, Chomsky introduced transformational grammar to the field of linguistics. His theory states that languages are innate and that the differences we see are only due to parameters developed in our brains over time. This helps to explain for example, why children are able to learn different languages more easily than adults. One of his most famous contributions to linguistics is what is called the Chomsky Hierarchy, which is a division of grammar into groups, moving up or down, based on expressive abilities. These ideas have had a significant impact in the fields of modern psychology and philosophy, both answering and raising questions about human nature and how we process and express information in our brains, in the form of language. Chomsky has said: "It's perfectly obvious that there is some genetic factor that distinguishes humans from other animals and that it is language-specific. The theory of that genetic component, whatever it turns out to be, is what is called universal grammar. In fact, by universal grammar I mean just that system of principles and structures that are the prerequisites for acquisition of language, and to which every language necessarily conforms. Human language appears to be a unique phenomenon, without significant analogue in the animal world." Chomsky further states: "Language is a process of free creation; its laws and principles are fixed, but the manner in which the principles of generation are used is free and infinitely varied. Even the interpretation and use of words involves a process of free creation."

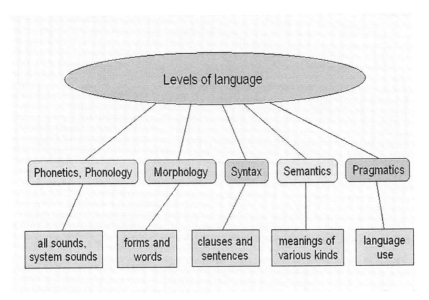

(119) The hierarchical structure of Human language

How did language evolve in our human ancestors? There are three major theories. Let us examine each one. One theory states that the earliest proto-languages were not for speaking, but for singing. This idea was proposed by Charles Darwin and claims that during this early singing stage of human evolution, the survival and reproductive capabilities of our early ancestors would have depended on producing "singing sounds" to serenade and attract mates, sooth infants, and maintain social bonds. Other paleoanthropologists believe the first forms of proto-language involved bodily-facial gestures and pantomime; and only at a later stage in the evolution of hominines was language developed using the vocal chords. This theory comes from the work of Jane Goodall and her observations of our closest relatives, chimpanzees, which in the wild exhibit over 70 gestures, compared to only 4 varieties of sounds. This idea does not explain how the gesture form of communication evolved to the use of speech-dominated language. Other researchers are convinced that the hierarchical syntax language structures as proposed by Noam Chomsky, emerged much later, and preceding this was a proto-language based on symbolic sounds in the form of individual words, but no complex-structured sentences. According to this idea, our pre-linguistic ancestors talked more like babies, such as "Water! Thirsty! or like the popularly cultural image of cavemen – "Me hunt mammoth. Me strong!"

I think the reality of the developmental evolution of language amongst our early human ancestors does represent successive stages, first with grunting, singing and facial/bodily gestures; then simple words; and of course ultimately structured sentences leading to more complex and creative language. The development of language was the single greatest catalyst that accelerated the spread of human knowledge and collective learning from one culture to another and from one generation to another.

(120) The ethnic and cultural diversity of modern Humans

9

THE DEVELOPMENT OF AGRICULTURE

A LONG WITH THE INVENTION AND USE OF STONE TOOLS, ONE OF the greatest achievements of our early human ancestors, was the domestication of plants and animals, resulting in the development of farming and agriculture. Agriculture first arose in the Fertile Crescent of the Middle East, alongside the fertile banks of the Euphrates, Tigris and Nile Rivers, towards the end of the last Ice Age, around 12,000 years ago. The Fertile Crescent comprised what is modern day southern Iraq, Syria, Lebanon, Jordan, the Palestinian territories, Israel, Egypt and western Turkey.

The emergence of agriculture is the seventh threshold of increasing complexity in the story of Big History. The ingredients were a significant accumulation in the knowledge of plants, animals, seasonal climate patterns, and soil fertility brought about by collective learning passed down mainly orally from one generation to another; and an

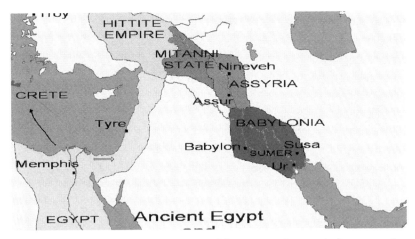

(121) The Fertile Crescent in the Middle East – where agriculture started

increasing ability for humans to manipulate and extract resources from the environment and other organisms. These ingredients resulted in human communities settling near fertile river systems; and what emerged for the first time was an exponential increase in the capacity of humans to extract energy and food from the domestication of selective species of plants and animals; leading to larger and denser communities, increased social complexity and an even greater acceleration in collective learning.

The Neolithic Revolution, also referred to as the Agricultural Revolution, represented the transition in human prehistory from small nomadic groups of hunter-gatherers, to larger agricultural settlements and the earliest civilizations. There was no single factor that led humans down the path of farming, approximately 12,000 years ago. It seems that the causes of this Neolithic Revolution differed from region to region. Archaeological evidence indicates that the domestication of various types of plants and animals evolved in separate locations worldwide. Interestingly, the Agricultural Revolution greatly reduced the diversity of foods available to our early human ancestors, compared to the nomadic foraging hunters and gatherers who preceded them. This actually led to a downturn in the nutritional value of foods in diets and a deterioration in the general wellbeing of our human ancestors, during the Agricultural-Neolithic Revolution.

There are a number of theories as to why humans stopped foraging and started farming. The most prominent of these theories are summarized here. The Oasis Theory which was originally proposed by Raphael Pumpelly in 1908, states that as the climate became drier at the end of the last Ice Age, human communities contracted to smaller-sized oases where they were forced into close association with animals, which were then domesticated, along with the planting of seeds. In recent years, this theory has lost support amongst archaeologists because more accurate climate data has been obtained, suggesting that the Middle Eastern Fertile Crescent region was actually getting wetter rather than drier, 12,000 years ago. The Hilly Flanks Theory proposed by Robert Braidwood in 1948, suggests that agriculture began in the hilly flanks of the Taurus (Iraq) and Zagros (Iran) mountains, supporting a variety of plants and animals which were suitable for domestication. The Feasting Model proposed by Brian Hayden in 1990, suggests that agriculture was motivated by overt displays of power by ruling tribal leaders, such as the giving of feasts, in order to assert their dominance over a population. These feasts required assembling large quantities of food in one place, driving agricultural innovation.

The Demographic Theory proposed by Carl Sauer in 1927, states that the population of foraging hunters and gatherers in some locations reached a point where there just wasn't enough food available, and as a result farming techniques were developed in order to produce more food for a growing population. The Evolutionary/Intentionality Theory developed by David Rindos in 1980, views agriculture as an evolutionary adaptation of plants and humans. Under this theory, agriculture started with the human domestication of selected wild plants, and then led to specialized domestication of additional particular plants and animals, depending on the location. In 2009, Peter Richerson, Robert Boyd and Robert Bettinger outlined a case for the development of agriculture coinciding with an increasingly stable climate at the beginning of the Holocene, 12,000 years ago. The Younger Dryas Impact Hypothesis was proposed in 2007, suggesting

(122) The domestication of animals and plants led to agriculture

that a comet or asteroid struck North America 12,900 years ago, and was in part responsible for a megafauna extinction and the heating of the planet leading to the end of the last Ice Age; providing the catalyst for humans to develop agricultural societies to replace extinct large mammals and plants, in order to survive. Leonid Grinin argued in 2007 that the independent invention of agriculture always took place in particular natural environments where suitable plants were available for domestication.

In his 1984 book, "A Reassessment of the Neolithic Revolution," Frank Hole outlines the relationship between plant and animal domestication. He suggests that domestication events could have occurred independently over different periods of time, in as yet, unknown locations within the Fertile Crescent; with the full domestication of goats, sheep, cattle and pigs, for example, happening much later than the domestication of a variety of plants. The available evidence seems to indicate that in the Fertile Crescent plants were domesticated for agriculture, before animals were. Once agriculture started gaining momentum, human activity resulted in the selective breeding of cereal grasses such as wheat and barley. Plants that possessed traits such as small seeds or a bitter taste would have been considered undesirable for domestication. In addition to the cereals, such as wheat and barley, mentioned above, some of the other "Neolithic foundation crops" would have included flax, pea, chickpea, bitter vetch and lentil.

In the 1980s, the British scientists, botanist Gordon Hillman and geneticist Stuart Davies, carried out experiments with wild wheat varieties to show that the process of domestication would have happened over a relatively short period of time, covering between twenty and two hundred years. Some of these pioneering attempts by our human ancestors would have failed at first and the crops abandoned, only to be taken up again and successfully domesticated thousands of years later. A good example of this is rye, which was tried and abandoned in Neolithic Anatolia (modern day Turkey), then made its way to Europe as weed seeds, and was successfully domesticated in Europe thousands of years after first being tried in Anatolia. Wild lentils presented a different challenge that needed to be overcome;

which is that most of the wild seeds do not germinate in the first year. The earliest evidence of successful lentil domestication, breaking this first year of dormancy, was found in the early Neolithic Period at the settlement of Jerf el Ahmar in modern Syria; and quickly spread south to the Netiv HaGdud settlement in the Jordan Valley. This process of domestication allowed the abovementioned foundation crops to adapt over time and eventually become larger, more easily harvested, more dependable in storage, and more useful as a source of food for our early human ancestors.

(123) Irrigation in the Fertile Crescent of Mesopotamia

Once early farmers perfected agricultural innovations like irrigation and soil fertilization, their crops would yield surpluses that required storage. Most preceding hunter-gatherers could not easily store food for long due to their nomadic migratory lifestyle; whereas those early ancestors engaged in farming with sedentary shelters could store their surplus grain. Eventually, granaries were developed which allowed villages to store their seeds and produce for longer periods of time. So, with the availability of more food, the human population expanded and communities developed specialized workers engaged in all sorts of fields other than farming, like artisans, traders, priests, administrators, and construction; and also more advanced tools were invented. The Fertile Crescent region of the Middle East was the centre of domestication for three types of cereals; einkorn wheat, emmer wheat and barley. Four varieties of legumes being lentil, pea, bitter vetch and chickpea were also grown. The climate, which consisted of a long dry season and a short period of rain, was ideal for the farming of these plants. The Fertile Crescent also had a large area which consisted of diverse geographical environments and altitudes, which was favourable for the earliest farmers.

Some of the earliest evidence for the domestication of plants, including large quantities of seeds and a grinding stone, have been found at the Paleolithic site of Ohalo II, near the Sea of Galilee in Israel around 12,000 years ago, suggesting that humans grew the grain and then processed it, before consuming it. This excavated site consisted of half a dozen brush huts. Archaeologists discovered that one of the huts contained over 150,000 charred remains of seeds and fruits, including many types like almonds, grapes and olives, that would later become productive crops. A stone blade found at Ohalo II appears to have been used as a sickle to harvest cereal crops. A stone slab was used to grind the seeds.

Another early site in the Fertile Crescent was Tell Aswad in Syria, where emmer wheat was grown; and Jericho in the Jordan valley where two-row barley crops were grown. Beneath a rocky slope in central Jordon lie the remains of a 10,000- year-old village called Ain Ghazi, whose inhabitants lived in stone houses with timber roof beams, the walls and floors gleaming with white plaster. Hundreds of people living in this village, worshipped in circular shrines and made haunting, wide-eyed sculptures that stood three feet high. They buried their beloved dead under the floors of their houses, decapitating the skulls and using them for ornamental decorations. As fascinating as this culture was, something else about Ain Ghazi excited archaeologists even more; it was one of the first farming villages to have emerged at the dawn of the Agricultural Neolithic Revolution. Surrounding this village settlement, the Ain Ghazi farmers grew barley, wheat, chickpeas and lentils. Other villagers would leave for months at a time to herd sheep and goats in the surrounding hills. Archaeologists have been able to establish that the peoples of Ain Ghazi migrated into East Africa, bringing crops and animals with them. As a result, many East Africans such as Somalians, can trace their DNA to the Ain Ghazi ancestry of central Jordan.

(124) The earliest trade in the Fertile Crescent involved bartering

Growing populations of farmers began linking to one another via trade networks, initially within the Fertile Crescent and ultimately beyond. People would move along these trade routes and married into other groups, having children in the process. This expansion driven by trade seems to have branched out in all directions. For example, the early farmers in what is modern-day Turkey, moved across the western part of the country, crossed the Bosporus river and travelled into Europe, around 9,000 years ago. When they first arrived, they encountered no farmers there. Europe had been the home of groups of ancestral human hunter–gatherers for more than 30,000 years, as we have seen in the last chapter. When the farmers arrived, they seized much of the fertile territories and converted them into farming land, without initially interbreeding with the groups of hunter-gatherers these farmers dispossessed. The hunter-gatherers in Europe continued with their existence for a number of centuries, but were eventually absorbed into the agrarian lifestyle by rapidly expanding farming communities. Many modern day Europeans can trace their ancestry to both of these groups. On the other hand, the early Neolithic farmers in what is now Iran, expanded eastward. Eventually, their descendants ended up in what is present-day India and Afghanistan, and their DNA makes up a substantial portion of the genomes of many modern-day Indians.

Sometime after the start of the Agricultural Revolution in the Fertile Crescent, Northern China appears to have introduced the domestication of foxtail millet and broomcorn millet, 9,000 years ago. These species were also cultivated in the Yellow and Yangtze River basins; along with rice, soybean, orange and peach, all introduced at later times. The earliest recorded evidence of the farming of wheat and lentil plants in Europe was around 9,000 years ago, as we have seen above, in the fertile Carpathian Basin, located in modern day Hungary,

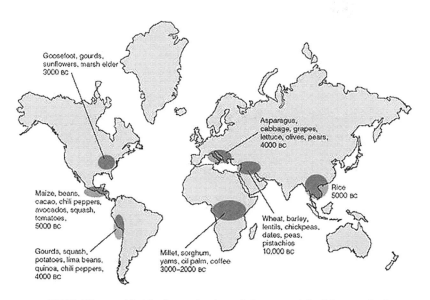

(125) The worldwide domestication of plants used for Human food

Slovakia, the Czech Republic, Romania, Serbia, Croatia and the Ukraine. This was the place where our early human ancestors survived the last Ice Age, and in particular, the territory between the Danube and the Tisza rivers, was the powerhouse of early human agricultural innovation and knowledge in Europe.

As far as the African continent is concerned, agriculture appears to have developed in four separate areas; the Nile River Valley, the Ethiopian Highlands, the Sahel, and West Africa. Farming commenced in the Nile River Valley of Egypt around the same time as it did in other regions of the Fertile Crescent described above, approximately 12,000 years ago, during the early Neolithic Revolution. The earliest crops domesticated in Egyptian cultures were the emmer wheat grain, chickpea, lentil, lettuce, onion, garlic, sesame, corn and barley; a legacy of the exceptional fertility of the Nile River. Many grinding stones have been uncovered as part of the excavation of artefacts from the early Egyptian Sebilian and Mechian cultures. The most famous crop domesticated in the Ethiopian Highlands was coffee, as well as khat, ensete, noog, teff and finger millet. In the Sahel region the main crops were sorghum and pearl millet; and in West Africa it was predominately rice, yam and palm oil.

Developing entirely independently of farming in Afro-Eurasia (the name we use in Big History for the continuous landmass comprising Africa, Europe and Asia), was the cultivation of maize corn, beans, potatoes and squash in Central America, starting from around 10,000 years ago. In what is now the eastern United States, Native Americans domesticated sunflower, sump-weed and goosefoot, at around the same time. Drainage ditches that have been excavated at Kuk Swamp on the borders of the Western and Southern Highlands of Papua New Guinea show evidence of the planting, digging and staking of taro, banana, sugar cane, sago and yam, also starting from approximately 10,000 years ago.

When hunter-gatherers began to be replaced by sedentary or settled food production in small villages, it became more profitable to keep animals nearby and readily available as a food source. So the presence of animals became a permanent part of early human village settlements. Only animals of a particular temperament, size, diet, mating pattern, and life span were selected for domestication. For example, wild animals that could not be tamed, were not considered suitable; on the other hand, animals that provided milk, such as cows and goats, offered a source of protein that was renewable, and therefore were considered to be valuable. The animal's ability to be used as a source of work, such as for transportation, ploughing and towing, also had to be taken into account. Besides being a direct source of food, some animals could provide additional resources such as leather, wool, hides and fertilizer. Some of the earliest domesticated animals included dogs, sheep, goats, cows and pigs.

The Middle East served as the source of many animals that could be domesticated, such as sheep, goats and pigs. This area was also the first region to domesticate camels. The presence of these animals, in addition to the wide variety of domesticated plants, gave the Fertile Crescent a significant advantage to progress economically and culturally, compared to other regions. As the climate in the Fertile Crescent of the Middle East changed and became drier, many of our early ancestral farmers were forced to leave, taking their domesticated animals with them. It was this largescale emigration from the Middle East that would help distribute these animals and farming practices to the rest of Afro-Eurasia. This emigration occurred mainly on an east–west axis of similar climates, from the Middle East to Europe, because crops usually have a narrow optimal climatic range, outside of which they cannot grow, particularly if there are changes to rainfall patterns. For example, wheat does not normally grow in tropical climates; just like tropical crops such as bananas do not grow in colder climates. The historian Jared Diamond, has suggested that this east–west axis is the main reason why plant and animal domestication in the form of agriculture, spread so quickly from the Fertile Crescent to the rest of Eurasia and North Africa. In his 1997 book, "Guns, Germs and Steel", Jared Diamond had this to say about the early Agricultural Revolution: "Twelve thousand years ago, everybody on earth was a hunter-gatherer; now almost all of us are farmers or else are fed by farmers. The spread of farming from those few sites of origin usually did not occur as a result of the hunter-gatherers' elsewhere adopting farming; hunter-gatherers tend to be conservative. Instead, farming spread mainly through farmers' outbreeding hunters, developing more potent technology, and then killing the hunters or driving them off all lands suitable for agriculture."

The Agricultural Revolution had social consequences for our early human ancestors. As we have seen, the nutritional standards of Neolithic human populations were generally inferior to that of the earlier hunters and gatherers. Their life expectancy was believed to be shorter as well, partly due to diseases and also due to the much harder

(126) The first cities were in Mesopotamia

work required to engage in farming. Paleoanthropologists estimate that nomadic hunters and gatherers devoted around 20 hours per week obtaining their food; whereas agriculture took up at least twice that time, with no guarantees of a successful crop or enough animal produce. The hunter gatherers' diet was also more varied and nutritious, as we have seen. Average height went down from 5'10" (178cm) to 5'5" (165cm) for men, and from 5'6" (168cm) to 5'1" (155cm) for women. It took until the twentieth century for the average height of men and women to come back to pre-Neolithic Revolution, hunter and gatherer levels.

However, this decrease in human nutrition levels was accompanied by a significant increase in the human population as a direct result of agriculture. Agricultural food production supported a higher density population, which in turn supported larger village communities, the accumulation of goods and tools, and specialization in the forms of new jobs and labour outside of farming. Not all humans had to be farmers; they could engage in a variety of other productive activities such as artisan trades, hand-making and selling goods, construction; and later, soldiers, priests, scribes, government officials and tax collectors. The development of larger communities led to many villages becoming towns and some towns becoming cities. This required decision-making, writing for record-keeping, hierarchical structures, the introduction of laws and forms of government. Food surpluses made possible the development of a social elite who were not engaged directly in agriculture, industry or commerce; but instead accumulated wealth and power, monopolized decision-making and dominated their communities. Along with these societal structures, agriculture also brought about deep social divisions and encouraged inequality between the sexes, with patriarchal societies emerging, where men dominated women.

The English archaeologist Andrew Sherratt has argued that following from the Neolithic Revolution, there was a second phase of discovery that he refers to as the "Secondary Products Revolution." The Secondary Products Revolution came about when our early human ancestors realized that animals were not just a source of meat,

but also provided communities with a number of other beneficial
secondary products such as; hides and skins from undomesticated
animals; manure for soil fertilizing from all domesticated animals; wool
from sheep, llamas, alpacas and Angora goats; milk from cattle, goats,
yaks, sheep, horses and camels; transportation from oxen, donkeys,
horses, camels and dogs; and guarding and herding assistance from
dogs. Sherratt argued that this phase in the Agricultural Revolution
enabled humans to harness the energy of their animals in new and
innovative ways, allowing for permanent intensive subsistence farming
and crop production, which opened up heavier and more diverse soils
for farming. It also made possible forms of nomadic pastoralism in
semi-arid areas along the margins of deserts which amongst other
things, eventually led to the domestication of camels.

In time, our prehistoric ancestors were able to stockpile food for
their survival during lean times, and also trade unwanted surpluses with
other communities, opening up the fields of commerce, accounting
and economics. Once a secure food supply and trade was established,
populations could grow and further labour specialization could occur.
With more artisans for example, new technology developed, such as
the use of different metals not only for tools, but also weapons. This
level of social complexity would have required some societal organization, cohesiveness and unity, such as that provided by the earliest
religions. Also as populations grew and wealth was accumulated, ruling
families emerged, who supported and secured their power base with
large professional armies of soldiers. Concepts of property ownership
also began to develop and became increasingly important to more
and more people. Societies needed to maintain ownership rights,
safety, order and stability, so laws were also developed. Ultimately, the
Australian archaeologist Vere Gordon Childe argued that this growing
social complexity, with its attendant accelerated collective learning,
led to an Urban Revolution in which the first cities were built. We
will examine the rise of city states which ultimately led to the earliest
civilizations, later in the book.

One of the biggest challenges our Neolithic human ancestral
farmers had to contend with, more so than the earlier hunters and

gatherers, was the spread of diseases originating from animals. A combination of poor sanitary practices and the domestication of animals explained the rise in deaths and sicknesses from diseases during the Neolithic Revolution. Many viral diseases such as influenza, smallpox and measles, came from domesticated animals and rapidly spread within human populations. As a result of natural selection, the humans who first domesticated bigger mammals quickly built up immunities to the diseases, increasing their chances of survival over many generations. In their approximately 10,000 years of shared proximity to domesticated animals such as cows and sheep, Eurasians and Africans became more resistant to those diseases, compared with the indigenous populations humans encountered outside of these regions, who did not build up an adequate resistance. For example, the populations of most Caribbean Islands and several Pacific Islands were completely wiped out by mammal-sourced diseases; and over 90% of the populations of many cultures in the Americas were also wiped out by mammal-sourced diseases, thousands of years before these indigenous peoples encountered the European explorers of the fifteenth, sixteenth and seventeenth centuries. Some cultures in the Americas, like the Inca Empire, did have a large domestic mammal, the llama; but llama milk was not drunk, nor did llamas live in close proximity to humans, so the risk of disease contagion was limited.

In his book, "Guns, Germs and Steel" the historian Jared Diamond effectively argues that Europeans and East Asians benefitted from an

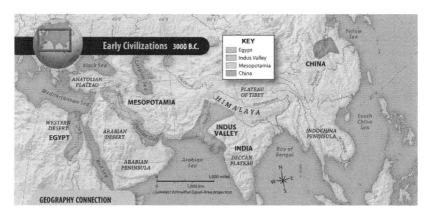

(127) The early agrarian civilizations situated alongside the great fertile river systems

attractive geographical location that gave them significant advantages during the Neolithic Revolution. Both regions shared a mild-temperate climate which was ideal for agriculture; and both regions also had the good fortune of possessing many plants and animals in natural environments that were ideally suited to domestication. Also, humans living in Western Europe and East Asia, were not as exposed to attacks from other civilizations; compared to those humans living in the land-locked central regions of Eurasia. Being among the first to fully adopt agriculture, establish towns and cities and exploit trade and commerce opportunities; the Europeans and East Asians were also among the first to benefit from new technologies such as bronze, iron and steel tools and swords. Due to their close relationship with domesticated animals over many generations, these groups of peoples also developed resistances to infectious diseases such as smallpox. Whereas, groups of indigenous peoples who had not lived in close proximity to large mammals, such as the Australian Aborigines and American indigenous peoples, were much more vulnerable to infection and were devastated by diseases when they encountered Europeans.

The dominant narrative about early human society is that the Neolithic Revolution directly led to the domestication of plants, animals and fixed-field agriculture which allowed humans to form sedentary villages and towns, ultimately leading to the formation of cities and states. The first city-states are typically viewed as a leap forward for humanity in the form of a linear progression that gave us civilization, public order, and increased health and leisure. The past two decades of archaeological research have produced evidence which contradicts this narrative. In his 2017 book, "Against the Grain: A Deep History of the Earliest States," historian James C. Scott, poses the central question; "How did Homo sapiens end up living in crowded, sedentary communities, packed with domesticated livestock and a variety of cereal grains, and governed by the ancestors of what we now call states, so very recently in its species history?" In the book, Scott focuses mainly on southern Mesopotamia, where the first city-states emerged around 8,000 years ago. Scott questions the established narrative which asserts that the early city-states brought into being a more comfortable existence for humanity; and instead he puts forward a counter-argument based on the current archaeological evidence.

It was long thought that hunting and gathering required mobility and dispersal, making the domestication of grain a precondition for a sedentary-village lifestyle. Yet the reality is, there were areas where hunter-gatherers lived in permanent settlements before the domestication of plants and animals. Neolithic villages in Syria, central Turkey, and western Iran, for example, existed in water-rich areas, surviving mainly on hunting, gathering and foraging. It also turns out that the domestication of grains and animals occurred around 4,000 years before any city-states were formed. This seems to suggest the first agrarian city-states were neither natural or inevitable. Another belief this book challenges is the idea that the first agrarian states arose from a need to mobilize and manage human labour for the building of irrigation works and intensifying agricultural production to support a growing population. This narrative is based on the dry-arid conditions which dominate the Mesopotamian sites today. However, more recent studies have revealed that the southern Mesopotamian sedimentary soil deposits around 6000 BCE were in the form of a vast delta-wetland, with the Persian Gulf extending further inland. Therefore, the first city-states emerged within an ecologically rich environment with an abundance of food and resources. In addition, these early sedentary settlements were situated near several different ecological environments, each providing diverse plant and animal food sources to draw from; removing the danger of overreliance on any one environment or food source. This suggests the traditional belief that the establishment of the earliest sedentary villages and towns were the direct result of irrigating fertile soils for agriculture, may not be the reality.

Given the fact that cereal plant cultivation emerged in these Mesopotamian areas, why did our human ancestors, who were leading an easy hunter-gathering lifestyle, suddenly engage in the hard-working, energy-intensive lifestyle of agriculture? Scott suggests that it began as "flood-retreat" agriculture, where seeds could be easily planted in the fertile silt deposited on the banks of the Euphrates, Tigris and Nile rivers, by annual riverine floods. The floods also served

The inhabitants might use the upper floor to store crops such as wheat and apples.

Many houses included a shrine. It was decorated with bulls' horns and sometimes wall paintings. A burial often lay sheltered beneath the shrine floor.

Kevin Jones Associates

History Interactive
For: Interactive illustration
Web Code: nap-0121

Food was prepared in a small clay oven or over a hearth.

(128) Early Human Neolithic stone housing in the Fertile Crescent of Mesopotamia

to clear the fields by removing other vegetation and depositing new layers of silt. This form of plant cultivation was less intensive than agriculture by ploughing. The next question that needs to be asked is: why did the "back-breaking" labour intensive ploughed fields and livestock farming subsequently come to dominate Mesopotamia and the wider Fertile Crescent? In his book, Scott does not provide a conclusive answer, but instead suggests that it may have been due to growing population pressure in sedentary village communities, making it harder to feed everybody and to move, combined with a decline in large game-hunted animals; making it harder for our early human ancestors to survive on a hunter-gatherer lifestyle. If this was the case, hunter-gatherers would have had no choice, other than to supplement their food resources with plant cultivation and the domestication of selected animals; that is, agriculture.

The arguments put forward in "Against the Grain" are the strongest when Scott tackles the widely accepted narratives which present the formation of early city-states as a new golden age in human history. Scott argues compellingly that early agrarian societies in fact produced poorer health and nutritional outcomes for our human ancestors. Their heavy reliance on a narrower food source, centred mainly on domesticated grains, severely limited their diets. Their primary food sources were highly vulnerable to climatic changes, parasitic plagues, and poor yields. So, the shift to a heavy reliance on domesticated grains led to physical changes not only in the artificially selected domesticated plants and animals, but also in the people themselves. Humans living in these societies became shorter in height, smaller in build-stature, physically weaker, and more prone to injury and disease than their hunter-gatherer counterparts. In addition, agriculture was extremely hard work, especially for such poor nutritional output. In contrast to this, hunter-gatherers enjoyed a more diverse and nutritious diet from a wide variety of sources; a healthier lifestyle; and more flexibility to respond to changes in climate or the environment.

The early city states were extremely fragile and subject to collapse because the sedentary communities they controlled had been built on a significant concentration of humans, domesticated animals and plants. Within these more densely populated areas a variety of vermin, parasites and diseases flourished. In fact, it was at this point in human history that many diseases which had crossed over from animals to humans appeared for the first time. The early city states relied heavily on coercion and force to sustain and reproduce themselves. There is archaeological evidence for the use of slave labour and the forcible resettlement of people into work camps. Inventories recording the spoils of war prioritized captives according to their levels of expertise to perform various work tasks. Walls at this time were not only built to keep invaders out, but also to keep the population within them. Therefore, contrary to the longstanding belief that the first city states were a major attraction to people with their promise of prosperity and a better quality of life; in fact, in many instances they were places to be feared, and relied on force to maintain and grow themselves.

However, there appears to be a major weakness in Scott's above-mentioned analysis. While he makes use of the archaeological evidence to present a picture of the physical conditions allowing for the establishment and expansion of the first city states; he does not take into account, the active role in human achievements in shaping these societies. While he does acknowledge that the people were not always willing participants and that they often resisted or sought to escape; in most of his narrative Scott treats the people as objects of history, upon whom the apparatus of city state formation was imposed, rather than as active antagonists struggling to shape their own destiny in what was a hostile environment. This problem of not giving the actions of people enough credit particularly comes through in Scott's analysis of how these oppressive city states emerged in the first place. A key observation he makes is that these city states were all based on some form of wheat grain-based agriculture. The reason, Scott argues, is that grains were the best suited crop for intensive cultivation, tax assessment and organised distribution. The nature of grains as seasonal crops growing above the ground which can be stored for long periods of time after harvest, made them ideal as crops that could be easily identified, accounted for and taxed, and then for the surplus to be distributed within the city state. Whereas legumes, for example, were much harder to control and tax, because they grew beneath the ground and could be left in the fields for long periods of time, hidden from tax collectors.

The rest of the people in the world during this early Neolithic Period were not dominated by emerging city states, but instead carried on their lifestyles of nomadic hunter–gatherers; and so therefore were not the subject of forcible control and taxation. Although this is a compelling hypothesis describing the conditions which existed in the early stages of city state formation, it does not explain why these city states were formed by our human ancestors. The implication from Scott's narrative is that the early city states resulted from the logical progression of sedentary societies that were relying on the cultivation of domesticated plants, centred on grain; and later, the domestication of animals. Scott does not talk about the inevitable class struggle that would have existed amongst our human ancestors in these early city states as a result of an imbalance of power and wealth between those few who controlled large quantities of agricultural produce and the majority who laboured in the fields. The formation of these city states required leadership and power, which surely would have been possessed by those few families that had accumulated wealth from the farming of domesticated plants and animals. These were the ruling elites, who would have exercised significant control over the working population. As we have seen, after Neolithic societies began producing agricultural surpluses, the need to manage these surpluses would have led to new divisions of labour and power hierarchies in city states; with the elite and wealthy rulers sitting at the top; their advisers, priests, and confidants at the next level; followed by the working people; and then slaves brought in from outside communities possible through conquest and also the debt-ridden, sitting at the bottom of the hierarchy. There is archaeological evidence that some groups of people did not willingly accept a class structural hierarchy imposed upon them. One example is the Neolithic site of Cayonu in modern day Turkey. The site suggests the emergence of a ruling class that was subsequently overthrown by the population, who then reformed the society along egalitarian (equal) lines.

Scott's book "Against the Grain" is a valuable contribution to the study of early agrarian societies, in that it summarizes much of the evidence regarding the emergence of the first city states in the Neolithic Period. The book also identifies common patterns in the emergence of early agriculture; the realities of daily life in agrarian

(129) James C. Scott's book – Against the Grain

societies; and the oppressive nature of the first city states in the Fertile Crescent of the Middle East. Scott's book however, does not shed light on the reasons why city states emerged. There is clearly much more scope for additional research and debate in this field of early human history.

There is no doubt that the Agricultural Revolution accelerated collective learning to levels not seen before, in the history of humans on Earth. The freeing up of most people's time, from what was literally a full time job of foraging, hunting and gathering, enabled specialization; and with specialization, people had more time to accumulate knowledge and think innovatively; progressing human culture and society in the process. Therefore, it is not surprising to see the emergence of city-states which then evolved into great civilizations, mainly centred around the great river systems of the world. Each one of these civilizations added to the accumulated collective learning of humanity; creating an explosion of knowledge. In the following chapters, we will examine a number of these great civilizations, starting with the first one to emerge along the banks of the Tigris and Euphrates rivers; the Sumerian Mesopotamians.

- New Farming Inventions
  - Plows
  - Sickles
  - Irrigation canals
- Required Cooperation
  - Shared canals used by many fields
- More food
  - Surpluses
- **Cities become first civilizations**

(130) The domestication of animals in Ancient Egypt

10

SUMERIAN CIVILIZATION OF MESOPOTAMIA

T HE SUMERIANS WERE THE PEOPLE OF SOUTHERN MESOPOTAMIA whose civilization flourished between 6500 and 1750 BCE. Their name comes from the region which is frequently, and incorrectly, referred to as a "country." Sumer was never one single political state. Instead it was a region of numerous independent city-states, each with its own power structure headed up by a king. The name "Sumer" literally translates to the "land of the civilized kings."

The Sumerians were responsible for many of the most important innovations, inventions and ideas in early ancient times. They effectively came up with the concept of measuring time, by dividing day and night into 12-hour periods, hours into 60 minutes, and minutes into 60 seconds. They also established the first schools, religious narratives, heroic epic stories, government bureaucracy, monumental architecture and irrigation techniques. Amongst their more significant inventions were the first written language (cuneiform), the wheel, chariots, arithmetic, geometry, irrigation, saws, sandals, beer, sailboats, the plough and metallurgy.

We do not know where the Sumerians originated from, but by 2900 BCE they were firmly established in southern Mesopotamia. Ancient historians divide the history of the Sumerians of Mesopotamia into six periods:

- The Ubaid Period from 6500 to 4100 BCE.
- The Uruk Period from 4100 to 2900 BCE.
- The Early Dynastic Period from 2900 to 2334 BCE.
- The Akkadian Period from 2334 to 2218 BCE.

- The Gutian Period from 2218 to 2047 BCE.
- The Ur III Period, also known as the Sumerian Renaissance from 2047 to 1750 BCE.

The Ubaid originated on the flat alluvial plains of southern Mesopotamia. This period was characterized by a distinctive type of pottery. It was during this period that the first identifiable villages developed in the region; where people farmed the land using irrigation; and fished the rivers and the sea of the Persian Gulf. Towns also began to develop to exploit trade with each other. One of the more prominent towns was called Uruk, which ultimately developed into a city. The first Sumerian kings, ruled the city-state of Uruk, including the legendary Gilgamesh, more of whom we will speak of later in this chapter. By around 3200 BCE, the largest settlement in southern Mesopotamia, if not the world, was Uruk; a true city dominated by monumental mud-brick buildings decorated with mosaics of painted clay cones embedded in the walls, and extraordinary works of art. Large scale sculpture in the round and relief carving appeared for the first time, together with metal casting using the lost-wax process. Simple pictographs were drawn on clay tablets to record the management of

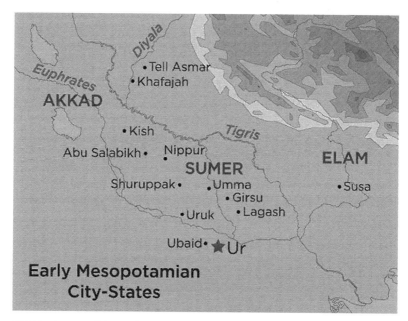

(131) A map of the early Mesopotamian City-States in the Fertile Crescent

goods and the allocation of workers' rations. These pictographs were the precursors of later cuneiform writing.

It was during the Early Dynastic Period that power structures were established giving rise to dynastic kings; the establishment of government and bureaucracy; and competition between the Sumerian city-states for land and water access-usage rights. From time to time, the Sumerian city-states were united under a single all-powerful king; such as in the case of King Enembaragesi of the city-state of Kish, who led a combined Sumerian army into the first recorded battle in history in 2700 BCE, against the opposing city-state of Elam; which the Sumerians of Kish won. Kish is believed to be the first city-state to have a dynasty of kings, starting with Jushur, an ancestor of Enembaragesi.

132) King Sargon of Akkad ruled the Akkadian Empire from 2334 to 2279 BCE

One of the most powerful kings and one of the earliest of the world's great empire builders was Sargon of Akkad, an ancient Mesopotamian ruler who reigned from 2334 to 2279 BCE. During his reign, he successfully conquered all of southern Mesopotamia as well as parts of Syria, Anatolia (Turkey), and Elam (in what is now western Iran). He established the region's first Semitic dynasty and was considered the founder of the Mesopotamian military tradition, with a permanent standing professional army. Sargon is known almost entirely from the legends and tales that followed his reputation through over 2000 years of cuneiform written Mesopotamian history recorded after his lifetime; and not from any written records during his lifetime. The lack of a contemporary record of this king can be put down to the fact that the capital city of Akkad, which Sargon built, has never been located and excavated. It was destroyed at the end of his family dynasty and was never again inhabited.

According to subsequent legend, Sargon originated from humble beginnings and rose to great power. Not only did he found an empire, but he also kept it operating smoothly with the innovative use of Akkadian bureaucrats who he installed in every city he conquered. Akkadians spoke a distinctly Semitic language and originated from northern Mesopotamia, whereas the Sumerians occupied the south. Sargon became the first person in recorded history to create an empire, ruling over a multi-ethnic group of people which united the north and south into one unified Mesopotamia. Sargon went on to become a legendary figure; his heroic and epic achievements were passed on as oral stories from one generation to the next, for thousands of years. Eventually they were recorded in writing and became major works of Sumerian literature. This is why he gained the title of Sargon the Great and the period of his rule became the Akkadian Golden Age.

Sargon was born to an unknown mother and father; but his mother was believed to be a temple priestess and possibly a member of the order of sacred prostitutes; and his father believed to be a wandering nomad. His mother was unable to keep and look after her infant son, so she placed him onto a reed basket and sent him down the Euphrates River. The infant Sargon was found and raised by a gardener to the King of the city-state of Kish, Ur-Zababa. Later, as a young man, Sargon became a cup-bearer to the king. The role of a cup-bearer was to bring the king his wine, and also to act as a trusted adviser. Lugalzagesi, the king of the city-state of Umma, came to conquer Kish. King Ur-Zababa, mistrusting Sargon and his personal political ambition, sent him to meet with King Lugalzagesi, supposedly with a message of peace. In reality, the message asked the Umma king to kill Sargon. Lugalzagesi instead asked Sargon to join his military campaign, which he did; and they proceeded to conquer Kish, at which time Ur-Zababa fled the city-state, going into exile. Soon after, Sargon fell out with Lugalzagesi, and they became arch-enemies in the process, vowing to destroy each other.

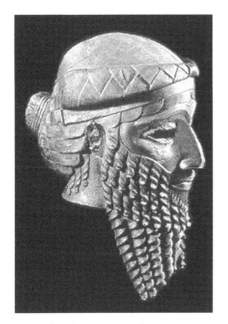

(133) King Lugalzagesi   (134) King Ur-Zababa

Lugalzagesi had already united the many city-states of Sumer under his own control. When Sargon captured him during a battle, he gained control of the Sumer Empire. Sargon then proceeded to place one of his trusted confidantes in each of the Sumerian cities he ruled in his name, while continuing to expand his empire with military conquests. This is when he ultimately conquered Elam in Iran, parts of Syria, Lebanon, Anatolia (Turkey) and Cyprus. After conquering all of Mesopotamia, Sargon built his own city, on the banks of the Euphrates river, and named it Akkad. He established a powerful dynasty that would rule the region for the next 150 years.

Sargon maintained his empire by strategically placing men he trusted into each conquered city or region, as we have seen. He would then deploy troops to protect and secure each of these areas. To consolidate his dynasty, he appointed his daughter Enheduanna, as high priestess of Inanna in Ur, where she would go on to influence religious and political affairs for the next 40 years. When Sumerians rebelled, the Sargon-led Akkadians would ruthlessly crush any opposition and execute rebel leaders. During his reign, Sargon built many roads and irrigation canals; extended trade routes; and encouraged the pursuit of science and the arts. He also created a postal system, ensuring privacy of the mail by innovating the use of clay envelopes for the tablets. Sargon was also fair and equitable in his taxing of the rich and poor. In spite of fighting in many battles, Sargon lived a long life and died of natural causes. After his death, his legendary status grew more and more from one generation to the next, culminating in Mesopotamians revering him like a god.

The Akkadian Empire created by Sargon, retained control of the Mesopotamian region until the area was invaded and conquered by the Gutians, who toppled the Akkadians and then ruled from 2218 to 2047 BCE. The Gutians were a mixture of tribes that descended from the Zagros Mountains (in modern day Iran and Turkey), apparently attracted to the fertile flatter plains of the city-states and their prosperity. The Sumerian Mesopotamians treated them as sub-human beings because of their unwillingness to conform to their customs and laws of what they considered to be civilized behaviour. Chronicles written around this time described the Gutians as barbarians, having the intelligence of dogs and the appearance of monkeys, while speaking a language resembling a confused babble. There is no doubt the Gutians were fierce warriors who practised hit-and-run tactics, often long gone by the time regular Sumerian troops could arrive to deal with a crisis situation. Their raids crippled the economy of Sumer. Travel became unsafe, as did working in the farming fields. The end result was famine. The Gutians though, proved to be poor rulers. Under their crude and cumbersome rule, prosperity declined and the enlightened progress previously made in innovation, culture and the arts went backwards. They were not accustomed to the complexities of ruling a civilization, such as organized planning, particularly in the construction of canal networks, irrigation channels, and monumental buildings. During the Gutian Period, a "dark age" swept over ancient

Mesopotamia. Ultimately, the Gutian rulers were expelled from
Mesopotamia by a coalition of rulers from the city-states of Uruk
and Ur, who defeated the last Gutian king, Tirigan. The chronicle
titled "Victory Stele of Utu-Hengal" describes the expulsion of King
Tirigan as follows: "By the envoys of Utu-Hengal, Tirigan and his wife
and children in Dabrum were captured. They placed fetters (restraints)
on his hands and put a cloth (blindfold) over his eyes. Utu-Hengal
made him lie at his (Utu's) feet, and on his neck he set his foot. Gutian,
the fanged snake of the mountain ranges, he made drink from the
cracks in the earth."

The Third Dynasty of Ur, refers to the Sumerian ruling dynasty
which was based in the city of Ur, the most prominent city of the
Sumerian Mesopotamian Empire at the time. The Third Dynasty of
Ur was the last Sumerian dynasty which came to dominant power in
Mesopotamia. It began in 2047 BCE after several centuries of control
by Akkadian and Gutian kings. In addition to Ur, it also controlled
the major cities of Isin, Larsa, Nippur and Eshnunna and extended as
far north as upper Mesopotamia. The land ruled by the Ur kings was
divided up into provinces that were each run by a governor (called
an ensi). In periods of imminent danger, military commanders would
assume the role of governor. Each province contained a government
bureaucracy where provincial taxes (called bala) would be collected
largely in the form of agricultural produce, livestock or land; and would
be transported to the capital city of Ur. The central government would
then distribute goods to each province based on their respective needs,
and would also give food rations to the numerous religious temples
spread throughout the kingdom.

The city of Nippur was one of the most important cities in the
Third Dynasty of Ur. Nippur was believed to be the religious centre of
Mesopotamia. It was home to the shrine of Enlil, who was the supreme
lord of all gods; the "Lord Wind," and ruler of the cosmos. At this
shrine was where the God Enlil spoke the king's name and was calling
the king to join him in the afterlife. Connecting the human king to
the god Enlil was used as an effective tool to secure and legitimize
the power of kings over the highly religious and superstitious people.

The city of Nippur was also a place where people could take and resolve their disputes in person, according to stone tablets excavated in the area. Nippur was never destroyed, surviving numerous conflicts because of its spiritual and religious significance.

Mesopotamia had a distinctive social and hierarchical class structure. There were three major classes. First, there was the Upper Class, comprising of the king and his family, other nobility, high-priests, high-level government officials (including provincial governors) and warriors-military leaders. The second group in the hierarchy were the Middle Class, which included scribes, lower-priests, lower-level government officials, tax collectors, artisans, merchants and traders. The Lower Class included common labourers and peasant farmers; some of whom were forced to work and others who were freer. There is evidence that certain groups performed their labouring or farming work under compulsion. This group worked in order to keep their property or receive food rations from the state. On the other hand, there was a group of labourers and farmers who were free men and women for whom social mobility was a possibility. Many of these families travelled together in search of work. These workers

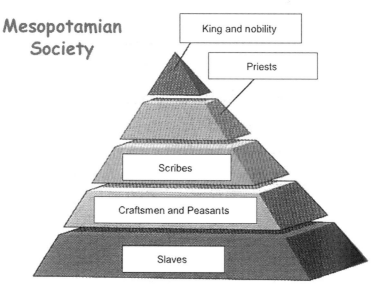

(135) The Hierarchical Structure in Mesopotamian Society

could accumulate their own private property and be promoted to government or military positions. Sitting at the bottom of the Lower Class, were slaves. Slaves made up a critical group of labour for the state. It is estimated around 40% of the slaves in Mesopotamia were not born slaves, but became chattel slaves; as a result of family members being sold into slavery for accumulating and not being able to repay debt. One surprising feature of this civilization though, is that slaves were able to accumulate some assets, including property, during their lifetimes, often resulting in them being able to buy their freedom from slavery.

One of the earliest known codes of law was established during the Third Dynasty of Ur. It was known as the Code of Ur-Nammu. This law established the king as the beacon of justice for his land, a role that previous kings did not play. The king was codified into the law as wanting justice for all groups, including slaves and unfortunate people like the poor, widows, disabled and orphans. Most legal disputes were dealt with in the local communities within a province, by government officials called mayors; although their decisions could be appealed and eventually overturned by the provincial governor. Sometimes, legal disputes were publicly prosecuted and defended with witnesses present at a location like the town square, or in front of the temple. However, throughout Mesopotamia the king

(136) The Epic of Gilgamesh was written in 2100 BCE

was seen as the supreme judge of the land; and this was reinforced in literary works and poems excavated from this time. The citizens would sometimes write letters of prayer to the king, asking him to be a just and fair arbitrator of the law of the land. The laws first outlined the crimes and then their punishment. Considering it is the oldest law-code known to history, it is remarkably advanced. For example, it includes fines of monetary compensation for physical injuries, as opposed to the later "eye for an eye" principles established in Babylonian law-codes. The Code of Ur-Nammu imposed a punishment of death for crimes of murder, robbery, adultery and rape.

Numerous Sumerian texts were hand-produced and circulated to the educated elite during the Third Dynasty of Ur. At this time, Semitic Akkadian was the common spoken language, but Sumerian was the dominant written language. Government officials would learn to write at special schools that only used Sumerian literature. A number of archaeologists and ancient historians believe that the classic Uruk text, "The Epic of Gilgamesh" was written down during this period into its classic Sumerian form. Interestingly, this text served as a powerful piece of propaganda because the Ur III Dynasty claimed to be from the same ancestral lineage as the earlier kings of Uruk. For example, the Ur III kings often claimed Gilgamesh's divine parents, Ninsun and Lugalbanda, as being from the same bloodline as themselves, drawing a direct connection with the epic hero. Another text from this period, known as "The Death of Urnammu", contains an underworld scene in which King Ur-Nammu showers "his brother Gilgamesh" with gifts.

There was nothing short of a Sumerian Renaissance during this time, which saw a rebirth of Sumerian culture, following the turmoil of the Akkadian and Gutian conquests. Sumerian cities grew rich from extensive trade. The relative stability of the cities encouraged cultural growth, innovation and invention. In his iconic 1956 book: "History Begins at Sumer", the world-renowned expert in Sumerian history, Samuel Noah Kramer, refers to 39 inventions and innovations which first appeared during the Sumerian Renaissance. I have listed them here, to give readers a deeper appreciation of the significant contri-

bution the Sumerian Mesopotamian Civilization made to collective
learning, human knowledge and progress:

1. The first schools
2. The first case of "apple polishing"
3. The first case of juvenile delinquency
4. The first "war of nerves"
5. The first bicameral congress
6. The first historian
7. The first case of tax reduction
8. The first "Moses"
9. The first legal precedent
10. The first list of pharmaceutical drugs
11. The first farmer's almanac
12. The first experiment in shade-tree gardening
13. The first study of cosmology
14. The first moral ideals
15. The first "Job"
16. The first proverbs and sayings
17. The first animal fables
18. The first literary debates
19. The first biblical parallels
20. The first "Noah"
21. The first tale of resurrection
22. The first "Saint George"
23. The first library
24. The first heroic literature
25. The first love song
26. The first library catalogue
27. The first golden age
28. The first "Sick Society"
29. The first literary laments
30. The first Messiahs
31. The first long-distance champion
32. The first literary imagery
33. The first sex symbolism

34. The first mater dolorosa (sorrowful mother)
35. The first lullaby
36. The first literary portrait
37. The first elegies (melancholic poems)
38. The first victory of labour
39. The first aquarium

Nobody has contributed more to our knowledge of the ancient Sumerian Mesopotamian Civilization than Samuel Noah Kramer. Born on September 28th, 1897 near Kiev in Ukraine, he was one of the world's leading Assyriologists and a world-renowned expert in Sumerian history and language. In his autobiography written in 1986, he summarized his accomplishments as follows: "First and most important, is the role I played in the recovery, restoration, and resurrection of Sumerian literature, or at least of a representative cross section. Through my efforts, several thousand Sumerian literary tablets and fragments have been made available to cuneiformists, a basic reservoir of unadulterated data that will endure for many decades to come. Second, I endeavoured to make available reasonably reliable translations of many of these documents to the academic community, and especially to the anthropologist, historian, and humanist. Third, I have helped to spread the name of Sumer to the world at large, and to make people aware of the crucial role the Sumerians played in the ascent of civilized man." Kramer died of throat cancer at age 93 on November 26th, 1990.

Although modern-day archaeologists have established Uruk as the oldest city in Mesopotamia, the Sumerians themselves believed the first city in the world was Eridu, presided over by their god of wisdom and water, Enki, who raised the city from watery marshes and established the concept of kingship and order in the land. The establishment of Eridu by Enki was seen as a kind of golden age, comparable to the biblical Garden of Eden, as the home of the gods and birthplace of the rules governing civilization (known by the Sumerians as the meh). The Austrian-British historian and Assyriologist, Gwendolyn Leick, had this to say: "The Mesopotamian Eden is not a garden but a city, formed from a piece of dry land surrounded by the waters. The first building is

a temple. This is how Mesopotamian tradition presented the evolution
and function of cities, and Eridu provides the mythical paradigm.
Contrary to the biblical Eden, from which man was banished forever
after the Fall, Eridu remained a real place, imbued with sacredness but
always accessible."

*(137) The Sumerian Mesopotamian city of Eridu – it was the most sacred city
in Sumer*

According to Sumerian mythology, the fall of Eridu had nothing
to do with humanity's sins, but with the cleverness of one of the most
popular Mesopotamian goddesses, Inanna. In the poem, "Inanna and
the God of Wisdom", the goddess travels from her city of Uruk to
Eridu, home of her father Enki, and invites him to sit and have a
few drinks with her and, as he drinks and becomes more and more
jovial, he gladly hands over the meh (rules governing civilization) to
his daughter. Once she has gathered all of the rules, she runs away and
brings them to Uruk, thus making her city preeminent and dimin-
ishing the importance of Eridu in the process. Modern-day scholars
believe this myth arose in response to the shift of Mesopotamia from
an agrarian culture (symbolized by Eridu) to an urbanized culture,
symbolized by Uruk, the most powerful city in the region.

Religion was fully integrated into the lives of the Sumerian people and was an important part of the ruling government and the society structure. The Sumerians believed that the gods had created order out of chaos and the individual's role in life was to provide labour as a co-worker with the gods, to make sure chaos would never appear again. The gods themselves however, would reverse all of their own work later, and return the world to chaos when humanity's noise and trouble became too great to bear. The Sumerian creation story was recorded in a text known as the "Eridu Genesis" which was composed in 2300 BCE and found amongst the ruins of Eridu. It is the earliest known version of the Great Flood legend, which was later retold in the "Atrahasis", "The Epic of Gilgamesh", and even later in the Jewish and Christian "Book of Genesis." The Eridu Genesis legend tells the story of how the gods destroyed humanity through a flood, except for one man, Ziusudra, who is saved when Enki tells him to build an ark and rescue two of every kind of animal. Afterwards, the gods relent and determine to control the human population, and limit their annoying tendencies by introducing death and disease into the world; in the process re-establishing order and setting a limit to human life and earthly ambition. This highlights how the ideas, legends and myths developed in the religions of the earliest civilizations, were utilized in later civilization religions; similar events, but key names being changed. So, in the "Book of Genesis" Ziusudra, became Noah.

The Sumerian gods expected human beings to use their lives to help maintain order, including finding ways to work together. The Sumerians took great pride in their individuality, as can be seen by the elevation of the patron deities of each city, and intermittent rivalry and conflicts; but were required by the gods to set this rivalry aside, in the interests of the common good. The historian Samuel Noah Kramer, had this to say: "While the Sumerians set a high value on the individual and his achievement, there was one overriding factor which fostered a strong spirit of cooperation among individuals and communities alike; the complete dependence of Sumer on irrigation for its wellbeing, indeed, for its very existence. Irrigation is a complicated process requiring communal effort and organization. Canals had to be dug and kept in constant repair. The water had to be divided equitably among all concerned. To ensure this, a power stronger than the individual landowner or even the single community was mandatory; hence the growth of governmental institutions and the rise of the Sumerian state."

The Sumerian King List, a document composed around 2100 BCE in the city of Lagash, lists all of the kings going back to the beginning of the Mesopotamian world, when the gods first established kingship in Eridu; but the first mortal kings ruled Uruk. The first king who can be verified archaeologically was Etana, described by Kramer as "he who stabilized all the lands;" and the list then continues in chronological order, often though with impossibly long dates for some monarchs (such as Gilgamesh), right up to the reign of the king in 2100 BCE. We have seen that the Sumerian city-state was governed by a king, or the Lugal (meaning "big man") who oversaw the cultivation of the land, among many other responsibilities, and was bound to the gods to ensure their will was done on earth. The Lugal (or king) was initially the head of a "household", a closely knit community which pooled their resources. This "household" concept would continue as the underlying power structure of the city-states. With the rise of the cities and the development of agricultural farming innovations, the Sumerians had forever changed the way human beings had previously lived, and would live into the future. The Austrian-British historian, Paul Kriwaczek described this monumental change as follows: "This was a revolutionary moment in human history. The Sumerians were consciously aiming at nothing less than changing the world. They were the very first to adopt the principle that has driven progress and advancement throughout history, and still motivates most of us in the modern times; the conviction that it is humanity's right, its mission and its destiny, to transform and improve on nature and become her master."

Over an extended period of time, the Sumerian city-states expanded, and when they needed more land and resources, they took them from others. During the Uruk Period, the culture developed rapidly, with the greatest invention culminating in the emergence of writing around 3600 BCE. Early writing developed in response to the need to record produce and livestock in a basic manner, such as: "two sheep – five goats – Kish", which was clear enough to the sender at the time, but lacked the ability to inform the recipient whether the two sheep and five goats were coming or going from the city of Kish, whether they were dead or alive, and what their purpose was. This early-crude writing system would develop by the time of the Early Dynastic Period into the more sophisticated writing system which would produce such great literary works as "The Epic of Gilgamesh" and "Enheduanna's Hymms to Inanna," amongst others. Sumerian became the language of communication of Mesopotamia, and laid the foundation for the establishment of the writing system known as "cuneiform" which would later be used to record other languages. Gwendolyn Leick described the development of writing as follows:

"The more homogenous cultural horizon of the alluvial plains of Sumer finds expression in the development of writing in a particular idiom. Why Sumerian came to be the language repre-sented by writing is still uncertain. Mesopotamia was never linguistically or ethnically homogeneous and the personal names in the early texts clearly show that languages other than Sumerian were spoken at the time."

Sumerian was well established as the Mesopotamian written language by the late 4th century BCE; and Sumerian culture, religion, architecture, and the other significant aspects of this civilization were as well. The literature of the Sumerians was to have a significant influence on later writers; notably the scribes who wrote the Old Testament Bible; as their tales of "The Myth of Adapa," "The Eridu Genesis," and "The Atrahasis," would inform the later biblical accounts of the Garden of Eden, Fall of Man, and the Great Flood. Enheduanna's works would become the models for later religious liturgy; Mesopotamian animal fables would later be popularized by Aesop, a slave and storyteller who lived in ancient Greece between 620 and 564 BCE; and "The Epic of

Gilgamesh" would inspire Homer's great literary works such as "The Iliad" and "The Odyssey" which were written between 700 and 750 BCE.

(138) A Sumerian Ziggurat – a temple designed to get closer to the gods in the heavens

The concept of the gods living in the city's temple, as well as the shape and size of the Sumerian ziggurat, is believed to have influenced the Egyptian development of the pyramid and their beliefs in their own gods and the afterlife. The first wheel was also believed to have been invented by the Sumerians; the wheel, their concept of measuring time, as well as their writing system, was also adopted by later civilizations. The Sumerian cylinder seal, an individual's sign of personal identification, remained in use in Mesopotamia until the fall of the Assyrian Empire in 612 BCE. There was literally no other subsequent civilization that the Sumerian culture did not make some contribution to.

In spite of its incredible achievements, the Sumerian culture began to decline long before its ultimate demise. The Sumerian civilization collapsed in 1750 BCE with the invasion of the region by the Elamites. King Shulgi of Ur had erected a great wall in 2083 BCE to protect his people from such an invasion, but as it was not anchored at either end, the wall could easily be walked around; which is exactly what the Elamite invaders did. By the time of the invasion, the Sumerian culture had already been weakened because it struggled to retain its autonomy ever since the Semitic Amorites had gained power in Babylon, a Mesopotamian city-state. The city of Babylon was founded more than 4000 years ago as a small port town on the Euphrates River. It grew into one of the largest cities in the ancient world under the rule of the Mesopotamian King Hammurabi.

Hammurabi was the sixth and best known ruler of the First Amorite Dynasty of Babylon. His reign was from 1792 to 1750 BCE. When Hammurabi succeeded his father, King Sin-Muballit in 1792 BCE, he was still young, but as was customary in Mesopotamian royal courts at the time, he had almost certainly already been given royal responsibilities before becoming king. Archaeological artefacts excavated show that he was engaged in the traditional activities of an ancient Mesopotamian king; such as building and restoring temples, city walls, and public buildings; digging irrigation canals; dedicating cult objects to the deities in the cities and towns of his kingdom;

(139) King Hammurabi   (140) The Hammurabi Law Code

and fighting wars for conquest of fertile land and other city-states. Hammurabi inherited one major political imperative, and that was to succeed in controlling the Euphrates River. This was critical for a civilization that depended greatly on irrigation agriculture alongside the fertile banks of the Euphrates for its survival. This policy had been begun by Hammurabi's great-grandfather, but was most forcefully and partially successfully pursued by his father. His years of reign were dominated by shifting power dynamics and coalitions among the main city-state kingdoms at that time; which were Babylon, Mari, Ashur, Elam, Eshnunna and Larsa. Hammurabi astutely took advantage of this situation by securing and fortifying a number of cities along his northern borders.

The last 14 years of Hammurabi's reign were dominated by continuous warfare. In 1764 BCE he fought a series of successful battles against a coalition of Ashur, Eshnunna, and Elam, the most powerful city-states east of the Tigris River, whose position threatened to block his kingdom's access to the metal-producing areas of what is modern-day Iran. On the back of this military success, Hammurabi took the initiative in moving against King Rim-Sin of Larsa in 1763 BCE. There is evidence to suggest that he successfully employed a strategy which involved damming up the water of a main water-course alongside the Euphrates, and then either releasing it suddenly to create a devastating flood, or simply withholding it from the people of Larsa, effectively starving them of a precious resource. In any event, Hammurabi prevailed and defeated Rim-Sin. He then proceeded to build a canal along the western branch of the Euphrates River in order to resettle an uprooted population who had previously lived there. He also secured control of the city-state of Larsa. Hammurabi's other battles during this 14-year period included the 1761 BCE conquest of Mari, led by his former long-time ally, King Zimrilim, to gain control of Mari's excellent crossroads trade location and to secure more water access upstream from Babylon alongside the Euphrates; and the final destruction of Eshnunna in 1755 BCE again, also for greater Euphrates water access. This last conquest turned out to have created problems for him, because it removed a buffer zone between his kingdom of

Babylon and the tribal group of Kassite peoples of the north-east, originating in the Zagros mountains, who ended up attacking and conquering Babylon 160 years later.

By this time, Hammurabi was a sick man and he died in 1750 BCE with the responsibility of power, rule and government already having been transferred to his son Samsuiluna, before he died. Hammurabi is most famous and remembered for his Code of Hammurabi; a well-preserved Babylonian code of law applying to ancient Mesopotamia and written in 1754 BCE. It is one of the oldest translated writings of significant length in human history. The Code consists of 282 laws, with scaled punishments, including "an eye for an eye, a tooth for a tooth"; based on the seriousness of the offense, and the social ranking, gender, and freeperson or slave status, of the person found guilty. Nearly half of the Code deals with matters of contract, labour and remuneration, such as establishing the wages to be paid to an ox driver or a surgeon, for example. Other sections set the terms for a trade transaction, the liability of a builder for a house that collapses, or property that is damaged while left in the care of another. A third of the Code is concerned with issues relating to household and family relationships such as inheritance, divorce, paternity and sexual relations.

Hammurabi is also known for completely reversing the Sumerian religious belief system by elevating one supreme male god, Marduk, over all of the others. Marduk was the god of creation, water, vegetation, judgment and magic. Under Hammurabi's reign, temples dedicated to goddesses were replaced by those for Marduk and, even though the goddesses' temples were not destroyed, they were marginalized from Babylonian society. By the time Hammurabi was ruling, women's rights, which had traditionally been equal to men's in ancient Mesopotamian culture, had declined, as did the great Sumerian cities. Overuse of farming land and excessive urban expansion, along with more frequent military conflicts, are the major reasons for the decline of these once great city-states.

I believe it is fitting to end this chapter with a synopsis of the great, and one of the earliest works of extended human literature, The Epic of Gilgamesh.

The oldest written epic story in human history was written 1500 years before Homer wrote the Illiad. It tells the story of the Sumerian Gilgamesh, the hero King of Uruk, and his adventures. This epic story was discovered in the ruins of the library of Ashurbanipal in Nineveh, by the archaeologist Hormuzd Rassam in 1853. It was written in cuneiform on 12 clay tablets. According to the story, Gilgamesh is a handsome, athletic young king of the Uruk city-state. His mother was the goddess Ninsun and his father the priest-king Lugalbanda, making Gilgamesh semi-divine. He is courageous and energetic, but also cruel and arrogant. He challenges other young men in the kingdom to physical contests and combat. He also proclaims his right to have sexual relations with all new brides. Gilgamesh's behaviour upsets the citizens of Uruk, and in exasperation, they cry out to Anu, the great god of heaven, for help with dealing with this "out of control" young king. Anu responds by sending a wild man to earth, Enkidu, to challenge Gilgamesh. At first, Enkidu lives in the rural wilds amongst animals. He is partially civilized by a temple priestess, Shamhat, who seduces him and teaches him how to eat and behave like a civilized human being. Enkidu then heads to the city of Uruk where he meets and fights with Gilgamesh. Gilgamesh wins the fight, and he and Enkidu then become the best of friends.

The first half of the epic concerns the adventures of Gilgamesh and Enkidu. They conquer and kill the monster Humbaba, who the gods had released over the Forest of Cedar. Gilgamesh rejects Ishtar/Inanna when she tries to seduce him. In revenge, Ishtar/Inanna asks the god Enlil for the Bull of Heaven, with which to attack Gilgamesh. However, Gilgamesh and Enkidu kill the Bull, which angers all the gods. The gods decide to punish Gilgamesh by killing Enkidu. The second half of this epic story has Gilgamesh searching for immortality as he deeply mourns the death of his close friend Enkidu, and worries about his own demise. He searches for Utnapishtim, an immortal man who survived the Great Flood, a precursor to the biblical Noah. Gilgamesh finally finds Utnapishtim, who tells him to accept his mortality as he cannot change it. Gilgamesh then returns to Uruk and becomes a good king. He rules for 126 years, according to the Sumerian King List; clearly a "super-human" achievement, only fit for a god.

Gilgamesh was not only an epic hero, but also a historical king of Uruk who appears in contemporary inscriptions found by archaeologists. From a human mortal king however, Gilgamesh became the semi-divine hero of Mesopotamia's greatest tale. The Epic of Gilgamesh conveys many themes that are important to our understanding of Mesopotamia and its kings. These themes encompass friendship, the role of kings, enmity, immortality, death, male-female relationships, city versus rural life, civilization versus the wild, and the relationship between mortal humans and the gods.

II

THE EGYPTIANS

THERE WERE SOCIETIES ON THE FERTILE BANKS OF THE NILE RIVER thousands of years before the Egyptian civilization emerged over 5000 years ago. These pre-Egyptian societies had already developed a number of technological skills such as the skilful use of stone tools, pottery, textiles, glassmaking, domesticized plant and animal farming, canal irrigation and wooden boat construction. They also developed their own style of rudimentary basic hieroglyphic writing, and standard weight length and measuring techniques. Few written records or artefacts have been found from this period, which is known as the Predynastic Period (6000-3100 BCE). The written history of the land began at some point between 3400 and 3200 BCE when hieroglyphic script was developed by the Naqada culture. In this culture each village had its own animal deity which was associated with the clan of the villagers. They buried their dead in graves with statuettes to keep them company in the afterlife. The Naqada also buried their dead with food, weapons, amulets, ornaments and decorated vases. By 3500

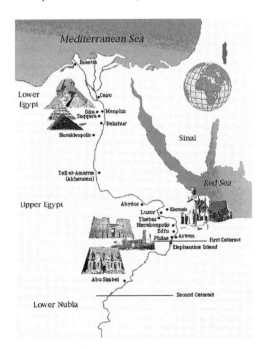

(141) Map of Ancient Egypt

(142) Naqada pot from early Egypt

BCE mummification of the dead was in practice at the city of Hierakonpolis, and large stone tombs were built at Abydos.

Organized farming began in the region around 6000 BCE and communities known as the Badarian culture began to flourish along the Nile River. They were the first in this region to engage in organized farming. The Badarian people also developed industry at about this time as evidenced by faience (glazed pottery) workshops and their use of metal, discovered at Abydos dating to 5500 BCE. The Badarian were followed by the Amratian (more advanced pottery), the Gerzean (more advanced metallurgy), and the Naqada (trading with Nubia) cultures; all of which contributed significantly to the development of what became Egyptian civilization.

Around 3400 BCE, two separate kingdoms were established in Egypt, near the Fertile Crescent, an area as we have seen, that was home to some of the world's oldest civilizations. The first kingdom was known as the Red Land, and was to the north, based in the Nile

(143) Upper and Lower Egypt was united under King Menes – the first Pharaoh

River Delta, and extended south along the Nile. The second kingdom was known as the White Land, and was further south. A southern king, known as Scorpion, made the first, but unsuccessful attempt to conquer the northern kingdom around 3200 BCE. Many city-states had already been established along the banks of the Nile River by the time the first pharaoh (King) Menes united these city-states into the first kingdom in 3100 BCE, which became known as the Old Kingdom. King Menes founded the First Dynasty of Egypt and created the city of Memphis. This version of the early history comes from "Aegyptica" (History of Egypt) written by the ancient Greek historian Manetho who lived in the third century BCE under the rulers of the Ptolemaic Dynasty. Although his chronology has been disputed by some later historians, it is still regularly consulted on dynastic succession and the early history of ancient Egypt. Manetho's work is the major source which cites Menes as the conqueror and unifier of the city-states into one kingdom.

Commencing from around 6000 BCE a belief in the gods defined the Egyptian culture. An early Egyptian creation myth tells the story of the god Atum who stood in the midst of swirling chaos before

(144) Osiris, Seth and Horus – three prominent Egyptian gods

the beginning of time, and made creation into existence. Atum was accompanied by the eternal force of heka (magic) personified in the god Heka, and by other spiritual forces which could animate the world. Heka was the primordial force which infused the universe and caused all things to operate as they did; it also allowed for the central values of the Egyptian culture; ma'at, meaning harmony and balance. All of the gods and their responsibilities went back to ma'at and heka. The Sun rose and set as it did, and the Moon travelled its course across the sky, and the seasons came and went in accordance with balance and order, which was possible because of these two forces. Ma'at was also personified as a deity, the goddess of the ostrich feather, to whom every pharaoh king promised his full abilities and devotion. The pharaoh king was associated with the god Horus in life, and Osiris in death, based upon a myth which became the most popular in Egyptian history. In Egyptian religion, Osiris and his sister-wife Isis were the original monarchs who governed the world and gave people the gifts of civilization. Osiris' brother Seth, grew jealous of him and murdered him, but Osiris was brought back to life by Isis who then bore his son Horus. Osiris was incomplete however, and so descended to rule the

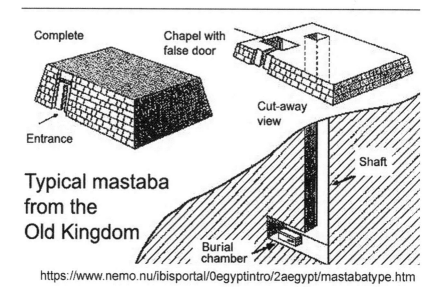

https://www.nemo.nu/ibisportal/0egyptintro/2aegypt/mastabatype.htm

(145) Mastaba burial tombs preceded the construction of pyramids

underworld while Horus, once he had matured, avenged his father's fate and defeated Seth. This myth highlighted how order triumphed over chaos, and would become a consistent core key belief in Egyptian religion, mortuary rituals, religious texts, in works of art, and in regular life. There was no period in Egyptian history in which the gods did not play a central role in the daily lives of the Egyptians, and this is clearly seen from the earliest times in their history.

The first 300-400 years of the Old Kingdom became known as the Early Dynastic Period (3100-2686 BCE). Each dynasty represented the rule of one family bloodline. During the Second Dynasty, Memphis was well and truly established as the capital city of the newly united Egypt; and this dynasty served as a bridge between the founding of a central government in the First Dynasty and the consolidation of Egyptian culture during the Third Dynasty. At this time, significant tomb building emerged in the form of block-shaped mastabas, the predecessors to pyramids. This period saw the development of the foundations of Egyptian society, including the ideological structure based on an all-powerful supreme ruler, called a king, and later to be given the title of pharaoh. To the ancient Egyptians, the king was a godlike being who was closely identified with the all-powerful god Horus. During this Early Dynastic Period, most of the people were farmers living in small villages; and agriculture based mainly on the cultivation of wheat and barley was the economic foundation of the kingdom. The annual flooding of the great Nile River provided the necessary irrigation and fertilization each year. Farmers sowed the wheat and barley after the flooding receded and harvested it before the season of high temperatures and drought returned. By the 27th century BCE the Egyptian kingdom was well on its way to becoming a significant and enduring civilization. The Third Dynasty brought much needed stability to the kingdom.

One of the great challenges of Egyptology (the study of ancient Egyptian history) is to devise an accurate list of the ruling pharaoh kings in chronological order and with their correct names. This is due to the wide variety of sources uncovered and the different names that were ascribed to the kings by these sources. For example, we have contemporary lists of kings from more recent dynasties inscribed on monuments that are still standing in Egypt; as well as stone tablets and papyrus-written records which have been excavated that record the names of kings from older dynasties. We also have the writings of ancient historians that are dated from the 3rd century BCE. Establishing the correct names of pharaoh kings is further complicated by the fact that during the Hellenic period of Egyptian rule many of the earlier pharaohs were given Greek names. We also have secondary sources coming from later Greek historians like Herodotus.

The Third Dynasty ushered in a period of more powerful pharaohs and would be the beginning of what was to become a stable and powerful Egypt and the beginning of the Old Kingdom (2686-2181 BCE). Around 2630 BCE, King Djoser asked his close advisor and confidante, the architect, priest and healer Imhotep, to design a funerary monument for him. The result was the world's first major stone building, the disproportionately shaped Step-Pyramid at Saqqara near Memphis. Imhotep also wrote one of the first medical texts describing the treatment of over 200 different diseases, and argued that the cause of disease could be natural, not the will of the gods. In the following century, the Fourth Dynasty pharaohs would advance the construction of pyramids to a more sophisticated level by developing a site at Giza, where three perfectly triangular pyramids were to be constructed under the direction of the pharaoh Khufu, who ruled from 2589 to 2566 BCE; each one to become the tomb of a different Fourth Dynasty pharaoh. At the same site, the Great Sphinx of Giza was also constructed in the image of the pharaoh Khafra (2558-2532 BCE), who we believe one of the pyramids was constructed for. The third pyramid was built for the pharaoh Menkaure (2532-2503 BCE).

The grandeur of the pyramids on the Giza plateau as they originally would have appeared, perfectly proportioned, visible from a great distance, and covered in gleaming polished white limestone, are a testament to the power and wealth of the rulers during this early period. Many theories exist regarding how these pyramids and monuments were constructed, but modern architects and Egyptologists are far from in agreement on any single theory. Considering the level

(146) The Great Pyramid of Giza and the Sphynx were completed by 2500 BCE

of technology of the day, some have argued that a monument such as the Great Pyramid of Giza should not have been able to be built, based on existing technology at that time. Others claim however, that the existence of such buildings and tombs suggest superior technology, the evidence for which has never been recovered, and instead lost in time. Most reputable scholars today reject the claim that the pyramids and other monuments were built by slave labour, although slaves of different nationalities certainly did exist in Egypt and were employed regularly in the mines, and as servants, for example. Egyptian monuments were considered public works created for the public benefit of the entire kingdom, and used both skilled and unskilled Egyptian workers in their construction, all of whom were paid for their labour. Workers at the Giza site, which was only one of many, were given a ration of beer three times a day, and ten loaves of bread each day. Egyptologists estimate that it took around 100,000 Egyptian labourers, many of whom were farmers, 30 years to complete the construction of the Great Pyramid of Giza.

It is easy to see why people are so fascinated with the Egyptian pyramids. There are a lot of mysteries surrounding their construction. Over a period of time, archaeologists have been able to piece together a viable hypothesis of how a pyramid such as the Great Pyramid of Giza was constructed. The stones themselves were mined from a quarry located not far south of the pyramid. It is believed the journey of transporting these massive stone blocks across the desert was made easier by wetting the sand first. Of course, this only explains how the stone blocks got from one location to another, not how they were then lifted high up into the air and deposited to form an enormous geometrically perfect triangular monument – a pyramid. Archaeologists believe this lifting of the stone blocks would have involved a ramp of some sort. This ramp would have needed to be very steep, at an incline of around 20 degrees; and that would have posed a significant challenge for a very heavy 2.5-ton block of stone.

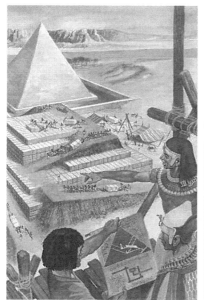

(147) & (148) Constructing the Great Pyramid of Giza – a massive project

In 2018, at Hatnub, a rock quarry located in Egypt's eastern desert, an Anglo-French team of archaeologists discovered a very unusual ramp carved into the ground, that hinted at some surprisingly advanced technological achievements. For one thing, it was very steep, but more significantly it was flanked on both sides by staircases. These stairs were marked with recurring holes that could have contained wooden posts, which of course could have rotted away long ago. According to the mission's co-director, Yannis Gourdon: "This kind of system has never been discovered anywhere else. Using a sled which carried a stone block and was attached with ropes to wooden posts, ancient Egyptians were able to pull up the alabaster blocks out of the quarry on very steep slopes of 20 percent or more." What's even more interesting, is that the ramp and stairs are dated to around 4500 years ago, well before the construction of the Great Pyramid of Giza commenced.

Roland Enmarch, another scholar who participated in the expedition, noted that the patterns of the post holes in the stairs suggested a particular kind of rope-and-pulley system. Since this specific ramp is cut into the rock itself, it would not have been used to build the Great Pyramid of Giza; but it does suggest that the ancient Egyptians had a firm grasp on the technology to move and hoist heavy stones upwards. Kara Cooney, a professor of Egyptian art and architecture at the University of California, referring to this discovery, said: "It's a stretch to take an alabaster quarry and say this is how the pyramids were built, because the pyramids weren't built out of alabaster. The way that the ancient Egyptians cut and moved stone is still very mysterious. Alabaster is a softer mineral, different from the heavy stone

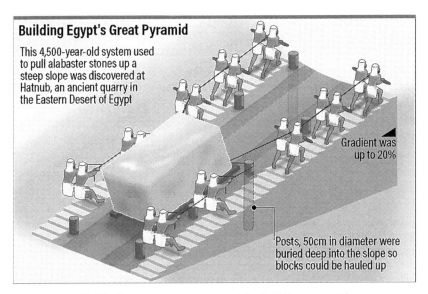

Building Egypt's Great Pyramid

This 4,500-year-old system used to pull alabaster stones up a steep slope was discovered at Hatnub, an ancient quarry in the Eastern Desert of Egypt

Gradient was up to 20%

Posts, 50cm in diameter were buried deep into the slope so blocks could be hauled up

(149) Ramp systems were used for building pyramids

blocks with which Egyptians built the outer structure of the pyramids. We actually don't know their mechanism of cutting hard stones like red granite. And we still don't know how the ancient Egyptians lifted blocks weighing hundreds of tons up the sides of pyramids."

Most Egyptologists already believe that the ancient Egyptians used ramp systems to build the pyramids, but there are different theories about what types they used. Professor Cooney says experts have theorized that the Egyptians could have used straight ramps that went up the outside walls of a pyramid; ramps that curved around these walls; or ramping systems inside the pyramid itself. So, although the ramp system discovery at the Hatnub alabaster quarry does tell us something about Egyptians' level of technological knowledge, it doesn't answer the big questions about how they actually built the pyramids. Cooney believes keeping this as a mystery is exactly the way the ancient Egyptians would have wanted it to be. "Just as any authoritarian regime is going to hide their secrets as long and as best as they can, the Egyptians purposefully left no record of how they built their pyramids," Cooney says. "The pyramids are there as mountains of stone proving the otherworldly nature of their god-kings. You stand in front of those pyramids and you feel it's impossible to build such a thing. That means that the propaganda is still working," says Cooney.

This period in the Old Kingdom was clearly one of affluence for the Egyptians. By this time, they had developed a superior level of technology that enabled them not only to design and construct perfectly geometrical pyramids; but also sophisticated engineering knowledge for their Nile River canal irrigation systems; detailed anatomical knowledge of the human body, for their mummification process; and more advanced boat-building, including the use of rudders and oars. During this time, the Egyptians also established lucrative trading networks with surrounding civilizations both within Africa, such as Nubia, and into the Middle East, such as Mesopotamia. This added great wealth to the Old Kingdom.

Over the course of the Fifth and Sixth Dynasties, the wealth of dynastic kings steadily depleted, partially due to the huge expense of building pyramids and other grand monuments; and also as a result of

Pepi II Neferkare

(150) King Pepi II ruled Egypt for 94 years and died around 2216 BCE

the growing influence and power of the nobility and priesthood who worshipped the Sun-god Ra. After the death of the Sixth Dynasty's King Pepi II, who ruled for 94 years, the Old Kingdom period was in utter chaos. It seems that conditions in Egypt at this time were becoming less favourable. For example, the annual flooding of the Nile appeared to have declined over a number of decades due to climatic changes. The annual flooding of the Nile was an integral aspect of ensuring there was a bountiful depositing of fertile nutrient-rich black soil for the growing and harvesting of wheat, barley and other produce to sustain the growing population of Egypt. So when the annual flooding was disrupted, there would have been political and societal upheaval because the people were not being adequately fed and as a result, widespread famines occurred. This civil unrest and the famines ultimately led to the collapse of the Old Kingdom and the beginning of the First Intermediate Period (2181-2055 BCE).

Immediately after the collapse of the Old Kingdom, the Seventh and Eighth Dynasties consisted of a rapid succession of Memphis-based rulers until about 2160 BCE, when the king's absolute central power completely dissolved, leading to civil war between various provincial governors. This chaotic situation was further intensified by the presence of Bedouin invasions, as well as famine and disease. The Bedouin tribes originated from Syria and North Arabia, and lived in the desert between the Nile River and Dead Sea during this time. From this era of conflict there emerged two different kingdoms. The Ninth and Tenth Dynasties produced 17 kings, based in Hierakonpolis, who ruled Middle Egypt between Memphis and Thebes; while another family of rulers rose to power in Thebes itself, and challenged the power centred around Hierakonpolis. Around 2055 BCE, the Theban prince Mentuhotep I managed to conquer Hierakonpolis, reunited Egypt, began the Eleventh Dynasty, and ended the First Intermediate Period.

After the last ruler of the Eleventh Dynasty, Mentuhotep IV, was assassinated the throne passed to his vizier, or chief minister, who became King Amenemhet I and the founder of the Twelfth Dynasty. It is believed Amenemhet was behind the assassination, in a naked grab for power. A new capital city was established at It-towy, south of Memphis, while Thebes remained as the major religious centre. During this Middle Kingdom (2055-1786 BCE), Egypt once again flourished as it had during the Old Kingdom. It is considered to be Egypt's "Classical Age" when art and culture reached great heights and Thebes became the most important and wealthiest city in the kingdom. The Twelfth Dynasty kings ensured the smooth succession of their bloodline by making each successor a co-regent (co-ruler), a custom that began with Amenemhet I (1991-1962 BCE). According to the historians Oakes and Gahlin, "the Twelfth Dynasty kings were strong rulers who established control not only over the whole of Egypt but also over Nubia to the south, where several fortresses were built to protect Egyptian trading interests. The first permanent standing army was created during the Middle Kingdom by Amenemhat I; the Temple of Karnak was begun under Senruset I (1971-1926 BCE); and some of the greatest Egyptian literature and art was created during this time.

Middle Kingdom Egypt pursued an aggressive foreign policy, colonizing Nubia (with its rich supply of gold, ebony, ivory and other resources) and repelling the Bedouins, who had infiltrated into Egypt during the First Intermediate Period, as we have seen. During this time the kingdom also built diplomatic and trade relations with Syria and Palestine; undertook building projects including military

(151) The Temple of Karnak was the largest religious complex in Ancient Egypt

fortresses and mining quarries; and returned to pyramid-building in the tradition of the Old Kingdom. The Middle Kingdom reached its peak under Amenemhet III (1842-1797 BCE); and its decline began under Amenemhet IV (1797-1790 BCE) and continued under his sister and regent, Queen Sobekneferu (1789-1786 BCE), who was the first confirmed female ruler of Egypt and the last ruler of the Twelfth Dynasty. The end of the long reign of Sobekneferu's father, Amenemhet III, brought her half-brother to the throne late in his life. When her brother died, the absence of a male heir resulted in Sobekneferu (her name translates to "the beauties of Sobek") being next in the line of dynastic succession.

As the first officially confirmed queen of Egypt, she deserves some recognition and praise. She certainly made full use of her naturally strong personality, determination, perseverance, sharp intellect, and tactical knowledge. Sobekneferu realised, that as the first female ruler of Egypt, she had to work hard to win over the respect and admiration of the people. During the nearly four years of her rule she was instrumental in the completion of a number of important buildings and monuments including Amenemhat III's funerary complex (called the "Labyrinth" by the historian Herodotus); she built the religious centre called Shedet, dedicated to the crocodile god Sobek; and a number of religious buildings in the city of Heracleopolis. The rulers of the Twelfth Dynasty established a religious and economic centre in the city of Fayoum, where crocodiles were nurtured and worshipped. Sobekneferu made no attempt to rule as a man, and did not conform to the gender rules of Egypt at that time. She insisted on being addressed by her female name, and did not adopt a male name, as was the common custom amongst women of the nobility. Several of her surviving portraits and statues show her wearing unusual clothes and symbols of power, indicating that she wanted her subjects to view her as an exceptionally powerful and unique ruler, and not to be identified as similar to "male" rulers. Nothing is known of Sobekneferu's death or burial. Some archaeologists have suggested that her place of burial might be one of the pyramids at Mazghuna, but this seems very unlikely because there is no evidence supporting this. Therefore, the destiny of one of the most powerful women in ancient history remains a mystery to us.

The Thirteenth Dynasty marked the beginning of another unsettled period in Egyptian history, during which time a rapid succession of kings failed to consolidate power. As a result, during this Second Intermediate Period (1786-1567 BCE), Egypt was divided into several spheres of influence. The official royal court and seat of government was relocated to Thebes, while a rival Fourteenth Dynasty, centred on the city of Sakha (known by the Greeks as Xois) in the Nile delta, seems to have existed at the same time as the Thirteenth Dynasty.

Around 1650 BCE, a line of foreign rulers known as the Hyksos took advantage of Egypt's instability to take control of Lower Egypt. The Hyksos were a mysterious people, most likely from the region of Syria and Palestine, who first appeared in Egypt around 1800 BCE and settled in the town of Avaris. While the names of the Hyksos kings are Semitic in origin, no definitive ethnicity has been established for them. The Hyksos grew in power until they were able to take control of a significant portion of Lower Egypt by 1720 BCE, resulting in the Theban Dynasty of Upper Egypt virtually becoming a vassal (subservient) state. I should explain at this stage that because the Nile River originates from the centre of the African continent and of course flows north and ultimately ends at the Mediterranean Sea; what is referred to as Lower Egypt is the northern region of Egypt all the way to the Nile delta which flows into the Mediterranean Sea; and what is referred to as Upper Egypt, is the southern region of the kingdom, the most southern portion of which borders Nubia.

The Hyksos people originated from Mesopotamia and spoke a Semitic language. They introduced chariots to Egypt, as well as elements of their own culture. The Hyksos rulers of the Fifteenth Dynasty adopted and continued many of the existing Egyptian traditions in government and culture. They ruled concurrently with the line of native Theban rulers of the Seventeenth Dynasty, who retained control of most of southern Upper Egypt, despite having to pay taxes to the Hyksos. We are unsure whether the Sixteenth Dynasty were Theban or Hyksos rulers. Conflict eventually flared-up between the two groups and the Thebans launched a war against the Hyksos around 1570 BCE, driving them out of Egypt. During the Middle Kingdom there were further advances in Egyptian technology and culture; such as more advanced glass making techniques, and the development of hieroglyphics language and writing to such a level that it would eventually evolve into the Phoenician alphabet, the precursor to the alphabet we use today. There were also advances in mathematics, science, human anatomy, and the identification and treatment of medical conditions.

The Theban king from Upper Egypt Ahmose I, was responsible for finally expelling the Hyksos Dynasty from Lower Egypt. Egypt was ridden of the foreign invaders and once again was reunited as one kingdom. Military commanders who had served the king faithfully were given high-ranking positions within the kingdom. This marked the beginning of the New Kingdom (1567-1085 BCE) in Egyptian history. Political, government, and military operations would be conducted out of the city of Memphis, whereas Thebes would act as the ceremonial and religious capital. It was to be near Thebes, where the renowned Valley of the Kings would become the new burial place for pharaohs. During the New Kingdom there were also significant

temples being built such as the Karnak Temple Complex and the
Mortuary Temple of Queen Hatshepsut (1479-1458 BCE), who was a
member of the Eighteenth Dynasty.

(152) & (153) Queen Hatshepsut was a gifted and exceptional leader

Hatshepsut demonstrated great leadership during her time in
power, and she reigned for over 20 years, commanding the great respect
of her political advisors and the common people. She dedicated herself
to the role of pharaoh to the extent where she dressed like a man with
a false beard and headdress, because only men were leaders during this
period in Egyptian history. This was another period of great accumulation of wealth and power for Egypt. Hatshepsut led one notable
trading expedition to the land of Punt, in the ninth year of her reign.
The Kingdom of Punt was located in north eastern Africa, in present
day Somalia. It was a land that was rich in resources and products
such as frankincense, wooden furnishings, sweet-smelling resin, ivory,
spices, gold, ebony, and aromatic trees. Scenes of this expedition can
be seen at her mortuary temple, Djeser-Djeseru at Deir el Bhari. One
of Hatshepsut's most important viziers and advisors was Senenmut.
He had been among the queen's servants and rose with her in power.
Some have speculated that he was also her lover.

While the pharaoh Queen Hatshepsut was looking after the
political and government affairs of the kingdom, her step-son and

regent (pharaoh-in-waiting), Thutmose III (1479-1425 BCE), was leading the army and looking for conquests beyond the borders of the Egyptian kingdom. Officially, Thutmose III ruled Egypt for almost 54 years, from the age of two; however, during the first 22 years of his reign, he was co-regent with his step-mother and aunt, Hatshepsut. In 2007, Egyptologists announced that Hatshepsut's mummy had been found in a tomb in the Valley of the Kings. The mummy showed signs of arthritis, many dental cavities, and root inflammation; as well as diabetes and bone cancer. She was just over 5 feet tall when she died, and there is evidence that she died obese. The lands of the Middle Eastern Levant (within Mesopotamia) were being contested for by the Mitannians, from northern Syria; and the Hittites from Anatolia, modern-day Turkey. The biggest military legacy of Thutmose III was the Battle of Megiddo in 1457 BCE, which was un uprising of a coalition of local rulers against the Egyptians themselves. Thutmose III was able to outwit the Levant coalition and through military conquests was able to expand Egypt to its greatest extent in its history, deeply into the Middle Eastern Levant and Asiatic lands.

After this period of great military success, one of the pharaohs who came to lead the New Kingdom was Amenhotep IV (1353-1336 BCE); who would achieve infamy for trying to alter the entire religious belief system of the Egyptians, from one of worshipping many gods (polytheism), to the exclusive worship of the Sun Disk God, Artem (monotheism). He undertook a religious and highly provocative and contro-versial revolution by disbanding the priest-hoods dedicated to Amon-Ra (a combi-

(154) The "Boy-King" Tutankhamun was nine years old when he ascended to the throne

nation of the local Theban god Amon and the sun god Ra) and forcing the exclusive worship of another sun-god, Aton. Renaming himself Akhenaton ("servant of the Aton"), he built a new capital in Middle Egypt called Akhetaton, known later as Amarna. This went totally against the polytheistic centuries old Egyptian religious worship of many gods. As we can imagine, this decision created great turmoil within Egyptian culture and society at the time. When Akhenaton died, the capital returned to Thebes, and under the leadership of the young "boy-king" pharaoh Tutankhamun (1336-1327 BCE), the Egyptians returned to worshipping a multitude of gods. He was originally named Tutankhaten to reflect the religious beliefs of his father but, upon assuming the throne he changed his name to Tutankhamun to honour the ancient god Amun. He restored the ancient temples, removed all references to his father's single deity, and returned the capital to Thebes. By the time Tutankhamun died, mummification techniques had advanced to a highly skilled level. It is incredible how extensive and deep their knowledge of the human body had progressed to.

The Nineteenth and Twentieth Dynasties saw the restoration of the weakened Egyptian Empire and an impressive amount of building, including additional great temples and cities. According to biblical chronology, the exodus of Moses and the Israelites from Egypt to the Promised Land, possibly occurred during the reign of Ramses II (1279-1213 BCE). All of the New Kingdom rulers (with the exception of Akhenaton) were laid to rest in deep, rock-cut tombs (not pyramids) in the Valley of the Kings, a burial site on the west bank of the Nile, opposite Thebes. Most of the tombs were raided and destroyed throughout the ancient and medieval historical periods, with the notable exception of the tomb and treasure of Tutankhamun. His tomb was discovered largely intact by the British Egyptologist Howard Carter, in 1922.

Ramses II is considered to be the greatest ruler of the New Kingdom, commencing the most elaborate building projects of any Egyptian ruler, and reigned so efficiently and effectively that he had the necessary resources to achieve his big ambitions. Although the famous Battle of Kadesh in 1274 BCE, between the forces of Ramses

II of Egypt, and Muwatalli II of the Hittites, is today regarded as having been a stalemate; Ramses II considered the battle to be a great Egyptian victory; celebrating himself as a champion of the people, and finally as a revered god, in his many public monuments. His temple of Abu Simbel, built for his queen Nefertari, depicts the Battle of Kadesh. Under the reign of Ramses II, the first peace treaty in the world, The Treaty of Kadesh, was signed in 1258 BCE; and Egypt enjoyed great prosperity and affluence as evidenced by the sheer number of monuments that were built or restored during his reign.

Ramses II's fourth son Khaemweset, is known as the "First Egyptologist" for his efforts in preserving and recording old monuments, temples, and their original owner's names. It is largely due to Khaemweset's initiative that Ramses II's name is so prominent at so many ancient sites in Egypt. Ramses II became known to later generations as "The Great Ancestor" and reigned for so long that he outlived most of his 96 sons, 60 daughters, and over 200 wives. In time, all of his subjects had been born knowing only Ramses II as their ruler, having no memory of any other. He enjoyed an exceptionally long life of 96 years, over double the average lifespan of an ancient Egyptian, and virtually unheard of in ancient times. At the time of his death, the sources record that many feared the end of the world had come, because he reigned for so long they expected that as a living god on earth he would reign for an eternity.

(155) King Ramses II and Queen Nefertari – a powerful ruling couple

One of the great pharaoh's successors, Ramses III (1186-1155 BCE), followed his policies but, by this time, Egypt's phenomenal wealth had attracted the attention of the "Sea Peoples" who began to make regular incursions along the Mediterranean coast. The Sea Peoples, like the Hyksos, are of unknown origin but are thought to have come from the southern Aegean region. Some archaeologists believe they may have been Mycenaean or even Minoans, the precursor civilizations to the ancient Greeks. Between 1276-1178 BCE the Sea Peoples were a threat to Egyptian security. Ramses II had defeated them in a naval battle early in his reign, as had his immediate successor Merenptah (1213-1203 BCE). After Merenptah's death however, the Sea Peoples increased their efforts, sacking Kadesh, which was then under Egyptian control, and ravaging the coast. Between 1180-1178 BCE Ramses III fought them off, finally defeating them at the Battle of Xois in 1178 BCE.

The Third Intermediate Period (1085-664 BCE) saw important changes in Egyptian politics, society and culture. Centralized government under the Twenty-First Dynasty pharaohs gave way to the resurgence of local officials including viziers and provincial governors, while foreigners from Libya and Nubia grabbed power for themselves and left a lasting imprint on the population of Egypt. The Twenty-Second Dynasty began around 943 BCE with King Sheshonq (943-922 BCE), a descendant of the Libyans who had invaded Egypt during the late Twentieth Dynasty and had settled there. Many local rulers were virtually autonomous during this period, and as a result, we have little records of the Twenty-Third and Twenty-Fourth Dynasties.

In the eighth century BCE, Nubian pharaohs beginning with Shabako (705-690 BCE), ruler of the Nubian kingdom of Kush, established their own Twenty-Fifth Dynasty at Thebes. Under Kushite rule, Egypt clashed with the rapidly growing Assyrian Empire. In 671 BCE, the Assyrian ruler Esarhaddon drove the Kushite King Taharqa out of Memphis and destroyed the city. He then appointed his own rulers from local governors and officials who were loyal to the Assyrians. One of them, Necho of Sais (671-664 BCE) ruled as the first king of the Twenty Sixth-Dynasty, before being killed by the Kushite leader Tanuatamun in a final unsuccessful grab for power.

Beginning with Necho's son Psammetichus I (664-610 BCE),
the Saite Dynasty ruled a reunified Egypt for less than two centuries.
In 525 BCE, Cambyses II (529-522 BCE), King of Persia, defeated
Psammetichus III (526-525 BCE), the last of the Saite kings, at the
Battle of Pelusium, and Egypt then became part of the Persian Empire.
Knowing the devotion Egyptians had for cats (who were thought to
be the living representations of the popular goddess Bastet), Cambyses
II ordered his men to paint cats on their shields, and to unleash cats
and other animals sacred to the Egyptians in front of the army as they
marched into battle towards Pelusium. The Egyptian forces surrendered and the kingdom fell to the Persians. Persian kings such as Darius
(522-485 BCE) ruled the country largely under the same terms as the
native Egyptian kings. Darius supported Egypt's religious cults and
undertook the building and restoration of its temples. The rule of the
Persian King Xerxes (485-465 BCE) sparked uprisings in Egypt under
his particularly cruel, ruthless rule and that of his successors. One of
these rebellions succeeded in 404 BCE, beginning one last period of
Egyptian independence under native rulers. These were the Twenty-
Eighth, Twenty-Ninth, and Thirtieth Dynasties.

In the middle of the fourth century BCE, the Persians again
attacked Egypt in 343 BCE, reviving their empire under Ataxerxes III
(359-338 BCE). Less than a decade later, in 332 BCE, the Macedonian
king Alexander the Great (336-323 BCE), defeated the armies of the
Persian Empire and conquered Egypt. Alexander was welcomed as a
liberator and conquered Egypt without a fight. He established the city
of Alexandria with its spectacular lighthouse and library, and moved on
to also conquer Phoenicia and the remainder of the Persian Empire.
After Alexander's death in 323 BCE his body was brought back to
Alexandria and Egypt was ruled by a dynastic line of Macedonian
kings, starting with one of Alexander's key generals and confidants,
Ptolemy I Soter (323-285 BCE), and continuing with his descen-
dants. The last ruler of Ptolemaic Egypt was the legendary Queen
Cleopatra VII (51-30 BCE), who surrendered Egypt to the armies of
Octavian (later to become the first Roman Emperor Augustus) in 31
BCE. Six centuries of Roman rule followed, from 31 BCE to 646 CE,

during which time Egypt was considered the personal property of the emperors; and Christianity became the official religion of Rome and the Roman Empire's provinces, including Egypt. The conquest of Egypt by the Arab Muslims under Caliph Umar in 646 CE, and the subsequent introduction of Islam was to do away with the last outward remnants of the ancient Egyptian culture and propel the country towards its ultimate status as a modern nation.

Cleopatra has been a fascinating character throughout history. She has been portrayed in books, theatre, art and cinema; including William Shakespeare's play "Antony and Cleopatra" and Jules Massenet's opera "Cleopatre." The most famous contemporary depiction of her was by the actress Elizabeth Taylor in the 1963 blockbuster film "Cleopatra" which won four Academy Awards, and was nominated for five more. This film also earned Elizabeth Taylor a Guinness World Record for the most costume changes in a single film. Taylor made 65 costume changes in this film. But who was the real Cleopatra? Let us examine her life in some detail.

(156) Queen Cleopatra
(157) Elizabeth Taylor as
Queen Cleopatra

Cleopatra VII Philopator, was an Egyptian Queen and the last pharaoh of Ancient Egypt. Her name means "glory of her father." As we have seen, she was a descendant of Ptolemy 1 Soter, a Macedonian Greek general, companion, confidante and bodyguard of Alexander the Great, who took control of Egypt after Alexander's death, and founded the Ptolemaic Kingdom. Embroiled in the internal politics of the Roman Republic, she was the lover of both Julius Caesar and then later, Marc Antony.

Cleopatra was born in 69 BCE. Her father Ptolemy XII, was forced out of Egypt due to earlier disastrous economic policies, personal bankruptcy, and for engaging in a covert scheme to handover Egyptian territories to the Romans, whom he was heavily indebted to. During his exile, he took his 11 years-old daughter Cleopatra with him and settled in the outskirts of Rome, and later in Ephesus (in modern day Turkey); while his throne of Egypt was forcibly taken over by his elder daughter, and Cleopatra's older sister, Berenice IV Epiphaneia. A Roman army fought a successful campaign to remove and execute Berenice IV, and successfully reinstated Ptolemy XII on the Egyptian throne. However, Ptolemy XII was still beholden to the Romans in part because of the debt he owed them. He died in 51 BCE when Cleopatra was 18, leaving her and her 12 years-old brother Ptolemy XIII as co-regents. As was the custom at the time, Cleopatra married her younger brother and together they ruled Egypt. However, they had a fallout in 50 BCE and Ptolemy XIII had her exiled, leaving himself in charge. Cleopatra was forced to flee Egypt for Syria in 49 BCE. While there, she raised an army of mercenaries and returned the following year to face her brother's forces in a costly and damaging civil war at Pelusium, on Egypt's eastern border with Syria. She was unsuccessful in regaining control of Egypt and had to go into hiding again.

In 48 BCE the Roman Republic was itself embroiled in a civil war between the two ambitious generals, Julius Caesar and Pompey. The Pompey family had close connections with the Ptolemies of Egypt; so when Pompey lost the civil war in the Battle of Pharsalus in Greece, he fled to the Egyptian capital Alexandria, where he was murdered and decapitated on the orders of Ptolemy XIII, who thought this murder and the bringing of Pompey's head to him would please Caesar. In fact, Caesar was horrified by this brutal murder of a noble Roman general, and had the assassins executed. Soon after arriving in Egypt, Caesar met and fell in love with Cleopatra. Caesar needed to fund his own return to power in Rome, and needed Egypt to repay the debts it owed to Rome. With Caesar's military strength and support, Ptolemy XIII was overthrown and killed, and Cleopatra was reinstalled as Queen. In 47 BCE Cleopatra gave birth to Caesarion, though Caesar never publicly declared him to be his son.

By 46 BCE, Caesar had to return to Rome, and he was openly accompanied by Cleopatra and her entourage who were allowed to live within the palatial estate of Julius Caesar on the banks of the Tiber River in close proximity to the central Roman Forum. During her stay in Rome, Cleopatra was officially designated as a "friend and ally of Rome", which gave her the status of a client ruler. Cleopatra reciprocated by endowing lavish gifts and meeting powerful influential senators like Cicero. Unfortunately for the couple, not everybody

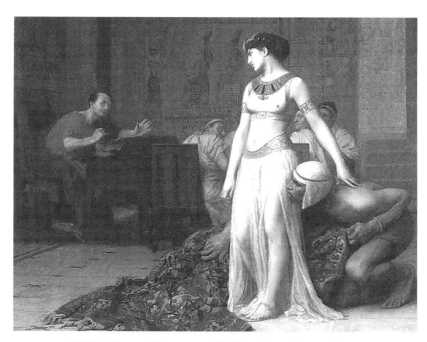

(158) Queen Cleopatra and Julius Caesar became lovers

was impressed by their opulent show of political and regal power. This was not helped by the fact that Caesar was still married to the highly respected Calpurnia, and Roman laws prohibited bigamy. The clearly unsympathetic Cicero had this to say in 45 BCE: "I detest the queen. For all the presents she promised were things of a learned kind, and consistent with my character, such as I could proclaim on the housetops, and the insolence of the Queen herself when she was living in Caesar's trans-Tiberine villa, the recollection of it is painful to me."

Beyond the opulence, it was the dictatorial nature of Caesar that caused his ultimate demise; his brutal assassination on the famous Ides of March, 15th March 44 BCE, on the floor of the Senate, when he was stabbed 27 times by a group of conspiratorial Senators led by Brutus and Cassius, who believed that by performing this "honourable act" they were protecting the Roman Republic from the tyranny of Julius Caesar; whom they were convinced was setting himself up to bring back the monarchy and become the supreme King of Rome. Amazingly, the confident and guileful Cleopatra remained in Rome for a month after her lover's murder, in what appeared to be a bid to have her son, the infant Caesarion, recognized as the legitimate heir to Caesar. But in his will, Julius Caesar had already named his great-nephew Octavian as his heir and prime beneficiary, a decision that was to sow the seeds of another civil war between the various factions of Caesar's supporters.

For a time after her return from Rome, Cleopatra's reign in Egypt brought relative stability to the region, and also peace and prosperity to a kingdom that had been bankrupted by civil war. Although she was brought up to speak fluent Greek like her family, she made an effort to learn Egyptian; and later only spoke in Egyptian. The years following Caesar's death led to a great power struggle between Caesar's cousin and loyal general, Marc Antony; and Caesar's great nephew and adopted son, Octavian. Despite being married to Octavian's sister, Octavia, Marc Antony began a relationship with Cleopatra, and together they had three children. Octavian was able to convince the majority of the Roman Senate that Marc Antony was no longer loyal to Rome and instead would give away Rome to the Egyptian Queen, who seemed to have Mark Antony under her spell. It was also seen as a family insult that Mark Antony was married to Octavian's sister, but yet was having an affair with Cleopatra. The historian Plutarch, had this to say about the Egyptian Queen's first formal encounter with Marc Antony in the city of Tarsos in Anatolia (modern day Turkey): "She came sailing up the river Cydnus in a barge with gilded stern and outspread sails of purple, while oars of silver beat time to the music of flutes and fifes and harps. She herself lay all along, under a canopy of cloth of gold, dressed as Venus in a picture, and beautiful young boys, like painted Cupids, stood on each side to fan her. Her maids were dressed like Sea Nymphs and Graces, some steering at the rudder, some working at the ropes." This first formal encounter sowed the seeds for what was to become a full-scale passionate love affair between the couple which resulted in Marc Antony moving to Alexandria; and Cleopatra giving birth to twins in 40 BCE, a son named Alexander Helios, and a girl named Cleopatra Selene II, both of whom were acknowledged by Antony as his own children. They later had another son, who they named Ptolemy Philadelphos. It should be noted that during his passionate relationship with Cleopatra, Antony was already married to Fulvia and later, to Octavian's sister, Octavia.

Over the course of the decade between 41 and 31 BCE, the relationship between Antony and his one-time ally Octavian worsened

(159) Elizabeth Taylor and Richard Burton as Cleopatra and Marc Antony

significantly due to the political fallout over the control of the Roman Republic. In effect, Octavian continued to claim his status as the "true heir" of Julius Caesar in Rome; while Antony, often pressured by Cleopatra, began to assert his ties with the Greek Eastern provinces and client states of the Roman Republic, which he controlled or influenced. Antony's political power in Rome took a big hit after he led a disastrous military campaign into Parthia, Rome's major super-power rival at the time; and when he officially divorced Octavia in 33 BCE, in a bid to marry Cleopatra. By 32 BCE, many senators and officials who still supported Antony had to flee Rome, under threat from Octavian and his loyal army. Finally, Octavian achieved his justi-fication for war when Mark Antony's Will was forcefully seized from the sacred Temple of Vesta by agents of Octavian, and then read to the Senate. It is believed the contents of Antony's Will revealed how he had plans to divide up the Roman territories in the east among his own sons. But the most contentious point in the Will relates to how Antony put forth Caesarion, the alleged son of Julius Caesar and Cleopatra, as the legitimate heir to Caesar, thereby sidelining Octavian who was widely perceived as the true heir of Caesar. Consequently, in 32 BCE, the Senate officially revoked the consulship of Marc Antony and declared war on Cleopatra's regime in Egypt.

This antagonism between Mark Antony and Octavian flared up into a full scale civil war when in 31 BCE, Cleopatra joined her Egyptian military naval forces with the Roman forces of Marc Antony, and fought against Octavian's forces in The Battle of Actium off the west coast of Greece. Cleopatra and Marc Antony were decisively defeated in this battle, but just managed to escape with their lives, heading back to Alexandria, despondent and dejected. However, Octavian's forces pursued the couple and captured the city of Alexandria. With no chance of escape, and their future rule and dynastic plans in tatters, Marc Antony and Cleopatra both could see the end was drawing near. On 1st August 30 BCE, hearing a false rumour that Cleopatra had killed herself, and knowing that Octavian had just arrived in Alexandria with his army and was about to capture him; Marc Antony, at the age of 53, dressed himself in his full Roman military general's uniform and died an honourable soldier's death by piercing his heart underneath his ribcage with his sword. The tragedy of this act was that Cleopatra had not in fact committed suicide; the rumour that Antony had heard was false. On 12th August 30 BCE, inconsolable with grief and after burying Marc Antony and meeting with the victorious Octavian, and finding out he intended to bring her and the children back to Rome with him, and parade them in a triumph like circus animals; Cleopatra, at the age of 39, shut herself within her chamber with two of her closest female servants, and then according to Plutarch (45-120 CE), used a poisonous snake known as an asp, a symbol of divine royalty, to commit suicide as well. Interestingly, the ancient historians had differing accounts of how Cleopatra died. According to Cassius Dio (164-235 CE), she injected herself with poison; while Strabo (63 BCE – 23 CE) claimed she used a toxic ointment. We will never know the truth of exactly how she died. According to her wishes, Cleopatra's body was buried with Antony's, leaving Octavian to celebrate his conquest of Egypt and his consolidation of power in Rome, having effectively removed his only remaining immediate rival, Marc Antony. Octavian later had Cleopatra's son Caesarion strangled, ending the Cleopatra-Ptolemaic dynasty, and any potential threat to his rule of the Roman Empire. Egypt became a province of the newly formed Roman Empire, and Octavian became the first Roman Emperor, Augustus Caesar. Cleopatra turned out to be the last of the Egyptian Pharaohs.

Many contemporary sources at that time spoke of the mystique of Cleopatra's beauty and allure. It was also rare for a woman to appear on an Egyptian coin, as Cleopatra did. The historian Plutarch, wrote this about Cleopatra in his biographical manuscript, "The Life of Mark Antony": "For as they say, it was not because her beauty in itself was so striking that it stunned the onlooker, but the inescapable impression produced by daily contact with her; the attractiveness in the persuasiveness of her talk, and the character that surrounded her conversation was stimulating. It was a pleasure to hear the sound of her voice, and she turned her tongue like a many-stringed instrument expertly to whatever language she chose."

(160) The death of Marc Antony depicted in the arms of Cleopatra

In contrast to the other members of the Ptolemaic family, Cleopatra did learn to speak Egyptian fluently and represented herself as the reincarnation of the Egyptian goddess, Isis. In her book, "Cleopatra: Last Queen of Egypt" published in 2008, author Joyce A. Tyldesley in describing Cleopatra's character and behaviour, said she was: "A woman who worshipped crude gods, dominated men, slept with her brothers and gave birth to bastards." Whereas, another contemporary author, Stacy Schiff, in her book, "Cleopatra: A Life" published in 2010, focuses more on Cleopatra's achievements in this quotation, when she says Cleopatra was: "A capable, clear eyed sovereign, she knew how to build a fleet, suppress an insurrection, control a currency, alleviate a famine."

The 1963 film Cleopatra was the pinnacle in the various shifts, perceptions, idealizations, stereotyping, glorifications, mythologizing and personification of the Cleopatra story. In this film though, as magnificent a spectacle as it was, Cleopatra's personal perspective and those of her contemporaries who lived with her, were largely overlooked and glossed over. Taking the works of authors, such as Shakespeare, and exaggerating them, Hollywood sexualized and glamorized the character of Cleopatra, which appealed to the modern masses. She was even depicted with a "lily-white" European skin complexion, rather than the reality of her darker North African complexion. Cleopatra's true identity, cleverly manipulated from the start, to the extent we can determine it from reliable sources, can be found buried underneath centuries of differing historical interpretations and artistic liberties. The competent, shrewd, strategic and highly educated Queen, has been distorted and undermined by the tale of an exotic, manipulative enchantress. To this day, Cleopatra still is the subject of movies, plays, books and works of art; and her image is constantly changing in the process. Over the next hundred years and beyond, what more will be added to her ever-changing myth and legendary status, and what pieces of future societies will she reflect?

The glory of Egypt's past was rediscovered during the 18th and 19th centuries CE and has had a profound impact on our current-day understanding of ancient history and the world. The historian Will Durant eloquently expressed this sentiment, felt by many:

"The effect of remembrance of what Egypt accomplished at the very dawn of history has influence in every nation and every age. 'It is even possible', as Faure has said, 'that Egypt, through the solidarity, the unity, and the disciplined variety of its artistic products, through the enormous duration and the sustained power of its effort, offers the spectacle of the greatest civilization that has yet appeared on earth.' We shall do well to equal it."

Egyptian culture and history has long held a universal fascination for people around the world; whether through the work of archaeologists in the 19th century CE, such as Champollion who deciphered the Rosetta Stone in 1822, or the famous discovery of the Tomb of Tutankhamun by Howard Carter in 1922. The ancient Egyptian belief in life as an eternal journey, created and maintained by divine magic, inspired later cultures and religious beliefs. Much of the iconography and beliefs of Egyptian religion found their way into the new religion of Christianity, and many of their symbols are recognizable today with largely the same meaning. It is an important legacy to the power of

the Egyptian civilization that so many works of the creative imagination, from films to books to paintings and even to religion, have
been and continue to be inspired by the elevating and profound vision
the Egyptians had of the universe and humanity's place in it.

On 19th July 1799, during Napoleon Bonaparte's Egyptian
military campaign, a French soldier discovered a black basalt slab
inscribed with ancient writing near the town of Rosetta, about 55
kilometres east of Alexandria. The irregularly shaped stone contained
fragments of passages written in three different language scripts; Greek,
Egyptian hieroglyphics and Egyptian demotic. The ancient Greek on
the Rosetta Stone indicated to archaeologists that it was inscribed by
priests honouring the King of Egypt, Ptolemy V, in the second century
BCE. Incredibly, the Greek passage announced that the three scripts
all had an identical meaning. This artefact therefore held the key to
solving the complex puzzle of deciphering hieroglyphics, a written
language that had been "dead" for thousands of years.

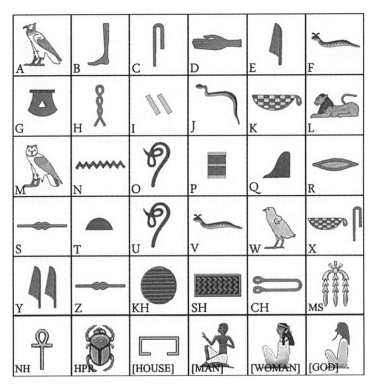

(161) Egyptian hieroglyphics were first deciphered in 1822

When Napoleon, an emperor known for his enlightened view of education, art and culture, invaded Egypt in 1798, he took along with him a group of scholars and instructed them to seize all important cultural artefacts for France. Many of these Egyptian artefacts are today displayed in the Louvre Museum in Paris. Pierre Bouchard, one of Napoleon's soldiers, was aware of this instruction from his leader when he found the basalt stone, which was almost four feet long and two-and-a-half feet wide, at a fort near Rosetta. When the British defeated Napoleon in 1801 they took possession of the Rosetta Stone, and today it is housed in the British Museum in London.

Several scholars, including the English archaeologist Thomas Young, made limited progress with the initial hieroglyphics analysis of the Rosetta Stone. It was the French Egyptologist Jean-Francois Champollion though, who taught himself ancient languages and ultimately cracked the code, and deciphered the hieroglyphics using his knowledge of ancient Greek as a guide. Hieroglyphics used pictures to represent objects, sounds and groups of sounds. Once the Rosetta Stone inscriptions were translated, the language, history and culture of ancient Egypt was suddenly opened up to scholars; and our knowledge of ancient Egypt has been significantly enhanced, continuing to the present day.

Late in 1922, the British archaeologist Howard Carter discovered the tomb of the Pharaoh Tutankhamun, who as we have seen, had died in 1323 BCE aged 18, in the Valley of the Kings, across the Nile River from Luxor in Egypt. As we have also seen, pharaohs had been buried in the Valley of the Kings from the 16th to the 11th centuries BCE. Most of the tombs were ransacked and plundered by tomb-raiders from earlier times. Tutankhamun's tomb was the first to be discovered intact and undisturbed. The Fifth Earl of Carnarvon, George Herbert, an enthusiastic amateur Egyptologist who was financing the excavation project, joined Carter and his team to enter the burial chambers, where they found the young pharaoh's mummified body, his spectacular gold funeral-mask, and a wide variety of precious jewellery, religious objects, wall paintings, statues, furniture, and inscriptions; all of which the young pharaoh wanted to take with him into the afterlife.

(162) Howard Carter's discovery of the tomb of Tutankhamun in 1922

The Tutankhamun discovery created a big sensation around the world in 1922. Stories also began to spread in the worldwide media in 1923 about a curse which afflicted anybody who had dared to break into a pharaoh's tomb. The Times in London and New York World magazine published the best-selling novelist Marie Corelli's prediction that "the most-dire punishment follows any rash intruder into a sealed tomb." It was not long before these articles were published that Lord Carnarvon had died in Cairo at age 56, from blood poisoning from an infected mosquito bite on his cheek, and the lights of the entire city went out, setting off a frenzy of speculation about a mummy's curse. This "Curse of the Pharaohs" supposedly promised death to anyone who disturbed the rest of the kings and queens buried in the Valley of the Kings. The "Sherlock Holmes" author Arthur Conan Doyle told the American press that "an evil elemental created by priests to protect the mummy could have caused Carnarvon's death."

No evidence of a curse had actually been found in the tomb, but the deaths in succeeding years of various members of Carter's excavation team, and those of real or supposed visitors to the burial site, created a lot of mystery and kept the "curse" story alive; especially

(163) The mysterious death of Lord Carnarvon in 1923

in the cases of death by violence or in unusual, unexpected circumstances. The alleged victims of the "mummy curse" included Prince Ali Kamel Fahmy Bey of Egypt who had visited the tomb and was shot dead by his wife in 1923; Sir Archibald Douglas Reid who apparently X-rayed the Tutankhamun mummy and then died in mysterious circumstances in 1924; Arthur Mace, a member of Carter's excavation team, who is believed to have died from arsenic poisoning in 1928; Carter's secretary, Richard Bethell, who supposedly died when he was smothered in his bed in the middle of the night in 1929; and Bethell's father who committed suicide in 1930. In addition to these cases, not long after the discovery of Tutankhamun's burial chamber; Lord Carnarvon's pet bird was eaten by a snake, his dog died in England at the exact time Carnarvon died in Egypt, and a wealthy American died of pneumonia shortly after visiting the tomb.

In reality however, most of the people who worked in or visited the tomb, lived long lives, but this did not undermine belief in the curse by those who wanted to believe it. Howard Carter himself, angrily dismissed the whole idea of a mummy's curse as being "tommy rot", but when he himself died alone and miserably unhappy, of Hodgkin's disease in his London flat in March 1939 at the age of 64, the story of the mummy's curse sprang back to life with a vengeance, and has persisted to this day.

(164) & (165) "Mummy-Mania" as depicted in popular culture in the movies

In 2002, the British Medical Journal conducted a study on the survival rates of 44 "non-Egyptians" whom Howard Carter identified as being in Egypt when the tomb was discovered and excavated. The reason only non-Egyptians were included in this study is that the curse was said not to affect native Egyptians. This study compared the mean (average) age of death for those 25 non-Egyptians who were present at the tomb, with the others who were not. It found no significant connection between potential exposure to the mummy's curse and survival; as well as no sign at all that those who were exposed to the mummy or burial chamber were more likely to die within 10 years.

Some contemporary medical researchers who have sought to find scientific explanations have stated that Lord Carnarvon's death may have been linked to toxins within Tutankhamun's burial chamber. A number of ancient mummies have been shown to carry potentially dangerous species of mould, and the tomb walls could have been covered in bacteria which is known to attack the respiratory system. The reality is that Lord Carnarvon was chronically ill before he even visited the Tutankhamun burial chamber. Also, he did not die until a number of months after he visited the burial chamber, so if he had been exposed to toxins, he surely should have died much earlier than he did.

One of the greatest achievements of the ancient Egyptians, in addition to the construction of perfectly geometrical pyramids, was their advanced level of medical knowledge and how they applied these skills to the mummification of bodies in order to prepare them for the afterlife. The ancient Egyptians believed that each individual possessed a "Ka", a life force that departed the body after death. Upon death, the "Ka" needed to continue to receive nourishment in the form of offerings of food, whose spiritual essence it still consumed. A person also had a "Ba", a set of spiritual characteristics which were unique to each person. These remained attached to a body after death and would return each night to receive new life. Because of the post-mortem importance of a body, the Egyptians believed that bodies had to be well preserved. While the most elaborate versions of this practice were only reserved for the highest levels of Egyptian society, mummification was a cornerstone of Egyptian religion. After death, a body begins to decompose, which can only be prevented by depriving the tissues of moisture and oxygen.

The earliest Egyptians buried their dead in shallow graves in the sands of the desert. The hot, dry sand quickly removed moisture from the dead body and created a natural mummy. This was not the case if they placed the body in a coffin. Instead, it would decompose and could not be preserved. In order to ensure the body was preserved, the ancient Egyptians began to use a process known as mummification to embalm the bodies, wrap them in linen, and then preserve them.

The practice of mummifying the dead began in ancient Egypt around 3500 BCE. The English word "mummy" comes from the Latin word "mumia" which is derived from the Persian "mum" meaning "wax", and refers to an embalmed corpse which was wax-like. The

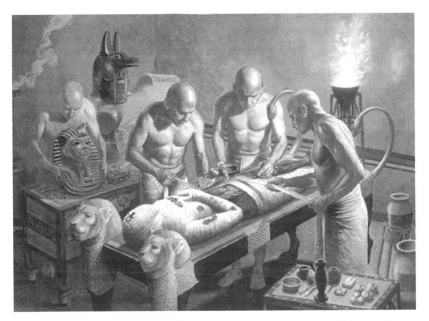

(166) Ancient Egyptian mummification was an elaborate and sophisticated process

idea of mummifying the dead seems to originate from how well corpses were preserved in the arid sands of the desert. During the Early Dynastic Period (3150-2613 BCE), the mastaba tomb had replaced the simple grave, and cemeteries became common. Mastabas were seen not as a final resting place, but as an eternal home for the body, but not the soul. The tomb was now considered a place of transformation in which the spiritual soul could leave the body to go on to the afterlife. It was thought however, that the body had to remain intact in order for the soul to continue its journey to the afterlife.

Once it was freed from the body, the soul would need to orient itself by what was familiar to it. It was for this reason that tombs were painted with stories and spells from "The Book of the Dead", to remind the soul of what was happening and what to expect, as well as with inscriptions known as "The Pyramid Texts" and "Coffin Texts" which recounted particular events from the dead person's life. Death was not the end of life to the Egyptians, but simply a transition from one state to another. With this in mind, the body had to be carefully prepared in order to be recognizable to the soul upon its awakening in the tomb, and its journey to the afterlife.

(167) The Journey of the Egyptian dead to the Afterlife

The Egyptian Book of the Dead was a collection of spells which enabled the soul of a deceased to navigate and enter the afterlife. The famous title was given to the work by western scholars. The actual title would more accurately translate to "The Book of Coming Forth by Day" or "Spells for Going Forth by Day." Although the work is often referred to as "the Ancient Egyptian Bible" it is no such thing; although the two works share the similarity of being ancient compilations of texts written at different times, eventually gathered together in book form. The Book of the Dead was never codified into a book-form, and no two copies of the work are exactly the same. The reason for this is that they were created specifically for each individual who could afford to purchase one as a manual to help them after death. The Egyptologist Geraldine Pinch, explains it as follows: "The Egyptian Book of the Dead is a term coined in the nineteenth century CE for a body of texts known to the Ancient Egyptians as the Spells for Going Forth by Day. After the Book of the Dead was first translated by Egyptologists, it gained a place in the popular imagination as the Bible of the Ancient Egyptians. The comparison is very inappropriate. The Book of the Dead was not the central holy book of Egyptian religion. It was just one of a series of manuals composed to assist the spirits of the elite dead to achieve and maintain a full afterlife."

The god Osiris was often depicted as a mummified ruler and regularly represented with green or black skin, symbolizing both death and resurrection. The Egyptologist Margaret Bunson writes:

"The cult of Osiris began to exert influence on the mortuary
rituals and the ideals of contemplating death as a 'gateway into
eternity'. This deity having assumed the cultic powers and rituals
of other gods of the necropolis, or cemetery sites, offered human
beings salvation, resurrection, and eternal bliss."

Eternal life was only possible though, if one's body remained intact. A
person's name and their identity represented their immortal soul; and
this identity was linked to one's physical form. The soul was an integral
part of the ancient Egyptian religious afterlife beliefs, and was thought
to consist of nine separate parts. These were:

1. The "Khat" which was the physical body.
2. The "Ka" which was one's double-form (the astral self).
3. The "Ba" which was a human-headed bird aspect that could
    speed between earth and the heavens; specifically, between the
    afterlife and one's body.
4. The "Shuyet" which was the shadow self.
5. The "Akh" which was the immortal, transformed self, after
    death.
6. The "Sahu" which was an aspect of the "Akh."
7. The "Sechem" which was another aspect of the "Akh."
8. The "Ab" which was the heart; the source of good and evil,
    and the holder of one's character.
9. The "Ren" which was one's secret name, to be used in the
    afterlife.

The "Khat" needed to exist in order for the "Ka" and "Ba" to
recognize itself, and to be able to function properly. Once released
from the body, these different aspects of the soul would be confused,
and would therefore need to orientate and centre themselves by some
familiar form, before undertaking the sacred journey into the afterlife.

When an ancient Egyptian person died, they were brought to the
embalmers, who offered three differing types of services. The historian
Herodotus (484-425 BCE) describes these three levels as follows: "The
best and most expensive kind is said to represent Osiris, the next best is
somewhat inferior and cheaper, while the third is cheapest of all." The

grieving family would be asked to choose which of the three services they preferred, and their answer was extremely important, not only for the deceased, but also for themselves. Obviously, the best service was the most expensive, but if the family could afford it and yet chose not to purchase it, they ran the risk of a "haunting." The dead person would know they had been given a cheaper service than they deserved, and would not be able to peacefully go into the afterlife; instead they would return to make their relatives' lives miserable until the wrong was rectified. Burial practices and mortuary rituals in ancient Egypt were taken so seriously because of the deep belief they held, that death was not the end of life. The individual who had died could still see and hear, and if wronged, they would be given leave by the gods, to seek out their revenge on earth.

(168) Mummification – The Opening of the Mouth ceremony

The mummification process took around 70 days and involved the following steps:

1. The body was washed.
2. A cut was made on the left side of the abdomen and the internal   organs; the intestines, liver, lungs, and stomach were removed. The heart, which the ancient Egyptians believed to

be the centre of emotion and intelligence, was left within the body, for use in the afterlife.

3. A hooked instrument was used to remove the brain through the nostrils of the nose. This was done because the brain was not considered to be important in the afterlife, so hence it was thrown away.

4. The body and the internal organs were then packed with natron salt for forty days, in order to remove all remaining moisture.

5. The dried organs were wrapped in linen and placed in special "canopic jars." The lid of each jar was shaped to represent one of the god Horus' four sons.

6. The body was thoroughly cleaned and the dried skin was rubbed with oil.

7. The body was packed with sawdust and rags, and the open cuts were sealed with wax.

8. The body was then wrapped with linen bandages. About 20 layers were used. This would take 15–20 days to complete.

9. A death mask was then placed over the bandaged face. The higher in rank the person was, the more elaborate and precious the mask was.

10. The bandaged body was placed in a shroud (a large sheet of cloth) which was secured with linen strips.

11. The body was finally placed in a decorated mummy case (sarcophagus) or coffin.

After the completion of the above process of mummification, the mummies were then secured into their tombs. Egyptologists and archaeologists are continuing to find these burial tombs at excavation sites throughout areas of ancient Egyptian settlement.

The Egyptologist, Salima Ikram, professor of Egyptology at the American University in Cairo, has studied mummification in great depth, and provides the following observations:

"The key ingredient in the mummification was natron, or netjry, divine salt. It is a mixture of sodium bicarbonate, sodium carbonate, sodium sulphate and sodium chloride that occurs naturally in Egypt, most commonly in the Wadi Natrun some sixty-four kilometres northwest of Cairo. It has desiccating and defatting properties and was the preferred dessicant, although common salt was also used in more economical burials."

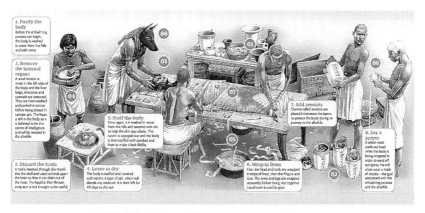

(169) The Mummification process in Ancient Egypt

The historian Herodotus described the work of the ancient Egyptian embalmers during the most expensive type of burial service, once the body was laid out on a table and washed:

"The brain was removed via the nostrils with an iron hook, and what cannot be reached with the hook is washed out with drugs; next the flank is opened with a flint knife and the whole contents of the abdomen removed; the cavity is then thoroughly cleaned and washed out, firstly with palm wine and again with an infusion of ground spices. After then it is filled with pure myrrh, cassia, and every other aromatic substance, excepting frankincense, and sewn up again, after which the body is placed in natron, covered entirely over for seventy days – never longer. When this period is over, the body is washed and then wrapped from head to foot in linen cut into strips and smeared on the underside with gum, which is commonly used by the Egyptians instead of glue. In this condition the body is given back to the family who have a wooden case made, shaped like a human figure, into which it is put."

In the second-most expensive type of burial, we can see from Herodotus, that less care was given to the body:

"No incision is made and the intestines are not removed, but oil of cedar is injected with a syringe into the body through the anus which is afterwards stopped up to prevent the liquid from escaping. The body is then cured in natron for the prescribed number of days, on the last of which the oil is drained off. The effect is so powerful that as it leaves the body it brings with it the viscera in a liquid state and, as the flesh has been dissolved by the natron, nothing of the body is left but the skin and bones. After this treatment, it is returned to the family without further attention."

The third and cheapest method of embalming was "simply to wash out the intestines and keep the body for seventy days in natron" according to Herodotus. The internal organs were removed in order to help preserve the corpse, but because it was believed the deceased would still need them, the viscera were placed in canopic jars to be sealed in the tomb. Only the heart was left inside the body as it was thought to contain the "Ab" aspect of the soul. Most Egyptians could not afford the more expensive abovementioned first and second

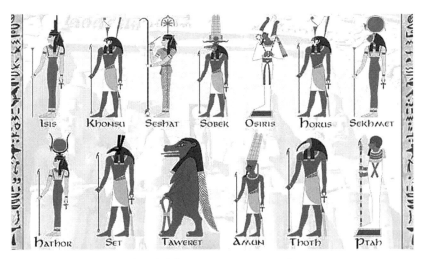

(170) The major Egyptian gods

embalming methods, so they had no choice other than to accept this cheapest third method. Of course, the pharaoh and his family and closest political and religious advisers would be embalmed in the most elaborate "first-method" fashion.

Ancient Egyptian funerals were a public affair at which, if one could afford them, women were hired as professional mourners. These women were known as the "Kites of Nephthys" and would encourage people to express their grief through their own emotional outpourings in the form of cries, screams and wailings. They would make references to how short life was, and how suddenly death came, but would also give assurances of the eternal aspects of the soul, and the confidence that the deceased would pass through the trial of the weighing of the heart in the afterlife by Osiris, to pass on to paradise in the "Field of Reeds." The ancient Egyptians believed that when they died, they would be judged on their behaviour during their lifetime, before they could be granted a place in the afterlife. This judgment ceremony was called the "Weighing of the Heart" and was recorded in Chapter 125 of the funerary text known as "The Book of the Dead." By weighing

(171) Magnificent Ancient Egyptian art

the heart of a deceased person against Ma'at (or truth), who was often represented as an ostrich feather, the god Anubis dictated the fate of souls. If the souls were heavier than an ostrich feather, they would be devoured by Ammit; and if the souls were lighter than an ostrich feather, it would ascend peacefully into the afterlife. In ancient Egyptian religion, the god Anibus (the dead) weighed the heart; the god Thoth (wisdom and judgement) recorded the results; and the god Ammit (divine retribution) stood by.

12

PHOENICIANS AND MINOANS OF THE MEDITERRANEAN

P HOENICIA WAS AN ANCIENT CIVILIZATION THAT CONSISTED OF A number of independent city-states which were located along the coast of the Mediterranean Sea; covering mainly what today is Syria, Lebanon and northern Israel. The Phoenicians were an advanced maritime people, who were renowned for their excellent ships, decorated with horses' heads in honour of their god of the sea Yamm, who was the brother of Mot, the god of death. The island city of Tyre and the coastal city of Sidon were the two most powerful city-states in Phoenicia; while

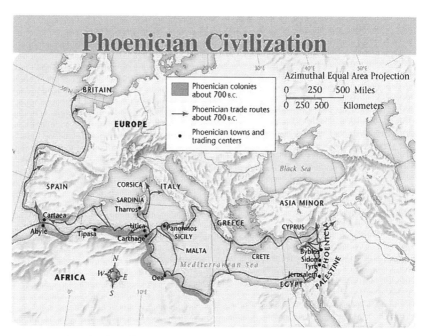

(172) The Phoenician civilization, colonies, and trade routes

Gebal/Bablos and Baalbek were the most significant religious spiritual centres. Phoenician city-states were first established around 3200 BCE and began to gain prominence around the Mediterranean by 2750 BCE. Phoenicia was an affluent maritime trading civilization and major manufacturing centre predominately from 1500 to 332 BCE; and was highly regarded for its superior skills in ship-building, glass-making, the production of dyes, and the skilful artisan hand-crafted manufacturing of a variety of luxury and common goods.

The purple dye manufactured and used in Tyre for the robes of the Mesopotamian royal families gave Phoenicia the name by which we know it today; the word "Phoinikes" means Tyrian purple, in Greek. This also explains why the Phoenicians were widely known as the "purple people" by the Greeks, as the historian Herodotus wrote; "because the purple dye would stain the skin of the workers." The ancient Greek historian also cites Phoenicia as the birthplace of the alphabet, stating that it was brought to Greece by the Phoenician scholar Kadmus, sometime before the 8th century BCE; and that prior to this time, the Greeks had no concept of an alphabet. In fact, the Phoenician alphabet is the foundation for Latin, which of course subsequently evolved into contemporary widely spoken European languages such as English, French, German, Italian and Spanish. The Phoenician city of Gebal (called "Byblos" by the Greeks) gave the Bible its name (from the Greek words "Ta Biblia", meaning "the books"). This is because the city of Gebal was a significant exporter of papyrus ("bublos" to the Greeks), which was the form of paper used in ancient Egypt and Greece.

Many archaeologists believe that most of the gods of ancient Greece were derived from the Phoenician religion, because there are so many indisputable similarities in some stories about the Phoenician gods, Baal and Yamm; and the Greek deities Zeus and Poseidon. It is also worth noting that the eternal battle between the Christian God and Satan as related in the biblical Book of Revelation seems to be a much later version of the same conflict, with many of the same details and events mentioned in the Phoenician Myth of Baal and Yamm.

During ancient times Phoenicia was known as Canaan, and is the land referenced in the Hebrew Scriptures to which Moses led the Israelites from Egypt, and which Joshua then conquered. These stories are referenced in the biblical books of Exodus and Joshua from Egypt, but are uncorroborated by other ancient texts, and also unsupported by the physical evidence so far excavated from the region. According to the British historian and archaeologist Richard Miles: "The people of this land shared an ethnic identity as Can'nai, inhabitants of the land of Canaan yet, despite a common linguistic, cultural, and religious inheritance, the region was very rarely politically united, with each city operating as a sovereign state ruled over by a king." We can see here that the basic political structure of Phoenician city-states was similar to those of the preceding Sumerian Mesopotamian city-states; each having a supreme dynastic king as their ruler.

The city-states of Phoenicia experienced their greatest economic prosperity through maritime trade with surrounding civilizations from 1500 to 332 BCE. This prosperity came to a sudden grinding halt when in 332 BCE the major cities were conquered by Alexander the Great, and after his death the region became a battleground in the fight between his generals for succession and control of Alexander's empire. We can see from artefacts recovered throughout the region that Phoenician luxury goods were highly valued and prized by the cultures with whom they traded. The Phoenicians were primarily known as excellent sailors who had developed a high level of skill in ship-building technology and were able to navigate the often turbulent waters of the Mediterranean Sea. To the layman, the Mediterranean Sea is often thought of as being a calm, tranquil and serene sea; but during a cruise in 2015 I experienced first-hand just how quickly the Mediterranean can change from one of calmness to a very turbulent-rough sea. In my case, it became turbulent quite suddenly, when a storm suddenly emerged, whipping up large waves crashing into the sides of the cruise ship, and lasting for about 8 hours through the night; a night in which I didn't get any sleep, and numbed my senses for most of the night engaged in jovial conversation with a number of fellow passengers around an adequately stocked bar, while a DJ played one sentimental 1960s and 70s hit-song after another, in the background.

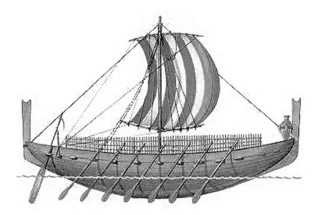

(173) The Phoenicians were advanced ship builders pioneering curved hulls

The Phoenicians appear to have perfected the craft of shipbuilding at Byblos, where the design of the very innovative curved hull was first developed. Richard Miles states: "…over the following centuries, Byblos and other Phoenician states such as Sidon, Tyre, Arvad, and Beirut created an important niche for themselves by transporting luxury goods and bulk raw materials from overseas markets back to the Near East. These new trade routes took in much of the eastern Mediterranean, including Cyprus, Rhodes, the Cyclades, mainland Greece, the Minoans of Crete, the Libyan coast, and Egypt." We also have evidence which indicates that Phoenician sailors travelled to Mesopotamian ports and as far west as Sicily, Italy and Britain. Recovered Phoenician shipwrecks have given archaeologists a valuable insight into the cargo these ships carried. According to Richard Miles: "There were ingots of copper and tin, as well as storage vessels which are thought to have contained unguents (medicinal ointments), wine and oil, glass, gold and silver jewellery, precious objects of faience (glazed earthenware), painted pottery tools, and even scrap metal."

Because its goods were so highly prized, Phoenicia was often spared the numerous violent military incursions suffered by other civilizations of the Near East during these ancient times. For the most part, the great military powers preferred to leave the Phoenicians to their trade, but that did not mean there was no envy on the part of their neighbours. The Old Testament of the Bible refers to the Phoenicians as the "princes of the sea" in a passage from Ezekiel 26:16, in which the prophet seems to predict the destruction of the city of Tyre, and seems to take a certain satisfaction in the humbling of those who had previously been so highly respected and admired. So extraordinary were the skills of the artisan glassmakers of Sidon, that it was thought the Sidonians had invented glass. They provided the model for the Egyptian manufacture of faience, and set the standard for high quality bronze and silver crafts. It appears that the Phoenicians were the first civilization to develop the concept of mass production of similar artefacts, fashioned in the same way, and in large quantities. Again, the British archaeologist Richard Miles notes: "Favourite motifs included Egyptian magic symbols such as the eye of Horus, the scarab beetle and the solar crescent, and these were thought to protect their wearers from evil spirits that prowled the world of the living."

The Phoenician purple dye, already mentioned above, became the standard adornment of royal families from Mesopotamia, Egypt and other civilizations right up to the Roman Empire. All of this was accomplished through intense competition between the city-states of the region, the skill of the sailors who transported the goods, and the high levels of artistic workmanship achieved by the artisans who manufactured these goods. The competition was particularly fierce between the cities of Sidon and Tyre, arguably the most well-known

(174) The Phoenician island city-state of Tyre

of the city-states of Phoenicia who, along with the merchants of
Byblos, carried and transmitted the cultural beliefs and societal norms
of the civilizations they traded with, to each other. This is why the
Phoenicians have been called the "ancient middlemen" of culture
by many scholars and historians; because of their critical role in the
transfers of cultures from one civilization to another.

The city of Sidon (modern Sidonia in Lebanon) was initially
the most prosperous Phoenician city-state, but steadily lost ground
to its sister-city of Tyre. Tyre formed an alliance with the newly
formed Kingdom of Israel, which proved to be very lucrative, and
further expanded its wealth and prosperity by decreasing the power
of the religious priests, and more efficiently distributing the wealth
to the ordinary citizens of the city. In the hope of forming an equally
prosperous trading relationship with Israel, Sidon attempted to cement
an alliance through marriage. Sidon was the birthplace of the princess
Jezebel, who was married to the King of Israel, Ahab, as has been
chronicled in the biblical Hebrew Books of I and II Kings. Jezebel's
refusal to give up her religion, dignity and cultural identity to her
husband's culture, did not sit well with many of his Israelite subjects,
most notably the Hebrew prophet Elijah, who regularly denounced
her. Ahab and Jezebel's rule was ended by an internal military coup,
inspired by Elijah, in which the general Jehu took control of the army
and took over the throne. Following this monumental event, the trade
relationship between Sidon and Israel suddenly ceased, However, Tyre's
trading relationship with Israel continued to flourish.

In 332 BCE, Alexander the Great conquered the Phoenician
city-state of Baalbek and renamed it Heliopolis. His army then
proceeded to march on and subdue the cities of Byblos and Sidon that
same year. Upon Alexander's arrival at Tyre, the citizens followed the
example set by Sidon and submitted peacefully to Alexander. Once in
Tyre however, Alexander wanted to offer a sacrifice in the holy temple
of Melqart, but the Tyrians did not allow this. The religious beliefs of
the Tyrians forbade foreigners of any sort from sacrificing, or even
attending ritual services in the most sacred temple; and so instead,
they offered Alexander a compromise whereby he could offer sacrifice

in the old city on the mainland, but not in the temple on the island complex of Tyre.

(175) & (176) Alexander the Great and the siege of Tyre

Alexander categorically rejected this proposal and sent military envoys to Tyre, demanding their complete surrender. Instead, the angry Tyrians killed the envoys and threw their bodies over the city walls. Hearing of this, the enraged Alexander ordered the immediate siege of Tyre and was so determined to take the city that his army built a causeway from the ruins of the old city on the mainland, full of felled trees and debris; to the island city, and after seven months, breached the walls and massacred most of the city's citizens; men, women and children. It is estimated that over 30,000 citizens of Tyre were massacred or sold into slavery, and only those wealthy enough to adequately bribe Alexander were allowed to leave the island and escape with their lives. After the utter total conquest and subjugation of Tyre by Alexander, the other Phoenician city-states certainly got the message and followed the earlier example of Sidon by surrendering peacefully to Alexander and his army, as it swiftly moved on one town and city-state after another. These events marked the end of the Phoenician Civilization and ushered in the beginning of the Hellenistic Age.

By 64 BCE the fragmented remains of Phoenicia were occupied by the Romans and by 15 CE they became colonies of the Roman Empire, with Heliopolis remaining an important sacred pilgrimage site, boasting the grandest religious building in all of the Roman Empire, the Temple of Jupiter Baal; the ruins of which remain preserved to this day. It is fair to say, the most famous legacy of the Phoenicians is undoubtedly the alphabet which they left to western civilization; but their contribution to the arts, and their role in transferring a diversity of cultures from one civilization to another in the ancient world, should never be overlooked.

Let us now examine another ancient civilization from the Mediterranean Sea; this one centred from the island of Crete. The Minoan civilization flourished during the Middle Bronze Age on the eastern Mediterranean island of Crete, from 2000 to 1500 BCE. With their unique art and architecture, and the spread of their ideas as a result of contact with other cultures across the Aegean, including the Phoenicians, the Minoans made a significant contribution to the development of Western European civilization. Labyrinth-like palace complexes, spectacularly vivid frescoes depicting scenes such as bull-leaping and processions, fine gold jewellery, elegant polished stone vases, and pottery with vibrant decorations of marine life, were all hallmarks of the Minoans.

The Minoans engaged in what was either a sporting event or a religious ceremony called "bull-leaping." Archaeologists recovered the "Bull-Leaping Fresco" from the Minoan palace at Knossos on the island of Crete. The fresco was painted around 1450 BCE and depicts a young man performing what appears to be a handspring or flip over a charging bull. Two young women are flanking the bull. Archaeologists and anthropologists have studied the Bull-Leaping Fresco for many years. Some say that this form of bull-leaping is purely decorative or metaphorical. There are other scholars who believe the fresco represents a cultural or religious ceremonial event, and not a display of athletic prowess. Others disagree and think it was a sporting event, suggesting a series of actions depicted in the fresco, including using the bull's horns for leverage, which would propel an athlete over a bull's back, as is actually shown in the fresco.

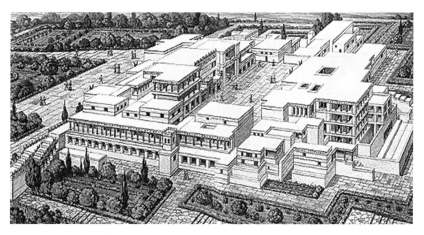

(177) The Minoan Palace at Knossos on the island of Crete

The archaeologist Sir Arthur Evans was first alerted to the possible presence of an ancient civilization on Crete when he examined excavated carved and sealed stones worn as charms by native Cretans. Conducting excavations at Knossos, Crete from 1900 to 1905 CE, Evans discovered extensive ruins which confirmed the ancient accounts, both literary and mythological, of a sophisticated culture in Crete, and the possible site of the legendary labyrinth and palace of King Minos. In ancient Greek mythology, Minos was a King of Crete, and the son of the god Zeus and goddess Europa. According to the legend every nine years he made King Aegeus pick seven young boys and seven young girls to be sent to the skilful architect and craftsman Daidolos' creation, the monumental labyrinth temple, to be eaten by the Minotaur. After his death, Minos became a judge of the dead in the underworld. In his book, "The History", translated by George Rawlinson in 1862, the Greek historian Herodotus writes: "Minos, according to tradition, went to Sicania, or Sicily, as it is now called, in search of Daidolos, and there perished by a violent death. Men of various nations now flocked to Crete, which was stripped of its inhabitants; but none came in such numbers as the Hellenes. Three generations after the death of Minos, the Trojan war took place, and the Cretans were not the least distinguished in this battle. But on this account, when they came back from Troy, famine and pestilence fell upon them, and destroyed both the men and the cattle. Crete was a second time stripped of its inhabitants, a remnant only being left; who form, together with the fresh settlers, the third Cretan people by whom the island has been inhabited."

It was Evans who coined the term "Minoan" in reference to this legendary Bronze Age King Minos. Seeing what he believed to be the rise and fall of a unified culture on Crete, Evans divided the development of the Minoan civilization into three distinct phases, based on different pottery styles. These were:

- The Early Minoan Age – from 3000 to 2100 BCE
- The Middle Minoan Age – from 2100 to 1600 BCE
- The Late Minoan Age – from 1600 to 1100 BCE

Minoan settlements, tombs, and cemeteries have been found all over the island of Crete; but the four major palace sites, in order of size, were: Knossos, Phaistos, Malia and Zakros. At each of these sites, large complex palace structures seemed to have acted as local administrative, trade, religious and political centres. The relationship between these palaces and the power structure within them, and over the island as a whole, is not entirely clear, due to a lack of archaeological and literary evidence. What is clear though, is that the palaces exerted localized control over the surrounding communities of cities, towns and villages; particularly in the gathering and storage of surplus produce such as wine, oil and grain, as well as precious metals and ceramics.

Small towns, villages and farms were spread around the territory under the control of a single palace. Roads connected these settlements to each other and to the main centre. There is a general agreement among archaeologists that the palaces were independent from each other, right up to 1700 BCE; and thereafter they came under the central control of Knossos, as evidenced by a greater uniformity in architecture, and the use of Linear A writing across various palace sites. Linear A, was a system of writing used by the Minoans from 1800 to 1450 BCE as a form of language. Linear A was the primary script used in palace and religious writings. It was discovered by the archaeologist Sir Arthur Evans. No texts written in the Linear A language have been successfully deciphered. The term "linear" refers to the fact that the script was written by using a stylus to cut lines into a clay tablet; as opposed to the Mesopotamian language script of cuneiform, which was written by using a stylus to press wedges into the clay.

The absence of fortifications in the settlements suggests a relatively peaceful co-existence between the different communities. However, the presence of weapons such as swords, daggers and arrowheads; and defensive equipment such as armour and helmets indicates that peace may not have always been enjoyed, and some conflict occurred. Minoan roads too, show evidence of the presence of regular guard-houses and watchtowers, indicating that there may have been roving groups of bandits and brigands who posed a threat to unsuspecting travellers.

The palaces themselves covered two periods. The first palaces were constructed around 2000 BCE and, following destructive earthquakes and or volcanic eruptions, were rebuilt again around 1700 BCE. These second rebuilt palaces survived until their final destruction, between 1500 and 1450 BCE; once again by either earthquake, volcanic eruption or possibly an invasion by outside forces.

(178) "The Last Days of Crete" – a painting by Daniel Blair Stewart

There is evidence to suggest that the Minoan civilization may have ultimately collapsed because of the economic, social and political repercussions of a massive volcanic eruption on the Aegean island of Santorini (then called Thera by the ancient Greeks) which is believed to have occurred in 1646 BCE. Excavations there have uncovered Akrotiri, a Minoan town which was buried during the eruption, one of the largest in recorded history. The volcanic eruption was only 110 kilometres from Minoan Crete. Recent evidence suggests that the Santorini eruption was up to 10 times more powerful than the eruption on the Indonesian island of Krakatoa in 1883. The Santorini eruption caused massive short term climatic disruption, and the explosive blasts were heard over 4800 kilometres away. We also have evidence to indicate that the Santorini eruption and the collapse of the volcanic cone into the sea caused tsunamis which devastated the coasts of many nearby Aegean islands, including Crete; destroying many Minoan coastal towns. Radiocarbon dating indicates that very large tsunamis struck Crete at the time of the Santorini eruption. A catastrophe of this magnitude could easily have sown the seeds for the ultimate collapse of the economic, political and cultural livelihood of the Minoans. Writing long after the disappearance of the Minoans, the Greek historian Herodotus, mentions the significantly diminished civilization of the Cretans, ravaged by what he says was pestilence and disease, a reference that is consistent with the aftermath of the violent volcanic eruption on Santorini.

The Minoan palaces were well-appointed monumental struc-tures with large courts, colonnades, ceilings supported by tapered wooden columns, staircases, religious crypts, light-wells, extensive drainage systems, large storage magazines, and even theatres for public spectacles or religious ceremonies. The palaces were magnif-icent structures reaching up to four stories high and spreading over several thousand square metres. The majestic complexity of these palaces in conjunction with the practice of bull-leaping; the worship of bulls as indicated by the presence throughout the palaces and beyond of sacred bulls' horns; and the depictions of double axes (or labrys) in stone frescoes; may all have combined to give birth to the legend of Theseus and the labyrinth-dwelling Minotaur, so popular in later Greek mythology.

Our knowledge of the religious beliefs and practices of the Minoans remains incomplete, but some details are revealed through art, architecture and artefacts. These include scenes depicting various religious ceremonies and rituals, such as the pouring of libations (wine or other liquids in honour of a deity); making food offerings; processions; feasts; and unique events such as bull-leaping. Supernatural forces and nature in general were manifested in artworks such as a voluptuous female mother-earth goddess figure and an accompanying male figure holding several animals, which seemed to have been revered. Palaces contained open courtyards for mass gatherings and many rooms had wells and channels for the pouring of libations. As we have seen, bulls were sacred animals and were prominently featured in Minoan art; while bull horns were an architectural feature of palace walls, as well as a general decorative feature widely used in jewellery, frescoes and pottery decoration. Dramatic ritual sites such as hilltops and caves often show evidence of cult rituals being performed there.

(179) The Minoan Ritual of Bull-Leaping

The sophistication of the Minoan culture and its trading prosperity is evidenced by the presence of writing; firstly, in the form of Cretan Hieroglyphics from 2000 to 1700 BCE, and then Linear A scripts; mainly found on various types of administrative clay tablets. Seal

impressions on clay were another important form of record-keeping.
Neither of these writing scripts have been deciphered, to date. A
further example of the Minoan culture's high degree of development
and sophistication is the variety and quality of the art forms practised.
Pottery finds reveal a wide range of vessels from wafer-thin cups
to large storage jars. Ceramics were initially hand-turned but then
increasingly made on potter's wheels. In the area of decoration, there
was a progression from flowing geometric designs to vibrant naturalistic depictions of flowers, plants and sea life. Common pottery shapes
included three-handled amphorae, tall beaked-jugs, squat round vessels
with a false spout, beakers, small lidded boxes, and ritual vessels with
figure-of-eight-shaped handles. Stone was also used to produce similar
vessel types and "rhyta", which were special ritual vessels for pouring
libations, often in the shape of animal heads.

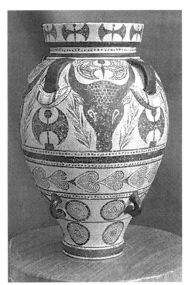

(180) Minoan pottery depicting dolphins and bulls

Large-scale figure sculpture has not survived, but there are many
figurines in bronze. Early types in clay show the dress-styles of the time,
with men wearing red belted white loincloths, and women wearing
long flowing white dresses and open-fronted jackets. A leaping male
acrobat made of ivory, and the faience snake goddess are notable works

which reveal to us the Minoan love of capturing figures in active striking poses. Magnificent frescoes from the walls, ceilings and floors of the palaces also reveal the Minoans' love of the sea and nature, giving us insights into their religious, communal and funeral practices. Subjects range in scale from miniature to larger-than-life size. The Minoans were one of the earliest cultures to paint natural landscapes without the presence of any humans; such was their admiration of the natural world. Animals such as monkeys, birds, dolphins and fish, were often depicted in their natural habitats. Although Minoan frescoes were often framed with geometrically designed decorative borders, the principle palace fresco itself occasionally went beyond conventional boundaries such as corners, and instead covered several walls of a single room, immersing and surrounding the viewer.

As a seafaring culture, the Minoans were also in contact with other civilizations throughout the Aegean, as evidenced by Near Eastern and Egyptian influences in their early art and writing, but also in their later export trade, in particular the exchange of pottery and foodstuffs such as olive oil and wine, in return for precious objects such as copper from Cyprus and Attica (Greece), and ivory from Egypt. Several Aegean islands, especially in the Cyclades, display the characteristics of a palace-centred political and economic structure, as did the royal palaces of Egypt and Mesopotamia. There is some evidence to show that the Minoan trade routes extended as far west as Italy and Sicily, and as far east as Phoenicia and Asia Minor including Syria.

The reasons for the ultimate decline and demise of the Minoan civilization continue to be debated amongst archaeologists. Palaces and settlements show evidence of destruction caused by fire, dated around 1450 BCE; but not at Knossos, which appears to have been destroyed a century later. The rise of the Mycenaean civilization around 1400-1500 BCE on the Greek mainland, and the evidence of their cultural influence on later Minoan art and trade, make the Mycenaean civilization a prime candidate for the Minoan downfall. However, as we have seen earlier in this chapter, another prime candidate which may have triggered the beginning of the demise of Minoan civilization was the massive volcanic eruption on the island of Santorini (Thera).

We know that this eruption triggered a wave of tsunamis and earthquakes throughout the Aegean. I believe the most likely scenario was
a combination of the above two events. I think the Santorini eruption
significantly weakened the Minoan economy and political system
over a 200-year period; so that by the time the Mycenaean civilization was at its peak, it was able to conquer the Minoans relatively
easy, destroying their culture in the process. Whatever the cause of the
collapse, most of the Minoan sites were abandoned by 1200 BCE and
Crete would not prominently feature in Mediterranean history until
the 8th century BCE, when it was colonized by the ancient Greeks.

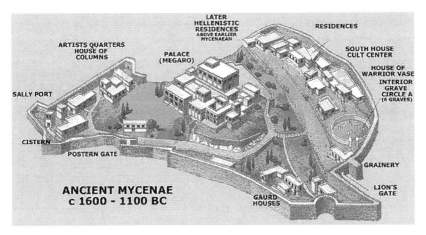

(181) Ancient Mycenae – 1600 to 1100 BCE

(182) A Mycenae religious mask

13

ASSYRIANS
AND PERSIANS

A ssyria was the region located in the ancient Near East which, under the Assyrian Empire, stretched from Mesopotamia (modern-day Iraq and Syria) through Asia Minor (modern Turkey) and down into Egypt. The empire began modestly at the city of Ashur, located in Mesopotamia north-east of Babylon, where merchants trading in Anatolia became increasingly wealthy; and that affluence allowed for the growth and prosperity of the city.

According to one interpretation of passages in the biblical Book of Genesis, Ashur was founded by a man named Ashur, son of Shem, and the grandson of Noah; after the Great Flood. Ashur then went on

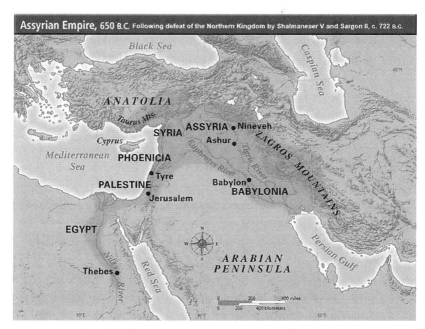

(183) The Assyrian Empire at around 650 BCE

to found other important Assyrian cities. A more likely account is that the city was named Ashur after the deity of that same name, sometime in the 3rd millennium BCE; the same god's name is the origin of "Assyria." The biblical version of the origin of Ashur appears later in the historical record, the Book of Genesis being dated to around 1450 BCE at the earliest; and seems to have been adopted later by the Assyrians after they had accepted Christianity.

The Assyrians were a Semitic people who originally spoke and wrote Akkadian before the easier to use Aramaic language became more popular. Historians have divided the rise and fall of the Assyrian Empire into three periods: The Old Kingdom, The Middle Empire, and The Late Empire (also known as the Neo-Assyrian Empire); although it should be noted that Assyrian history continued on past that point. There are still people of Assyrian ancestry living in the regions of Syria, Iran and Iraq, and elsewhere, in the present day. The Assyrian Empire is considered to be the greatest of the Mesopotamian empires due to its expansive size, and the development of the bureaucracy and military strategies which allowed it to grow and flourish.

Although the city of Ashur existed from the 3rd millennium BCE, the earliest ruins of that city date back to 1900 BCE, which is now considered to be the date the city was founded. According to early stone inscriptions, the first king was Tudiya, and those who followed him were known as "kings who lived in tents" suggesting agrarian farming villages, rather than an urban community. Ashur was certainly an important centre of commerce even at this early time. The king Erishum I built the temple of Ashur in the city in around 1900 BCE. The historian Wolfram von Soden writes: "Because of a dearth of sources, very little is known of Assyria in the third millennium. Assyria did belong to the Empire of Akkad at times, as well as to the Third Dynasty of Ur. Our main sources for this period are the many thousand Assyrian letters and documents from the trade colonies in Cappadocia, foremost of which was Kanesh (modern Kultepe).

The trade colony of Karum Kanesh (the Port of Kanesh) was among the most lucrative centres for trade in the ancient Near East, and definitely the most important for the city of Ashur. Merchants from Ashur travelled to Kanesh, set up businesses there, and then, after placing trusted employees (usually family members) in charge, returned to Ashur and supervised their business dealings from there. The wealth generated from trade in Karum Kanesh provided the people of Ashur with the stability and security necessary for the expansion of the city, and so therefore laid the foundation for the rise of the empire. Trade with Anatolia (modern day Turkey) was equally important in providing the Assyrians with raw materials from which they were able to perfect the craft of ironworking. The iron weapons of the Assyrian army would provide a decisive advantage in the military campaigns which were to ultimately result in the conquest of the entire region of the Near East.

King Adad Nirari I (ruled 1307-1275 BCE) is the first Assyrian king about whom anything is known with certainty because he left inscriptions of his achievements which have survived mostly intact. In addition, letters between the Assyrian king and the nearby Hittite rulers have also survived, and make it clear that initially, the Assyrian rulers were not taken seriously by those of other kingdoms in the region, until they proved themselves too powerful to resist. The historian Will Durant offers these comments on the rise of the Assyrian Empire at this early stage: "If we should admit the imperial principle that it is good, for the sake of spreading law, security, commerce and peace, that many states should be brought, by persuasion or force, under the authority of one government, then we should have to concede to Assyria the distinction of having established in western Asia a larger measure and area of order and prosperity than that region of the earth had ever, to our knowledge, enjoyed before."

Adad Nirari I completely conquered the rival Mitanni people and began what would become standard policy under the Assyrian Empire; the deportation of large segments of the population that were conquered. With the Mitanni people under Assyrian control, Adad Nirari I decided the best way to prevent any future uprising was to remove the former occupants of the land and replace them with Assyrians. This should not be misunderstood as the cruel treatment of the captives. Writing about this, the historian Karen Radner states:

"The deportees, their labour and their abilities were extremely valuable to the Assyrian state, and their relocation was carefully planned and organised. We must not imagine treks of destitute fugitives who were easy prey for famine and disease; the deportees were meant to travel as comfortably and safely as possible in order to reach their destination in good physical shape. Whenever deportations are depicted in Assyrian imperial art, men women and children are shown travelling in groups, often riding on vehicles or animals and never in bonded cruelty. There is no reason to doubt these depictions as Assyrian narrative art does not otherwise shy away from the graphic display of extreme violence."

Adad Nirari's son and successor, Shalmaneser I (1275-1245 BCE) completed the destruction of the Mitanni and absorbed their culture into Assyria. Shalmaneser I continued with his father's policies, including the relocation of populations, but his son, Tukulti-Ninurta I (1245-1208 BCE), went even further in expanding the empire, maintaining an iron grip of power, and reacting with spectacular cruelty to any sign of revolt. Tukulti-Ninurta I was also very interested in acquiring and preserving knowledge and cultures of the peoples he conquered, and developed a more sophisticated method of choosing which sort of individual or community would be relocated, and to which specific location. Scribes and scholars for example, were chosen carefully and sent to urban centres where they could help catalogue written works and help with the political and administrative bureaucracy of the Assyrian Empire. A highly educated and literate man, Tukulti-Ninurta I wrote an epic poem chronicling his victory over the Kassite king of Babylon, and the subjugation of that city and the areas under its influence; and wrote another poem on his victory over the Elamite people.

He also defeated the Hittites at the Battle of Nihriya in 1245 BCE, which effectively ended Hittite power in the region, and began the decline of their civilization. The Hittites were an Anatolian people who had played an important role in establishing an empire centred mainly in the region of Hattusa, in central-northern Anatolia, around 1600 BCE. When Babylon made incursions into Assyrian territory, Tukulti-Ninurta I punished the city severely by sacking it, plundering the sacred temples, and carrying the king and a portion of

(184) The Battle of Nihriya in 1245 BCE

the population back to Assur as slaves. With his plundered wealth, he renovated his grand palace in the city he had built across from Assur, which he named Kar-Tukulti-Ninurta, to which he seems to have retreated once the tide of popular opinion turned against him. His desecration and vandalising of the temples of Babylon was seen as an offense against the gods, because the Assyrians and Babylonians shared many of the same deities. His sons and court officials rebelled against him for putting his hand on the property of the gods. He was assassinated in his palace, probably by one of his sons, Ashur-Nadin-Apli, who then took the throne in 1208 BCE.

Following the death of Tukulti-Ninurta I, the Assyrian Empire went through a period of aimless drifting, in which it neither expanded nor declined. While the rest of the Near East fell into a "dark age" following the so-called Bronze Age Collapse of around 1200 BCE, the Ashur-based Assyrian Empire, although weakened, remained relatively intact. Unlike other civilizations in the region which suffered a complete collapse, the Assyrians seemed to have experienced something closer to a slowdown in momentum. This all changed with the rise of King Tiglath Pileser I to the throne, from 1115 to 1076 BCE. According to the historian Gwendolyn Leick: "He was one of the most important

Assyrian kings of this period, largely because of his wide-ranging military campaigns, his enthusiasm for building projects, and his interest in cuneiform tablet collections. He campaigned widely in Anatolia, where he subjugated numerous peoples, and ventured as far as the Mediterranean Sea. In the capital city Assur, he built a new palace and established an impressive library which held numerus tablets on all kinds of scholarly subjects. He also issued a legal decree, the so-called Middle Assyrian Laws, and wrote the first royal annals. He was one of the first Assyrian kings to commission parks and gardens stocked with foreign and native trees and plants."

The Late Empire (also known as the Neo-Assyrian Empire) is the one most familiar with students of ancient history, as it is the period of the largest expansion of the empire. It is also the era which most decisively gives the Assyrian Empire the reputation it has for ruthlessness and cruelty. The historian Paul Kriwaczek writes: "Assyria must surely have among the worst press notices of any state in history. Babylon may be a byname for corruption, decadence and sin but the Assyrians and their famous rulers, with terrifying names like Shalmaneser, Tiglath-Pileser, Sennacherib, Esarhaddon and Ashurbanipal, rate in the popular imagination just below Adolf Hitler and Genghis Khan for cruelty, violence, and sheer murderous savagery." While the reputation for decisive, ruthless, military tactics is understandable, the comparison with the Nazi regime is harsh. Unlike the Nazis, the Assyrians treated the conquered people they relocated well, and considered them Assyrians once they had submitted to central power and authority. There was no concept of a 'master race' in Assyrian policies; everyone was considered an asset to the empire, whether they were born Assyrian or were assimilated into the culture from outside." Kriwaczek further notes: "In truth, Assyrian warfare was no more savage than that of other contemporary states. Nor, indeed, were the Assyrians notably crueller than the Romans, who made a point of lining their roads with thousands of victims of crucifixion dying in agony. The only fair comparison between Germany in World War II and the Assyrians, if one can be made, is the efficiency of the military and size of the army; yet this same comparison could also be made with ancient Rome."

(185) The ancient city of Babylon and its luxurious lush gardens and temples

The rise of King Adad Nirari II (912-891 BCE) brought the kind of invigorating revival Assyria needed at that time. Adad Nirari II reconquered the lands which had been lost, including Eber Nari, and secured the borders. The defeated Aramaeans were executed or deported to regions within the heartland of Assyria. He also conquered Babylon but, learning from the mistakes of the past, refused to plunder the city, and instead entered into a peace agreement with the king, in which they married each other's daughters and pledged mutual support. Their treaty would secure Babylon as a powerful ally, instead of a perennial problem, for the next 80 years. Babylon was the capital city of the ancient Babylonian Empire, which itself is a term referring to either of two subsequent empires within Mesopotamia. These two empires achieved regional dominance between both the 19th and 15th centuries BCE, and again between the 7th and 6th centuries BCE. Babylon was renowned for its spectacular gardens, amongst other things. The Gardens of Babylon were a set of gardens in the city that were so spectacular, they were one of the Seven Wonders of the Ancient World.

The kings who followed Adad Nirari II continued the same policies and military expansion. Tukulti-Ninurta II who ruled from

891 to 884 BCE, expanded the empire to the north and gained further territory toward the south in Anatolia; while Ashurnasirpal II (884-859 BCE) consolidated rule in the Levant and extended Assyrian control throughout Canaan. Their most common method of conquest was through siege warfare which would begin with a brutal assault on the city. The historian Simon Anglim writes: "More than anything else, the Assyrian army excelled at siege warfare, and was probably the first force to carry a separate corps of engineers. Assault was their principal tactic against the heavily fortified cities of the Near East. They developed a great variety of methods for breaching enemy walls; sappers were employed to undermine walls or to light fires under-neath wooden gates, and ramps were thrown up to allow men to go over the ramparts or to attempt a breach on the upper section of a wall where it was the least thick. Mobile ladders allowed attackers to cross moats and quickly assault any point in defences. These operations were covered by masses of archers, who were the core of the infantry. But the pride of the Assyrian siege train were their engines. These were multi-storied wooden towers with four wheels and a turret on top and one, or at times two, battering rams at the base."

When it comes to Assyrian society, culture and religious beliefs, we can ascertain a number of factors from the archaeological artefacts and written sources. Schools were established throughout the Assyrian Empire, but were only for the sons of the nobility and the wealthy. Women were not allowed to attend school or hold positions of authority even though, earlier in Mesopotamia, as we have seen in a previous chapter, women had enjoyed almost equal rights to men. The decline in women's rights corresponds with the rise of Assyrian monotheistic religious beliefs; the belief in one god only, and hence sidelining the female gods, amongst others. As the Assyrian armies campaigned and conquered new lands, their supreme and now only god Ashur, went with them; but, as Ashur was previously linked with the temple of that city and had only been worshipped there, a new way of imagining the god became necessary in order to continue that worship in other locations. Paul Kriwaczek writes: "One might pray to Ashur not only in his own temple in his own city, but anywhere.

As the Assyrian Empire expanded its borders, Ashur was encountered in even the most distant places. From faith in an omnipresent
god to belief in a single god is not a long step. Since He was everywhere, people came to understand that, in some sense, local divinities
were just different manifestations of the same Ashur." This unity of
vision of a supreme deity helped to further unify and consolidate the
diverse regions of the empire. The different gods of the conquered
peoples, and their various religious practices, became absorbed into
the worship of Ashur, who was recognized as the one true god who
had been called different names by different people in the past, but
who now was clearly known and could be properly worshipped as the
universal deity of the Assyrian Empire.

(186) The Assyrian Supreme God Ashur

The Assyrian culture became increasingly cohesive with the
ongoing expansion of the empire, the new understanding of a single
deity, and the assimilation of the people from the conquered regions.
King Shalmaneser III (859-824 BCE) expanded the empire up
through the coast of the Mediterranean, and received tribute from
the wealthy Phoenician cities of Tyre and Sidon. He also defeated the
Armenian kingdom of Urartu which had long been a significant thorn

in the side of the Assyrians. Following his reign however, the empire erupted into civil war, as the King Shamshi Adad V (824-811 BCE) fought with his brother for control. Although the rebellion was put down, expansion of the empire halted during the reign of Shalmaneser III. The empire was revitalized by Tiglath Pileser III (745-727 BCE) who reorganized the military and restructured the bureaucracy of the government. According to the historian Anglim, Tiglath Pileser III "carried out extensive reforms of the army, reasserted central control over the empire, reconquered the Mediterranean seaboard, and even subjugated Babylon. He replaced conscription (in the military) with a manpower levy imposed on each province, and also demanded contingents (soldiers) from vassal states." Under his reign, the Assyrian army became the most effective military force in history, to that date, and would provide a model for future armies, in organization, tactics, training and efficiency.

King Sennacherib (705-681 BCE) campaigned widely and ruthlessly, conquering Israel, Judah, and the Greek provinces in Anatolia. His capture and sack of Jerusalem is detailed on the "Taylor Prism", a cuneiform stone block (discovered in 1830 CE) describing Sennacherib's military exploits, and the king's claim to have trapped the people of Jerusalem inside the city until he overwhelmed them; as well as his claim to have captured 46 additional cities during his conquests. Sennacherib's military victories further increased the majesty, prestige and wealth of the empire. He moved the capital city to Nineveh and built what was known as "the Palace without a Rival." He beautified and improved upon the city's original structure, planting orchards and gardens. The historian Christopher Scarre writes: "Sennacherib's palace had all the usual accoutrements of a major Assyrian residence; colossal guardian figures and impressively carved stone reliefs (over 2000 sculptured slabs in 71 rooms). Its gardens, too, were exceptional." Recent research by British Assyriologist Stephanie Dalley has suggested that these were the famous Hanging Gardens, one of the Seven Wonders of the Ancient World. Later writers placed the Hanging Gardens as being located in the city of Babylon; but extensive research has failed to find any trace of them. Sennacherib's proud account of the palace gardens he created at Nineveh fits that of the Hanging Gardens in several significant details.

The empire flourished under the reign of Sennacherib's son, Esarhaddon (681-669 BCE). He successfully conquered Egypt (which his father had tried and failed to do) and established the empire's borders as far north as the Zagros Mountains (modern-day Iran), and as far south as Nubia (modern Sudan), with a span from west to east of the Levant (modern-day Lebanon to Israel), and through Anatolia (Turkey). His successful military campaigns, and careful management of the government during his reign, provided the necessary stability for advances in many fields including medicine, literature, mathematics, astronomy, architecture, and the arts. The historian Will Durant writes: "In the field of art, Assyria equalled her preceptor Babylonia and in bas-relief surpassed her. Stimulated by the influx of wealth into Ashur, Kalakh, and Nineveh, artists and artisans began to produce - for nobles and their ladies, for kings and palaces, for priests and temples – jewels of every description, cast metal as skilfully designed and finely wrought as on the great gates at Balawat, and luxurious furniture of richly carved and costly woods strengthened with metal and inlaid with gold, silver, bronze, or precious stones."

In 612 BCE the Assyrian capital city of Nineveh was sacked and burned by a military coalition of Babylonians, Persians, Medes, and Scythians, among others. The destruction of the palace brought the flaming walls down on the great Library of Ashurbanipal and, although it was far from the intention, preserved the great library and its literary treasures of the history of the Assyrians, by baking hard and burying the clay tablet books. Historian Kriwaczek writes, "Thus did Assyria's enemies ultimately fail to achieve their aim when they razed Ashur and Nineveh in 612 BCE, only fifteen years after Ashurbanipal's death; the wiping out of Assyria's place in history." Nevertheless, the destruction of the great Assyrian cities was so complete that, within two generations of the empire's fall, no one knew where the cities had been. The ruins of Nineveh were covered by the sands and lay buried for the next 2000 years.

Thanks to the Greek historian Herodotus, who considered the whole of Mesopotamia "Assyria", scholars have long known the culture existed. Mesopotamian scholarship was traditionally known as Assyriology until relatively recently (though the term is still in use), because the Assyrians were so well known through the primary sources of Greek, Roman and Persian writers. As a result of their expanding empire, the Assyrians spread Mesopotamian culture to the other regions of the world, which have in turn, impacted cultures worldwide up to the present day. Will Durant writes: "Through Assyria's conquest of Babylon, her appropriation of the ancient city's culture, and her dissemination of that culture throughout her wide empire; through the long Captivity of the Jews, and the great influence upon them of Babylonian life and thought; through the Persian and Greek conquests which then opened with unprecedented fullness and freedom all the roads of communication and trade between Babylon and the rising cities of Ionia, Asia Minor, and Greece – through these and many other ways the civilization of the Land between the Rivers passed down into cultural endowment of our race. In the end nothing is lost; for good or evil, every event has effects forever."

Following the decline and rupture of the Assyrian Empire, Babylon assumed supremacy in the region from 605 to 549 BCE. Babylon then fell to the Persians under Cyrus the Great who founded the Achaemenid Empire (549-330 BCE), which fell to Alexander the Great and, after his death, was part of the Seleucid Empire. The region of Mesopotamia corresponding to modern-day Iraq, Syria, and part of Turkey, was the area at this time known as Assyria and, when the Seleucids were driven out by the Parthians, the western section of the region, formerly known as Eber Nari and then Aramea, retained the name Syria. The Parthians gained control of the region and held it until the coming of the Romans in 116 CE, and then the Sassanid Empire held supremacy in the area from 226 to 650 CE until, with the rise of Islam and the Arabian conquests of the 7th century CE, Assyria ceased to exist as a national entity.

Amongst the greatest achievements of the Assyrians was the Aramaic alphabet, imported into the Assyrian government by King Tiglath Pileser III from the conquered region of Syria. Aramaean was easier to write than Akkadian, and so older documents collected by kings such as Ashurbanipal, were translated from Akkadian to Aramaic, while newer ones were directly written in Aramaic. As a result, thousands of years of history and culture was preserved for future generations; a great legacy to the world and humanity.

The Persian Empire can trace its earliest roots to a collection of semi-nomadic tribes who raised sheep, goats and cattle on the Iranian plateau. Persia (mostly modern-day Iran) is among the oldest inhabited regions in the world. Archaeological sites in the country have established that this region has been occupied by human settlements dating back over 100,000 years, to the Paleolithic Age, with semi-permanent settlements established by 10,000 BCE. The ancient Kingdom of Elam in this region was among the most advanced of its time, before parts of it were conquered by the Sumerians, then by the Assyrians, and finally by the Medes. The oldest settlement recovered is the archaeological site of Chogha Bonut, which dates back to 7200 BCE.

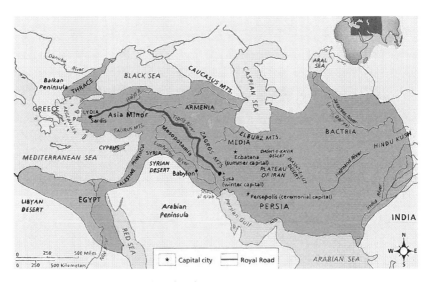

(187) The Persian Empire

The Median Empire (678-550 BCE) was followed by one of the greatest political and social kingdoms of the ancient world, the Persian Achaemenid Empire (550-330 BCE), which was conquered by Alexander the Great, and later replaced by the Seleucid Empire (312-63 BCE), Parthia (247 BCE – 224 CE), and the Sassanian Empire

(224-651 CE) in succession. The Sassanian Empire was the last of the Persian kingdoms to hold power in the region before the Muslim Arab conquest of the 7th century CE.

At some point between 1500-1000 BCE, the Persian spiritual visionary Zoroaster (also known as Zarathustra) claimed to have received a divine revelation from Ahura Mazda (the creator god), recognizing the purpose of human life as choosing sides in an eternal struggle between the supreme deity of justice and order, and his adversary Angra Mainyu, the god of discord and strife. Human beings were defined by whose side they chose to act out their lives. Zoroaster's teachings formed the foundation of the religion of Zoroastrianism, which would later be adopted by the Persian empires and inform their culture and way of life.

The Persians settled primarily across the Iranian plateau and were established by the 1st millennium BCE. The Medes united under a single king named Dayukku (727-675 BCE), and founded their kingdom in Ecbatana. Dayukku's grandson, Cyaxares (625-585 BCE) would extend Median territory into what is modern-day Azerbaijan. In the late 8th century BCE, under their king Achaemenes, the Persians consolidated their control of the central-western region of the Bakhityari Mountains, with their capital city at Anshan. The indigenous people of this area were known as the Elamites. Under King Thiepes (675-640 BCE), the son of Achaemenes, the Persians settled to the east of Elam in the territory known as Persis, which would give this group the name they were known by; Persians. They later extended their control of the region into Elamite territory, intermarried with the Elamites, and absorbed their culture. Sometime prior to 640 BCE, Thiepes divided his kingdom between his sons, Cyrus I (625-600 BCE), and Ararnamnes. Cyrus ruled the northern kingdom from Anshan, and Ararnamnes ruled in the south. Under the subsequent rule of King Cambyses I (580-559 BCE) these two kingdoms were united and continued to be controlled from Anshan.

The Medes were the dominant power in the region at this time, and the Kingdom of Persia was only a small vassal state. The situation completely reversed itself after the fall of the nearby Assyrian Empire

(188) Cyrus the Great ruled Persia from 550 to 530 BCE

in 612 BCE, hastened by the campaigns of the Medes and Babylonians who led a coalition of others against the weakened Assyrian state, as we have seen. The Medes at first maintained control until they were overthrown by the son of Cambyses I of Persia and grandson of Astyages of Media; Cyrus II, also known as Cyrus the Great, who ruled from 550 to 530 BCE, and was the founder of the Achaemenid Empire.

Cyrus II overthrew Astyages of Media in 550 BCE and then began a systematic campaign to bring other principalities under his control. He conquered the wealthy Kingdom of Lydia in 546 BCE, Elam in 540 BCE, and Babylon in 539 BCE. By the end of his reign, Cyrus II had established an empire which stretched from the modern-day region of Syria, across Iran and down through Turkey all the way to the borders of India. This was the Achaemenid Empire, named after Cyrus II's ancestor Achaemenes. Cyrus II is unique among ancient conquerors for his humanitarian vision and policies, as well as encouraging technological innovations. Much of the land he conquered suffered from a lack of adequate water supply, and so he had his engineers revive an older means of tapping underground aquafers known as qanat, a sloping channel dug into the earth with vertical shafts at intervals down to the channel which would bring the water up to ground level. Although Cyrus II is often credited with inventing the qanat system, it is believed to have been invented during the earlier reign of Sargon II of Assyria (722-705 BCE).

Cyrus II's humanitarian efforts are well known through the Cyrus Cylinder, a record of his policies and proclamation of his vision, that everyone under his reign should be free to live as they wished to, as long as they did so peacefully with others. After he conquered Babylon, he allowed the Jews, who had been taken from their homeland by King Nebuchadnezzar (605-562 BCE) in the so-called Babylon Captivity, to return to Judah, and even provided them with funds to rebuild their temple. The Lydians continued to worship their goddess Cybele, and other ethnic groups their own deities as well. All Cyrus II wanted, was that the citizens of his empire live peacefully with each other, serve in his armies when required, and pay their taxes.

In order to maintain a stable empire, Cyrus implemented a governmental hierarchy with himself at the top, surrounded by key-advisors who relayed his decisions and decrees to secretaries, who then passed these on to regional governors (satraps) in each province (satrapy). These governors only had authority over bureaucratic and administrative matters, while a military commander in the same region oversaw all military, security and police matters. By dividing the responsibilities of government in each satrapy, Cyrus II significantly reduced the chance of any official or rival amassing enough money, power and influence to attempt to overthrow him in a coup. The decrees of Cyrus II, and any other news, travelled along a network of roads linking the major cities. The most famous of these roads would become the Royal Road which ran from Susa to Sardis. Messengers would leave one city on horseback, and find a watchtower and resting station within two days, where they would be given food, drink, a bed, and then be provided with a new horse to continue their journey. The Persian postal system was considered by Herodotus to be a marvel of his day, and became the benchmark for future postal systems.

Cyrus died in 530 BCE, possibly during a battle, and was succeeded by his son, Cambyses II (530-522 BCE) who went on to extend Persian rule into Egypt. Between 522 and 486 BCE a very powerful and effective king ruled the Persian Empire, his name was

Darius I, who was also known as Darius the Great. He would extend the empire even further, and initiate some of its most famous building projects, including the great city of Persepolis, which became one of the empire's capitals. Even though Darius I continued Cyrus II's policy of tolerance and humanitarianism, unrest broke out during his reign. This was not an uncommon occurrence, as it was standard practice in this region for some provinces to rebel after the death of a monarch; going right back the Akkadian Empire of Sargon the Great in Mesopotamia (2334-2279 BCE). The Ionian Greek colonies of Asia Minor were among those provinces that rebelled, and since their efforts were backed by the city-state of Athens, Darius launched an invasion of Greece which halted at the Battle of Marathon in 490 BCE, when the Athenians repulsed the first Persian invasion of Greece in a single afternoon of fierce fighting.

After the death of Darius, he was succeeded by his son Xerxes I (486-465 BCE), who is said to have raised the largest army in history up to that point, for his unsuccessful invasion of Greece in 480 BCE. Afterwards, Xerxes I occupied himself with grand building projects, notably adding to the city of Persepolis; and his successors continued to do the same. The Achaemenid Empire remained stable under later rulers, until it was conquered by Alexander the Great during the reign of Darius III (336-330 BCE). Darius III was assassinated by his confidante and bodyguard Bessus, who then proclaimed himself ruler, as Artaxerxes V (330-329 BCE), but he didn't last long until he was executed by the forces of Alexander, who styled himself as the successor of Darius.

Following the death of Alexander in 323 BCE, his vast empire was divided up among his generals, as we have seen. One of these, Seleucus I Nicator (323-281 BCE), took control of Central Asia and Mesopotamia, expanded these territories, and founded what was to become the Seleucid Empire. Seleucus I retained the Persian model of government and religious toleration, but filled the top administrative positions with Greeks. Even though Greeks and Persians intermarried, the Seleucid Empire favoured Greeks, and Greek became the language of the court. Seleucus I began his reign by putting down rebellions

(189) King Xerxes I of Persia   *(190) Seleucus I – founder of*
the Seleucid Empire

in some regions, and conquering others, but always maintaining
the Persian government policies which had worked so well in the
past. Even though this same practice was followed by his immediate
successors, regions continued to rise in revolt, and some like Parthia
and Bactria, broke away into separate kingdoms. In 247 BCE, Arsaces I
of Parthia (247-217 BCE) established an independent kingdom which
would become the Parthian Empire.

Parthia continued to grow, as the Seleucid Empire shrank. The
Seleucid King Antiochus IV Epiphanes (175-164 BCE) focused
entirely on his own self-interests and his successors would also follow
this pattern. Eventually though, the Seleucids were finally reduced to a
small buffer kingdom in Syria after their defeat by the Roman general
Pompey the Great (106-48 BCE) while, at around the same time, in
63 BCE, the Parthian Empire was at its height, following the reign
of Mithridates II (124-88 BCE) who had expanded the empire even
further. The Parthian army was the most effective fighting force of that
time, primarily due to its cavalry, and the perfection of a technique
known as the Parthian shot, which consisted of mounted archers
feigning retreat, who would then turn and shoot back at advancing
adversaries. This tactic of Parthian warfare came as a complete surprise,

and was very effective, even after the opposing forces became aware of it. Under the leadership of Orodes II (57-37 BCE), the Parthians easily defeated the Roman general Crassus at the Battle of Carrhae in 53 BCE, and killed him by pouring molten gold down his throat (a literal symbolic reference for his unquenchable thirst for accumulating wealth); and later also defeated Roman general Mark Antony in 36 BCE; delivering two very severe blows to the awesome might and morale of the Roman Army.

In spite of these military setbacks, the power of the Roman Empire was on the rise, starting from its first Emperor, Augustus (27 BCE to 14 CE). By 165 CE the Parthian Empire had been severely weakened by a series of Roman military campaigns under the leadership of various emperors. The last Parthian king, Artabanus IV (213-224 CE) was overthrown by his vassal Ardashir I (224-240 CE), a descendant of Darius III, and a member of the royal Persian dynasty. Ardashir I was mainly preoccupied with building a stable kingdom founded on the religious beliefs of Zoroastrianism, and also keeping the kingdom safe from further Roman attack and influence. With this in mind, he made his son Shapur I (240-270 CE) his co-regent in 239 CE. When Ardashir I died a year later, Shapur I became the "king of kings" and initiated a number of military campaigns to enlarge his territory and protect his borders. Shapur I was a devout Zoroastrian, like his father was; but he adhered to the policy of religious tolerance, just like the earlier Achaemenid Empire had. Jews, Christians, and members of other religious faiths were free to practice their beliefs, build houses of worship, and participate in the government. The religious visionary Mani, founder of Manichaeism, for example, was an honoured guest at Shapur I's court. His Zoroastrian vision cast him and the Sassanians as the forces of light, serving the great supreme god Ahura Mazda, against the forces of darkness and disorder epitomized by the Roman Empire. Shapur's campaigns against Rome were almost universally successful, even to the point of capturing the Roman Emperor Valerian (253-284 CE), and using him as a personal servant and footstool. Shapur saw himself as a warrior king and lived up to that vision, taking full advantage of

Rome's weakness during the Crisis of the Third Century (235-284 CE) to enlarge his empire.

Shapur I laid the foundation for the Sassanian Empire which his successors would build on. The greatest of these was King Kosrau I (531-579 CE); who reformed the tax laws so they were more equitable; divided the empire into four sections, each under the defence of its own general for quick response to external or internal threats; secured his borders tightly; and elevated the importance of education throughout the empire. The Academy of Gondishapur, founded by Kosrau I, was the leading university and medical centre of its day, with scholars from India, China, Greece and elsewhere within the ancient world at that time, in attendance. Kosrau I continued the long-held policies of religious tolerance and inclusion as well as the ancient Persian antipathy towards slavery. Prisoners of war taken by the Roman Empire became slaves; but those taken by the Sassanian Empire became paid servants. It was illegal to beat or in any way injure a servant, no matter one's social class; and so the life of a servant under the Sassanian Empire was far more superior than that of a slave in the Roman Empire or anywhere else.

The Sassanian Empire is considered to be the height of Persian rule and culture during the times of antiquity, as it was built upon the best aspects of the Achaemenid Empire, and then further improved upon them. The Sassanian Empire, like most in the ancient world, ultimately declined because of weak rulers who made poor choices and decisions; but the decline was further accelerated by the corruption of the religious clergy; and the onslaught of a Plague which struck the region in 627-628 CE. It was significantly weakened by the time it was conquered by the Muslim Arabs in the 7th century CE. Even so, Persian technological, architectural, scientific and religious innovations would come to inform and enlighten the culture of the conquerors, and their newly-found religion of Islam. The high civilization of ancient Persia continues in modern Iran today with direct, unbroken ties to its past through the Iranian culture. Although modern-day Iran corresponds to the heartland of ancient Persia, The Islamic Republic of Iran is a multicultural nation. To say that one is Iranian, is to state one's nationality, while to say one is Persian, is to define one's ethnicity and culture; these are not the same things. Even so, Iran's multicultural heritage comes directly from the paradigm of the great Persian empires of the past, which had many different ethnicities, all living under the Persian banner; and that past is reflected in the diverse character of modern Iranian society.

14

THE GREEKS

THE ANCIENT CLASSICAL AND HELLENISTIC ERAS OF GREECE ARE without a doubt, a period in history when human collective learning flourished, providing the foundation of what we call "western civilization." However, the two millennia that preceded these eras are also an integral part of the history of Greece, and have left just as rich a cultural and intellectual footprint on this timeless land. Much of the ancient Greek civilization has survived to this day through primary sources of mainly archaeological evidence and works of literature. The ancient Greek dialects are also influential to this modern day, with much Greek vocabulary embedded in the Modern Greek and

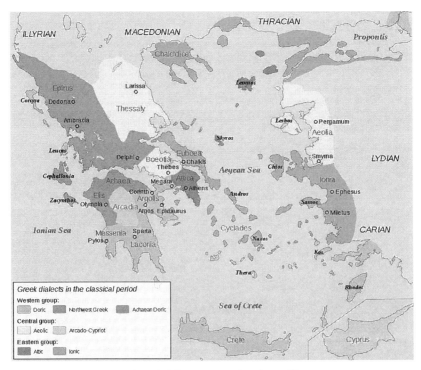

(191) The Ancient Greek world

English languages, amongst others. Likewise, the art and architecture of Ancient Greece has remained relevant and influential up to our time, throughout many cultures worldwide. The much celebrated European Renaissance was guided in large part by the re-discovery of ancient Greek ideas and philosophy, through text and art, which had been previously suppressed, particularly during the Middle Ages, by the absolute power of the Catholic Church and its believers.

History is a discipline that was first conceived in Ancient Greece. Herodotus (484-425 BCE) is considered to be the "Father of History" as he was the first to attempt to record events and human actions for the sole purpose of preserving them for future generations. The very first lines of his "Histories" read: "Herodotus of Halicarnassus here displays his inquiry, so that human achievements may not become forgotten in time, and great and marvellous deeds – some displayed by Greeks, some by barbarians – may not be without their glory." Being the first to attempt such a feat for humanity, Herodotus was not spared from harsh criticism for including in his "Histories" (written between 431 and 435 BCE), myths, legends and what some called, outrageous tales. Not long after Herodotus, Thucydides (460-395 BCE) put his stamp on the discipline of History, with his monumental work, "History of the Peloponnesian War", by attempting to present historical events in an objective way, and to make correlations between human actions and events. Their approach and methods of recording historical events became the shining and guiding light for historians over the next two thousand years.

(192) The Greek historian Herodotus produced the first narrative of ancient history

Greek history is generally divided into the following eras:
- Cycladic – 3300 to 2000 BCE
- Minoan – 2600 to 1200 BCE
- Helladic – 2800 to 1600 BCE
- Mycenaean – 1600 to 1100 BCE
- Dark Ages – 1100 to 700 BCE
- Archaic – 700 to 480 BCE
- Classical – 480 to 323 BCE
- Hellenistic – 323 to 30 BCE

Each of the above eras had its own unique and distinctive cultural characteristics, and the transition between them was often tumultuous and dramatic.

The Bronze Age, a period that lasted around three thousand years, saw major advances in social, economic and technological advances that made Greece the central hub of activity in the Mediterranean. Historians have identified three distinct civilizations which existed independent of each other during this time, even though they interacted culturally, economically and sometimes, militarily; these civilizations overlap in time and coincide with the major geographic regions of Greece. The Cycladic civilization developed on the islands of the Aegean, but mainly centred on the Cyclades islands, between Greece and modern day Turkey. They lived a subsistence life of basic farming and fishing in small villages. The Minoans, as we have seen in a previous chapter, occupied the large island of Crete. At the same time, the civilization which existed on the Greek mainland was known as the Helladic-Mycenaean. Towards the end of the 11th century BCE, the Helladic civilization experienced an "Age of Heroes" because it was the source of the great Greek mythological heroes and epics like Hercules; and Homer's, the Iliad and the Odyssey. All three of these civilizations of the Bronze Age had many characteristics in common, while at the same time were also distinct in their culture and ways of life. The Minoans are considered to be the first advanced civilization of Europe; whilst the Helladic-Mycenaean culture had a great deal of influence with its legends and Greek language; and what later became the splendour of Classical Greece.

The Mycenaeans are often referred to as "the first Greeks." Either by fortune or force, the Mycenaeans outlasted both the people of Cyclades and the Minoans; and by the end of the 10th century BCE, they had expanded their influence over the Greek mainland; the islands of the Aegean and Ionian seas; Crete; and the coast of Asia Minor. However, after 1100 BCE, the Mycenaean civilization mysteriously ended; either as a result of internal conflict; or outside invasions (the Dorians are often cited as the probable invaders); or through a combination of both events; we just do not know for certain. What is known, is that the extensive damage done to the Mycenaean civilization took over three hundred years to reverse. This period is called "the Dark Ages" partly because the people of Greece fell into a period of basic subsistent existence, barely surviving off the land, with no significant evidence of cultural and intellectual development; and partly because of the incomplete historical record which does not shine much light for us on what actually happened during this mysterious period.

One thing we do know about the Dark Ages of Greece, is that during this period, the old major settlements were abandoned (with the notable exception of Athens), and the Greek population dropped dramatically. Within these 300-400 years, the people of Greece lived mainly in small rural communities, moving constantly in accordance with their new agrarian-herding-farming lifestyle and livestock needs. They left no written record behind, which leads us to believe they were illiterate. Later in the Dark Ages, between 950 and 750 BCE, the Greeks re-learned how to write once again, but this time instead of using the Linear B language script used by the Mycenaeans, they adopted the alphabet used by the Phoenicians, innovating in a fundamental way by introducing vowels as letters. The Greek-Phoenician version of the alphabet eventually went on to form the basis of the alphabet used for the English language today.

The main difference between the Linear A script developed by the Minoans and the Linear B script later developed and refined by the Mycenaeans is that Linear A was a pictorial form of writing. It consisted of 100 symbols, each one representing a different phonetic syllable. Various combinations of different symbols would create words from sounds. Written from left to write, Linear A script seems to have developed from earlier Egyptian hieroglyphic forms of writing. Linear B script on the other hand, was phonetic writing consisting of 90 symbols. It is generally seen as more simplified and a less pictorial version of Linear A script. Linear B formed the foundation for the first Mycenaean Greek written language. Although Linear A script has never been deciphered, as we have seen; in 1952 the scholar Michael Ventris finally deciphered Linear B. Scholars then began to study the many clay tablets discovered in Greece that were written in Linear B. Mycenaean Linear B writing survives mainly in the form of records and accounts. These include rectangular clay tablets recording trade transactions and calculations, and clay 'palm leaf' tablets which were elongated tablet lists. The script was also found on vases, probably representing an ownership mark, or even the signature of the object's creator.

Linear B tablets have been found to record many different things, including:

- Personal records
- Records of domestic animals owned such as sheep
- Records of crops harvested such as corn
- The distribution of food and olive oil
- The distribution of religious offerings
- The registering and distribution of manufactured items such as metalwork, fabric, vases, weapons and chariots

Life was undoubtedly harsh for the Greeks of the Dark Ages. However, in retrospect, we can identify one major benefit of the period. The deconstruction of the old Mycenaean economic and social structures, with the strict class hierarchy and hereditary rules, were forgotten; and eventually replaced with new innovative, progressive socio-political institutions that eventually were to lead to the rise of Democracy in Athens, in the 5th century BCE. There were some notable events during the Dark Ages; including the great poems of Homer (the Iliad and the Odyssey) and the very first Olympic Games held in 776 BCE.

(193) The gods descending into battle as depicted in Homer's "Iliad"

The next period of Greek history is described as the Archaic, and lasted for around two hundred years, from 700 to 480 BCE. During this era, the Greek population recovered, and were organized politically into city-states (the Polis). These city-states were comprised of citizens, foreign residents, and slaves. This kind of complex social organization required the development of an advanced legal structure that ensured the smooth coexistence of the different hierarchical classes; and the equality of the citizens irrespective of their economic status. This was a required precursor for the democratic principles that were to develop in the city-state of Athens, two hundred years later.

Greek city-states of the Archaic Era spread throughout the Mediterranean region as a result of aggressive colonization. As the major city-states grew in size, they spawned a large number of mainly coastal towns in the Aegean, the Ionian, Anatolia (Turkey), Phoenicia (the Middle East), Libya, southern Italy, Sicily, Sardinia, and as far as southern France, Spain and the Black Sea. These city-states, towns, and trading posts numbered in the many hundreds, and became part of an extensive commercial trading network that involved all of the advanced civilizations in the region at that time. As a consequence, Greece came into contact with, and assisted in, the exchange of goods and ideas throughout ancient Africa, Asia, and Europe. Through domination of commerce in the Mediterranean, aggressive expansion abroad, and competition at home, several very strong city-states began to emerge as dominant cultural centres; the most notable of these city-states were Athens, Sparta, Corinth, Thebes, Syracuse, Miletus and Halicarnassus.

The rapid development and expansion of the Archaic Era was followed by a period of maturity which came to be known as Classical Greece. Between 480 and 323 BCE, Athens and Sparta dominated the Hellenic world with their cultural and military achievements. These two cities, with the involvement of the other Hellenic states, rose to power through alliances, reforms, and a series of victories against the invading Persian armies. They eventually resolved their rivalry in a long, and particularly nasty war that concluded with the demise of Athens first, Sparta second, and the emergence of Macedonia as the dominant power of Greece. Other city-states like Miletus, Thebes, Corinth and Syracuse among many others, played a major role in the cultural achievements of Classical Greece. Early in the Classical Era, Athens and Sparta coexisted peacefully, in spite of an underlying suspicion of each other in the middle of the 5th century BCE.

The political and cultural structures of the two city-states were at the opposite ends of the spectrum. Sparta was a closed society, governed by an oligarchic system of government led by two kings who were advised by a council of elders, and occupied the harsh southern end of the Peloponnese region, organizing its affairs around a powerful military that protected the Spartan citizens from both external invasion and internal revolt. In Sparta, all boys were removed from their families at age seven, and underwent vigorous military training to prepare them to become members of the prestigious Spartan army. Athens on the other hand, grew into an intellectually progressive, open society, governed by a direct-democratic form of elected government that thrived on widespread commercial activity. The period of leadership by Pericles in Athens, is described as the "Golden Age." It was during this period that massive public building projects were undertaken, including the construction of the Acropolis.

Pericles, born in 495 and died in 429 BCE was a brilliant general, orator, patron of the arts and politician. He was considered to be 'the first citizen" of democratic Athens, according to the historian, Thucydides. He was an Athenian statesman who played a significant role in devel-

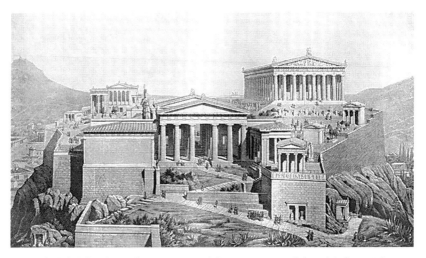

(194) The Acropolis in ancient Athens – a powerful and influential
Greek city-state

oping direct democracy in Athens; expanding the direct participation of the citizens in politics, for the first time in human history; helping to make Athens the political and cultural centre of Ancient Greece. Pericles transformed his city's alliances into an Athenian Empire, and was responsible for the construction of the famous Parthenon in the Acropolis; a spectacular building that still stands today within the very heart of Athens. Built to the highest standards of aesthetics, engineering and mathematics, this massive white marble structure, standing high up on a hill, was decorated with intricate statues, carved by the city's greatest sculptors. Between 463 and 461 BCE, Pericles emerged as the leader of Athens and promoted the new ideology of direct democracy within the city-state. In 454 BCE he took on the role of a great general, by leading a successful military campaign in Corinth, and sponsoring the establishment of Athenian colonies in Thrace, and along the Black Sea coast. As a direct result of his spectacular military accomplishments, in 443 BCE Pericles was elected Strategos, one of Athens' leading generals, a position which he held, with the exception of one short interruption, for the remainder of his life.

Pericles' social innovations were equally important to Athens. He worked to democratize the fine arts by subsidizing theatre admission for the poorer citizens, and enabled civic participation by offering

(195) Pericles, the founder of direct democracy, speaking to the citizens of Athens

pay for jury duty and other civil service. Pericles maintained close friendships with the leading intellectuals of his time. The playwright Sophocles and the sculptor Phidias were among his friends. Pericles' consort Aspasia, a well-known and highly respected woman in Athens, taught rhetoric to the young philosopher Socrates. Pericles himself was a master orator. Here is an example of his superb oratory skills, in a speech he delivered, praising his fellow Athenians: "We cultivate refinement without extravagance and knowledge without effeminacy; wealth we employ more for use than for show, and place the real disgrace of poverty not in owning to the fact but in supporting the struggle against it. Our public men have, besides politics, their private affairs to attend to, and our ordinary citizens, though occupied with the pursuits of industry, are still fair judges of public matters; for, unlike any other nation, regarding him who takes no part in these duties not as unambitious but as useless, we Athenians are able to judge at all events if we cannot originate, and instead of looking on discussion as a stumbling-block in the way of action, we think it an indispensable preliminary to any wise action at all."

As Athens grew in power under the leadership of Pericles, Sparta was becoming more and more of a threat, and began to demand concessions from the Athenians, whom they envied, felt threatened by, and believed had amassed too much power and influence among the Greek city-states. Pericles refused outright, and as a result, in 431 BCE conflict between Athens and Sparta's ally Corinth, pushed the Spartan King Archidamus II to invade Attica, just north of Athens. In response, Pericles adopted a strategy that played to the Athenians' advantage as a naval force, by evacuating the Attic countryside to deny the superior Spartan army anybody to fight against. When the Spartans arrived at Attica, they literally found it to be deserted. With all of his people safely within the walls of Athens, Pericles was free to engage in a series of naval "tactical hit and run" assaults against city-states that were allied with Sparta. Although financially costly, this naval-assault strategy worked very well during the war's early years; but a Plague struck the densely populated Athens, taking many lives and stirring civilian discontent and unrest. As a result, Pericles was briefly deposed as leader in 430 BCE, but after the Athenians' effort to negotiate some kind of ceasefire with Sparta failed, he was quickly reinstated. In 429 BCE, the two legitimate sons of Pericles died of the Plague. A few months later, Pericles himself succumbed to the pestilence. According to the historian Thucydides, his untimely death was a disaster for Athens. His military strategies were quickly abandoned, and the leaders who followed lacked Pericles' foresight, strategic vision, tactical military skills and courage. Instead, "committing even the conduct of state affairs to the whims of the multitude." The downside of democracy; succumbing to the will of the masses?

(196) The Ionian Revolt – a Greek rebellion against Persian rule

The biggest threat to the collective Greek city-states were the Persians. The Athenian adventurous spirit, and loyalty to their Ionian allies obliged them to assist the Greek colonies that were feuding with the powerful Persian Empire in Asia Minor. To aid the Ionian Revolt of 499 BCE, led by the city-state of Miletus, the Athenians landed a small military garrison in Ionia to join the fight against the Persians, and to spread the revolt. The combined Greek forces enraged the Persians by burning the capital city of Lydia, Sardis in 498 BCE; but the Greeks were finally defeated in 494 BCE, and were unsuccessful in removing the domination of these Ionian cities by the Persians. The sacking of Sardis and the defiance and provocation of the Athenians invoked the wrath of the Persian King Darius, who vowed revenge. In 490 BCE, Darius landed his substantial Persian forces thirty kilometres north of Athens, at Marathon. While the Spartans were preoccupied with a religious festival, the heavily outnumbered Athenians under the leadership of Miltiades, mounted an ambitious surprise attack, and routed the unprepared and dumbfounded Persians at the Battle of Marathon, preserving Greek independence in the process. It took ten years, but the Persian King Xerxes, determined to succeed where his father Darius had failed, amassed what the historian Herodotus described as the greatest army ever put together to that date, in order to launch an attack on Greece again. The Athenians, expecting a huge attack from the Persians, prepared for that moment as well. Under the leadership of Themistocles, the Athenians cashed in their silver extracted from the newly dug mines of Lavrion in southern Attica, and built a formidable navy of trireme ships. King Xerxes and his massive army crossed the Hellespont in 480 BCE and began annexing Greece through land and sea. The first line of defence for the Greek alliance of city-states was at the narrow passage of Thermopylae, where the Spartan King Leonidas, with an army of 300 Spartans and 700 Thespians, held back the entire mighty Persian army of around 80,000 soldiers, for three days, before the Spartans were finally overwhelmed and defeated.

King Leonidas of Sparta would have been in his sixties by the time of Thermopylae and undoubtedly was an experienced military commander. Following Xerxes' Persian land invasion of mainland Greece in 480 BCE, Leonidas was selected to lead a small contingent of elite Spartan hoplite soldiers; 300 hand-picked men with male heirs, to defend the narrow pass of Thermopylae, and hold the massive invading Persian force until more Spartan and other Greek troops could be mustered. The Spartans at this time were involved in the sacred Karneia religious festival and so, theoretically could not go to war until it was over. Sparta could have fielded up to 8000 soldiers, but not during the Karneia. The 300 Spartans were joined by troops from various other city-states to make up a force of 7000 men, which was woefully inadequate to halt the Xerxes led Persian army of 80,000.

(197) The Battle of Marathon in 490 BCE – Miltiades of Athens against Darius I of Persia

At the same time, the Athenian ships fought the Persian navy to a stalemate at nearby Artemision, before the Athenians withdrew to the Straits of Salamis not far from the city of Athens. The Athenians vacated the entire non-combat population from their city, so when the Persians arrived they met no resistance. The Persians took vengeance on the buildings and temples of Athens by burning them to the ground, and anchored their fleet at Faliron, in pursuit of the Athenian-Greek navy that was sheltered at nearby Salamina Island. While the joint leadership of the Hellenes argued in typical Greek fashion if they should withdraw to the Peloponnese, and where to engage the Persians next, Themistocles, seeking an advantageous quick battle, provoked the Persian fleet into attacking, just as the Greek ships were faking an early morning escape from Salamina. As the Persians pursued what they thought was a fleeing foe, the Greek triremes turned around and engaged the surprised Persians in a fierce naval battle, inflicting massive casualties and decimating the Persian navy. With his navy utterly destroyed, Xerxes feared that the Greek triremes would rush to the Hellespont to cut off his only way home, so he quickly withdrew back to Asia Minor, leaving his capable general Mardonious to continue fighting the Greeks. The next year in 479 BCE, this Persian army was defeated at Plataea by an alliance of Greek city-states, under the leadership of the Spartan general Pausanias, putting a permanent end to any further Persian ambitions to conquer Greece, once and for all.

The victory of the Greek forces at Marathon and Salamis are considered to be a significant turning point in the development of western civilization. The reason being that if the Persians had been victorious in conquering Greece; all of the subsequent achievements

(198) The Spartan King Leonidas led the Spartans against King Xerxes of Persia

of Greece, and especially Athens; particularly in the fields of philosophy, politics, the arts, languages, drama and literature; would not have transpired and would not have found their way from the Greek world, to the Roman world; and then thanks mainly to thousands of medieval monks who transferred much of Greek literature onto manuscripts; ultimately into the very foundations of western civilization. Instead, western civilization would have been heavily influenced by the Persian culture. Our modern Western world would have been entirely different as a result; for the better or for the worst, we shall never know.

Following the successful defence of their homeland, the Greek city-states commenced a high level of cultural development. Athens especially, emerged as a major superpower, and led a defensive alliance of numerous other Greek city-states, called The Delian League; which was initially established to counter any further threat from the Persians. The tributes (money and resources) collected from its allies helped Athens expand and maintain a formidable, though not easy to manage, empire in the Aegean world. At the same time, the other powerful city-state of Sparta, formed and led The Peloponnesian League; an alliance of city-states mostly from the Peloponnesian region, that felt threatened by the growing power and influence of Athens. The competitive spirit, suspicion, and animosity toward each other that characterized many of the Greek city-states, re-emerged once the external threat of the Persians had subsided; and with the two most dominant empires occupying opposite ends of the political and cultural spectrum; it was not long before the underlying differences and mistrust spilled over into a particularly long and nasty conflict; The Peloponnesian War. While Sparta and Athens were clearly the primary adversaries, just about every other Greek city-state took part in this conflict, at one time or another. With Sparta possessing the strongest and most formidable army, and Athens dominating the sea with its navy of triremes; the war lasted from 431 until 404 BCE; with the temporary Peace of Nicias interrupting it briefly in 421-418 BCE. After surviving a devastating Plague in 430-429 BCE and a spectacular defeat in Sicily by the Spartan-allied Syracuse in 413 BCE, Athens was drained of resources and morale, and finally capitulated to the Spartans in 404 BCE.

The Classical Greek Period produced remarkable cultural and scientific achievements. As we have seen, the city-state of Athens introduced the world to Direct Democracy. The rational and logical approach to exploring and explaining the world as reflected in Classical Art, Philosophy, and Literature, became the well-grounded springboard that western civilization and culture used to leap forward, beginning with the subsequent Hellenistic Age. The great philosophers of the Classical Greek Period have since dominated thought for thousands of years, and have remained relevant to this day. The teachings of Socrates, Plato, and Aristotle among others, have been used as a reference point of countless western (and other) thinkers over the last two thousand years. Hippocrates became the "Father of Modern Medicine", and the Hippocratic Oath is still used today when doctors enter the medical profession. The dramas of Sophocles, Aeschylus, Euripides, and the comedies of Aristophanes, are considered to be amongst the masterpieces of western culture.

Let us now briefly examine the controversial life of "the father of philosophy." Socrates (469-399 BCE) was a classical Greek philosopher who is credited with laying the foundation of modern Western philosophy. He is known for creating Socratic irony and the Socratic method of rational logical argument. He is also recognized for inventing the teaching practice of pedagogy; where a teacher engages with a student through a series of questions in a manner that ultimately draws out the correct response. He has had a profound influence on Western philosophy along with his students Plato and Aristotle. Scholars and historians who try to gather

(199) The Athenian Socrates was one of the founders of Western philosophy

accurate information about the life of Socrates face a peculiar problem, known as "the Socratic problem." This problem arises because there is no proof that Socrates ever wrote anything, philosophical or biographical. Whatever information we derive about Socrates is mainly from the works of four scholars; Xenophon, Plato, Aristotle, and Aristophanes. The writings of these Greek scholars are in an artistic and creative style, therefore creating some doubt as to whether they are truth or fiction, or a combination of both.

Much of what we know of the life of Socrates, comes from Plato's literary works, like "The Apology." Socrates was primarily known for his ideas, communication skills, and public teachings. In Plato's work, Socrates' father was Sophroniscus and his mother, a midwife, was Phaenarete. Socrates married Xanthippe, who was much younger than him; and he had three sons, Lamprocles, Sophroniscus, and Menexenus. Very little is known on what Socrates did for a living. Xenophon suggests that he dedicated his life to philosophical thought and discussion. Aristophanes' writings describe Socrates as running a sophist school and getting paid for it. Xenophon and Plato disagree with this, saying that Socrates did not accept any payment for his teaching; with his poverty acting as proof of this fact. In Plato's dialogues he portrays Socrates as a soldier who served in the Athenian army, and fought in the battles of Potidaea, Amphipolis, and Delium.

The Socratic Method is described in Plato's "Socratic Dialogues." According to Plato, the Socratic Method clarified the concepts of Good and Justice. If you have any problem, break it down to a series of questions, and you will find your required answer in those responses. This style of philosophy earned Socrates the crown of "father of political and moral philosophy" and a leader in mainstream Western philosophy. The Socratic Method was designed to help a person closely and rationally examine their own beliefs, and then evaluate their worthiness. Socrates was morally, intellectually, and politically against the Athenians. When he was on trial for corrupting the minds of young Athenians he explained that while they were primarily concerned about their families and careers, they would have been better to be concerned about "the welfare of their souls." He also vigorously contested the Sophistic doctrine, which stated that virtue could be taught to anybody; and argued that successful fathers do not necessarily produce successful sons, and that moral excellence was more a matter of divine bequest than parental upbringing.

Socrates believed that wisdom was in direct parallel to one's ignorance. A person's deeds were a result of this level of intelligence and ignorance. He constantly connected the "love of wisdom" with "art or love." He also drew a clear line of distinction between wisdom and ignorance. Socrates believed that a person must concentrate more on self-development than the acquisition of material possessions. He encouraged people to develop friendships and love amongst themselves. Humans possess certain basic philosophical or intellectual virtues, and those virtues were the most valuable possessions a person could have in life. To act Good and to be truly Good from within, is different; and virtue relates to the Goodness of the soul. Socrates believed that ideals belong in a world that only the wise man can understand. He had no particular political beliefs or ideologies, but did object strongly to democracy, in its Athenian form. Basically, he objected to any form of government that did not run on the basis of his ideas of perfect governance. Socrates refused to enter politics because he believed he could not tell other people how to live their lives, when he didn't know how to live his own. He thought he was a philosopher of truth, which he had not fully discovered, believing it was a lifetime work in progress. Towards the end of his life, Athenian democracy had been replaced by the Thirty Tyrants (an Oligarchy) for around one year, before being restored again. For Socrates, the oligarchical Thirty Tyrants were no better, and arguably worse rulers than the democracy they sought to replace.

The death of Socrates happened at the climax point in his career, and was well depicted in Plato's works. His death could have been avoided if he had deserted his philosophy and gone back to minding his own business. Even after Socrates was convicted, he could have escaped with the help of his friend Crito, who argued that by not escaping, Socrates was letting down his students and family. His complete lack of cooperation seemed to be a combination of a strong belief in his

(200) The Death of Socrates from hemlock poisoning in 399 BCE

philosophical views; and an expression of political infighting with representatives of the Athenian power structure. At that time, Athens was in political turmoil, undergoing a change from authoritarian rule to democracy, and as we have seen, Socrates was against direct democracy. He didn't believe the common people were virtuous and wise enough to be entrusted with major city-state decisions. Despite his loyalty to Athens, his attitude of defending his truth clashed strongly with the Athenian politics and society of that time. Even the Oracle of Delphi had agreed that there was nobody wiser than Socrates; but Socrates refused to believe this. Eventually Socrates was put on trial and found guilty of refusing to recognise the Gods recognised by the State; of introducing his own divinities (meaning his Socratic philosophy); and corrupting the youth of Athens. He was sentenced to death by poison (in the form of hemlock). His death narrative can be found in Plato's "Phaedo." After drinking the poison, Socrates was made to walk until his legs felt heavy. The man who gave him the hemlock pinched his foot, but Socrates only felt a numbness. The numb feeling eventually spread to his heart, and he died. Shortly before dying, Socrates spoke his last words to Crito saying: "Crito, we owe a cock to Asclepius. Please don't forget to pay the debt."

Plato attributed the following quotations to Socrates:

- The only true wisdom is in knowing you know nothing.
- The unexamined life is not worth living.
- I cannot teach anybody anything. I can only make them think.
- There is only one good, knowledge, and one evil, ignorance.
- Be kind, for everyone you meet is fighting a hard battle.
- Wonder is the beginning of wisdom.
- Strong minds discuss ideas, average minds discuss events, weak minds discuss people.
- To find yourself, think for yourself.
- Education is the kindling of a flame, not the filling of a vessel.
- By all means marry; if you get a good wife, you'll become happy; if you get a bad one, you'll become a philosopher.
- He who is not contented with what he has, would not be contented with what he would like to have.
- Be slow to fall into friendship, but when you are in, continue firm and constant.
- If you don't get what you want, you suffer; if you get what you don't want, you suffer; even when you get exactly what you want, you still suffer because you can't hold onto it forever. Your mind is your predicament. It wants to be free of change. Free of pain, free of the obligations of life and death. But change is law and no amount of pretending will alter that reality.
- The children now love luxury; they have bad manners, contempt for authority; they show disrespect for elders and love chatter in place of exercise. Children are now tyrants, not the servants of their households. They no longer rise when elders enter the room. They contradict their parents, chatter before company, gobble up dainties at the table, cross their legs, and tyrannize their teachers.
- Sometimes you put walls up not to keep people out, but to see who cares enough to break them down.
- No man has the right to be an amateur in the matter of physical training. It is a shame for a man to grow old without seeing the beauty and strength of which his body is capable.

- The secret of happiness, you see, is not found in seeking more, but in developing the capacity to enjoy less.
- Know thyself.
- Let him who would move the world, first move himself.
- Death may be the greatest of all human blessings.
- Contentment is natural wealth; luxury is artificial poverty.
- Employ your time in improving yourself by other men's writings so that you shall come easily by what others have laboured hard for.
- The hour of departure has arrived, and we go our separate ways, I to die, and you to live. Which of these two is better, only God knows.
- Do not do to others what angers you if done to you by others.
- I am not an Athenian or Greek, but a citizen of the world.
- I examined the poets, and I look on them as people whose talent overawes both themselves and others, people who present themselves as wise men and are taken as such, when they are nothing of the sort. From poets, I moved to artists. No one was more ignorant about the arts than I; no one was more convinced that artists possessed really beautiful secrets. However, I noticed that their condition was no better than that of the poets, and that both of them have the same misconceptions. Because the most skilful among them excel in their specialty, they look upon themselves as the wisest of men. In my eyes, this presumption completely tarnished their knowledge. As a result, putting myself in the place of the oracle and asking myself what I would prefer to be – what I was or what they were, to know what they have learned or to know that I know nothing – I replied to myself and to the god: I wish to remain who I am. We do not know – neither the sophists, nor the orators, nor the artists, nor I – what the True, the Good, and the Beautiful are. But there is this difference between us: although these people know nothing, they all believe they know something; whereas, I, if I know nothing, at least have no doubts about it. As a result, all this superiority in

wisdom which the oracle has attributed to me, reduces itself to the single point that I am strongly convinced that I am ignorant of what I do not know.

- Every action has its pleasures and its price.
- Prefer knowledge to wealth, for the one is transitory, the other perpetual.
- We cannot live better than in seeking to become better.
- We can easily forgive a child who is afraid of the dark; the real tragedy of life is when men are afraid of the light.
- Beware the barrenness of a busy life.
- Understanding a question is half an answer.
- Life contains but two tragedies. One is not to get your heart's desire; the other is to get it.
- The hottest love has the coldest end.
- Envy is the ulcer of the soul.
- Be as you wish to seem.
- Thou should eat to live; not live to eat.
- I know that I am intelligent, because I know that I know nothing.
- True wisdom comes to each of us when we realize how little we understand about life, ourselves, and the world around us.
- From the deepest desires often come the deadliest hate.
- Regard your good name as the richest jewel you can possibly be possessed of – for credit is like fire; when once you have kindled it you may easily preserve it, but if you once extinguish it, you will find it an arduous task to rekindle it again. The way to a good reputation is to endeavour to be what you desire to appear.
- If all our misfortunes were laid in one common heap whence everyone must take an equal portion, most people would be content to take their own and depart.
- In all of us, even in good men, there is a lawless wild-beast nature, which peers out in sleep.
- The greatest blessings granted to mankind come by way of madness, which is a divine gift.

- Those who are hardest to love need it the most.
- Esteemed friend, citizen of Athens, the greatest city in the world, so outstanding in both intelligence and power, aren't you ashamed to care so much to make all the money you can, and to advance your reputation and prestige; while for truth and wisdom and the improvement of your soul, you have no care or worry?

The art of Classical Greece began the trend towards a more naturalistic depiction of the world, thus reflecting a shift in philosophy from the abstract and supernatural, to more immediate earthly concerns. Artists stopped merely "suggesting" the human form, and instead began "describing" it with the greatest of accuracy. Man became the focus, and "measure of all things" in daily life, through pursuits like democratic politics and culture. Rational thinking and Logic became the driving force behind this radical cultural revolution, at the expense of the emotions and impulse. The most striking example of this "Logic over Emotion" approach is frozen on the faces of the statues of the Temple of Zeus at Olympia. In this complex array of sculptures, it is easy to determine who is "Barbarian" and who is a "Civilized Hellene", through the expressions on their faces. Barbarian Centaurs exhibit an excess of emotion, while Lapithae women and Apollo remain calm, collected and emotionless, even in the direst of situations.

(201) The Temple of Zeus at Mount Olympia – completed in 457 BCE

Even after its defeat in the Peloponnesian War, Athens remained a guiding light and inspiration for the rest of Greece; but slowly, over a period of time, the influence of Athens began to wane. After winning the Peloponnesian War, Sparta had emerged as the unequalled dominant power in the Greek world; but politically, Sparta was nowhere near as astute as it was militarily. Soon after the conflict ended, and while Sparta was still fighting to subdue a number of city states all over Greece, Athens commenced a major reconstruction program, which amongst other things involved rebuilding her walls, her navy, army, buildings, and infrastructure. Eventually, the military might of Sparta began to diminish, especially after two major defeats at the hands of the Thebans first in Leuctra in 371 BCE, and then nine years later at Mantinea. This power vacuum was quickly and decisively filled by the emerging power of the Macedonians, who under the leadership of Philip II, emerged as the only major military authority in Greece, after their significant victory against the Athenians at Chaeronea in 338 BCE.

As a result of shrewd diplomacy, and superior military might, Philip II, who became King of Macedonia in 359 BCE, managed to consolidate the region around northern Greece, under his control; and right up to his assassination in 336 BCE, he had added central and southern Greece to his growing empire. The excuse for his military incursions into southern Greece was to protect the Oracle of Delphi from the Phocaeans (Ionian Greeks). Just before his assassination, Philip II had his ambitions set on expanding his empire beyond the borders of Greece. His desire was to lead a united Greek military expedition into Persia, and attack the Persian Empire in order to avenge their earlier destructive attacks on Greece. After the assassination, this ambition was to be fulfilled by his son Alexander the Great, who took over as king after his father's death.

The assassination of King Philip II of Macedonia is clouded in controversy, mystery, and intrigue. Why was he killed? Who was responsible for the assassination? Did his former wife Olympias and even his son Alexander, play any role in planning the killing? Philip had been the ruler of Macedonia for twenty-three years, and was married to

*(202) The assassination of King Philip II of Macedonia in 336 BCE
by Pausanias*

his seventh wife when he was assassinated. He had turned Macedonia into a formidable force to be reckoned with, by revolutionizing his army into a fierce and efficient fighting machine. He had subdued Greece and conquered surrounding territories. He had also fathered a number of children from his seven wives. His eldest son Alexander, was from wife number four Olympias, who was in fact Greek. Even though Alexander was the oldest; the oldest son did not always get the throne; it was pretty much up to Philip, who would succeed him. He had a young son named Caranus, from his then current seventh wife. There had been rumbling complaints amongst his entourage and royal court that his heir should be a pure Macedonian, like Caranus was; and not have any Greek blood, like Alexander had. Alexander was highly offended by this suggestion, and angry words were exchanged between father and son, which resulted in Alexander and his mother Olympias being temporarily exiled. Olympias and her son Alexander went into voluntary exile in Epirus, staying at the palace of her brother Alexander, who was the King of Epirus at that time.

In October 336 BCE, Philip was celebrating the wedding of his daughter Cleopatra, to King Alexander of Epirus. They were also celebrating Philip's imminent invasion of Persia. Philip had arranged lavish musical competitions and a huge feast, with food and wine sourced from throughout his empire, in honour of the married couple. Everyone who had any kind of fame in the Greek world showed up to be a part of the wedding celebrations. At the beginning of one of the competitions, after a night of hard drinking, Philip went to the theatre of Aegae in a procession with twelve statues of the gods. His bodyguards appear to have been dismissed from duty that evening. This was not an uncommon practice for Philip, who believed he was so powerful that nobody would dare to harm him. According to the historical account given by Diodorus of Sicily, Philip was having an affair with one of his bodyguards, a man named Pausanias, who had the same name as another bodyguard he had also had an affair with. This was not unusual in Greek society at that time. The first Pausanias was upset that the King's eye had been caught by a second Pausanias. The first Pausanias had earlier insulted the second Pausanias, who then complained to his friend, and the king's aide, Attalus. All manner of unsavoury things then happened, which resulted in the second Pausanias being killed, and the first Pausanias being raped. The humiliated and violated Pausanias was full of hatred and anger, and decided to get his revenge on both Attalus and Philip. When he saw Philip was without his bodyguards, Pausanias rushed forward and stabbed him in the chest with a dagger. Then he immediately fled, while half of the bodyguards who were present at the wedding went after him in hot pursuit. He was killed by one of the bodyguards. Alexander then became King.

Was this really all that happened on that fateful evening? Many historians believe Olympias had something to do with the assassination of Philip. She and Philip had a highly passionate and volatile love-hate relationship. Olympias had even urged her Greek military commander brother to declare war on her ex-husband. This brother of Olympias was none other than King Alexander of Epirus, the man who was marrying Philip's daughter, Cleopatra! Olympias certainly had a motive to conspire to murder Philip. Even though the Macedonians wanted a "pure heir", they were less likely to choose a child as the next king, over a grown man who had proven battle experience; her son Alexander. Therefore, the death of Philip would put her son Alexander

(203) Olympia presenting her son, the young Alexander
to the philosopher Aristotle

on the throne, and place herself in a position of great power within
the Kingdom of Macedonia. In fact, shortly after her son became king,
Olympias forced Philip's seventh wife, Cleopatra Eurydice, to commit
suicide shortly after Olympias had also arranged for her young children
fathered by Philip, to be killed. There was no doubting that Olympias
was ruthless in securing her power.

Historians are divided as to whether Alexander was involved in
the assassination or not. He of course, also had a clear motive, and
had recently had a big upsetting argument with his father Philip. Also,
the bodyguards who went after Pausanias and killed him were close
friends with Alexander. It is possible they were silencing him before
he could be tortured and then declare that he was simply a "pawn"
in a wider plot involving Alexander. There is a possibility that the
bodyguards may have carried out this plan without consulting with
Alexander. What we do know is that Philip was buried in a lavish
tomb, after a monumental funeral ceremony; and Alexander went on
to conquer the known world. A treasure-filled tomb was found in
1977 at the site where Philip was believed to have been buried; but

historians are undecided whether it was Philip's tomb, or the tomb of Alexander's successor and half-brother, Philip III Arrhidaios. The recovered skeleton did have a notch in the eye socket, which was consistent with a battle wound that left Philip's face disfigured.

With a copy of the Iliad always in his possession, and a dagger in his hand, Alexander continued the centuries-old conflict between East and West by leading a united Macedonian-Greek army into Persia and other regions of Asia. His success on the battlefield and the amount of land he conquered became legendary throughout the ancient world, earning him the epithet "the Great." In addition to lightning fast, breathtakingly brilliant military tactics, Alexander possessed strong and decisive leadership skills, and an abundance of charisma, that made his huge united army unbeatable in numerous battles against many opponents; pushing Alexander and his army all the way to Egypt, India and Bactria (modern day Afghanistan). Alexander always led battles from the front, placing himself on a white horse at the point of attack, so therefore engaging in fierce face to face combat right alongside his soldiers. He spared no mercy when in battle leading his army; obliterating all manner of opposition in his path. In the process, Alexander the Great amassed the largest empire known to the ancient world up to that time.

In 334 BCE, Alexander led his army across the Hellespont into Asia, and scored successive decisive victories against the Persian Empire. His first success came at Granicus River in northwest Asia Minor, where his cavalry smashed the outnumbered Persian mercenaries, who fought under the leadership of General Memnon of Rhodes. In 333 BCE, Alexander's outnumbered army defeated the Persians at Issus, and forced King Darius to flee for his life. He then went on to conquer the city of Tyre in 332 BCE and Egypt in 331 BCE, giving the Greeks control of the entire eastern shores of the Mediterranean; and allowing Alexander the strategic path to initiate attacks deep into the heart of the Persian Empire. In Egypt, Alexander was proclaimed to be the son of the supreme god Ammon (the equivalent of the Greek god, Zeus), and he established himself as the King of Asia after his victory at the Battle of Gaugamela in 331 BCE, which sealed the fate of the Persian Empire.

Establishing himself in the great city of Babylon, where he was warmly greeted as a conquering hero by its citizens, when he announced he had plans to make this city the capital of his Eastern Empire; Alexander proceeded to lead his army towards the heart of South Asia, subduing all resistance, and establishing new cities along the way. During his conquests he named more than 70 cities after himself. Despite the objections of his generals and officers, he integrated conquered soldiers into his own army, adopted local customs, and married a Bactrian (Afghan) woman, Roxane. His continuing march eastward eventually came to a grinding halt on the edge of India, not far from the Himalayan Mountains, mainly as a result of objections from his fatigued, demoralised and home-sick army. Understanding that his soldiers were at the end of their tether, he returned to Babylon for his soldiers to rest. While there, Alexander was planning a military expedition southward, towards Arabia; but in 323 BCE, he suddenly died of an "apparent fever" at the age of 32, putting an end to his brilliant military career; and leaving his vast conquered empire without an appointed heir.

(204) Alexander the Great and his conquering army first attacked Persia in 334 BCE

The exact cause of Alexander's death is a mystery. Historians have debated the issue for centuries; attributing it to poison, malaria, typhoid fever, or other diseases. What is agreed upon from the sources is that the Macedonian king died in early June 323 BCE while suffering from a high fever that lasted ten days. The following description of the death of Alexander and how he spent the last ten days of his life, was written by Arrian, a Greek historian who wrote this account around 350 years after the event. Arrian based his account on the Royal Diaries, contemporary chronicles of Alexander's campaign recorded by his scribes. We join Arrian's account as Alexander begins to feel ill. This is a fascinating insight not only into his death, but also the lifestyle that Alexander had during his last days in the great city of Babylon, where he was treated as the king of the world; and the power of religion and the gods in the lives of ancient Greeks, including Alexander.

"A few days later he (Alexander) had performed the divine sacrifices (those prescribed for good fortune and others suggested, by the priests) and was drinking far into the night with some friends. He is said to have distributed sacrificial victims and wine to the army by detachments and companies. Some state that he wanted to leave the drinking-party and go to bed, but then Medius met him, the most-trusty of his companions, and asked him to a party, for he promised that it would be a good one.

Day 1

The Royal Diaries tell us that he drank and caroused with Medius. Later he rose, had a bath and slept. He then returned to have dinner with Medius and again drank far into the night. Leaving the drinking, he bathed, after which he had a little to eat and went to sleep there. The fever was already on him.

Day 2

Each day he was carried on his couch to perform the customary sacrifices, and after their completion he lay down in the men's apartments until dusk. During this time, he gave instructions to his officers about the coming expedition and sea-voyage, for the land forces to be ready to move on the fourth day, and for those sailing with him to be prepared to cast off a day later. He was carried thence on his couch to the river, where he boarded a boat and sailed across to the garden where he rested again after bathing.

Day 3

The next day, he again bathed and performed the prescribed sacrifices. He then entered his room, lay down and talked to Medius. After ordering the officers to meet him in the morning, he had a little food. Carried back to his room, he lay now in continual fever the whole night.

Day 4

In the morning he bathed and sacrificed. Nearchus and the other officers were instructed to get things ready for sailing two days later.

Day 5

The following day, he again bathed and sacrificed, and after performing them, he remained in constant fever. But in spite of that he summoned the officers and ordered them to have everything quite ready for the journey. After a bath in the evening, he was now very ill.

Day 6

The next day, he was carried to the house by the divine place, where he sacrificed, and in spite of being very poorly, summoned the senior officers to give them renewed instructions about the voyage.

Day 7

The next day he was carried with difficulty to perform the sacrifices, and continued to give orders just the same to his officers about the voyage.

Day 8

The next day, though very weak, he managed to sacrifice. He asked the generals to stay in the hall, with the brigadiers and the colonels in front of the doors. Now extremely sick, he was carried back from the garden to the Royal Apartments. As the officers entered, he clearly recognized them, but he said not a word to them.

Days 9 and 10

He had a high fever that night; another day as well, and all the next day as well.

The above information comes from the Royal Diaries, where we also learn that the soldiers wanted to see him, some hoping to see him before he died and others because there was a rumour that he was already dead, and they guessed that his death was being kept back by his personal guard, or so the historian Arrian thought. Many pressed into the room in their grief and longing to see Alexander. They say that he remained speechless as the army filed past him. Yet he welcomed each one of them by a nod with his head or a movement of his eyes. The Royal Diaries say that Peithon, Attalus, Demophon, Peucestas, Cleomenes, Menidas and Seleucus spent the night in the temple of Serapis and asked the god whether it would be better and more profitable for Alexander to be carried into the temple to pray the god for his recovery. A reply came from the god that he should not be brought into the temple, but that it would be better for him to remain where he was. The Companions brought this news, and, shortly after, Alexander died; for this was what was better. That is the end of the account given by Aristoboulos and Ptolemy.

The conquests of Alexander the Great changed the course of ancient history. The centre of gravity in the Greek world moved from the self-contained city-states (the polis), to a vast territorial empire that spanned the entire coast of the Eastern Mediterranean, Egypt and reached far and deep into Asia. Alexander's conquests placed a diversity of cultures and peoples under Macedonian-Greek control and influence, paving the way for the development of the distinctive Hellenistic culture that followed his death.

The Hellenistic Age marked the transformation of Greek society from the localized and introverted city-states to an open, cosmopolitan, and exuberant culture that permeated the entire Eastern Mediterranean and Southwest Asia. While the Hellenistic world comprised a diverse range of ethnic peoples; Greek thinking, language, culture, traditions, knowledge, and ways of life, dominated the public affairs of the time.

All aspects of culture took on a Greek perspective, with the Greek language being established as the official language of the Hellenistic world. The art and literature of the era was also transformed. Instead of the previous preoccupation with the Ideal, Hellenistic art focused on the Real. Depictions of humanity in both art and literature mainly revolved around the exuberant and often amusing themes that for the most part, explored the daily life and emotions of humans, gods, and heroes.

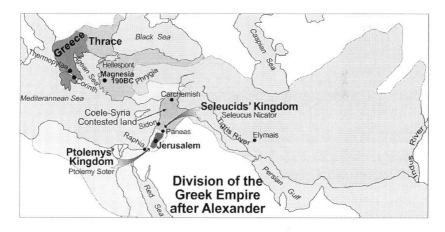

(205) The division of Alexander's Empire after his death in 323 BCE

The autonomy of individual city-states of the Classical Era gave way to the emergence of large kingdoms that were led by one ruler-king. As Alexander left no apparent heir, his most senior generals took control of different pieces of the vast empire. After fighting common enemies and against each other, as they attempted to establish their power; eventually three major kingdoms emerged from the strife, turmoil, and unrest; following the untimely death of Alexander the Great in 323 BCE. Egypt and parts of the Middle East came under the rule of Ptolemy; Seleucus controlled Syria and what remained of the Persian Empire; while Macedonia, Thrace, and parts of northern Asia Minor came under the rule of Antigonus and his son Demetrius. Several smaller kingdoms were established by other generals in Hellenistic Greece. The most notable of these was the Attalid Kingdom which was formed around Pergamum in eastern Asia Minor; and the independent

Kingdom of Bactria, which was created after Diodotos led a rebellion of Greeks there, against Seleucid rule. In the meantime, most of the Classical Greek cities south of Thessaly and on the southern shores of the Black Sea remained independent.

Several Greek city-states became dominant during the Hellenistic Era. The Classical Greek cities like Athens, Corinth, Thebes, Miletus and Syracuse in particular, continued to flourish; while others emerged as major political, cultural or trading centres throughout the three major kingdoms. Pergamum, Ephesus, Antioch, Damascus, and Trapezus were a few of these cities. Although it remained as an independent city-state, Sparta was significantly weaker during this time, only a shadow of its powerful position during the Classical Era. During this time though, no city was more influential and powerful than Alexandria in Egypt. Alexandria was founded by its namesake, Alexander the Great himself in 331 BCE, and very quickly became the centre of commerce and culture of the Hellenistic world under the Ptolemy rulers. Alexandria hosted the Tomb of Alexander the Great, the Lighthouse of Alexandria (one of the Seven Wonders of the Ancient World), and the famous Library of Alexandria which aspired to be the repository of the entire knowledge of the known world, and housed the biggest collection of manuscripts in the ancient world at that time.

(206) The School of Athens by the Italian Renaissance artist Raphael

Many famous thinkers and artists of the Hellenistic Era created works that remained influential for centuries. Philosophical Schools of Thought such as the Stoics, the Sceptics, and the Epicureans, continued the substantial enlightened thinking for which the Greeks were renowned. Art, literature, and poetry reached new heights of innovation and creativity through the masterful works of great Greeks like Kalimachus, Apollonious of Rhodes, Menander, and Theocritos. The sculptures of Polykleitos remained influential and were copied throughout the Hellenistic and Roman Eras, and even centuries later during the Italian Renaissance. Great works of art were created during the Hellenistic Era. In architecture, the classical styles were further refined and developed with new design concepts and ideas like the Corinthian Order Columns which were first used on the exterior of the Temple of Zeus in Athens. Public buildings and monuments were constructed on a larger scale in more ambitious designs, configurations, and complexity. The Mausoleum of Pergamum, for example, merged architectural space and sculpture by the placement of heroic statues in the close proximity of a grand staircase. Hellenistic Greece also became a time of substantial maturity and progress in the sciences. In geometry, Euclid's elements became the standard all the way to the 20th century CE; and the work of Archimedes in mathematics, along with his practical inventions, became influential and legendary. Eratosthenes calculated the circumference of the Earth within an accuracy of 2400 kilometre, by simultaneously measuring the shadow of two vertical sticks, one placed in Alexandria, and the other in Syene. During the Hellenistic times it was common knowledge that the Earth was a sphere.

The Hellenistic Era was by no means free of conflict; even after the major kingdoms were established following the death of Alexander the Great. These great Hellenistic kingdoms were threatened from within and also from outside forces. The size of the three great kingdoms in particular, made securing them next to impossible; and life outside the usually orderly cities was very dangerous and precarious, because of roaming bandits, brigands and pirates. Internal conflicts and power struggles caused the borders of the kingdoms to shift several times, as the rulers of the major and nearby minor kingdoms engaged in continuous conflict with each other. One big external threat at that time came from the Celtic Gauls, who invaded Macedonia and reached into southern Greece in 279 BCE, attempting to plunder the treasures of Delphi at Mount Olympus in the process; which were saved when King Attalus of Pergamum raised an army and defeated and evicted the Gauls, after they crossed into Asia Minor.

During the Hellenistic Era, Rome had risen to become a formidable power by 200 BCE; by which time it occupied not only Italy, but also the entire coastal Adriatic Sea and Illyria. During the Second Punic War against Carthage (218-201 BCE) when the Carthaginian General Hannibal managed to establish a successful campaign against the Romans, after crossing the Alps and invading Italy; Philip V of Macedonia allied himself with Hannibal, and successfully annexed and occupied Illyria; thereby starting a series of wars with Rome that ultimately led to the eventual conquest of Greece by the Romans. In the end, large parts of the Hellenistic kingdoms disintegrated as a result of constant incursions by various barbarian tribes coming from the outer fringes. Large areas of these former kingdoms were simply given to Rome, through bequests made in the Wills of deceased kings and rulers. Other remnants of these once three great Hellenistic kingdoms won independence by inciting people's revolutions. In 31 BCE, Octavian, later to become the first Roman Emperor Augustus, defeated the rulers of Egypt, Antony and Cleopatra, in the naval Battle of Actium, which completed the demise of the Hellenistic Era in Greece. In fact, the Battle of Actium is considered to be the pivotal event that led to the end of Ancient Greece. After the Battle of Actium, the entire Hellenic world virtually came under the control of Rome.

Greek Mythology has always excited the imaginations of so many people from a diverse range of civilizations and cultures all through history, right up to modern times. The stories of the Gods are full of drama, courage, heroism, intense passion, love, hatred, jealousies, and admiration; and the full range of fallacies that we call the human condition. There was a direct connection between Greek drama, highlighting courage and tragedy, and the pantheon of Gods

(207) The Greek gods were led by Zeus and his wife Hera

themselves. We mustn't forget that even though the stories of the Greek Gods are fascinating myths for us; for the ancient Greeks the stories of the Olympian Gods were the heart and soul of their beliefs, and an integral part of their religion and everyday life. Let us examine the major Greek Gods, to give us a taste of the full scale drama that is, Greek Mythology.

Zeus was the God of the sky and lightning. He was the supreme deity of the Ancient Greek Pantheon and King of Mount Olympus. He was actually the youngest of his siblings. His father was the Titan Cronus, who had swallowed all of his previous children in fear of being overthrown. Zeus escaped this terrible fate with the help of his mother, the Titan Rhea, who tricked Cronus. He grew up in the island of Crete, the home of the Minoans as we have seen; and managed to free his siblings from Cronus' stomach when he grew up. In the great battle that followed, known as the Titanomachy, the Olympian Gods defeated the Titans and became the rulers of the world. The Olympic Games in Ancient Greece were held in honour of Zeus.

(208) & (209) The Greek gods Zeus and Poseidon

Poseidon was also the son of the Titans, Cronus and Rhea, and the brother of Zeus. In Greek mythology, Poseidon was the god of the sea, earthquakes, and horses. Poseidon's weapon was a trident, and by striking it he would cause natural disasters, storms, earthquakes, and volcanic eruptions. The ancient Greeks had a period of the calendar dedicated exclusively to Poseidon, which they named after him; this was the months of November and December. During this period, the Greeks would hold a festival in his honour, called Poseidea. In Greek mythology, Hades was the King of the Underworld; the place down below where the souls would end up after death. Hades was the brother of Zeus and Poseidon, and was a frightening figure for living mortals. For many Greeks, just mentioning his name would incite great fear and dread. When the Greeks were praying to him, they would clap their hands on the ground to make sure he heard them. Black animals such as sheep, were sacrificed in his honour. Every hundred years, great festivals were also held in his honour.

In Greek mythology, Hera was a sister and wife of Zeus, and the daughter of the Titans, Cronus and Rhea. She was the goddess of marriage and the protector of all women. Hera was often portrayed holding a sceptre as a symbol of sovereignty, or pomegranate as a symbol of fertility. She was jealous of her husband, and well known for her vindictiveness! In honour of Hera, festivals were held in many ancient Greek cities, and were called Heraia. The most famous Heraia was held in the locations of Argos, Samos, and Olympia. Demeter was one of the greatest and oldest goddesses of ancient Greek mythology. She was the goddess of vegetation and agriculture, and particularly protected cereal crops and farmers. Demeter and her daughter Persephone, were central characters in the Eleusinian Mysteries, and were probably deities worshipped at the Olympian Pantheon. The capture of Persephone by the god Hades, resulted in the demise of Demeter. She abandoned Mount Olympus and began wandering the Earth, wearing black, and in silence among the mortals, in search of her dear Persephone. Nature and crops began to wither as she became weaker and weaker. When her daughter returned to her, everything blossomed once again! Every year, ancient Athenians would celebrate this renaissance with the mystical Eleusinian festivals.

According to Greek mythology, Hestia was the goddess of family, home, and hearth, the eternal flame that would keep the family's home warm. The ancient Greek houses had a sanctuary in the centre of their homes, and the woman of the house had the responsibility of keeping Hestia's flame burning. The goddess Hestia was the eldest daughter and the first child of Cronus and Rhea. She always stayed on Mount Olympus, taking care of her sacred flame. Hephaestus was a clever and inventive Olympian god. He was the master of construction and metallurgy, a skilful and capable artisan craftsman. Hephaestus was such an ugly baby, that as soon as Hera, his mother, saw him, she threw him out of Olympus without a second thought, and he ended up in the depths of the ocean! Hephaestus was a workaholic whose symbol was his hammer. The Olympian god Ares was the son of Zeus and Hera. He was the god of war, and was provocative and impulsive in nature. He represented the most violent aspects of war, in contrast to the goddess of war, Athena. According to Greek mythology, Ares had a huge body, and was powerful and impetuous. He always rushed into battle with scary eyes, and screamed as loud as up to ten thousand people fighting. He was also always thirsty for blood, while ignoring the rules of war, and recognized no law as applicable to him.

(210) & (211) The Greek goddesses Hestia and Athena

Athena was the goddess of wisdom and strategic warfare. The Temple of the Parthenon in Athens was the most famous temple dedicated to her. Athena was Zeus' favourite daughter. The story of her birth is very interesting and unique, in that she stormed out of Zeus' forehead in full armour. She was symbolized by an owl, and held a spear and a shield made of goat's skin. There were many festivals in honour of Athena, but the Great Panathenaea in Athens, was probably the largest celebration in honour of her throughout the Greek world. She battled with Poseidon over the patronage of the city of Athens, and won. This is also the mythical explanation of the naming of the city of Athens. Protector of heroes and the wisest among the gods, Athena was considered to be one of the most powerful and important Olympian Gods. The goddess Aphrodite emerged from the sea and was the synonym of feminine beauty in ancient times. She was the goddess of love and desire. Gods and mortals were all entangled in her erotic and amorous nets. According to Greek mythology, she was born at the coast of Cyprus, from the foam of the sea, after Cronus cut off the genitals of Uranus, her father, and threw them into the sea. Driven by Zephyrus, the Greek god of the westerly wind, into the sea, she was then transferred to Mount Olympus where she was introduced to Zeus and the other gods. Aphrodite had many lovers, but her greatest love was the god of war, Ares.

The Olympian God Hermes was the god of trade, wealth, thieves, and travellers. He was known as the Messenger God, as he was the one delivering messages between the Gods and the mortals with his golden flying sandals. Hermes was also known as the soul barer, the one that would lead the soul of the dead to the Gates of the Underworld. He is considered to be the first teacher of the human race, introducing the letters and sciences to mankind, and teaching them to use their intellect. Apollo was the god of light, music, and harmony. Born on the sacred island of Delos, he climbed Mount Olympus from the very first day and joined the other Olympians. He was the son of Zeus and Leto, and the twin brother of Artemis. Apollo was one of the most important Olympian Gods and the teacher of the nine Muses. His sanctuary at Delphi became the centre of the ancient world. The Oracle of Delphi was conveying Apollo's words, and the oracles would influence the political scene of the then known world. The Pythian Games in Delphi were held in honour of Apollo, and were one of the four Panhellenic Games of Ancient Greece, another of which was the Olympic Games. Apollo's twin sister Artemis, was the goddess of wildlife and hunting. She was also a helper of midwives, being the goddess of births. Among the elements of her character was her relationship with nature and hunting, war, dance, singing, and her virginity, in combination with her fertility power. The Temple of Artemis in Ephesus, was one of the Seven Wonders of the Ancient World. The Festival of Artemis at Brauron, near Athens took place every four years in honour of the Olympian Goddess.

Hecate was one of the main deities worshipped in the households of Athens. She was considered to be one of the greatest goddesses in ancient Greek religion and mythology. Hecate was the goddess of witchcraft, the Moon, the night, and necromancy (communicating with the dead). Her sacred animals, among others, were the dog, the horse, and the snake. Possessing three heads, she was able to look in three different directions at the same time, and she was considered to be the protector of the crossroads. As we have seen, Persephone was the daughter of the goddess Demeter, the goddess of agriculture and fertility. Hades fell in love with Persephone and then proceeded to kidnap her to the Underworld. Saddened by her loss, the goddess Demeter searched for her daughter day and night without luck. Her sorrow caused the flora of the Earth to wither. The Supreme God Zeus, fearing the decrease in mortals' sacrifices to the Gods, ordered his brother Hades to free Persephone. Hades offered Persephone to eat six seeds of pomegranate before she returned to the world of the living, binding her to the Underworld for six months of every year. The myth of Persephone essentially explains the changing seasons and the circle of life. It is also the origin of the Eleusinian Mysteries, the most sacred religious rites in Ancient Greece.

Tyche was the goddess of fortune, and the personification of prosperity and wealth. She was depicted as a woman holding the cornucopia, which was a symbol of wealth. Tyche had the power to take away the fortunes of mortals if they were not appreciative of their good fortune and of her. Eris was the daughter of Night and the goddess of jealousy. She was always causing divisions, arguments, squabbles, and quarrels. According to Greek mythology, she was also a sister of the god Ares. Eris usually showed up gossiping or humming, but sometimes she beautified her appearance to achieve more success in her scheming and manipulating. It was Eris that tossed the Apple of Discord in the midst of the feasts of the gods. Aeolus was the god of winds, and lived on the enchanting island of Aeolia, where he kept all the winds inside his huge shepherd's bag, letting them free only under the instructions of the major Olympic Gods. He was the one appointed by Zeus as the "treasurer of the winds." Themis was the goddess of divine order, natural law, justice, and fairness. Her law was sacred and applied equally to the gods, superior even to their will. She was usually depicted holding a sword, which was believed to represent her ability to cut through fact from fiction.

The god Pan was an idealistic, human-centred, secondary deity, that his father Hermes took with him at Mount Olympus, and everybody absolutely adored him. He was the god of Nature, but also a personification of the genetic power of life. Pan was depicted as having the lower legs of a goat, and horns in his head. He was the protector of breeders, hunters, fishermen, and shepherds, with permanent residence in nature. The erotic adventures he had with the various Nymphs are numerous, and they defined almost every myth written about him. Eileithyia was the goddess of birth and labour pains. She helped women give birth and withstand the pains of childbirth, and caring for newborns. Before and after childbirth, the midwives offered her various special gifts of nurturing. However, because so many women died during childbirth in ancient times, she was also considered to be the goddess that caused death to women at childbirth. Dionysus was the god of wine and viticulture. The Cult of Dionysus was the most popular in Ancient Greece. He was a god of contrasts, of supreme exaltation, and also of horror. The son of Zeus and Semeli; when he grew up, his father taught him how to grow grapes and make wine. Dionysus did not want to hold on to this skill for himself, so he began to go from city to city teaching winemaking to the mortals. He had many followers, hybrid creatures that accompanied him everywhere, like satyrs and maenads.

Many celebrations were dedicated to Dionysus, with the Dionysia Festival being one of the largest celebrations in Ancient Greece.

In Greek mythology, Eros was the winged god of love. According to the legend, if he hit two people with his arrows,

(212) The Greek god Dionysus

they would fall in love. Eros was one of the most important gods of antiquity and, according to Hesiod's Theogony, one of the three deities that created the world: "First there was Chaos, then Eros, then Earth." Eros did not bless someone with birth, but instead encouraged and facilitated birth and creation. Eros was the son of Aphrodite and the god of war, Ares. Asclepius was the god of medicine who was worshipped all over Greece in ancient times. Although Asclepius was not considered a major Greek god, he was still considered to be a central figure in the archetype of gods-heroes-healers. Asclepius was the ideal conception of the healing powers of nature. One of the greatest festivals in his honour was the Prominent Askleipia, which was originally held every four years in Epidaurus, and lasted for seven days. The goddess Nyx was the personification of the night. Nyx was a sovereign, primordial, and cosmogenic entity that even the God Zeus himself respected and feared. According to Hesiod's Theogony, Nyx was born out of Chaos.

Nemesis was the goddess of divine retribution. The symbols of this goddess were the forearm and the bridle. These symbols were very indicative of her key function, which was to measure human thoughts, emotions, actions, and to set a limit on the ruthless actions of people's selfishness. Therefore, the mortal human's attitude towards the Cosmic Laws and the indifference to the Common Good, are swept away by the actions of Nemesis; giving eternal justice. Phobos was the personification of fear. The ancient Greeks considered him to be the son of the god of war Ares, and the goddess of love Aphrodite. He was represented with a lion's head and was similar looking to the god Pan. For the Spartans, the Sanctuary of Phobos was a symbol of the discipline and cohesion of their military forces. The god Deimos was the personification of terror. He was the brother of Phobos. Together with his brother, who as we have seen, was the personification of fear, Deimos would accompany his father Ares, to war. For this reason, he was offered sacrifices before a battle. His figure, along with that of his brother, would adorn the famous shields of Agamemnon and Achilles. Hebe was the goddess of youth and vitality. She was the daughter of Zeus and Hera. Hebe had undertaken the task of supplying the gods with nectar and ambrosia, the food that always kept them young. Hebe married Hercules when the demi-god hero ascended to Mount Olympus. She was the gift that Zeus and Hera offered to Hercules, for succeeding while he was on the mortal Earth. Finally, we have Enyo (meaning "terror") who was a minor god of war, often appearing with the major god of war Ares. She is often regarded as his daughter. Although Enyo has also sometimes been referred to as the sister or mother of Ares. In antiquity, Enyo was depicted as being bloodthirsty and often involved in violent acts. Enyo was first mentioned by the epic poet Homer, who characterized her as a deity worthy of respect, for she was the one who looted and dominated conquered cities, towns and villages.

Let us conclude with the sheer mystery, divinity, and power of The Oracle of Delphi. We have seen that the world of Ancient Greece was dominated by men, who filled the highest positions in society, fought on the battlefields, and ruled the mightiest city-states and empires. However, all of these men, from the lowliest humble peasants to the kings themselves, sought the counsel and advice of one person; and that person was a woman. The city of Delphi had a long tradition of being at the centre of the ancient world. It was said that the Supreme God Zeus himself, named it the navel of Gaia. According to legend, a huge serpent named Python, guarded the spot before it was slain by the infant god Apollo. When Apollo's arrows pierced the serpent, its body fell into a fissure and great fumes arose from the crevice as its carcass rotted. All those who stood over the gaping fissure fell into sudden, often violent trances. In this state, it was believed that Apollo would possess the person and fill them with divine presence.

These peculiar occurrences attracted Apollo-worshipping settlers during the Mycenaean Era, and slowly but surely the primitive sanctuary grew into a shrine, and then by the 7th century BCE, a temple. It would come to house a single person, chosen to serve as the bridge between this world and the afterlife. Named after the fabled serpent, this chosen divinely sacred person was named Pythia – the oracle. The ability to communicate with a god was truly unique and exceptional, so not just anybody could be allowed or trusted to serve

*(213) The Oracle of Delphi – Pythia was the name of the divinely chosen
sacred female*

this divine position. It was decided that a pure, chaste, and honest
young female virgin would be the most appropriate mortal vessel for
such a sacred role. However, there was one drawback; beautiful young
virgins were prone to attracting negative attention from the men
who would seek their counsel, which resulted in oracles being raped
and otherwise violated. Older women of at least 50 began to fill the
position, and as a reminder of what used to be, they would dress in the
virginal garments of old. These older women were often chosen from
the priestesses of the Delphi Temple, but could also be any respectful
woman from the settlement of Delphi. Educated noble women were
prized, but even peasants could fill the role. Those Pythia who were
previously married were required to relinquish all family responsibilities and even their individual identities. To be an oracle was to take
up an ancient and vitally important role – one that transcended self,
and entered into legend. Pythia were so important to Greek civilization that it was essential that they were a blank slate, so children,
husbands, lovers, and any other human links to their previous life, had
to be severed in favour of total worship and dedication to the divinity
of Apollo.

The reason for the growing importance of the oracles was quite simple – the Pythia provided answers on a whole range of life and state matters. For an ambitious and religious civilization, this very visual and vocal link to the gods was treated with the utmost respect. For the nine warmest months of each year, on the seventh day of each month, the Pythia would accept any questions from all members of Greek society. This corresponded with the belief that Apollo deserted the temple during the three winter months of the year. After being "purified" by fasting, drinking holy water, and bathing in the sacred Castalian Spring, the Pythia would assume her position upon a tripod seat, clasping laurel reeds in one hand, and a dish of spring water in the other. Positioned above the gaping fissure, the vapours of the ancient vanquished serpent would wash over her and she would enter the realm of the divine. The exact origin of these magical vapours – assuming they weren't actually being given off by the rotting remains of Python – remains something of a mystery. Excavation work of the temple ruins in the 19th century did not uncover the sort of cave or hole in the ground that archaeologists had expected to find; so for much of the 20th century, scholars believed the Delphic fissure-fault was strictly mythological. That was until the late 1980s, when a new team of curious scientists decided to investigate the ruins for themselves. The rocks they discovered beneath the temple were oily bituminous limestone and were fractured by two faults that crossed beneath the temple. This had to be more than a coincidence. The scientists theorised that tectonic movements and ancient earthquakes had caused friction along the faults. Combined with the spring water that ran beneath the temple; methane, ethylene, and ethane gas would rise through the faults to the centre and directly into the temple. The low room with its limited ventilation and lack of oxygen would help amplify the effect of the gasses, and induce the trance-like symptoms experienced by the oracles.

Back in Ancient Greece, once the story of the woman who could communicate with the gods became widespread, people began to flock to Delphi to speak with her. Many of those who wished to ask the oracle a question would travel for days or even weeks to reach

Delphi. Once they arrived, they underwent an intense grilling from the priests, who would determine the genuine cases from the frivolous ones, instructing the genuine people in the correct way to frame their questions. Those who were approved then had to undergo a variety of traditions, such as carrying laurel wreaths to the temple. It was also encouraged for these people (called consultants) to provide a monetary donation, as well as an animal to be sacrificed to the gods. Once the animal had been sacrificed, its guts and intestines would be studied. If the signs were seen as unfavourable, the consultant could be sent home. Finally, the consultant was allowed to approach the Pythia to ask his or her question. In some accounts, it seems the oracles gave the answers, but others report the Pythia would utter incomprehensible words that the priests would then "translate" into narrative verse. Once the consultants received an answer, they would journey home and act upon the advice of the oracle in their everyday lives.

The oracle received a multitude of visitors in the nine days she was available; from fathers desperate to know the outcome of the harvest, to kings asking if they should wage war on their enemies; and her answers were not always clear. Responses, or their translations by the temple priests, often seemed deliberately phrased so that, no matter the outcome, the oracle would always be right. It was essential for the consultant to carefully consider her words, or else risk a bad harvest, or even the defeat of an entire army in battle. When Croesus, the King of Lydia, asked the oracle if he should attack Persia, he received the response: "If you cross the river, a great empire will be destroyed." He viewed this response as a good omen, and went ahead with the invasion. Unfortunately, the great empire that was destroyed was his own. In this way, the oracle, just like the gods, was infallible, and her divine reputation grew. To question the oracle was to question the gods – and that was absolutely unthinkable.

Soon, no major decisions were made before first consulting the Oracle of Delphi. It wasn't just Greek people, but also foreign dignitaries, leaders, and kings who travelled to Delphi for a chance to ask the oracle a question. Those who could afford it would pay great sums of money for a fast pass through the long lines of pilgrims and commoners. Using these donations, the temple grew in size and prominence. Rapidly, Delphi seemed to be fulfilling its own prophecy of being the centre of the ancient world; attracting many visitors to the Pythian Games, a precursor of the Olympic Games. Because it was so revered throughout the Greek world, Delphi became a powerful city-state in its own right. The oracle sat at the centre of not just the city of Delphi, but the Greek World itself. No important decision was made without her consultation, and so for nearly one thousand years, the position of perhaps the greatest political and social influence in the ancient world was occupied by a woman.

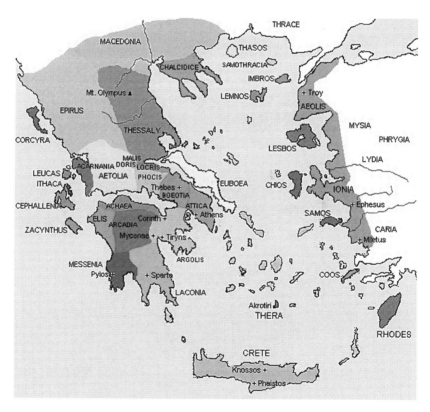

(214) The major city-states of Ancient Greece

15

THE ROMANS

ACCORDING TO LEGEND, ANCIENT ROME WAS FOUNDED BY TWO brothers, Romulus and Remus, on 21st April 753 BCE. The legend claims that in an argument over who would rule the city, Romulus killed Remus, and named the city after himself. Aeneas of Troy is also featured in this legend, and also famously, in Virgil's "Aeneid", as a founder of Rome and the ancestor of Romulus and Remus, therefore linking Rome with the grandeur and might which was once the city of Troy.

Originally a small town on the banks of the Tiber River surrounded by seven hills, Rome grew in size and strength early on through trade. The location of the city provided merchants with an easily navigable waterway on which to traffic their goods. The city was initially politically structured as a monarchy, and was ruled by seven consecutive kings, starting with Romulus himself, and ending with Tarquin. Greek culture and civilization, which came to Rome in the form of Greek

(215) Romulus and Remus, the founders of Rome, and the she-wolf

(216) A map of ancient Italy

colonies of city-states situated in southern Italy and Sicily, provided the early Romans with a model on which to build their own culture. From the Greeks they borrowed literacy and religion, as well as the fundamentals of architecture.

The Etruscans to the north, provided a model for the love of the arts and urban luxury. Etruria was also well situated for trade and the early Romans learned the necessary skills of commerce and trade from the Etruscans. As well as the Greeks, the Etruscans also played a key role in the development of the Roman culture. Right from the beginning, the Romans demonstrated a talent for borrowing and improving upon the skills, concepts, tradi-

(217) An Etruscan statue

tions, and belief systems of other cultures. The Kingdom of Rome grew rapidly from what began as a trading town and ultimately developed into a prosperous city, between the 8th and 6th centuries BCE. When the last of the seven kings of Rome, Tarquin the Proud, was deposed in 509 BCE, his rival for power, Lucius Junius Brutus, reformed the political system of government and was instrumental in establishing the Roman Republic. Interestingly, this Brutus was a distant ancestor of Marcus Junius Brutus, who was a senator of Rome, and one of the leaders in the conspiracy to assassinate Julius Caesar in 44 BCE, nearly 450 years later.

The story of Tarquin the Proud and why he was overthrown is a fascinating one. Lucius Tarquinius Superbus, also known as Tarquin the Proud, was the last Etruscan King of Rome from 535 to 509 BCE, succeeding Servius Tullus. He was overthrown by the people of Rome in 509 BCE following his son Sextus Tarquinius' rape of the noble-woman Lucretia. Lucius Tarquinius Superbus was the son of Lucius Tarquinius Priscus, the fifth king of Rome; and he married the two daughters of the sixth king, Servius Tullius. He was the father of Sextus Tarquinius, and two other sons. With encouragement from his wife Tullia, Tarquin enlisted the support of Rome's patricians (powerful elite families) in overthrowing and murdering King Servius Tullius. As a result, Tarquin became King of Rome in 535 BCE, and refused to bury the dead Servius Tullius; he also put several pro-Servius senators to death. He never replaced the slain senators; significantly reduced the size of his government; and never sought any advice from the remaining Senate. He had the nearby Latin King Turnus Herdonius killed after discovering a plot by Turnus to murder him; and in the process Tarquin took control of the Latin kingdom and integrated the Latins into his army, before going to war with the Volsci tribes east of Rome, and defeating them, gaining even more territory.

Tarquin built a magnificent temple dedicated to the supreme god Jupiter, on the Capitoline Hill in the centre of Rome. He used his son Sextus as part of a defection ploy to capture the Latin city of Gabii, 18 kilometres east of Rome; with Sextus "defecting" to Gabii and then killing or banishing all of its leading men, and compelling the city to surrender. Tarquin was very active militarily, politically and diplomatically, demonstrating a ruthless and yet an effectively efficient way of securing and further enhancing his power, and that of Rome. He made peace with the Italic tribes of the Aequi; renewed a peace agreement with the Etruscans; defeated the Sabine tribes; and estab-lished colonies at the Latin cities of Signia and Cerceii. In 509 BCE, Tarquin embarked on a military campaign against the Rutuli tribes, mainly as a distraction to the unhappiness of the people of Rome, who were becoming wary of his ruthless and cruel manner of ruling. However, while Tarquin was away from Rome leading his army into battle; his son Sextus Tarquinius raped Lucretia, the highly admired and revered wife of his cousin Lucius Tarquinius Collatinus. As a result of this unforgiveable brutal act, Lucretia was so humiliated that she committed suicide. Her grieving husband Collatinus, Lucretia's father Spurius Lucretius Tricipinius, Marcus Junius Brutus, and Publius Valerius Publicola, all swore a sacred oath to the gods that they would expel the king and his entire family from Rome, and do away with the monarchy in the process. Sextus was assassinated after fleeing to Gabii, and the Romans, having had an absolute "gutful" of monarchy, decided to create a Roman Republic system of government; electing Brutus and Collatinus as its first Consuls.

In antiquity, the rape of Lucretia was one of the most significant occurrences of violence against women. Her rape would mark the beginning of the Roman Republic, forever affecting the future of Rome and all of its people. For centuries, the rape of Lucretia was to become the topic of many poems, artworks and operas. To many modern historians, Lucretia is considered to be a mythological figure. Despite this belief, her existence was written about by the ancient historians Livy, in "Ab Urbe Condita Libri" and by Dionysius of Halicarnassus in "Roman Antiquities Book IV"; suggesting Lucretia has historical merit. According to the account of Dionysius; the story of Lucretia begins with Sextus, the eldest son of King Tarquin (Tarquinius), who was sent by his father to a city called Collatia to perform military services. In Collatia he stayed at the house of Tarquin's cousin, Lucius Tarquinius Collatinus. Lucretia was the wife of Collatinus. Sextus saw Lucretia as excelling above all the Roman women in beauty and in virtue, and decided he would seduce her. While lodging at Lucretia's home, Sextus woke late in the night and went to the room where he knew Lucretia slept. Careful not to awake her slaves who slept by her door, he entered her room with his sword in his hand. He woke her, told her his name and insisted she be silent and remain in the room; threatening to kill her if she attempted either to escape or to cry out. With no immediate source of aid, Lucretia was forced to listen to the strange proposition of Sextus as he said; "If you will consent to gratify me, I will make you my wife, and with me you shall reign, for the present, over the city my father has given me, and, after his death, over the Romans, the Latins, the Tyrrhenians, and all the other nations he rules; for I know that I shall succeed to my father's kingdom, as is right, since I am his eldest son."

(218) The end of the Roman Monarchy was triggered by Lucretia's rape

Sextus found Lucretia to not be moved to act by either fear or death or the declaration of his love. He then vowed to kill her and then slay one of her slaves, and having laid both their bodies together, said that he would state that he had caught Lucretia misbehaving with the slave and had punished her to avenge the dishonour of her behaviour to her family. He went on to threaten her by saying that her death would be attended with shame and her body would be deprived of both a burial and every other customary rite. In Ancient Rome, wool-weaving represented the archetype of ideal feminine behaviours, held by all Romans as a symbol of their devotion to sexual virtues such as chastity and modesty. Lucretia, being a chaste and sexually moral woman as represented by the symbolic meaning of wool-weaving she partook in, allowed Sextus to make his advances. By succumbing to his advances, Lucretia was acting in a moral manner. This is because the idea of being portrayed as being an unchaste woman was a fate worse than sexual abuse and death. In the morning, Sextus returned to his camp in nearby Ardea to continue his military efforts. Lucretia travelled to Rome to meet her father, carrying a dagger hidden under her cloak. She greeted him and demanded he summon all those he could. Then, in response to his hasty and urgent summons, the most prominent men of Rome had come to his house as she desired; she began at the beginning and told them all that had happened. After her sorrowful story, she drew the dagger she was keeping concealed under her robes, and plunged it deep into her breast with a single stroke which pierced her heart, and quickly died in her father's arms. This dreadful scene struck the Romans who were present with so much horror and compassion that they all cried out with one voice that they would rather die a thousand deaths in defence of their liberty than suffer such outrages committed by tyrants. Publius Valerius, a Roman aristocrat, was then sent to inform Lucretia's husband of the misfortune.

In this account given by the historian Dionysius, Lucretia asked those around her to avenge her by the means they could find necessary. She did not give directions on how to avenge her, and thus those that she left behind were responsible for the ensuing acts. The ensuing acts were therefore carried out in high honour of Lucretia, rather than any commands made by her. Lucretia found herself to not have sinned, but still be worthy of punishment; a trait likely symbolic of her dedication to chastity and a high moral code. Brutus, the nephew of King Tarquin and close friend of Collatinus, who was Lucretia's husband, led the battle against Sextus to avenge Lucretia's death. Brutus was present when Collatinus was told of Lucretia's death. With this war led by Brutus and Collatinus, alongside numerous Romans deeply disturbed and upset by Lucretia's sexual assault, the tyrant-led monarch of Rome was conquered. Sextus was exiled to Gabii where he was murdered by the civilians there for his past wrongdoings to them. Collatinus and

Brutus were met with rejoice for having successfully exiled Sextus and all his tyrant family members. The Monarchy of Rome collapsed with the revolt powered by Lucretia's death. The collapse of the Roman Monarchy was the beginning of the Roman Republic. Lucretia's death changed the entire political and social structure of Rome.

That same year, Tarquin approached the city-states of Tarquinii and Veli for help with restoring him to the throne; but the Etruscan army was defeated by the Romans at the Battle of Silva Arsia in 509 BCE, during which Tarquin's son Aruns was killed. Lars Porsena, the King of Clusium, decided to assist Tarquin, briefly occupying the city of Rome before deciding to withdraw. In 496 BCE, Tarquin launched one last attempt to retake Rome, but his army of Latins and Etruscans was defeated at the Battle of Lake Regillus. Defeated for the last time, Tarquin fled to the court of King Aristodemus of Cumae, where he died a year later.

The Republic System of Roman government was unique. Two Consuls were from a small group of candidates nominated by the Senate, and elected by the citizens; they governed Rome for one year, in consultation with the Senate. The idea of having two Consuls who only governed for one year, was to ensure that neither of them gained too much power; they were there to govern on behalf of the people. Usually one Consul was a highly skilled military general and the other was a competent government administrator. The 300 members of the then highly respected and all-powerful Senate were members of the wealthy ruling class, called patricians. The remainder of the citizens of Rome were called plebeians. Then there were slaves, most of whom came from conquered territories and performed most of the hard physical or menial work such as farm labourers, construction, the mines, and servants. During the time of the Republic, approximately 5% of citizens were patricians and 95% were plebeians.

Though the city of Rome owed its prosperity to trade in the early years, it was Roman warfare which would make it a powerful force to be reckoned with in the ancient world. The wars with the North African city-state of Carthage, known as the Punic Wars (264-146 BCE), consolidated Rome's power and helped the Romans grow in size, wealth, and prestige. Rome and Carthage were rivals in securing resources, trade, and territory in the Western Mediterranean. With Carthage finally defeated, Rome held virtually complete dominance over the region; though there were still ongoing incursions and raids by pirates which prevented Rome from having complete control of the sea. As the Roman Republic grew in power and prestige, the city of Rome began to suffer from the effects of corruption, greed, and the over-reliance on foreign slave labour. Gangs of unemployed Romans, put out of work by the influx of slaves brought in as a result of ever-increasing territorial conquests, hired themselves out as mercenaries and thugs, to do the violent bidding for whichever wealthy senator or patrician would pay them. The patrician wealthy elite of the city became even richer at the expense of the lower working class plebeians. Working class plebeians were so outraged by wealth and income inequality that they threatened a violent revolt unless the Senate agreed to share power with them. Eventually, after unrelenting pressure and violence, the Senate agreed to the establishment of the Plebeian Assembly, whereby plebeians could be elected by the people as Tribunes, and share power with the Senate. The Plebeian Assembly could propose new laws before the Senate, and could veto any laws or policies initiated by the Senate; so it did actually possess real and effective power.

(219) Hannibal meets Scipio during the Second Punic War which started in 218 BCE

In the 2nd century BCE, the Gracchi brothers, Tiberius and Gaius, two elected plebeian Roman Tribunes, led a movement for land and political reforms; effectively wanting to redistribute portions of fertile farming land in Italy from the large estates held by wealthy patricians, to smaller plots of land for plebeian farmers and soldiers. Though both of the brothers were killed during this noble cause, their efforts did lead to legislative reforms, and as a result, the rampant excessive corruption of the Roman Senate was curtailed. With relative political stability, the Republic underwent a period of flourishing prosperity. Even so, Rome found itself deeply divided across class lines. The ruling class of patricians called themselves "optimates" (the best men) while the lower plebeian classes were known as the "populares" (the people). These names were applied simply to those who held a certain political ideology; they were not strict political parties, nor were all of the patricians optimates, and all of the plebeians populares.

In general, the optimates held on to conservative traditional political and social values which favoured the power of the Senate of Rome and the prestige and superiority of the patrician ruling class. The populares on the other hand, generally favoured progressive reform and the further democratization of the Roman Republic. These sharply opposing ideologies were to clash in the form of three men who would unwittingly bring about the end of the Roman Republic. Marcus Licinius Crassus and his political rival, Gnaeus Pompeius Magnus (Pompey the Great) joined with another younger upcoming politician and military officer, Gaius Julius Caesar, to form what historians call the First Triumvirate of Rome. Crassus and Pompey both held the optimate political ideology, while Caesar was a populare. The three men who were equally ambitious and vying for power, were able to keep each other in balance and check, while helping Rome to continue its prosperity. Crassus was by far the richest man in Rome, and was corrupt to the point of forcing wealthy citizens to pay him "safety protection money." If the citizen paid, Crassus would not burn down that person's house, but if no money was paid, the fire would be lighted and Crassus would then charge a fee to send his "personal fire brigade" to put the fire out. Although the motive behind this

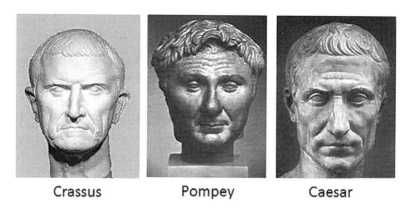

Crassus   Pompey   Caesar

(220) The First Triumvirate of the Roman Republic ruled from 60 to 53 BCE

practice was clearly corrupt, Crassus did create and finance the first fire department in Rome, which ultimately proved to be of great value to the city.

Both Pompey and Caesar were outstanding generals who, through their respective military conquests, (Pompey in the east and Caesar in the west) brought enormous additional territory, prestige, and wealth to Rome. Pompey even married Caesar's daughter Julia, a move designed to tighten their political alliance. Though the richest man in Rome, Crassus was envious of Pompey and Caesar, and craved the same military success and respect. In 53 BCE he led a sizeable army and attacked Rome's only superpower rival at that time, the eastern Kingdom of Parthia; but he was badly defeated at the Battle of Carrhae in modern-day Turkey, where he was brutally killed by having molten gold poured down his throat, when ceasefire negotiations broke down. With Crassus dead, the First Triumvirate disintegrated, and Pompey and Caesar then declared war on each other; igniting a civil war in the process. Pompey tried to eliminate his rival through legal means and had the Senate order Caesar back to Rome to stand trial on charges of treason and corruption. Caesar had been in Gaul (modern-day France and Spain) with his loyal army for eight years, conquering large amounts of territory, accumulating resources and slaves for the benefit of Rome, even though he substantially enriched himself as well. Instead of returning to the city to face the charges and be humiliated, Caesar and his sizeable, fiercely loyal army crossed the Rubicon River in northern Italy and entered the city of Rome in 49 BCE. He refused to answer the charges laid against him by the Senate and instead directed his focus towards eliminating Pompey as a rival. Pompey and Caesar met in the Battle of Pharsalus in Greece in 48 BCE, where Caesar's numerically inferior army defeated Pompey's larger army. Pompey himself survived and fled to Egypt, expecting to find sanctuary there, but instead, he was assassinated upon his arrival by being beheaded along the shores of Alexandria. His assassins incorrectly believed they were gaining favour with Caesar by carrying out this brutally heinous act of violence; and presented Pompey's head to Caesar when he arrived in Alexandria shortly after. Caesar was outraged by this barbarous act inflicted upon a great general of Rome; and arranged for Pompey's murderers to be brutally executed. News of Caesar's great victory against overwhelming odds at Pharsalus had spread quickly, and many friends and allies of Pompey swiftly sided with Caesar, believing he was favoured by the gods.

Julius Caesar was now the most powerful man in Rome. He effectively ended the period of the Republic by having the Senate proclaim him Dictator of Rome. His popularity among the people was enormous and his efforts to create a strong and stable central government meant even greater prosperity for Rome. In 44 BCE, on the Ides (middle) of March, at the height of his power, a day before he was to lead a large army to attempt to conquer the Kingdom of Parthia in the east, he was assassinated by a group of Roman senators. The conspirators, led by Brutus and Cassius, feared that Caesar was becoming so powerful he was planning to introduce the dreaded monarchy back to Rome, and that he would then become the King and abolish or significantly reduce the power of the Senate. Following his death, his right-hand man and cousin Mark Antony (Marcus Antonius) joined forces with Caesar's great-nephew and heir Octavian (Gaius Octavius), and Caesar's friend, Marcus Lepidus, to defeat the forces of Brutus and Cassius at the Battle of Philippi in 42 BCE.

Octavian, Antony and Lepidus then proceeded to form the Second Triumvirate of Rome, but as with the first, these men were also weary

Octavian   Antony   Lepidus

(221) The Second Triumvirate of the Roman Republic ruled from 43 to 32 BCE

of each other and equally ambitious. Lepidus was effectively neutralized when Antony and Octavian agreed he should have Hispania (modern day Spain and Portugal) and the northern African provinces to rule over, thereby keeping him away from the centre of power in the city of Rome. It was further agreed that Octavian would rule the western Roman lands including the city of Rome, and Antony would rule the resource-rich eastern Roman provinces. Antony's involvement with the Egyptian Queen Cleopatra VII however, upset the balance of power Octavian had hoped for; so as a result, after a build-up of ongoing animosity and friction, the two went to war. Antony and Cleopatra's combined forces were defeated at the crucial Battle of Actium in 31 BCE. Shortly after, in the city of Alexandria, as we have seen in the earlier chapter on Egypt, Cleopatra and Mark Antony took their own lives. Out of this, Octavian emerged as the sole power in Rome. In 27 BCE he was granted extraordinary powers by the Senate, and took the name of Augustus, the first Emperor of Rome. At this point the Roman Republic ceased to exist and the period of the Roman Empire began.

Probably the two most powerful senators at this time were Marcus Tullius Cicero and Marcus Porcius Cato (also known as Cato the Younger), who both arguably did more than any other senator to try and preserve the optimate patrician traditional conservative principles, and protect the Roman Republic from the threat of monarchy or dictatorship; and particularly from the threat they believed the powerful Julius Caesar posed, with his radical populare beliefs, and his widespread popularity among the plebeian common people. The best way to gauge the impact these two senators had at the time, is to look at some of their quotations. As much as they led the opposition to Julius Caesar in the Senate Chamber, there is no evidence to suggest they knew about, or were involved in, the assassination of Julius Caesar.

*(222) The great orator and lawyer Cicero in the Roman Senate:
from the series - ROME*

Marcus Tullius Cicero was a Roman politician, lawyer, and orator who lived from 106 to 43 BCE. He was one of the very few "new men" in Rome, meaning the first man in his family to become a senator, and gain the highest office of consul. He is best known for his philosophical works, love of Greek literature, and his devotion to the Republic. Cicero was invited to join the powerful First Triumvirate formed by Caesar, Crassus and Pompey, but he refused, and instead became an opponent of Caesar and everything he represented. Cicero had this to say about Caesar's impact on the people: "Do not blame Caesar, blame the people of Rome who have so enthusiastically acclaimed and adored him and rejoiced in their loss of freedom and danced in his path and gave him triumphal processions. Blame the people who hail him when he speaks in the Forum of the 'new wonderful good society' which shall now be Rome, interpreted to mean more money, more ease, more security, more living fatly at the expense of the industrious." Regarding Caesar's motives, Cicero said: "Unscrupulous behaviour of Gaius Caesar who disregarded all divine and human law for the sake of pre-eminence on which he had deludedly set his heart. If a man insists on outcompeting everyone else, then it is hard for him to respect the most important aspect of justice; equality. Men of this type put up with no restraint by way of debate or due process; they emerge as spendthrift faction leaders, because they wish to acquire as much power as possible and would sooner gain the upper hand through force than fair dealing."

(223) The death of Cicero in 43 BCE was ordered by
Mark Antony and Octavian

Cicero died a violent death when he fell out with Mark Antony during the time of the Second Triumvirate. Cicero was not only a long-standing critic of Julius Caesar, but also of Mark Antony; believing that both men did not respect the Roman Republic and possessed personal ambitions to become dictatorial tyrants. Mark Antony had insisted that Cicero be put on a proscriptions list, which was a list of Romans who the triumvirate leaders wished to assassinate either because of their political opposition to them, or to confiscate their wealth and assets, in order to financially support their armies and personal ambitions. According to the historian Plutarch, Octavian had argued for two days with both Mark Antony and Lepidus that Cicero be left alone, but to no avail; he yielded to Antony on the third day, thereby sealing Cicero's fate. Plutarch also claims that Cicero had decided to flee Rome, but had second thoughts on his way to Macedonia and returned to Rome. Fearful that he would be taken captive and tortured, he ordered his servants to take him to his villa in the coastal city of Caieta, south of Rome. Cicero's day of reckoning was on the 7th December 43 BCE. The historian Plutarch, describes what happened on that day: "But meantime his assassins came to the villa, Herennius a centurion, and Popillus a tribune, who had once been prosecuted for parricide and defended by Cicero; and they had helpers. After they had broken in the door, which they found closed, Cicero was not to be seen, and the inmates said they knew not where he was. Then, we are told, a youth who had been liberally educated by Cicero, and who was a freedman of Cicero's brother Quintus, Philologus by name, told the tribune that the litter was being carried through the wooded and shady walks towards the sea. The tribune, accordingly, taking a few helpers with him, ran round towards the exit, but Herennius hastened on the run through the walks, and Cicero, perceiving him, ordered the servants to set the litter where they were. Then he himself, clasping his chin with his left hand, as was his wont, looked steadfastly at his slayers, his head all squalid and unkempt, and his face wasted with anxiety, so that most of those that stood by covered their faces while Herennius was slaying him. For he stretched his neck forth from the litter and was slain, being then in his sixty-fourth year. Herennius cut off his head, by Antony's command, and his hands – the hands with which he wrote the Philippics. For Cicero himself entitled his speeches against Antony 'Philippics', and to this day the documents are called Philippics." After his death, Cicero's head and hands were sent back to Rome, where Mark Antony inspected the head to satisfy himself that Cicero was in fact dead, and ordered that

Cicero's hands be fastened to the Rostra, a large platform in the centre of Rome where orators spoke.

Marcus Porcius Cato, also known as "Cato the Younger", born 95 BCE and died 46 BCE, was a conservative Roman Senator in the period of the late republic. He was a highly effective orator and a follower of the Stoic branch of Greek philosophy; which stated that the best way to live life was with minimal material possessions and extravagance, and control of all emotions. He is remembered for his stubbornness and tenacity, especially in his lengthy conflict with Julius Caesar, as well as his complete immunity to bribes and other forms of corruption. He lived and breathed moral integrity, and was a fierce protector of the values of the Republic, distinguishing himself in the Senate Chamber by wearing a black toga, instead of the customary white the others wore. So distressed was he, when it became apparent to him that the Roman Republic was about to fall under Caesar; he committed suicide by sticking a dagger into his chest, and tearing the wound with his own hands. According to the historian Plutarch, Cato's last words, while committing suicide, were: "Now, I am the master of myself." Referring to what he believed was the extreme dictatorial power of Julius Caesar, he stated: "I would not be beholden to a tyrant, for his acts of tyranny. For it is but usurpation in him to save, as their rightful lord, the lives of men whom he has no title to reign." Deeply disappointed at the Senate appeasement of Caesar, he said: "In doing nothing, men learn to do evil." Cato made this statement, when the Senate was debating whether to charge Caesar with treason for his actions while engaged in his eight-year conquest of Gaul: "I know not what treason is, if sapping and betraying the liberties of a people be not treason." On the vital importance of the responsibility of every citizen, in preserving the integrity of the Roman Republic, he said: "Some have said that it is not the business of private men to meddle with government, a bold and dishonest saying, which is fit to come from no mouth but that of a tyrant or slave. To say that private men have nothing to do with government is to say that private men have nothing to do with their own happiness or misery; that people ought not to concern themselves whether they be naked or clothed, fed or starved, deceived or instructed, protected or destroyed."

(224) The Stoic senator Cato the Younger: from the series - ROME

The assassination of Julius Caesar was a major turning point in the history of Rome. How did it unfold in total secrecy; who was involved in the meticulous planning of it; and why were the conspirators so determined to end the life of Julius Caesar?

In February 44 BCE, with the full backing of a fearful and compliant Senate, Julius Caesar declared himself The Dictator of Rome. This act, along with his continual effort to adorn himself with the trappings of power, turned many in the Senate against him, in spite of his immense popularity among the common people. A conspiratorial group of up to 60 senators within the 300 strong Senate concluded that the only resolution to the problem was to assassinate Caesar. Nicolaus of Damascus wrote this account of the murder of Caesar, only a few years after the event. Even though he was not present when the assassination took place, he had the opportunity to speak with many of those who were present. He was a friend of King Herod the Great of Judaea, and gathered his information from what most historians believe to be reliable sources.

With regard the planning of the assassination, Nicolaus of Damascus said: "The conspirators never met openly, but they assembled a few at a time in each other's homes. There were many discussions and proposals, as might be expected, while they investigated how and where to execute their design. Some suggested that they should make the attempt as he was going along the Sacred Way, which was one of his favourite walks. Another idea was for it to be at the elections during which he had to cross a bridge to appoint the magistrates in the Campus Martius; they should draw lots for some to push him from the bridge and for others to run up and kill him. A third plan was to wait for a coming gladiatorial show. The advantage of that would be that, because of the show, no suspicion would be aroused if arms were seen prepared for the attempt. But the majority opinion favoured killing him while he sat in the Senate, where he would be by himself since non-Senators would not be admitted, and where the many conspirators could hide their daggers beneath their togas. This plan won the day."

Good friends of Julius Caesar were hearing constant rumours of an attempt to assassinate him, and tried to persuade him to get more security, and wanted to stop him from going to the Senate House on that fateful day. This is what Nicolaus had to say about Brutus (one of the lead-conspirators) persuading Caesar to ignore his apprehensions: "His friends were alarmed at certain rumours and tried to stop him going to the Senate-house, as did his doctors, for he was suffering from one of his occasional dizzy spells. His wife, Calpurnia, especially, who was frightened by some visions in her dreams, clung to him and said that she would not let him go out that day. But Brutus, one of the conspirators who was then thought of as a firm friend, came up and said, 'What is this, Caesar? Are you a man to pay attention to a woman's dreams and the idle gossip of stupid men, and to insult the Senate by not going out, although it has honoured you and has been specially summoned by you? But listen to me, cast aside the forebodings of all these people, and come. The Senate has been in session waiting for you since early this morning.' This swayed Caesar and he left."

Before going to the Senate, Julius Caesar made what was unbeknownst to him, his last religious sacrifice. Nicolaus of Damascus had this to say: "Before he entered the chamber, the priests brought up the victims for him to make what was to be his last sacrifice. The omens were clearly unfavourable. After this unsuccessful sacrifice, the priests made repeated other ones, to see if anything more propitious might appear than what had already been revealed to them. In the end they said that they could not clearly see the divine intent, for there was some transparent, malignant spirit hidden in the victims. Caesar was annoyed and abandoned divination till sunset, though the priests continued all the more with their efforts."

Meanwhile, in spite of the bad omen, Caesar was going to meet his destiny. Nicolaus of Damascus continues: "Those of the murderer's present were delighted at all this; though Caesar's friends asked him to put off the meeting of the Senate for that day because of what the priests had said, and he agreed to do this. But some attendants came up, calling him and saying that the Senate was full. He glanced at his friends, but Brutus approached him again and said, 'Come, good sir, pay no attention to the babblings of these men, and do not postpone what Caesar and his mighty power has seen fit to arrange. Make your own courage your favourable omen.' He convinced Caesar with these words, took him by the right hand, and led him to the Senate which was quite near. Caesar followed in silence."

Nicolaus of Damascus then proceeds to describe the frenzied attack in the Senate, as it was recounted to him by many witnesses that day: "The Senate rose in respect for his position when they saw him entering. Those who were to have a part in the plot stood near him. Right next to him went Tillius Cimber, whose brother had been exiled by Caesar. Under pretext of a humble request on behalf of this brother, Cimber approached and grasped the mantle of his toga, seeming to want to make a more positive move with his hands upon Caesar. Caesar wanted to get up and use his hands, but was prevented by Cimber and became exceedingly annoyed. That was the moment for the men to set to work. All quickly unsheathed their daggers and rushed at him. First Servilius Casca struck him with the point of the blade on the left shoulder a little above the collar-bone. He had been aiming for that, but in the excitement he missed. Caesar rose to defend himself, and in the uproar Casca shouted out in Greek to his brother. The latter heard him and drove his sword into the ribs. After a moment,

Cassius made a slash at his face, and Decimus Brutus pierced him in the side. While Cassius Longinus was trying to give him another blow he missed and struck Marcus Brutus on the hand. Minucius also hit out at Caesar and hit Rubrius in the thigh. They were just like men doing battle against him. Under the mass of wounds, he fell at the foot of Pompey's statue. Everyone wanted to seem to have had some part in the murder, and there was not one of them who failed to strike his body as it lay there, until, wounded thirty-five times, he breathed his last."

(225) The assassination of Julius Caesar in 44 BCE
by a conspiracy of 40 senators

Following the Battle of Actium in 31 BCE, Gaius Octavian, Julius Caesar's great nephew and heir, became the first Emperor of Rome and took the name Augustus Caesar. Although Julius Caesar is often regarded as the first emperor of Rome, this is not correct; he never held the title of "Emperor" but instead he was "Dictator", a title the Senate had no choice other than to give to him because of the immense

military and political power Caesar held at that time. In contrast, the
Senate willingly and eagerly granted Augustus the title of emperor,
lavishing praise, power, and prestige on him at the same time, because
he had been instrumental in destroying Rome's enemies, and brought
much-needed stability. Augustus ruled the Roman Empire from 31
BCE until 14 CE when he died. In that time, as he himself said, he
"founded Rome a city of clay but left it a city of marble." Augustus
reformed the laws of the empire; secured Rome's borders; and initiated
vast building projects, many of which were overseen by his loyal,
faithful, close friend since childhood, General Marcus Agrippa, who
amongst other things, oversaw the building of the Pantheon in the
centre of Rome. Augustus secured the empire a lasting name as one of
the greatest political, military, and cultural powers in history. The "Pax
Romana" (Roman Peace) which he initiated, was a time of peace and
prosperity that would last for over 200 years.

(226) Emperor Augustus   (227) Emperor Nero

After the death of Augustus, power passed to his heir, stepson
Tiberius, who ruled from 14 to 37 CE, and continued many of
Augustus' policies, but lacked the political will, strength of character
and vision which so defined Augustus. This trend of weaker and
distracted emperors would continue with those that followed:

Caligula (37-41 CE), Claudius (41-54 CE), and Nero (54-68 CE). These first five Roman emperors are referred to as the Julio-Claudian Dynasty, in recognition of the two family names they descended from, either by birth or adoption; Julius and Claudius. Although Caligula has become notorious in history for his depravity, blood-thirstiness, and apparent insanity, his early rule was admirably competent, as was that of his successor Claudius, who expanded Rome's power and territory in Britain. Nero's rule was tyrannical and cruel. Caligula and Claudius were both assassinated while in power; Caligula by his elite Praetorian Guard, and Claudius apparently by his wife. Nero's suicide ended the Julio-Claudian Dynasty and initiated the period of social unrest in Rome known as "The Year of the Four Emperors."

(228) The assassination of Emperor Caligula by his Praetorian Guards in 41 CE

These four emperors were Galba, Otho, Vitellius, and Vespasian. After Nero's suicide in 68 CE, Galba assumed rule in 69 CE and immediately proved to be unfit to cope with the immense responsibility and power bestowed upon him. He was assassinated by the Praetorian Guard. On the very day of his death, Otho took over as emperor, and according to the ancient sources, was expected to become a capable and competent ruler. His General Vitellius however, desperately wanted power for himself and so orchestrated a brief civil war which resulted in Otho's suicide and Vitellius' ascent to the imperial throne. Vitellius proved no more fit to rule than Galba had been; as it appears he constantly engaged in luxurious and debauched entertainments and feasts, instead of concentrating on his duties and responsibilities of leading and managing the empire. The Roman Legions sparked an insurrection, disgusted at the immorality of Vitellius, and marched into Rome, demanding that their much-admired general Vespasian, take over as emperor. Whilst in Rome, a small group of soldiers murdered Vitellius, and Vespasian became emperor, ruling from 69 to 79 CE. Vespasian founded the Flavian Dynasty which was characterized by massive building projects, economic prosperity, and an ongoing expansion of the empire. It was under the rule of Vespasian that construction of the famous Coliseum amphitheatre of Rome commenced; which his son Titus, who ruled from 79 to 81 CE, would complete.

(229) *Emperor Vespasian*   (230) *The Coliseum was completed*
in 80 CE

It was during the reign of Titus that the volcanic eruption of Mount Vesuvius occurred in 79 CE, utterly destroying and burying the cities of Pompeii and Herculaneum. The ancient sources are unanimous in their praise of Titus' handling of the Vesuvius disaster as well as the great fire of Rome in 80 CE. Titus died of a fever in 81 CE and was succeeded by his brother Domitian, who ruled from 81 to 96 CE. Domitian further expanded and secured the border frontiers of Rome; repaired the damage to the city caused by the great fire; continued the spectacular building projects initiated by his brother; and further improved the economy and standard of living of the empire. Nevertheless, his dictatorial autocratic methods and "common-people" policies made him unpopular with the Senate, and he was assassinated in 96 CE.

Domitian's successor was his advisor Nerva, who founded the Nervan-Antonin Dynasty which ruled Rome from 96 to 192 CE. This period was marked by increased prosperity owing to the rulers known as "The Five Good Emperors of Rome." Between 96 and 180 CE, five exceptionally gifted men ruled in sequence, and brought the Roman Empire to the peak of its power and glory. They were:

- Nerva (96-98 CE)
- Trajan (98-117 CE)
- Hadrian (117-138 CE)
- Antoninus Pius (138-161 CE)
- Marcus Aurelius (161-180 CE)

Under their leadership, the Roman Empire grew stronger, more stable, and expanded in size and scope. Lucius Verus and Commodus were the last two emperors of the Nervan-Antonin Dynasty. Verus was co-emperor with Marcus Aurelius until his death in 169 CE, and from most accounts was quite docile and ineffective. Commodus, who reigned from 180 to 192 CE, was Aurelius' son and successor, and was one of the most disgraceful emperors in the history of Rome; universally depicted as excessively indulging himself and his whims, at the expense of the empire and its people, in a cruel manner. He was strangled by his wrestling partner in his bath in 192 CE, ending the Nervan-Antonin Dynasty; leading to the rise in power of the prefect Pertinax, who appears to have masterminded the assassination of Commodus.

In my opinion, one of the greatest Roman emperors of all was Marcus Aurelius. Let us examine his achievements. Marcus Aurelius was chosen by Emperor Hadrian to be his eventual successor. In 161 CE Aurelius took control of the Roman Empire, along with his brother Verus. At this time, war and disease threatened Rome. Marcus Aurelius was born on 26th April 121 CE in the city of Rome. Known for his philosophical interests, Aurelius became one of the most respected emperors in Roman history. He was born into a wealthy and politically prominent family. He grew up as a dedicated and motivated student, learning Latin and Greek. But his greatest intellectual pursuit was Stoicism, a philosophy that emphasized fate, modesty, reason, and self-control. A philosophical treatise called "Discourses" written by a former slave and Stoic Greek philosopher Epictetus, had a significant influence over Marcus Aurelius. His serious, intense, and hard-working nature came to the attention of Emperor Hadrian. After his earlier choice for a successor died, Hadrian adopted Titus Aurelius Antoninus (who would be known as Emperor Pius Antonius) to succeed him as emperor. Hadrian also arranged for

(231) Emperor Marcus Aurelius

Antoninus to adopt Marcus Aurelius and the son of his earlier successor. At around the age of 17, Marcus Aurelius became the son of Antoninus. He worked alongside his adopted father while learning the ways of government and public affairs. In 140 CE, Aurelius became a leader of the Senate. As the years passed, he received more responsibilities and official powers, and evolved into a strong source of support and counsel for Antoninus. Aurelius also continued his philosophical studies and developed an interest in law.

In addition to his flourishing career, Aurelius appeared to have a contented personal life. He married Faustina, the emperor's daughter, in 145 CE, whom he genuinely loved. Together, they had many children, though some did not live for long; the two best known were their daughter Lucilla and son, Commodus. After his adoptive father died in 161 CE, Aurelius continued his rise to power by becoming joint emperor of Rome. While some sources indicate that Antoninus selected only him as his successor, Aurelius insisted that his adopted brother Lucius Verus, served as his co-ruler. Unlike the peaceful and prosperous rule of Antoninus, the joint reign of the two brothers was marked by war and disease. During the 160s, they battled with the Parthian Empire for control over lands in the East. Verus oversaw the war effort while Aurelius stayed in Rome. Unfortunately, Roman soldiers returning from the East carried with them a dreadful pestilent disease which spread throughout the empire. As the Parthian War ended, the two rulers had to face another military conflict with Germanic tribes in the late 160s. Germanic barbarian tribes crossed the Danube River and attacked a Roman city, close to the northern border. After raising the necessary funds and troops, Aurelius and Verus went off to fight the invaders. Verus died in battle in 169 CE, so Aurelius persevered alone, attempting to drive the Germans back outside the border of the empire.

In 175 CE, he faced another challenge, this time for his very position as emperor. After hearing a rumour that Aurelius had died in battle, a Roman general Avidius Cassius, claimed the title of emperor for himself. This forced Aurelius to travel to the East with his army to regain control. But he did not have to fight Cassius because he was murdered by his own soldiers. Instead, Aurelius toured the eastern provinces with his wife, consolidating his authority and control in the process. Unfortunately, his wife Faustina died during this trip. While once again battling the Germanic tribes, Aurelius appointed his son Commodus as his co-ruler in 177 CE. Together they fought the northern enemies of the empire, hoping to even further extend the northern borders. This was not to eventuate however, because three years later on 17th March 180 CE, Marcus Aurelius died. His son Commodus became emperor and soon ended the northern military efforts. Unfortunately, Commodus was to taint the legacy of Aurelius because he turned out to be one of the nastiest, cruellest, depraved, and self-indulgent emperors in the history of Rome.

Marcus Aurelius is not so much remembered for the wars he waged, and for protecting the northern borders from barbarian invasions; but instead for his fair and balanced contemplative nature, and his compassionate rule and simple lifestyle driven by modesty, reason, and logic. A collection of his thoughts were published in a work called "The Meditations." This book has survived to this day, and is filled with his personal diary notes on how to best pursue a stoic life. Here are some of the most inspiring quotations from "The Meditations".

- The happiness of your life depends upon the quality of your thoughts.
- Dwell on the beauty of life. Watch the stars, and see yourself running with them.
- Everything we hear is an opinion, not a fact. Everything we see is a perspective, not the truth.
- If you are distressed by anything external, the pain is not due to the thing itself, but to your estimate of it; and this you have the power to revoke at any moment.
- When you arise in the morning think of what a privilege it is to be alive, to think, to enjoy, to love.
- The best revenge is to be unlike him who performed the injury.
- It is not death that a man should fear, but he should fear never beginning to live.
- Our life is what our thoughts make it.

- If someone is able to show me that what I think or do is not right, I will happily change, for I seek the truth, by which no one was ever truly harmed. It is the person who continues in his self-deception and ignorance who is harmed.
- If it is not right do not do it; if it is not true do not say it.
- I have often wondered how it is that every man loves himself more than all the rest of men, but yet sets less value on his own opinion of himself than on the opinion of others.
- Very little is needed to make a happy life; it is all within yourself in your way of thinking.
- The object of life is not to be on the side of the majority, but to escape finding oneself in the ranks of the insane.
- When you wake up in the morning, tell yourself: the people I deal with today will be meddling, ungrateful, arrogant, dishonest, jealous and surly. They are like this because they can't tell good from evil. But I have seen the beauty of good, and the ugliness of evil, and have recognized that the wrongdoer has a nature related to my own – not of the same blood and birth, but the same mind, and possessing a share of the divine. And so none of them can hurt me. No one can implicate me in ugliness. Nor can I feel angry at my relative, or hate him. We were born to work together like feet, hands and eyes, like the two rows of teeth, upper and lower. To obstruct each other is unnatural. To feel anger at someone, to turn your back on him; these are unnatural.
- Concentrate every minute like a Roman – like a man - on doing what's in front of you with precise and genuine seriousness, tenderly, willingly, with justice. And on freeing yourself from all other distractions. Yes, you can – if you do everything as if it were the last thing you were doing in your life, and stop being aimless, stop letting your emotions override what your mind tells you, stop being hypocritical, self-centred, irritable. You see how few things you have to do to live a satisfying and reverent life? If you can manage this, that's all even the gods can ask of you.

Pertinax governed for only three months before he too was assassinated. He was followed in rapid succession by four others in the
period known as "The Year of the Five Emperors", which culminated
in the rise of Septimus Severus to power. Severus, who ruled from
193 to 211 CE, founded the Severan Dynasty; defeated the Parthians
and conquered more territory in the east in the process. His military
campaigns in north Africa and Britain were extensive and very costly;
contributing to Rome's later financial and economic difficulties. He
was succeeded by his sons Caracalla and Geta; until Caracalla had his
brother murdered. Caracalla ruled until 217 CE, when he was assassinated by his bodyguard. It was under Caracalla's reign that Roman
citizenship was expanded to include all free men within the empire.
This law was said to have been enacted as a means of raising more
tax revenue, simply because after its passage, there were more people
recorded on the census that the central government could tax. The
Severan Dynasty continued, mainly under the guidance and manipulation of Julia Maesa; often referred to as the "empress" she was the
grandmother of emperor Alexander Severus, who ruled from 222 to
235 CE, and a very heavy influence on him; until he too was assassinated in 235 CE, plunging the empire into the utterly chaotic period
known as "The Crisis of the Third Century" which lasted from 235
to 284 CE.

This period, also known as The Imperial Crisis, was characterized
by constant civil war as various military leaders fought for control
of the empire, creating massive instability. The crisis has been further
noted by historians for widespread social unrest; economic instability
accelerated by the devaluation of the Roman currency; and finally, the
dissolution of the empire, which broke-up into three separate regions.
The empire was reunited by Aurelian (270-275 CE) whose sound
policies were further developed and improved upon by Diocletian
(284-305 CE), who established the Tetrarchy (the rule of four), to
maintain stability and order throughout the empire. Even so, the empire
was so vast that Diocletian divided it in half in 285 CE, by elevating
one of his military officers, Maximian (286-305 CE) to the position of
co-emperor. In doing this, he created the Western Roman Empire and

the Eastern Roman Empire (also known as the Byzantine Empire).
Since a leading cause of the Imperial Crisis was a lack of clarity in
succession, Diocletian decreed that successors must be chosen and
approved from the outset of an individual's rule. Two of these successors
were the generals Maxentius and Constantine. Diocletian voluntarily
retired from rule in 305 CE, and the tetrarchy dissolved as rival regions
of the empire vied with each other for dominance. Following the
death of Diocletian in 311 CE, Maxentius and Constantine plunged
the empire into another civil war.

By this stage the Christians were becoming more integrated into
the Roman Empire and were not being as overtly persecuted as they
had been prior to the rule of Constantine. The Catholic Church was
starting its long march towards the supreme control it was to ultimately
achieve after the fall of the Western Roman Empire. A major battle
was about to convert Constantine to Christianity and greatly legitimize Christianity within the empire.

In 312 CE, Constantine defeated Maxentius at the Battle of
Milvian Bridge, and became the sole emperor of both the Western

(232) The first Christian Emperor Constantine and the Battle of Milvian
Bridge in 312 CE

and Eastern Empires, ruling from 306 to 337 CE. Believing that Jesus Christ was responsible for his victory, Constantine initiated a series of laws, such as the "Edict of Milan" in 313 CE which mandated religious tolerance throughout the empire and specifically, tolerance for the faith which came to be known as Christianity. In the same way that earlier Roman emperors had claimed a special relationship with a deity to reinforce their authority and standing with the people; such as Caracalla with the god Serapis, and Diocletian with the god Jupiter; Constantine chose the figure of Jesus Christ. At the First Council of Nicea in 325 CE, he presided over a gathering to codify the Christian faith and decide on important issues such as the divinity of Jesus; and which manuscripts would be collected to form the holy book of the Bible. He stabilized the empire, revalued the currency, and reformed the military; as well as founding the city he called "New Rome" on the site of the former city of Byzantium, which came to be known as Constantinople (modern-day Istanbul).

He is known as Constantine the Great because of later Christian writers who saw him as a mighty champion of their faith, as has been noted by many historians, this honourable title could easily also be attributed to his religious, cultural, and political reforms; as well as his skill in battle and his large-scale building projects. After his death, his sons inherited the empire and fairly quickly embarked on a series of conflicts with each other which threatened to undo all the good-work Constantine had accomplished. His three sons, Constantine II, Constantius II, and Constans, divided the Roman Empire between them, but soon fell to greedy infighting over which of them deserved more. In those conflicts, Constantine II and Constans were killed. Constantius II died later, after naming his cousin Julian as his successor and heir. Emperor Julian ruled for only two years (361–363 CE) and in that time, tried to return Rome to her former glory through a series of reforms aimed at increasing efficiency in government.

As a Neo-Platonic philosopher, Julian rejected Christianity and blamed the faith and Constantine's endorsement of it, for the decline of the empire. While officially proclaiming a policy of religious tolerance, Julian systematically removed Christians from influential government positions; banned the teaching and spread of the religion; and barred Christians from military service. His death, while on campaign against the Persians, ended the family dynasty Constantine had created. Julian was the last pagan emperor of Rome, and came to be known as "Julian the Apostate" for his opposition to Christianity. After the brief rule of Jovian, who re-established Christianity as the dominant religion of the empire and replaced Julian's edicts, the responsibility of emperor fell to Theodosius I, who ruled from 379 to 395 CE. He took Constantine's and Jovian's religious reforms to their ultimate conclusion, and outlawed pagan worship throughout the empire. Theodosius also converted pagan temples into Christian churches after formally proclaiming Christianity to be the state religion of Rome in 380 CE. Theodosius I devoted so much time to promoting Christianity that he seems to have neglected his other duties as emperor, and ultimately was to become the last to rule the combined Western and Eastern Roman Empires.

(233) Emperor Julian the Apostate rejected Christianity

From 376 to 382 CE, Rome was engaged in a series of fierce battles against invading barbarian tribal Goths, known today as the Gothic Wars. At the Battle of Adrianople on 9th August 378 CE, the Roman Emperor Valens (364–378 CE) was defeated, and histo-

(234) Gothic Wars – The Battle of Adrianople in 378 CE

rians identify this event as a pivotal turning point in the ultimate decline of the Western Roman Empire. Various theories have been put forward over the centuries as to the cause of the empire's fall but even today, there is no universal agreement on what the specific factors were. The historian Edward Gibbon has famously argued in his monumental 1776 book, "The History of the Decline and Fall of the Roman Empire" that Christianity played a major role, because the new religion undermined the social normality and uniformity of the empire which paganism provided. The theory that Christianity was a root cause in the empire's fall was debated long before Gibbon. In the 5th century CE the theologian Orosius argued that Christianity did not contribute to Rome's inevitable decline. In 418 CE, Orosius claimed it was paganism itself and anti-social Pagan practices which brought about the fall of Rome. Historians have argued the following other factors as contributing to the fall of Rome:

- Political instability due to the substantial size of the empire
- The competing self-interests of the two halves of the empire
- Invasions by barbarian tribes
- Government and Emperor corruption
- Mercenary armies being engaged to fight for the Roman army

- An over-reliance on slave labour
- Massive unemployment and inflation

The ungovernable vastness of the empire, even though it was divided into two, still made it difficult to manage. The Eastern Empire flourished economically while the Western Empire struggled, and neither gave much thought to helping the other. Eastern and Western Rome more and more saw each other as rival competitors rather than part of one political unit, with both halves primarily pre-occupied with their own self-interests. The growing strength of the Germanic tribes and their constant incursions into Rome could have been dealt with much more effectively if the emperors and government had not been so self-indulgent and corrupt. Corruption was particularly rife among provincial governors who often saw their position as an opportunity to enrich themselves personally, so they often did not take barbarian tribal threats, grievances, and incursions, seriously.

By this stage, the Western Roman Army was manned largely with barbarian mercenaries who had no ethnic ties to Rome, and could no longer safeguard the borders as efficiently as the army once had. In addition, the government was encountering greater difficulties in collecting all taxes from the provinces. Added to this volatile mix, was the ongoing devaluation of the Roman currency because of rampant widespread inflation. This resulted in massive unemployment and extreme poverty among the citizens of the empire. The arrival of the Visigoths in the 3rd century CE, who were fleeing south from the invading Huns; and their subsequent rebellions, also contributed to the ultimate decline. The Western Roman Empire officially ended on 4th September 476 CE when Emperor Romulus Augustulus was deposed by the Germanic King Odoacer. Meanwhile, the Eastern Roman Empire continued on as the Byzantine Empire for another thousand years, until 1453 CE. There would be an attempt to reinvent the Western Roman Empire with the establishment and domination of Europe by The Holy Roman Empire from 962 to 1806 CE, but as we shall see in a future chapter, that construct was far removed from the Roman Empire of antiquity.

The Roman Empire has left numerous legacies that flow through into our modern world. The inventions and innovations which were generated by the Romans profoundly altered the lives of ancient people and continue to be used in cultures around the world today. Advancements in the construction of roads and buildings, indoor plumbing, aqueducts, and fast-drying cement are some examples. The calendar used in the West is derived from the one created by Julius Caesar, and the names of the days of the week and months of the year, also come from Rome. Apartment complexes (known by the Romans as 'insula'), public toilets, locks and keys, newspapers, and even socks were developed by the Romans; as were shoes, an advanced postal system (modelled after the Persians), cosmetics, the magnifying glass, and the concept of satire in language. During the time of the Roman Empire, significant advances also occurred in the fields of medicine, law, religion, government, and warfare. As we have seen, the Romans were highly adept at borrowing from, and improving upon, those inventions, innovations and concepts, which they found among the peoples and cultures that they conquered.

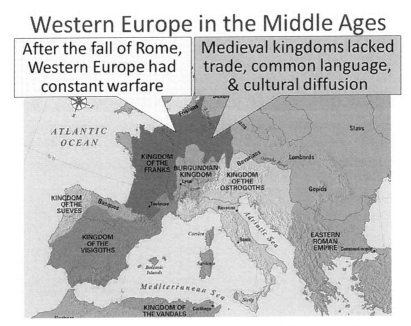

(235) European kingdoms after the collapse of the Western Roman Empire in 476 CE

16

ANCIENT CHINA

I T HAS BEEN GENERALLY ACCEPTED THAT THE CHINESE "CRADLE OF Civilization" was the Yellow River Valley which gave rise to villages sometime around 5000 BCE. In 2001 CE, archaeologists uncovered two skeletons buried in a collapsed house which was covered with a thick layer of silt deposits from the Yellow River. In the layer of deposits, archaeologists found more than 20 skeletons, an altar, a village square, pottery, and stone and jade utensils. This site was only one of many prehistoric villages in the area. From these small villages and farming communities grew centralized government; the first of which was the prehistoric Xia Dynasty (2070-1600 BCE). For many years, the Xia Dynasty was considered to be more of a myth than fact, until excavations in the 1960s and 1970s CE uncovered sites, which clearly showed evidence of the existence of this dynasty. Bronze works and tombs excavated, point to an evolutionary period of development between Stone Age villages and a recognizable structured and organized civilization.

(236) Emperor Yu of the Xia Dynasty

The Xia Dynasty was founded by Yu the Great who worked tirelessly for 13 years to control the flooding of the Yellow River which would routinely destroy the crops of farmers. He was so focused on his work that it was said he did not return home once in all those years, even though he seemed to have passed by his house on at least three occasions. This fiercely passionate dedication and work ethic inspired others to follow him. After he had finally controlled the flooding, Yu conquered the Sanmiao tribes, and was named successor by the then ruler Shun, reigning until his death. It was Yu who established the Chinese hereditary system of succession, and the concept of dynastic rule. The ruling class and the elite lived in urban clusters, while the peasant population which supported the elite's lifestyle remained largely agrarian farmers living in rural villages. Yu's son Qi, ruled after him, and power remained in the hands of the family until the last Xia ruler Jie, was overthrown by Tang, who established the Shang Dynasty which lasted from 1600 to 1045 BCE.

Tang was from the Kingdom of Shang. In 1600 BCE, he led a revolt against Jie and defeated his forces at the Battle of Mingtiao. The extravagance of the Xia court and the resultant burden on the population is thought to have led to this people's uprising. We will see in this chapter that peasant uprisings and rebellions are an integral part of the story of the ancient Chinese civilization. After his military victory, Tang then assumed leadership of the kingdom; lowered taxes; suspended the extravagant building projects begun by Jie, which were financially draining the resources of the kingdom; and ruled with such wisdom and efficiency that art and culture were freely allowed to flourish; as well as bronze metallurgy, architecture, and religion. Under the Shang Dynasty writing also developed in China for the first time.

Prior to the Shang Dynasty, the people worshipped many gods, with one supreme god Shangti, as head of the pantheon. This was a similar pattern to other cultures in existence during the same time, as we have seen. Shangti was considered to be "the great ancestor" who presided over victory in war, agriculture, the weather, and good government. The supreme god was considered to be remote and busy, prompting the people to worship lesser intermediary gods, for their day to day needs; and so the practice of ancestor worship began. When somebody died, it was thought they attained divine powers and could be called upon for assistance in times of need, similar to the Roman belief in the "parentes" ancestral spirits. This practice led to highly sophisticated and elaborate rituals dedicated to appeasing the spirits of the ancestors, which eventually included ornate burials in grand tombs filled with all the luxuries one would need to enjoy a comfortable afterlife; similar to the Egyptian burial belief system. The king, in addition to his political and military secular duties, also served as chief officiate and mediator between the living and the dead, and his rule was considered ordained by divine law. Although the famous Mandate of Heaven was developed by the later Zhou Dynasty, the idea of linking a just ruler with divine will has its roots in the religious beliefs developed during the Shang Dynasty.

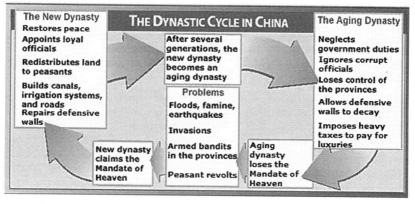

(237) The Dynastic Cycle in China and the Mandate of Heaven

In around the year 1046 BCE, King Wu (who reigned from 1046 to 1043 BCE) of the province of Zhou, rebelled against King Zhou of Shang, and defeated his army at the Battle of Muye, establishing the Zhou Dynasty in the process, which lasted from 1046 to 256 BCE. Wu rebelled against the ruling Shang after King Zhou of Shang killed Wu's older brother, cruelly and unjustly. The Mandate of Heaven was invoked by Wu and his family to legitimize the revolt, because he felt the Shang were no longer acting in the interests of the people, and so had forfeited the mandate between the monarchy and the god of law,

order, and justice, Shangti. The Mandate of Heaven was defined as
the gods' blessing given to a fair and just ruler, in the form of a divine
mandate. When the government no longer served the will of the gods,
that government and its king would be overthrown. Furthermore, it
was stipulated that there could only be one legitimate ruler of China,
and that his rule should be legitimized by his proper conduct as a
steward and custodian of the lands entrusted to him by the heavenly
gods. Rule could be passed from father to son, but only if the child
possessed the necessary virtue and dignity to rule. This Mandate of
Heaven would later be often manipulated by various rulers entrusting
succession to unworthy heirs, as we shall see.

(238) The Chinese philosopher and politician - Confucius

Under the Zhou Dynasty, culture flourished and civilization spread.
Writing was codified (formalized) and widely used; and iron metal-
lurgy became increasingly sophisticated. The greatest and best-known
Chinese philosophers and poets, Confucius, Mencius, Mot Zu, Lao
Tzu, Tao Chien, and the military strategist Sun Tzu, all came from the
Zhou period in China, and the time known as the Hundred Schools
of Thought. During the period from 481 to 221 BCE, seven provincial
states broke away from the kingdom's central rule, and fought with
each other for control of the kingdom. The seven states were Chu,
Han, Qi, Qin, Wei, Yan, and Zhao; all of whom considered themselves

to be sovereign, but none of which felt confident in claiming the Mandate of Heaven, still held by the Zhou Dynasty. All seven of the feuding states used the same tactics and observed the same rules of conduct in battle, and so as a result, none could gain advantage over the others. This situation was exploited by the pacifist philosopher Mo Ti, a skilled engineer, who made it his mission to provide each state with equal knowledge of fortification and siege ladders in the hope of neutralizing any one state's advantage over the others, and so hopefully finally ending the war. However, his efforts were unsuccessful, and between 262 and 260 BCE, the state of Qin gained supremacy over Zhao, finally defeating them at The Battle of Changping.

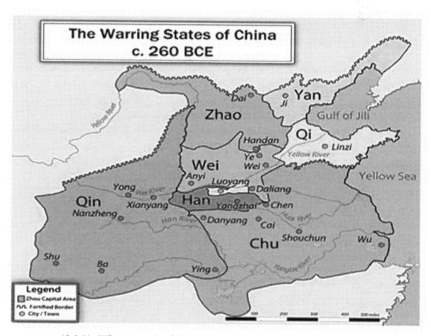

(239) The period of the Seven Warring Kingdoms lasted from 481 to 222 BCE

A Qin statesman by the name of Shang Yang, who was a great believer in efficiency and the rule of law, had recast the Qin understanding of warfare to focus on victory at any cost. Shang Yang was inspired by the famous work by Sun Tzu, "The Art of War" which detailed a radical new philosophical approach to military strategic and tactical warfare. Prior to the reforms initiated by Shang, Chinese warfare was considered to be a nobleman's game of skill with very clearly set out rules, dictated by courtesy and the perceived will of heaven. One did not attack the weak or the unprepared, and one was expected to delay engagement until an opponent had mobilized and formed ranks in the field of battle. On the other hand, Shang advocated total war in pursuit of victory, and believed in taking the enemies' forces by whatever means lay at hand. Shang's military principles were known in the province of Qin, and made use of at The Battle of Changping (262-260 BCE), where 450,000 captured Zhao soldiers were taken prisoner and then executed after the battle, giving the Qin the advantage they had been waiting for. Still, they did not make further effective use of these tactics until the rise of Ying Zheng, the King of Qin. Utilizing Shang's military strategies and directives, and with an army of significant size using iron weapons and driving chariots, Ying Zheng emerged from the Warring States conflict as the supreme leader by 221 BCE; subduing and unifying the six other states under his total rule, and proclaiming himself with the title "Shi Huangdi", the First Emperor of China; and creator of the Qin Dynasty.

Shi Huangdi (formerly Ying Zheng) established the Qin Dynasty which lasted from 221 to 206 BCE. This period was also known as the Imperial Era in China. He ordered the complete destruction of the walled fortifications which had separated the seven different states, and commissioned the building of a Great Wall along the northern border of his kingdom. Though little remains today of Shi Huangdi's original wall, construction of The Great Wall of China was begun under his rule. The historians Chris Scarre and Brian Fagan in their 1997 book "Ancient Civilizations" stated this about the Great Wall: "It stretched for over 5000 kilometres across hill and plain, from the boundaries of Korea in the east to the troublesome Ordos Desert in the west. It was an enormous logistical undertaking, though for much of its course it incorporated lengths of earlier walls built by the separate Chinese kingdoms to defend their northern frontiers in the fourth and third centuries." Shi Huangdi also strengthened the infrastructure through road building which helped to increase trade through the ease of travel.

Scarre and Fagan state: "Five trunk roads led from the imperial capital of Xianyang, each provided with police forces and posting stations. Most of these roads were of rammed-earth construction and were 15 metres wide. The longest ran southwest over 7500 kilometres to the frontier region of Yunnan. So precipitous was the countryside that sections of the road had to be built out from vertical cliff faces on projecting timber galleries."

Shi Huangdi also expanded the boundaries of his empire, built the Grand Canal in the south for better flows of irrigation channels, redistributed land to farmers, and initially was a fair and just ruler. But as his period of rule progressed, although he made great strides in building projects and military campaigns; his rule became increasingly heavy handed, particularly in his domestic policies and treatment of the educated classes. Claiming the Mandate from Heaven, he suppressed all philosophies with the exception of his favoured Legalism, which had been developed by Shang Yang and, taking the counsel of his chief advisor Liu Siu, he ordered the destruction of any history or philosophy books which did not correspond to Legalism, his family ancestry, the Kingdom of Qin, or himself. Legalism was a philosophical belief that human beings are more inclined to do wrong than right because they are motivated entirely by self-interest and require strict laws to control their impulses. The American historian William Durant states: "Since books were then written on strips of bamboo fastened with swivel pins, and a volume might be of some weight, the scholars who sought to evade the order were put to many difficulties. A number of them were detected; tradition says that many of them were sent to labour on the Great Wall, and that four hundred and sixty were put to death. Nevertheless, some of the literati memorized the complete works of Confucius and passed them on by word of mouth to equal memories."

This act, along with Shi Huangdi's suppression of general freedoms, including freedom of speech, made him progressively more unpopular with the people. The ancestor worship of the past and the land of the dead began to interest the emperor more than the welfare of the people in his kingdom; and Shi Huangdi became increasingly engrossed in what this other world consisted of, and how he might avoid travelling

(240) Emperor Shi Huangdi's terracotta army was built from 247 to 208 BCE

there. He seemed to have developed an obsession with death, and became increasingly paranoid regarding his personal safety. He was seeking to unlock the key to immortality. His desire to provide for himself an afterlife commensurate with his luxurious mortal life, led him to commission a palace to be built for his tomb, and an army of 8000 terracotta warriors created to faithfully serve and protect him in the eternal afterlife. This ceramic army buried with him, also included terracotta chariots, cavalry, a commander in chief, and assorted birds and animals. He is said to have died in 210 BCE while on a "quest of immortality"; and his key advisor Li Siu, hoping to gain control of the kingdom, kept his death a secret until he could alter his will to name Li Siu's son, Hu Hai, as heir. This plan proved to be an abject failure because the young prince was erratic and mentally unstable, executing many for no reason and initiating a widespread rebellion in the process. Shortly after the death of Shi Huangdi, the Qin Dynasty quickly collapsed as a direct result of the political and self-serving intriguing of people close to the centre of power, like Hi Hai, Li Sui, and another advisor, Zhao Gao.

With the fall of the Qin Dynasty, China was plunged into a period of chaos known as the Chu-Han Contention (206-202 BCE). Two generals emerged amongst the various fractured forces that rebelled against the Qin; Liu Bang of Han, and Xiang Yu of Chu; who fought for control of the kingdom and government. Xiang Yu, who had proven himself to be the most formidable opponent of the Qin, awarded Liu Bang the title of "King of the Han" in recognition of Liu

Bang's decisive defeat of the Qin forces at their capital city of Zianyang. It wasn't long though, before the two former allies became bitter enemies in the power struggle known as the Chu-Han Contention; until Xiang Yu negotiated the Treaty of Hong Canal, which delivered a temporary peace. Under this peace accord, Xiang Yu proposed that China be divided under the rule of the Chu in the east, and the Han in the west; but Liu Bang wanted a united Chinese Empire under Han rule, and after breaking the peace treaty, resumed military hostilities. At the Battle of Gaixia in 202 BCE, Liu Bang's great general Han Xin, trapped and defeated the forces of Chu under Xiang Yu; and Liu Bang was proclaimed as emperor under the title of Emperor Gaozu of Han. Xiang Yu was so humiliated and distressed by this outcome that he committed suicide, but his family was allowed to live, and even serve in government positions.

The new Emperor Gaozu treated all of his former adversaries with respect and magnanimously united the land under his rule. He pushed back the nomadic Xiongnu tribes, who had been making rampaging incursions into China; and made peace with the other states which had risen in rebellion against the failing Qin Dynasty. The Han Dynasty, deriving its name from Liu Bang's home in Hanzhong province, would now rule China, with the exception of a brief interruption, for the next 400 years from 202 BCE to 220 CE. The resultant peace initiated by Gaozu brought the stability that was necessary for the Chinese culture to again thrive and grow. Trade with other civilizations in central, western Asia, and the Romans began during this time, and the arts and technology increased in sophistication. The Han are considered to be the first Chinese dynasty to write their history down, but as Shi Huangdi destroyed so many of the written records of those who came before him, this claim is often disputed by historians. There is no doubt however that great advances were made under the Han in every area of culture.

One of the great achievements of the Han Dynasty was the construction of the network of trade routes called the Silk Road, which formally opened up reciprocal trade between China, central Asia, the Middle East, and Europe. Trade along the Silk Road

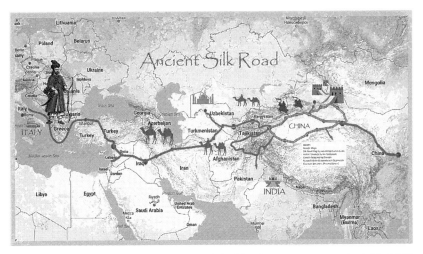

(241) *The Ancient Silk Road was established by the Han Dynasty in 130 BCE*

economic belt included a wide range of goods including fruits and vegetables, livestock, grain, leather and hides, tools, religious objects, artwork, precious stones and metals; and cultural exchanges in the form of language, religious beliefs, traditions, philosophy, and science. Commodities such as paper and gunpowder, both invented by the Chinese during the Han Dynasty, had obvious and lasting impacts on culture and history in the West. They were among the most traded items between the East and West. Paper was invented in China during the 3rd century BCE, and its use spread via the Silk Road, arriving first in Samarkand in around 700 CE, before moving to Europe through the then Islamic ports in Sicily and Spain. In addition to paper, the rich spices of the East quickly became popular in the West, and changed cuisine across Europe. Similarly, silk, a Chinese specialty, became the fashionable material of choice for the elite and nobility of the Roman Empire. Techniques for making glass migrated eastward to China from the Islamic world. The "Canon of Medicine", China's earliest written record on medicine was codified into written form during the Han Dynasty.

Emperor Gaozu reduced taxes, easing the burden of farmers and workers, and disbanded his army who nevertheless rallied to support him without delay, if they were called upon in the event of a crisis. After his death in 195 BCE, his wife, Empress Lu Zhi, installed a number of puppet kings, beginning with the crown prince Liu Ying, who reigned as Emperor Hui from 195 to 188 BCE, and very much served her interests, whilst also pursuing his own policies. These policies maintained stability and culture, enabling the greatest of the Han emperors, Wu Ti (also known as Wu the Great) to reign in the kingdom from 147 to 87 BCE. Emperor Wu the Great embarked on an ambitious enterprise of military expansion, economic growth, public works, and cultural initiatives. He sent his diplomatic emissary Zhang Qian to the West in 138 BCE, which resulted in the official opening of the Silk Road in 130 BCE. Increases in overall wealth within the kingdom led to a rise in large property estates held by the elite and nobility, with large farms and significant numbers of peasant farm labourers working this land. For the peasants who worked the land, life became increasingly difficult. In 9 CE, a high level government official and the acting regent Wang Mang, forcibly took control of the government, claiming the Mandate of Heaven for himself, and declaring an end to the Han Dynasty. Wang Mang founded the Xin Dynasty which lasted from 9 to 23 CE, and undertook extensive land reform for the benefit of peasant farmers, and a more equitable redistribution of wealth.

Emperor Wang Mang initially had enormous support from the peasant population and was opposed by the wealthy landowners. His programs and policies however were poorly conceived and executed, resulting in widespread poverty, unemployment, and resentment. Rebellious uprisings and extensive flooding of the Yellow River further destabilized Wang Mang's rule, and he was ultimately assassinated by an angry mob of peasants. This was tragic because it was on their behalf that he had seized the government and initiated his reforms in the first place. Emperor Guangwu, who ruled from 25 to 57 CE, returned the lands to the wealthy estate owners and restored order in the kingdom, maintaining the policies of the earlier Han rulers. In reclaiming lands lost under the Xin Dynasty, Emperor Guangwu was forced to spend much of his time putting down more peasant rebellions and re-establishing Chinese rule in the regions of modern-day Korea and Vietnam.

The Trung Sisters Rebellion of 39 CE, led by two peasant sisters, required over ten thousand soldiers and four years to put down. Even so, the emperor consolidated his rule and further expanded his boundaries, providing much need stability which gave rise to an increase in trade and prosperity. By the time of Emperor Zhang, who ruled from 75 to 88 CE, China was so prosperous that it was a trading partner with all of the major civilizations of that day. The Romans, under Emperor Marcus Aurelius in 166 CE, considered Chinese silk to be more precious than gold, and literally paid China whatever price it asked.

Confucianism was incorporated as the official doctrine in governments and schools throughout the empire, to foster literacy and teach its ideology. Let us now briefly examine the life of Confucius. Not a lot is known about the childhood of Confucius. He was born in the state of Lu in 551 BCE. His father was a soldier named Kong He who died when Confucius was three years old. The rest of his childhood was spent in poverty, and he was raised by his mother. In spite of his years of poverty as a child; Confucius' family was part of a growing middle class of people in China called "shi." They were not part of the nobility, but were considered above the common peasants; a type of middle-class. This gave him a different outlook on life than the majority of people. He thought that people should be promoted and rewarded based on their talents, not on what family they were born into. Confucius didn't start out as a wise teacher, he worked in a number of jobs first; including as a shepherd, and a clerk. Eventually, he came to work for the government; starting out as a governor of a small town, and working his way up until he became an advisor at the highest levels of government.

It was while he held these higher level government positions that Confucius developed his own philosophy which he taught to others. Today, his philosophy is known as Confucianism. His ideas did not become popular until years after his death, when they became the basic philosophy of the Chinese culture for over 2000 years. Some of the basic ideas of Confucianism include:

- Treat others kindly

- Have good manners and follow daily rituals
- A man should have good morals and ethics
- Family is important and ancestors are to be respected
- A true man has the qualities of integrity, righteousness, altruism, goodness, and loyalty
- One should practice moderation in all things
- A strong and organized central government is important for the welfare of the common people

Confucius quit his government job at the age of 51. He was disappointed that the leaders were not following his teachings. He then travelled throughout China for many years, teaching his philosophy in villages, towns and cities. Some of his followers wrote down his ideas in a book that would later be called "The Analects of Confucius." He died in 479 BCE of natural causes, spending his last few years in his hometown of Qufu, teaching his disciples. Confucius' teachings became the state philosophy of China during the Han Dynasty. His teachings formed the foundation of the government prestigious and rigorous civil service exams, which every citizen had to sit for and pass, if they wished to secure a government position. The government actively promoted the ideology of Confucianism because it taught the people to respect authority, and that a strong central government was

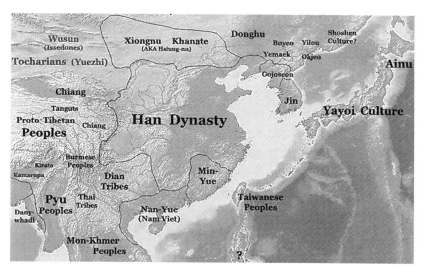

(242) The Han Dynasty ruled China from 202 BCE to 220 CE

important for political, cultural, and economic stability. The teachings of Confucius remain just as important a part of Chinese culture and government today, in the 21st century.

Some of the more memorable direct quotations from Confucius are:

- By three methods we may learn wisdom: First, by reflection, which is noblest; Second, by imitation, which is easiest; and Third by experience, which is the bitterest.
- Everything has beauty, but not everyone sees it.
- It does not matter how slowly you go as long as you do not stop.
- Wheresoever you go, go with all your heart.
- He who knows all the answers has not been asked all the questions.
- Life is really simple, but we insist on making it complicated.
- The man who moves a mountain begins by carrying away small stones.
- If you make a mistake and do not correct it, this is called a mistake.
- Study the past if you would define the future.
- The funniest people are the saddest ones.
- Before you embark on a journey of revenge, dig two graves.
- To be wronged is nothing, unless you continue to remember it.
- Respect yourself and others will respect you.
- You cannot open a book without learning something.
- The man who asks a question is a fool for a minute, the man who does not ask is a fool for life.
- Silence is a true friend who never betrays.
- Attack the evil that is within yourself, rather than attacking the evil that is in others.
- What the superior man seeks is in himself; what the small man seeks is in others.
- I hear and I forget. I see and I remember. I do and I understand.

- The hardest thing of all is to find a black cat in a dark room, especially if there is no cat.
- Music produces a kind of pleasure which human nature cannot do without.
- Give a bowl of rice to a man and you will feed him for a day. Teach him how to grow his own rice and you will save his life.
- The gem cannot be polished without friction, nor man perfected without trials.
- The way out is through the door. Why is it that no one will use this method?
- Only the wisest and stupidest of men never change.
- It is more shameful to distrust our friends than to be deceived by them.
- Real knowledge is to know the extent of one's ignorance.
- We have two lives, and the second one begins when we realize we only have one.
- If what one has to say is not better than silence, then one should keep silent.
- Forget injuries, never forget kindness.
- To put the world in order, we must first put the nation in order; to put the nation in order, we must first put the family in order; to put the family in order; we must first cultivate our personal life; we must first set our hearts right.
- Education breeds confidence. Confidence breeds hope. Hope breeds peace.
- In a country well governed, poverty is something to be ashamed of. In a country badly governed, wealth is something to be ashamed of.
- When a wise man points at the moon the imbecile examines the finger.
- It is easy to hate and it is difficult to love. This is how the whole scheme of things works. All good things are difficult to achieve; and bad things are very easy to get.
- Tzu Chang asked Confucius about jen. Confucius said, "If you can practice these five things with all the people, you can

be called jen." Tzu Chang asked what they were. Confucius said, "Courtesy, generosity, honesty, persistence, and kindness. If you are courteous, you will not be disrespected. If you are generous, you will gain everything. If you are honest, people will rely on you. If you are persistent, you will get results. If you are kind, you can employ people.

The Han Dynasty was now a memory and other short-lived dynasties, such as the Wei, Jin, Wu Hu, and the Sui, assumed control of the government in turn, and initiated their own policies from around 208 to 618 CE. The Sui Dynasty (589-618 CE) finally succeeded in reuniting China in 589 CE. The importance of the Sui Dynasty was in its implementation of a highly efficient bureaucracy which streamlined the operations of government and led to greater ease in maintaining the vast empire. Under the Emperor Wen, and then his son Yang, the Grand Canal was completed; the Great Wall was enlarged and portions rebuilt; the army was increased to the largest recorded in world history up to that time; and coinage was standardized throughout the kingdom.

During this time literature flourished, and it is believed that the famous "Legend of Hua Mulan", about a young girl who takes her father's place in the army and saves the country, was developed at this time. Unfortunately, both Wen and Yang were not content with domestic stability and organized massive military expeditions against the peoples of the Korean peninsula. Emperor Wen had already bankrupted the treasury as a result of his major building projects and military campaigns; and Yang followed his father's example, and equally failed in his attempts at military conquest. The Emperor Yang was assassinated in 618 CE, which then sparked an uprising led by Li Yuan, who took control of the government and called himself Emperor Gao Tzu of Tang, and reigned from 618 to 626 CE.

The Tang Dynasty lasted from 618 to 907 CE and is considered to be the "golden age" of Chinses civilization. Gao Tzu prudently and wisely maintained and improved upon the bureaucracy initiated by the Sui Dynasty, while dispensing with extravagant military operations and building projects. With minor modifications, the bureaucratic policies and methodology of the Tang Dynasty are still in use within

(243) The Great Wall of China was built mainly by the Qin Dynasty from 221 to 206 BCE

the modern-day Chinese government. Despite his highly effective rule, Gao Tzu was deposed by his son Li Shimm, in 626 CE. Having assassinated his father, Li Shimm then also killed his brothers and others of the noble house and assumed the title Emperor Taizong (626-649 CE). After the bloody coup however, Taizong decreed that Buddhist temples were to be built at various battle sites where the violent military coup took place, and that the fallen should be memorialized. Continuing and building upon the concepts of ancestor worship and the Mandate of Heaven, Taizong claimed a divine will in his actions and intimated that those he killed were now his counsellors in the afterlife. As he proved to be a remarkably efficient, strong ruler, as well as a skilled military strategist and warrior, his coup went unchallenged, and he set about the task of earnestly governing his vast empire. Amongst his many achievements was the introduction of an extensive and expanded legal code, and a series of military conquests which further expanded the kingdom.

Taizong was succeeded by his son Gaozong (649-683 CE) whose wife Wu Zetian, would become China's first and only exclusive female monarch. Empress Wu Zetian (690-704 CE) initiated a number of policies that improved the living conditions of peasant farmers and labourers; and made ample use of a highly efficient, ruthless secret police force, which kept her one step ahead of her foreign and domestic enemies. During this time, trade flourished within the empire, and along the Silk Road with other civilizations. The Roman Empire had fallen, and as a result, the Byzantine Empire became the foremost buyer of Chinese silk. By the time of the rule of Emperor Xuanzong (712-756 CE), China was the largest, most populous, and most prosperous civilization in the world. As a result of the large population, armies of hundreds of thousands of men could be conscripted into the military at short notice. A number of military campaigns against Turkish nomadic tribes and domestic rebels were swift and highly successful. Art, science, and technology all flourished under the Tang Dynasty and some of the most impressive pieces of Chinese sculpture and silverwork came from this period.

As successful as the Tang Dynasty was, the central government was not universally admired and regional uprisings were a regular concern. The most significant of these uprisings was the An Shi Rebellion of 775 CE. General An Lushan, a favourite of the imperial court, recoiled against what he saw as excessive extravagance in the government. With a force of over 100,000 troops, he rebelled and declared himself the new emperor, citing the Mandate of Heaven as his authority. Although his revolt was put down by 763 CE, the underlying causes of the insurrection and further military actions continued to plague the government up to 779 CE. The most apparent consequence of An Lushan's rebellion was a dramatic reduction in the population of China. It has been estimated that around 36 million people died as a direct result of the rebellion, either in battles, in reprisals, or through disease and famine. Trade suffered, taxes went uncollected, and the government, which had fled the then capital city of Chang'an when the revolt began, was ineffective in maintaining any kind of control, stability, and order. The Tang Dynasty continued to suffer from domestic revolts and, after the Huang Chao Rebellion from 874 to 884 CE, never recovered. Soon after, the Chinese kingdom then broke apart into the period known as The Five Dynasties and Ten Kingdoms, which lasted from 907 to 960 CE; with each regime claiming legitimacy for itself, until the ultimate rise of the Song Dynasty.

Under the Song Dynasty, which ruled China from 960 to 1279 CE, China once again achieved stability and order; and during this time institutions, laws, and customs were further codified and integrated into the culture. Neo-Confucianism became the most popular philosophy of the kingdom, influencing these laws and customs, and shaping the culture of China right through to the modern day. Nevertheless, in spite of all these advances, the age-old conflict between the wealthy landowners and the peasants who worked the land, continued on through the centuries. We have seen in so many other civilizations referred to in this book, that this was a universal human dilemma and struggle. Periodic peasant revolts in China were always crushed as quickly as possible, but no remedies for the people's grievances were ever offered. Each military action continued to deal with the symptoms of the problem rather than the root cause, a widening inequality of political power and wealth. In 1949 CE, Mao Zedong led the People's Revolution in China, toppling the government and instituting the People's Republic of China on the premise that finally, everybody would be equally affluent. We will examine the impact and consequences of this dramatic 20th century event in a future chapter.

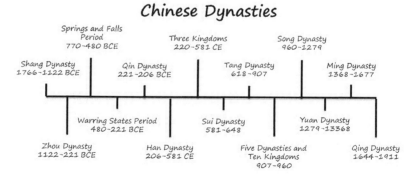

Chinese Dynasties

Springs and Falls Period 770–480 BCE

Three Kingdoms 220–581 CE

Song Dynasty 960–1279

Shang Dynasty 1766–1122 BCE

Qin Dynasty 221–206 BCE

Tang Dynasty 618–907

Ming Dynasty 1368–1677

Warring States Period 480–221 BCE

Sui Dynasty 581–648

Yuan Dynasty 1279–13368

Zhou Dynasty 1122–221 BCE

Han Dynasty 206–581 CE

Five Dynasties and Ten Kingdoms 907–960

Qing Dynasty 1644–1911

(244) The Major Chinese Dynasties from 1766 BCE to 1911 CE

17

INDUS VALLEY AND SOUTH EAST ASIAN CIVILIZATIONS

THE INDUS VALLEY CIVILIZATION WAS A CULTURAL AND POLITICAL group of peoples which flourished in the northern region of the Indian subcontinent between 7000 and 600 BCE. Its modern name comes from the fact it was located in the valley of the Indus River. It is also known as the Harappan Civilization, named after the ancient city of Harappa located in the region, and the first city from the civilization to be excavated. The Indus Civilization is categorized by archaeologists into the following historical periods:

- Pre-Harappan – 7000 to 5500 BCE
- Early Harappan – 5500 to 2800 BCE
- Mature Harappan – 2800 to 1900 BCE
- Late Harappan – 1900 to 1500 BCE
- Post Harappan – 1500 to 600 BCE

The Indus Valley Civilization is often compared by archaeologists with the more well-known cultures of Egypt and Mesopotamia. The discovery of Harappa in 1829 CE was the first indication that any such civilization existed in India. By this time in the early nineteenth century, Egyptian hieroglyphics had been deciphered; various Egyptian and Mesopotamian sites had been discovered and excavated; and cuneiform writing would soon be translated by the scholar George Smith. The two best known excavated cities of this culture are Harappa and Mohenjo-daro (located in modern day Pakistan), both of which are thought to have once had populations of around 50,000 people, which is quite incredible when we consider that most ancient cities had an average of 10,000 people living in them. The total population of the Indus Civilization is thought to have been around 5 million, and its territory stretched over 1500 kilometres along the banks of the Indus River, and then in all directions outwards. Archaeological sites have been found as far away as the border of Nepal, in Afghanistan, and around the city of Delhi.

Between 1900 and 1500 BCE, the civilization began to decline for unknown reasons. In the early 20th century CE, archaeologists thought that this decline had been caused by an invasion of light-skinned peoples from the north known as Aryans, who conquered a dark-skinned people called Dravidians. This claim, known as the Aryan Invasion Theory, has since been discredited. The Aryans, whose ethnicity is connected with the Iranian Persians, are now believed to have migrated to the region peacefully, and blended their culture with that of the indigenous people; while the term Dravidian, is now understood to refer to anyone of any ethnicity, who speaks one of the Dravidian languages. Why the Indus Valley Civilization declined and fell is unknown, but most scholars now believe it may have had

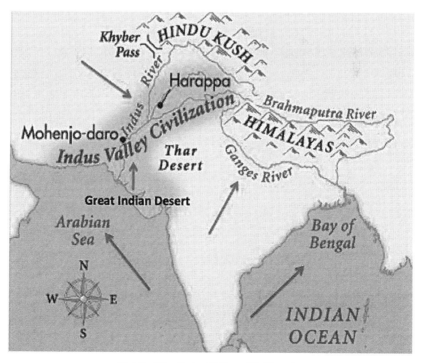

(245) The Indus Valley Civilization

something to do with a combination of factors triggered by climate change; which resulted in the drying up of the Indus and Sarasvati Rivers; an alteration in the path of the monsoons that provided valuable water to crops and livestock; overpopulation of the cities; and a decline in trade with Egypt and Mesopotamia. In the present day, excavations are continuing at many of the sites discovered so far. Hopefully, more clues as to the reasons why the civilization declined may be found at these sites, and future sites that may be discovered.

The discovery of the Indus Valley Civilization is a fascinating story in itself. James Lewis (also known as Charles Masson) born 1800 and died 1853 CE, was a British soldier serving in the artillery of the British East India Company Army when in 1827, he deserted with another soldier. In order to avoid detection by the authorities, he changed his name to Charles Masson and embarked on a series of travels throughout India. Masson was an avid coin collector who was especially interested in old coins and, while following up various leads, he ended up excavating ancient sites on his own. One of these sites was Harappa, which he discovered in 1829. He seems to have left the site very quickly, after making a written record of it in his dairy notes, but, having no knowledge of the significance of his discovery, and who could have built the city; he incorrectly attributed it to having been created by Alexander the Great during his military campaign in India in 326 BCE. When Masson returned to Britain after his adventures, he published a book, "Narrative of Various Journeys in Balochistan, Afghanistan and the Punjab" in 1842. His book attracted the attention of the British establishment and authorities, especially Alexander Cunningham. Sir Alexander Cunningham (1814-1893) was a British Engineer in India, who had a passion for ancient history, and had founded the Archaeological Society of India ("ASI") in 1861; an organization dedicated to maintaining a professional standard of excavation and to the preservation of historical sites. Cunningham led a team that began excavations at the site, and published his interpretation of artefacts and other findings in 1875. In this book, he identified and named a previously unknown form of writing which he named Indus Script; but this was incomplete and lacked definition because Harappa remained isolated with no connection to any known past civilization which could have built it. In 1904, a new director of the ASI was appointed, John Marshall (1876-1958), who later visited Harappa and concluded the site represented an ancient civilization previously unknown. He ordered the site to be fully excavated and, at around the same time, heard about another site a few miles away, which the local people referred to as Mohenjo-daro ("the mound of the dead") because of animal and human bones found there, along with various artefacts. Excavations at Mohenjo-daro began in 1924 and the similarities of the two sites as belonging to the same civilization, were soon recognized. The Indus Valley Civilization had been discovered.

The Hindu religious texts known as the Vedas, as well as other great works of Indian culture and tradition such as the "Mahabharata" and "Ramayana", were already well known to Western scholars, but they did not know what culture created them. Endemic racism at the time prevented scholars from attributing these great works to the people of India, and the same racist thinking at first, led archaeologists to conclude that the city of Harappa was a colony of the Sumerian Mesopotamians or perhaps the Egyptians. It soon became apparent that the artefacts recovered at Harappa did not conform to either Mesopotamian or Egyptian architecture; and there was no evidence of temples, palaces, monumental structures, names of kings or queens, or royal statuary. The city of Harappa spread over 370 acres of small brick houses with flat roofs made of clay. There was a citadel, walls, and streets laid out in a grid pattern, clearly demonstrating a high degree of skill in urban planning. In comparing the two sites, it was apparent to the excavating archaeologists that they were dealing with a highly advanced culture.

The houses in both cities had flush toilets, a sewer system, and fixtures on either side of the streets which appeared to be part of an elaborate drainage system, which was more advanced than even that of the Romans, who appeared later. Devices known to have originated in Persia, called "wind catchers" were attached to the roofs of some buildings, which provided air conditioning for homes and

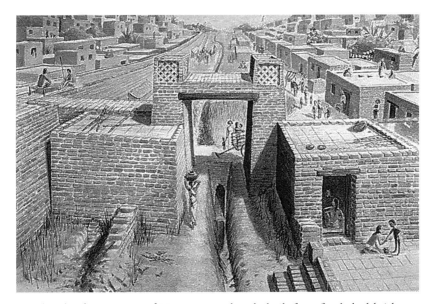

(246) The city-state of Harappa was largely built from fire-baked bricks

administrative buildings. At Mohenjo-daro, there was a large, intri-
cately designed public bathing complex, which was surrounded by a
courtyard with steps leading down into it. As other sites were unearthed,
the same degree of sophistication and skill came to light, as well as
a clear understanding that all of these cities had been pre-planned.
Unlike those of other cultures, which usually developed from smaller
rural communities, the cities of the Indus Valley Civilization had been
well thought out; sites deliberately chosen; and buildings purposefully
constructed prior to being fully inhabited by people. Furthermore,
these cities exhibited conformity to what appeared to be a single
vision, which further suggested a strong central government, possibly
ruled by dynastic kings with an efficiently functioning bureaucracy
that could plan, fund, and build such cities.

The British historian John Keay, had this to say about the Indus
Valley excavations: "What amazed all these pioneers, and what remains
the distinctive characteristic of the several hundred Harappan sites now
known, is their apparent similarity. Our overwhelming impression is
of cultural uniformity, both throughout the several centuries during
which the Harappan civilization flourished, and over the vast area it
occupied. The ubiquitous bricks, for instance, are all of standardized

dimensions, just as the stone cubes used by the Harappans to measure weights are also standard and based on the modular system. Road widths conform to a similar module; thus, streets are typically twice the width of side lanes, while the main arteries are twice or one and a half times the width of streets. Most of the streets so far excavated are straight and run either north-south or east-west. City plans therefore conform to a regular grid pattern and appear to have retained this layout through several phases of building."

Excavations at both sites continued between 1944-1948 under the direction of the British archaeologist, Sir Mortimer Wheeler (1890-1976) whose racist ideology made it difficult for him to accept that dark-skinned people had built the cities. Even so, he managed to establish a historical chronology for the differing stages of the development of the Indus Valley Civilization. Wheeler's work provided subsequent archaeologists with the means to recognize approximate dates from the foundations of the civilization through to its decline and ultimate fall. The chronology is primarily based on physical evidence from Harappan excavation sites, but also from knowledge of their trade contacts with Egypt and Mesopotamia. A semi-precious stone called "lapis lazuli" for example, was one product that was immensely popular in both cultures and, although scholars knew it came from India, they did not know from precisely where, until the Indus Valley Civilization was discovered. Even though this semi-precious stone would continue to be imported after the fall of the Indus Valley Civilization, it is clear that initially, some of the exports came from this region.

Each of the Indus Valley Civilization periods, can be categorized and described as follows:

- The Pre-Harappan from 7000 to 5500 BCE existed in the Neolithic Period, and was best exemplified by sites like the excavated village settlement of Mehrgarth, which shows evidence of agricultural development including the domestication of plants and animals; and the production of tools and ceramics.
- The Early Harappan from 5500 to 2800 BCE saw a period

when extensive trade was firmly established with Egypt, Mesopotamia, and possibly China. Ports, docks, and warehouses were constructed near waterways by communities living in small villages and towns.

- During the Mature Harappan period from 2800 to 1900 BCE, the construction of great cities and widespread urbanization occurred. The cities of Harappa and Mohenjo-daro were both flourishing by 2600 BCE. Other cities such as Ganeriwala, Lothal, and Dholavira were built, using the same city planning models, and the development of the land continued with the construction of hundreds of other cities, until there were over 1000 cities throughout the regions of the civilization.

- By the Late Harappan period from 1900 to 1500 BCE, the decline of the civilization had occurred, coinciding with a wave of migration of Aryan people from the north, most likely originating from the Iranian Plateau. Physical evidence suggests that significant climate change caused widespread flooding, drought, and famine throughout the civilization. A significant drop in trade with Mesopotamia and Egypt also contributed to the decline.

- During the Post Harappan period from 1500 to 600 BCE, the cities were abandoned, and the people moved further south. The Indus Valley Civilization had already fallen by the time the Persian King Cyrus the Great (who reigned from 550–530 BCE) invaded India in 530 BCE.

The people of the Indus Valley Civilization seem to have been primarily artisans, farmers, and merchants. There is no evidence of a standing army, no palaces, and no religious temples. The Great Bath at Mohenjo-daro is believed to have been used for ritual purification rites that were related to a religious belief system; as well as a public recreation pool. Each of the cities seem to have had their own governors, who it is believed all were accountable to a centralized government led by a king, in order to achieve the uniformity of the cities. British historian John Keay comments: "Harappan tools, utensils, and materials confirm this impression of uniformity. Unfamiliar with

(247) & (248) Harappan statue and ceremonial burial pottery

iron – which was nowhere known in the third millennium BC – the Harappans sliced, scraped, bevelled, and bored with effortless competence using a standardized kit of tools made from chert, a kind of quartz, or from copper and bronze. These last, along with gold and silver, were the only metals available. They were also used for casting vessels and statuettes and for fashioning a variety of knives, fishhooks, arrowheads, saws, chisels, sickles, pins, and bangles."

Among the thousands of artefacts discovered at the various sites were small, soapstone seals, a little over 3 centimetres in diameter, which archaeologists interpret to have been used for personal identification in trade. Like the cylinder seals of Mesopotamia, these seals are thought to have been used to sign contracts, authorize land sales, and authenticate point-of-origin, shipment, and receipt of goods in long distance trade. The people had developed the wheel; carts drawn by cattle, flat-bottomed boats wide enough to transport goods for trade; and may have also developed the sail. In agriculture, they understood and made use of irrigation techniques and canals; various farming implements; and established different areas for cattle grazing and the growing of crops. Fertility rituals may have been observed for a full harvest, as well as pregnancies of women, as evidenced by a number of excavated figurines, amulets, and statuettes in female form. It is thought that the Indus people may have worshipped a Mother Goddess deity and possibly, a male consort depicted in the form of a horned figure in the company of wild animals. However, archaeologists do not know much about the religious beliefs of the Indus people.

Let us now examine some of the great civilizations of South East Asia. The Khmer Empire was a powerful state formed by the people of the same name, and lasted from 802 CE to 1431 CE. At its peak, the empire covered much of what today is Cambodia, Thailand, Laos, and southern Vietnam. By the 7th century CE, the Khmer people inhabited territories along the Mekong River, the seventh largest river in the world. There were several kingdoms at constant war with each other at this time. The art and culture of Khmer was heavily influenced by India, due to long established sea trade routes with that subcontinent. Hinduism mostly, but Buddhism as well, were important religions in the region, along with animist and traditional cults. Important cities from that time included Angkor Borei, Sambor Prei Kuk, Banteay Prei Nokor and Wat Phu. A king called Jayavarman II, who appears to have originated from the Indonesian island of Java, led a series of successful military campaigns, subjugating most of these smaller kingdoms, and resulting in the founding of a large territorial state. In 802 CE he took the title "chakravartin" meaning "universal ruler", and effectively started the Khmer Empire.

(249) The Angkor Wat Temple was built in the 12th century CE by King Suryavarman II

Using the city of Angkor as its capital, the Khmer Empire expanded its territorial base, mostly to the north and the west. To the east, the outcomes were different; several times the Khmer fought wars against two neighbouring powerful kingdoms; the Cham (in today's central Vietnam) and the Vietnamese (in today's northern Vietnam). Despite some victories, such as in 1145 CE, when Cham's capital city Vijaya was taken, the empire was never able to conquer those lands. In fact, the Chams and Vietnamese enjoyed some victories of their own, the most spectacular of which was Cham's humiliating revenge on the Khmer, when it attacked and looted the capital of Angkor in 1177 CE, and pushed the empire close to the tipping point of its destruction. Throughout the empire's history, Khmer's kings were repeatedly concerned with putting down rebellions initiated by ambitious nobles trying to achieve independence, or fighting conspiracies against the king. This was particularly true each time a king died, as successions were usually contested, and male heirs not necessarily taking over.

The Khmer were exceptional builders, filling the landscape with monumental temples, huge reservoirs, and canals; and laying an extensive road network with many varieties of bridges; the main highways being hundreds of kilometres long. The most stunning temple, Angkor Wat, was a microcosm of the Hindu universe, and defied imagination as the world's largest religious complex, covering 200 hectares. Today, it is a UNESCO World Heritage Site and a major tourist, cultural, and religious destination, with amazing ruins that until recently were covered for centuries by thick jungle vegetation. Western explorers discovered the site in the 19th century CE, cleared the jungle overgrowth and began to restore it. Its original construction took around 30 years, and was commenced by one of the great Khmer kings, Suryavarman II, in around 1122 CE. Angkor Wat was dedicated to the Hindu god Vishnu. Its name means "City of the Temple", and it was created as a physical manifestation of human interaction with the realm of the gods. The spires represent the mountains of eternity and the moat, the eternal waters. Although he is remembered as a great ruler, King Suryavarman II was also a ruthless, power-hungry agitator, who assassinated his great uncle Dharanindravarman I (who ruled from 1107 to 1113 CE), in order to take the throne for himself. He is said to have compared the coup to destroying a serpent; although we are unclear with what he actually meant by this statement. He then proceeded to legitimize his rule through the sheer grandeur of his spectacular accomplishments; and of course, was immortalized as a result of the construction of the grand temple complex of Angkor Wat under his reign, which he dedicated to his personal protector-god Vishnu. King Suryavarman II had amassed substantial wealth through trade and taxes, and spared no expense in the creation of his temple. The British archaeologist Christopher Scarre had this to say: "The Khmer's unique form of kingship produced, instead of an austere civilization like that of the Indus, a society that carried the cult of wealth, luxury, and divine monarchy to amazing lengths. This cult reached its apogee in the reign of Suryavarman II who built the temple of Angkor Wat."

It is possible that the temple of Angkor Wat was intended to also serve as a tomb for the burial of his body; a funerary temple in the tradition of the pharaohs of Egypt. When Suryavarman II died though, the temple was still not completed, and he was instead cremated and buried elsewhere. It is more likely however, that Suryavarman II had the temple purposefully built to honour his god, particularly when we consider how devoted he was to Vishnu. The king practised a form of Hinduism known as Vaishnavism, which is a devotion to the god Vishnu above all others. Although Hinduism is generally regarded as a polytheistic religion by westerners, it is actually henotheistic, which means there is only one god; but with many different forms and aspects of that god. In a henotheistic belief system, a single god is considered too immense to be grasped by the human mind, and so appears in a multiplicity of personalities, all of which focus on a single different aspect of human life.

In Hinduism, Brahma is the supreme deity who created the world; while in his form as Vishnu, he preserves life; and as Shiva, he takes life away and rewards humans for their toil, with death, which then continues the cycle of rebirth, or leads to union with the over-soul. Angkor Wat reflects the course of life, death, and eternity according to Vaishnavism, removing Brahma as the supreme god, and replacing

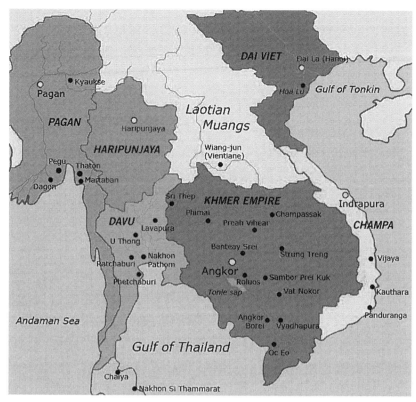

(250) The Khmer Empire and surrounding Southeast Asian kingdoms

him with Vishnu. The rise of Vaishnavism in the Khmer Empire was a direct result of the conflicts between the Khmers and the neighbouring Champa. King Suryavarman I, who reigned from 1006 to 1050 CE, extended the frontiers of his kingdom into what is modern day Thailand, and came into direct conflict with the cities of the Champa Kingdom in the process. The Champa's religion was Buddhism (which was also the faith of the Khmer elite at that time), which was viewed with hostility by most Khmer who saw it as a threat to their faith. Vishnu, as a protector god, rose in popularity through these conflicts and the backlash against Buddhism.

One of the most popular stories of Vishnu's kindness and cleverness in the interests of human beings is "The Churning of the Ocean" in which he tricks the demons into surrendering the "amrita" (meaning "immortality"), which will then make the gods immortal, and preserve eternal order. This story is among the most famous recorded in stone

inscriptions found at Angkor Wat and seems to support the claim that
the building was originally conceived of as a temple of worship rather
than a funerary site.

(251) The Churning of the Ocean of Milk: A Struggle Against Evil –
Angkor Wat

Angkor Wat was rededicated as a Buddhist temple in the 14th
century CE and statues of the Buddha and Buddha-related stories
were added to the already impressive iconography. As the Buddhists
respected the beliefs of the Hindus who still worshipped there, all
of the original statuary and artwork was left in place. The Buddhist
craftsmen added to the intricate story of the temple while taking
nothing away. By the early 16th century CE, use of the temple had
waned, even though it was still occupied by Buddhist monks, and it
became the subject of stories and legends. It was said to have been
built by the gods in the distant past, and a popular story emerged
that the god Indra had built it as a palace for his son, and that it rose
from nothing during the course of a single night! The temple was
protected from the surrounding thick jungle by an immense moat and

so, unlike other ancient temples and cities, it was never completely lost to humanity. Even though local people still visited the site, it became increasingly associated with hauntings and evil dark spirits. The great enthusiasm of devotees who used to visit the temple, it was said, needed to continue to infuse the area with positive energy, in order to try and drive away the negative energy of the evil spirits. Once worship at the site significantly fell off, it was believed that the dark spirits, attracted by the afterglow of the high energy, moved in and made the place their home. Dark energy was now thought to emanate from the empty galleries, porches, and entranceways; and fewer and fewer people went to visit. With only a few monks to care for it, the once magnificent buildings began to decay, and though it was never completely taken by the jungle, natural growth made headway up the walls and through the cracks between the stones.

However, the temple did not sit quietly for long. According to western documentary sources, Angkor Wat was first visited by the Portuguese monk Antonio da Madalena in 1586 CE, whose notes on the complex clearly convey his sense of awe and wonder. He made no efforts to restore the site, but did record its location. The next European to visit was the French archaeologist Henri Mouhot in 1860 CE. Mouhot is always cited as the man who "discovered" Angkor Wat but, in reality, it was never lost. Anybody who lived in the area knew of the site, and it was known in legends and stories by people from far-away places. Mouhot was the first westerner to take an active interest in Angkor Wat and publicize its existence within the western world. He was so impressed by the temple that he devoted the remainder of his life to its renovation and restoration. Mouhot believed the temple had been built by some ancient civilization now lost, and considered the stories of how Indra had raised the structure in a night to be evidence of the ancient culture's lost advanced technology. He refused to accept that it could have been built by the ancestors of the Cambodians he interacted with on a daily basis; in the same way that westerners rejected the notion that the great cities, monuments, and temples of Mexico and Central America could have been built by the Maya, as we shall see in later chapters; or that the monumental buildings of the

Indus Valley Civilization were built by the ancestors of Indians, as we saw earlier in this chapter. Again, these views reflected the scourge of the Eurocentric racism of that time.

Since Mouhot's time, Angkor Wat has become world famous. Millions of people visit the site every year and restoration projects are ongoing. In 1992, UNESCO declared it a World Heritage Site and, even though it was not chosen as one of the New Seven Wonders of the World in 2007, it gained great attention as a finalist. In 2016, a New York Times article reported on the ongoing efforts of archaeologists who continue to make discoveries in the surrounding jungle, and have located the sites of the workers who built the temple and of others who lived around the complex. The temple itself has undergone major restoration and is considered to be one of the most popular archaeological sites in the world. Those who visit Angkor Wat today are literally following in the footsteps of millions of people from the past who have emerged from the surrounding jungle to find themselves at the site King Suryavarman II created as the gateway between earth and heaven.

(252) & (253) King Jayavarman VII ruled the Khmer Empire from 1181 to 1215 CE

Another great king was Jayavarman VII, who reigned from 1181 to 1215 CE. He expelled the Chams who had taken Angkor, restored the kingdom from anarchy, and then invaded and occupied the Champa Kingdom. The scale of his construction program was unprecedented in its visionary scope. He built more temples, monuments, highways, over a hundred hospitals, and the spectacular Angkor Thom complex, a city within a city in Angkor. Under King Jayavarman VII, the Khmer Empire reached its greatest territorial size and power. After decades of turmoil, in 1181 CE King Jayavarman VII restored order to the Khmer Empire by embracing Buddhism and introducing amongst other things, a public healthcare program. On the foundation stone of a 12th century CE hospital temple built on the banks of the Mekong River, was written the following, in reference to the highly revered first Buddhist king: "He suffered the illnesses of his subjects more than his own; because it is the pain of the public that is the pain of kings rather than their own pain." Medical knowledge had a long history in the Khmer Empire. Along with a central hospital temple, stone inscriptions state that King Jayavarman VII founded an additional 102 hospitals throughout the empire. These were probably open to all, although how such healthcare was paid for is unknown. People may have had to make a donation, but it is also possible that healthcare was offered free of charge. Each hospital would have contained a shrine to Bhaisajyaguru, the Buddha of medicine and healing. Before entering Buddhahood, Bhaisajyaguru made a vow to help the physical and mental health of all living beings. The hospital edicts state that just hearing the name Bhaisajyaguru was enough to cure all ills. In addition to ritual and spiritual healthcare, these hospitals also offered healing in a manner similar to "modern" medicine. The inscriptions give an inventory of staff and supplies. Each hospital was home to "achar" or priests, and a large team of medical workers, including two doctors, two apothecaries (similar to modern day pharmacists), eight nurses and six assistants. There were also guards, cooks, rice-makers, and servants. The medical team offered diagnostic testing, most likely by reading the pulse of patients. The hospitals also prescribed medicines, including honey, butter, oil, and molasses. It is highly probable that medical alchemy formed a large part of the healthcare service in the Khmer Empire. An inscription at the Ta Prohm temple lists royal donations of metals and apparatus that would have been used for alchemy, including mercury sulphide and a gold cauldron.

The Khmer were festive people who enjoyed many celebrations all year round. Wrestling, horse races, cock fights, fireworks, music, and dances were all an integral part of their culture. Most of the kingdom's trade and commerce appeared to have been driven mainly by women. The king and noble elite were transported on palanquins – large lavishly appointed boxed enclosures, carried on two horizontal poles, by four to six men; and covered from the sun with large umbrellas. There were several religious beliefs, with Hinduism initially being the primary religion favoured by the kings; and then Buddhism later in the life of the empire, as we have seen. At its pinnacle of power, the Khmer Empire was divided into 23 provinces, each headed up by a king's direct representative (governor), and a sophisticated government administration, with an extensive bureaucratic structure going right down to local village level. Although high level government officials were essential for the proper management and maintenance of the prosperity of the empire; it was often also some of these same high level officials that were involved in conspiratorial plots to overthrow the king.

The decline and fall of the Khmer Empire is deeply connected with the great Thai migration of the 12th to 14th centuries CE. The Thais inhabited an area to the north of the empire, approximately where China ends and South East Asia begins, in the modern day; this northern area was then called the Yunnan. It is a mountainous, harsh land, where a Thai kingdom called Nanchao existed. For reasons that are unknown, Thai populations started migrating south into the Khmer Empire, in small groups first. Thais first appear in the records as hired mercenaries for the empire; and their numbers rose as they began to establish themselves as settlers in marginal areas. The migration intensified when Mongol military campaigns shook China, and when the Mongols conquered Yunnan in 1253 CE. Eventually the Thai peoples created their own small kingdoms, the most important of those were in the western side of the empire. As these kingdoms grew in power, they started to attack, annex and conquer territories within the Khmer Empire. By this time, the empire's economy appears to have significantly deteriorated as a result of a drop in agricultural production arising from the increased silting of the massive water works and canals that the Khmer depended on to irrigate their farms. The Thai Kingdom of Ayutthaya took control of Angkor in 1431 CE, which effectively led to the end of the Khmer Empire.

(254) The Mataram Sultanate of Java lasted from 1587 to 1755 CE

Let us conclude this chapter by briefly examining some other significant South East Asian civilizations and kingdoms. I have placed these kingdoms in what I consider to be their order of influence in the region; geopolitically and culturally; so the dates jump around somewhat. The Mataram Sultanate (1587-1755 CE) dominated the Indonesian island of Java at a time when the Dutch and Portuguese were first beginning to establish long-term trading relationships with various cultural groups in Asia. Mataram had a number of vassal states along the coasts of Java, which it preferred to work through when dealing with the Europeans, Chinese, and Japanese. This kept the mainly inland Sultanate of Mataram safe initially; but it eventually ended up being the Muslim state's undoing, as the vassals quickly learned the geopolitics and diplomacy of dealing with the "outsiders", eventually shutting out the Mataram, and enriching themselves in the process.

The Rattanakosin Empire (1782-1932 CE) arose out of the ashes of the Ayutthaya Empire in the late 18th century, and is largely responsible for keeping Thailand unoccupied by European powers throughout the 19th century, when European Imperialism was at its height. The Rattanakosin Empire not only played the Europeans off each other;

it also managed to keep numerous neighbouring ethnic groups under central Thai control; maintained Bangkok as the capital city of the Thai world; and undertook a number of modernization programs in order to counteract the growing influence of the Europeans in the region. The modernization program included the abolition of slavery; building schools and hospitals; and creating a nation state with a constitution, a large consumer middle class, and none of the ethnic baggage that afflicted many other post-colonial states in South East Asia.

(255) King Chulalongkorn and family – from the Rattanakosin Empire of Siam

The Aceh Sultanate (1496-1903 CE) was a powerful Muslim commercial empire located below Thailand and above the densely populated island of Java. Aceh gained most of its historical infamy by strongly resisting Dutch efforts to conquer Aceh territory in an endeavour to incorporate it into "Dutch East India." Unfortunately, the war with the Dutch depleted much of their resources and ruined this once-wealthy commercial sultanate. The Kingdom of Pagan (850-1300 CE) was responsible for the spread of Buddhism (which had originated in India) throughout South East Asia in the 9th century CE. Over 10,000 Buddhist temples were built under the rule of Pagan kings, much to the displeasure and offence of the traditional noble Hindu elite. There was not much the Hindus could do though, as Buddhism gained enormous popularity among the common people and spread rapidly in the region. The Kingdom of Pagan fell into decline in the 13th century CE, after it was invaded by the Mongols.

The Vietnamese Le Dynasty (1427-1789 CE) was situated in north Vietnam and was the longest–ruling in the history of Vietnam. It was responsible for removing the Chinese rulers from Vietnam and re-establishing an independent kingdom. When the Le Dynasty was finally eliminated by local rivals in 1789, the French imperialists stepped in to try and fill the power vacuum and introduce ordered stability to the region. It did not go well. The French colony of Indochina was constantly threatened by instability, conflict, and insurrection for the entire period it controlled the Vietnamese, from 1887 to 1954, when the French were finally militarily defeated and evicted from Vietnam.

The Toungoo Empire (1501-1752 CE) was originated in modern-day Burma and was Buddhist in nature. At its greatest, it included much of Burma, Thailand, Laos, and parts of Cambodia. Like all empires, Toungoo's power rose and fell throughout the centuries; but by 1752 the French had convinced one of Toungoo's most influential vassal states to rebel, bringing the Buddhist empire to its knees, once and for all. The Kingdom of Ayutthaya (1351-1767 CE) was known as "Siam" in the West. The Ayutthaya Kingdom reigned over much of what now is Thailand and maintained the independence of Thai culture and politics from outside influence. This was a Buddhist monarchy that managed to survive for so long by trading with not only the Chinese and Japanese, but also the Europeans, Persians, Ottomans, and Indians. Ayutthaya avoided falling under Chinese influence by using a rival Burmese kingdom as a buffer. The downside to this strategy was that Ayutthaya was often at war with the Burmese, who captured Ayutthaya's capital city in 1767, and completely laid waste to the kingdom.

The Rajahnate of Butuan (1001-1756 CE) first appeared in Chinese records in 1001 CE, but Butuan seemed to have been in existence before then. Butuan was a Hindu-dominated kingdom, and a testament to the power of Hinduism in South East Asia. It was a precolonial kingdom situated mainly on the Philippine island of Mindanao. It was known for its gold mining and its extensive trading network with surrounding kingdoms throughout South East Asia. The last independent ruler was Rajah Siagu, who was formally subjugated into the Spanish Empire after he and his brother, Rajah Kolambu, made a blood compact with the explorer Ferdinand Magellan in 1521, to place their kingdom as a vassal state of the Spanish.

(256) The fashion styles of Vietnamese Le Dynasty Women:
1427 to 1789 CE

18

THE AMERICAS CIVILIZATIONS

M ANY THOUSANDS OF YEARS BEFORE CHRISTOPHER COLUMBUS' ships arrived in North America, a different group of people discovered America; they were the nomadic ancestors of modern Native Americans who crossed the Baring Straits land bridge from Asia to what is now Alaska, commencing from around 20,000 years ago. By the time European explorers arrived in North America in the 15th century CE, anthropologists estimate that more than 50 million people were already living in the Americas. Of this total, around 10 million lived in the area that would ultimately become Canada and the United States of America. Over a period of time, these migrants and their descendants moved south and east from what today is Alaska, adapting to their different environments as they migrated. In order to keep track of these diverse groups, anthropologists have divided them into 10 separate cultural areas. These are the Arctic, the Subarctic, the Northeast, the Southeast, the Plains, the Southwest, the Great Basin, California, the Northwest Coast and the Plateau. Let us briefly examine each of these distinctive cultural groups.

The Arctic culture area was a cold, flat, treeless region, effectively a frozen desert, near the Arctic Circle in modern-day Alaska, Canada, and Greenland; and was the home to the Inuit and Aleut peoples. Both groups spoke dialects descended from the Eskimo-Aleut language family. Because it was such an inhospitable landscape, the Arctic's population was comparatively small and scattered. Some of its peoples, especially the Inuit in the northern part of the region, were nomads who hunted and followed seals and polar bears as they migrated across the frozen tundra. In the southern part of the region, the Aleut were more settled, living in small fishing villages along the coastline. The

Inuit and Aleut had many common aspects to their cultures. Many lived in dome-shaped houses made of sod or timber, or ice blocks in the northern Arctic region. They used seal and otter skins to make warm weatherproof clothing, aerodynamic dogsleds, and long open fishing boats that were called kayaks in Inuit and baidarkas in Aleut. By the time the United States purchased Alaska from Russia for $7.2 million in 1867, decades of oppression and exposure to European diseases had taken its toll; with the native population dropping to just 2500. The descendants of these survivors still live in the region today.

The Subarctic culture area mostly consisted of swampy taiga pine forests, and waterlogged tundra, that stretched across much of inland Alaska and Canada. Anthropologists have categorised this region's people into two distinctive language groups; the Athabaskan speakers at its western end, among them the Tsattine (Beaver), Kuchin, and the Ingalik; and the Algonquian speakers at its eastern end, including the Cree, Ojibwa, and Naskapi. In the Subarctic region travel was difficult, with toboggans (sleds), snowshoes, and lightweight canoes being the primary source of transportation; and the population was sparse. The peoples of this region did not form large permanent settlements; instead they lived in small close-knit family groups as they wandered after herds of caribou reindeers. They lived in small easy-to-move tents, and when it grew too cold to hunt game they would stay warm in underground dugouts. The growth in the European fur trade in the 17th and 18th centuries disrupted the Subarctic way of life; instead of hunting and gathering for subsistence, the Indians focused on supplying fur pelts to the European traders. This eventually led to their displacement and the extermination of many of these indigenous communities.

The Northeast (Eastern Woodands) culture area was one of the first to have ongoing contact with the Europeans. It stretched from the present-day Atlantic coast of Canada, to North Carolina, and inland to the Mississippi River Valley. Its peoples were members of two main groups; the Iroquoian speakers, who included the Cayuga, Oneida, Erie, Onondaga, Seneca, and Tuscarora tribal groups, most of whom lived alongside inland rivers and lakes in fortified, politically stable villages; and the more numerous Algonquin speakers, who included

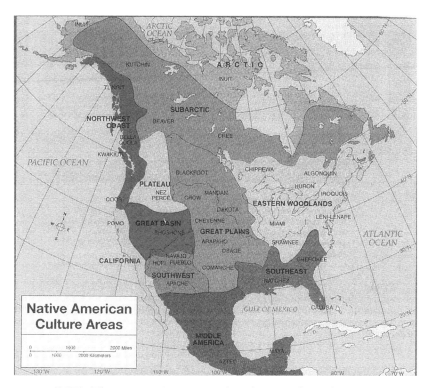

(257) The Native American cultural areas of North America

the Pequot, Fox, Shawnee, Wampanoag, Delaware, and Menominee tribal groups, who lived in small farming and fishing villages along the ocean. There, they would grow crops such as corn, bean, and a variety of vegetables. Daily life in the Northeast culture was often dangerous and included conflict. The Iroquoian tribal groups were aggressive and warlike; so tribes and villages outside of their allied confederacies were never safe from their lightning raids. Conflict became even more pronounced when the European colonizers arrived. Colonial wars repeatedly forced the region's natives to take sides, often pitting the Iroquois groups against their Algonquian neighbours. Meanwhile, with white settlers constantly moving westward, both of these indigenous groups were displaced from their native lands.

The Southeast culture area was situated north of the Gulf of Mexico and south of the Northeast culture. This area was a humid, fertile agricultural region; and many of its natives were expert farmers who grew staple crops like maize, beans, squash, tobacco, and sunflower. They organized their lives around small ceremonial and market villages that were known as hamlets. The major tribal groups in this region were the Cherokee, Chickasaw, Choctaw, Creek, and Seminole; often referred to as the Five Civilized Tribes. They spoke various Muskogean language dialects. By the time the United States had won its independence from Britain in 1776, the Southeast culture area had already lost many of its native people to disease and displacement. In 1830 the US federal Indian Removal Act forced the relocation of what remained of the Five Civilized Tribes, so that white settlers could take possession of their lands as they journeyed westward. Between 1830 and 1838, federal officials forced nearly 100,000 Indians out of the southern states and into "Indian Territory" which later became the state of Oklahoma, west of the Mississippi River. The Cherokee people called this often deadly trek, the "Trail of Tears."

The Plains culture area comprised the vast prairie region between the Mississippi River and the Rocky Mountains, from present-day Canada to the Gulf of Mexico. Before the arrival of European explorers and traders, its inhabitants, who mainly spoke Siouan, Algonquian, Caddoan, Uto-Aztecan, and Athabaskan languages, were well settled game hunters and farmers. After European contact, and particularly after the Spanish colonists brought horses to the region in the 18th century, the peoples of the Great Plains became much more nomadic. Tribal groups like the Crow, Blackfeet, Cheyenne, Comanche, and Arapaho, skilfully mastered the use of horses to pursue and hunt great herds of buffalo across the prairies. The most common dwelling for these tribal Indian groups was the cone-shaped tepee, a bison-skin tent that could be folded up and carried anywhere. Plains Indians were also renowned for their elaborately feathered war bonnet headdress. As European traders, cattle-ranchers, and settlers in their wagon trains moved westward across the Plains region, they caused much damage in the form of commercial goods such as knives and kettles, which native people came to depend on; guns and ammunition; and disease. By the end of the 19th century, white sport hunters had nearly extermi-nated the buffalo herds of the region. With settlers and cattle ranchers encroaching on their lands, and no way to make a living, the Plains natives were forced onto federal government reservations.

(258) A Native American Indian Tribal Chief with his family

The Southwest culture area was a large desert region spanning Arizona, New Mexico, along with parts of Colorado, Utah, Texas, and northern Mexico. The peoples of this region had developed two distinct ways of life. Firstly, there were the sedentary farmers such as the Hopi, Zuni, Yaqui, and the Yuma, who grew crops such as beans and squash. Many lived in permanent settlements that were known as pueblos, built of stone and adobe. These pueblos featured great multi-storey dwellings that resembled apartment houses. At their centres, many of these villages also had large ceremonial pit houses, or kivas. The second group, such as the Navajo, and the Apache tribes, were more nomadic. They survived by hunting, gathering, and raiding their more established neighbours for their crops. Because these groups were always on the move, their homes were much less permanent than the pueblos. For example, the Navajo tribes built their iconic eastward-facing round houses, known as hogans, out of materials such as mud and tree-bark. By the time the Southwest territories had become a part of the United States after the Mexican War in 1848, many of the region's native people had already been exterminated. For example, Spanish colonists and missionaries ended up enslaving many of the Pueblo Indians, working them to death on vast Spanish ranches known as encomiendas. Also, during the second half of the

19th century, the US federal government resettled most of the region's Indian natives onto reservations.

The Great Basin culture area was an expansive bowl formed by the Rocky Mountains in the east, the Sierra Nevada's in the west, the Columbia Plateau in the north, and the Colorado Plateau in the south. It was mainly a barren wasteland of deserts, salt flats, and brackish lakes. Most of the people here were from the Bannock, Paiute, and Ute tribes, and spoke Shoshonean or Uto-Aztecan dialects. They survived by foraging for roots, seeds, and nuts; and hunted snakes, lizards, and

(259) Cherokee Native American   (260) Apache Native American

small mammals. Because they were always on the move, they lived in compact, easy-to-build wikiups, which were huts made of willow poles or saplings, leaves, and brush. Their settlements and social groups were transient, and communal leadership when it existed, was informal. After European contact, some Great Basin groups got horses and formed equestrian hunting and raiding bands. After white prospectors discovered gold and silver in the region in the mid-19th century, most of the Great Basin people lost their land, and many were exterminated.

Before European contact, the temperate and hospitable California culture area had more people than any other North American culture at that time, an estimated 300,000 people in the mid-16th century. This culture was also the most diverse; with over 100 different tribes who spoke more than 200 dialects; languages that were mainly derived from the Penutian (spoken by the Maidu, Miwok, and Yokut tribes); the Hokan (spoken by the Chumash, Pomo, Salinas, and Shasta tribes); the Uto-Aztecan (spoken by the Tubabulabal, Serrano, and Kinatemuk tribes); and the Athapaskan (spoken mainly by the Hupa tribes). Despite this great diversity, many native Californians lived a very similar lifestyle. They did not practice much agriculture. Instead, they organized themselves into small, family based bands of around 150 hunter-gatherers known as tribelets. Inter-tribelet relationships, based on well-established systems of social connections, trade and common rights, were generally peaceful. Spanish explorers infiltrated the California region in the middle of the 16th century. In 1769, the priest Junipero Serra established a mission at San Diego, leading to a particularly brutal period in which forced labour, disease, and assimilation nearly exterminated the cultural area's native population.

The Northwest Coast cultural area covered the Pacific coast from British Columbia to the top of northern California. It had a mild climate, and an abundance of natural resources. In particular, the ocean and the region's rivers provided almost everything its people needed, including plenty of salmon, whales, sea otters, seals, and a wide variety of fish and shellfish. As a result, unlike many other hunter-gatherers who struggled to make a living and were forced to follow animal herds from place to place; the Indians of the Pacific Northwest were secure enough to build permanent villages that accommodated hundreds of people in each one. Those villages operated according to a rigidly stratified social structure, more sophisticated than any outside of Mexico and Central America. A person's status was determined by his closeness to the village's chief, and reinforced by the number of possessions he had at his disposal, including blankets, shells, skins, canoes, and even slaves in some instances. Goods like these played an important role in the "potlatch", an elaborate gift-giving ceremony designed to affirm these hierarchical class divisions. Prominent tribal groups in this region included the Athapaskan Haida, and Tlingit; the Penutian Chinook, Tsimshian, and Coos; the Wakashan Kwakuitl, and Nuu-chah-nulth or Nootka; and the Salishan Coast Salish.

The Plateau culture area was located within the Columbia and Fraser River Basins at the intersection of the Subarctic, Plains, Great Basin, California, and Northwest Coast cultures; in what is present-day Idaho, Montana, eastern Oregon, and Washington. Most of its people lived in small, peaceful villages along streams and riverbanks; and survived by fishing for salmon and trout, and hunting and gathering wild berries, roots, and nuts. In the southern Plateau region, the great majority of people spoke languages derived from the Penutian; and included the Klamath, Klikitat, Modoc, Nez Perce, Walla Walla, and Yakima tribal groups. North of the Columbia River, most tribal groups such as the Skitswish, Salish, Spokane, and Columbia, spoke Salishan dialects. In the 18th century, other Native American Indian groups brought horses to the Plateau. The region's inhabitants quickly integrated horses into their economy; expanding the radius of their hunts, and acting as traders and emissaries between the Northwest and the Plains cultures. In 1805, the American explorers Lewis and Clark passed through the area, drawing increasing numbers of disease-spreading white settlers. By the end of the 19th century, most of the remaining Plateau Indians had been cleared from their lands and resettled in federal government reservations.

As soon as the English colonists arrived in Jamestown Virginia in 1607, they shared an uneasy and suspicious relationship with the Native American Indians who as we have seen, had thrived on the North American continent for thousands of years. At that time in 1607, millions of indigenous people were scattered across North America in hundreds of different tribes. Between 1622 and the late 19th century, a series of wars known as the American-Indian Wars took place between native Indians and American settlers, mainly over control of land and resources. On 22nd March 1622, in an attempt to regain their confiscated land, Powhatan Indians attacked and killed colonists in eastern Virginia. Known as the Jamestown Massacre, this bloodbath gave the English government an excuse to justify their

(261) The Jamestown Virginia English Colony was founded in North America in 1607

efforts to attack Indians and confiscate more of their land. In 1636, the Pequot War broke out over settler expansion of trading routes; between the Pequot Indians and English settlers of Massachusetts Bay and Connecticut. The colonists' Indian allies joined them in battle and helped defeat the Pequot.

It wasn't just the British, who engaged in these battles. A series of battles took place from 1636 to 1659 between the New Netherlands (Dutch) settlers in New York and several Indian tribes, including the Lenape, Susquehannocks, Algonquians, and Esopus. Some of these battles were particularly violent and gruesome, and resulted in many settlers fleeing back to the Netherlands. The Beaver Wars between 1640 and 1701 occurred between the French and their Algonquin and Huron Indian allies, and the powerful Iroquois Confederacy. This fierce fighting started over territory and the control of the fur trade around the Great Lakes, and ended with the signing of the Great Peace Treaty. These earlier battles unfortunately set the scene for continuous conflict between the Indians and European settlers.

When the American Revolution broke out in 1775, Indians had to choose sides or try to stay neutral. Many tribes such as the Iroquois, Shawnee, Cherokee, and Creek, fought with the British; and others including the Potawatomi and the Delaware, sided with the American revolutionary patriots. No matter which side they fought with though, the Native Americans were adversely affected. They were left out of peace talks and lost additional land and resources. After the revolutionary war, some Americans retaliated against those Indian tribes who had supported the British. The Cherokee Chief, Dragging Canoe, led bands of Indians against white settlers in the Southern American States from 1776 to 1794. At the Battle of the Bluffs, he led 400 warriors to destroy Fort Nashborough in Tennessee, but a pack of unleashed hunting dogs forced them back into retreat during the battle. In 1830, President Andrew Jackson signed the Indian Removal Act, allowing the United States government to relocate Indians from their land east of the Mississippi River. In 1838, the government forcibly removed around 15,000 Cherokee from their homeland, and made them walk over 1900 kilometres west. Over 3000 Indians died on this gruelling route, which was known as the Trail of Tears. These involuntary relocations fuelled the Indians' anger and hatred towards the US government.

One of the most famous battles between the European Americans and Native American Indians was the Battle of the Little Bighorn on 25th June 1876. In this defining battle, the American General George Armstrong Custer led 600 men into the Little Bighorn Valley, where they were outnumbered and overwhelmed by 3000 Sioux and Cheyenne warriors, led by Chief Crazy Horse. Custer and his men were all killed in the battle, known as Custer's Last Stand. Despite the decisive Indian victory, the US government forced the Sioux to sell the Black Hills and vacate the land. George Armstrong Custer's apparent last words have been mythologized. While being swarmed and overrun by hostile Lakota, Cheyenne, and Arapaho Indian warriors at Little Bighorn, the colourful and charismatic US 7th Cavalry Commander reportedly shouted inspirational encouragement to his men. "Hurrah boys! Let's get these last few reds then head on to camp." This quotation is almost certainly a fabrication that was dreamed up sometime later

(262) The Battle of Little Bighorn – 25th June 1876

to mythologize Custer as an American hero and patriot. Custer and his entire force were wiped out, so nobody present survived to report anything that he said. General Custer was known to express supreme confidence and optimism on the battlefield, with statements like: "There are not enough Indians in the world to defeat the Seventh Cavalry. The Seventh can handle anything it meets" Before the start of the Battle of Little Bighorn he was heard to say: "Where did all these damn Indians come from?" And in a moment of empathy to the plight of the Indians, he said: "If I were an Indian, I would greatly prefer to cast my lot among those of my people who adhere to the free open plains, rather than submit to the confined limits of a reservation."

We can best gauge the thinking of Chief Crazy Horse with these quotations that were attributable to him. On the subject of European colonization: "We did not ask you white men to come here. The Great Spirit gave us this country as a home. You had yours. We did not interfere with you. We do not want your civili-

(263) *General George Custer*   (264) *Chief Crazy Horse*

zation!" In his war cry to his warriors before the Battle of Little
Bighorn: "Hokahey! Today is a good day to die." While smoking
a Sacred Pipe with Chief Sitting Bull, four days before Sitting
Bull's assassination, Crazy Horse asserted: "Upon suffering beyond
suffering, the Red Nation shall rise again, and it shall be a blessing
for a sick world. A world filled with broken promises, selfishness
and separations. A world longing for light again." Speaking to his
parents on his deathbed in 1877, he said: "We preferred hunting
to a life of idleness on the reservations. At times we did not get
enough to eat, and we were not allowed to leave the reservation
to hunt." He made the following request for the handling of his
body upon death: "At my death, paint my body with red paint and
plunge it into fresh water to be restored back to life, otherwise
my bones will be turned into stone and my joints into flint in my
grave, but my spirit will rise." Before his death, Chief Crazy Horse
made this prediction to the leader of the Oglada Lakota tribe: "I
will return to you in stone." In 1948, work began on a sculpture
in South Dakota to honour Crazy Horse. The Polish American
sculptor Korczak Ziolkowski designed the sculpture, thinking it
would take 30 years to complete. As of early 2021 over 70 years

later, it is not even close to finished. When completed, it will be the world's largest monument to a human.

In the late 19th century, Indian "Ghost Dancers" believed a specific dance ritual would reunite them with the dead, and bring them peace and prosperity. On 29th December 1890, the US Army surrounded a group of Ghost Dancers at Wounded Knee Creek, near the Pine Ridge reservation of South Dakota. During the ensuing Wounded Knee Massacre, fierce fighting broke out and 150 Indians were slaughtered. This battle was the last major conflict between the US government and the Plains Indians. By the early 20th century, the American-Indian Wars had effectively ended at an enormous cost. Even though the Indians had predominately helped the European colonial settlers to survive and flourish in the New World; helped the American revolutionaries gain their independence; and given up vast amounts of land and resources to the pioneering settlers as they expanded westward; hundreds of thousands of Indian and European lives were lost to war, disease, and famine; and the traditional culture and way of life of many hundreds of Native American Indian tribes was destroyed and eliminated. Tragically, this was to be a common pattern of European colonization throughout the world; including the Americas, Australia, New Zealand, other Pacific Islands, Asia, and Africa.

In addition to North American cultures, there were also civilizations that emerged in Central America (Mesoamerica). The history of Mesoamerica is usually divided into specific periods which, taken together, reveal the development of culture in the region; particularly that of the Mayan and Aztec Civilizations. The specific periods start with the Archaic Period, which lasted from 7000 to 2000 BCE. During this time a hunter-gatherer culture began to cultivate crops such as maize (corn), beans, and other vegetables; and the domestication of animals (such as dogs and turkeys) and plants became widely practiced. The first villages of the region were established during this period which included sacred spots and temples dedicated to various gods. The villages excavated so far are dated from 2000-1500 BCE.

(265) The Mayan and Aztec civilizations of Central America

The Olmec Period came next, lasting from 1500 to 200 BCE. During this period, the Olmec thrived; they were the oldest culture in Mesoamerica. The Olmec settled along the Gulf of Mexico and began building great cities of stone and brick. The famous Olmec heads strongly suggest highly sophisticated skills in sculpture, and the first indications of Shamanic religious practices, which date from this period. The enormous size and scope of Olmec archaeological ruins gave birth to the idea that the land was once populated by giants. Though nobody knows where the Olmec came from, nor what happened to them, they laid the foundation for all future civilizations in Mesoamerica. The remaining periods overlap somewhat with their dates, but are briefly outlined.

The Zapotec Period lasted from 600 BCE to 800 CE. In the region surrounding modern-day Oaxaca in Mexico, the cultural centre now known as Monte Alban was founded, which became the capital city of the Zapotec Kingdom. The Zapotec were clearly influenced by, and possibly related to the Olmec, and through them some of the most important cultural developments of the region occurred, such as writing, mathematics, astronomy, and accurate calendars; all of which the later Maya were to refine.

The Teotihuacan Period lasted from 200 to 900 CE. During this period, the great city of Teotihuacan grew from a small village to a metropolis of enormous size and influence. Early on, Teotihuacan was a rival of another city called Cuicuilco but, when that community was destroyed by a volcano in around 100 CE, Teotihuacan became the most dominant city-state in the region. Archaeological evidence suggests that Teotihuacan was an important religious centre which was devoted to the worship of a Great Mother Goddess and her consort the Plumed Serpent. The Plumed Serpent god Kukulkan (also known as Gucamatz) was the most popular deity among the Maya. Like many of the cities which lie in ruin throughout the southern Americas, Teotihuacan was abandoned sometime around 900 CE.

The El Tajin Period lasted from 250 to 900 CE. This period is also known as the Classic Period in Mesoamerican and Mayan history. The name 'El Tajin' refers to the great city complex in the Gulf of Mexico which has been recognized as one of the most important sites in Mesoamerica. During this time, the great urban centres rose across the land and the Maya people numbered in the millions. The very important ball game which came to be known as Poc-a-Toc was developed, with more ball courts being found in and around the city of El Tajin than anywhere else in the region. Who, precisely, the people were who inhabited El Tajin remains unknown, as there were over fifty different ethnic groups represented in the city, and dominance has been ascribed to both the Maya and the Totonac.

The Classic Maya Period lasted from 250 to 950 CE. This is the period which saw the consolidation of power in the great cities of the Yucatec Maya, such as Chichen-Itza and Uxmal. Direct cultural influences can be seen in some sites, from the Olmec and the Zapotec, and in the cultural values of Teotihuacan and El Tajin, but in other respects, a wholly new culture emerged; such as in Chichen-Itza for example, where there was a significantly different style of art and architecture. This period was the height of the Mayan Civilization in which they perfected mathematics, astronomy, architecture, and the visual arts; and also refined and perfected the calendar. The city-states of the Mayan

Civilization stretched from Piste in the north all the way down to modern-day Honduras.

The Post-Classic Period lasted from 950 to 1524 CE. At this time, the great cities of the Maya were abandoned. So far, no explanation for the mass exodus from the cities to outlying rural areas has been determined, but climate change and overpopulation have been strongly suggested, among other possibilities. The Toltec, a new tribe in the region, took over the vacant urban centres and repopulated them. At this time, Tula and Chichen-Itza became dominant cities in the region. The widely popular conception that the Maya were driven out of their cities by the Spanish Conquest is simply not true, because the cities were already vacant by the time of the Spanish invasion. In fact, the Spanish conquerors had no idea the natives they encountered in the region were responsible for the enormous complexes of the cities. The Quiche Maya were defeated by the Spanish at the Battle of Utatlan in 1524 CE, and this date traditionally marks the end of the Mayan Civilization.

(266) Artist's impression of a Mayan city

The Maya resided in independent city-states, much like in Ancient Greece, each of which was a city surrounded by farmland and ruled by a supreme king. Each city, with pyramids, palaces, and temples, served as the economic, religious, and political centre to its region. Tikal and

Copan were two of the largest Mayan city-states; at Tikal's peak, its population was somewhere between 50,000 and 100,000 inhabitants. The king was at the centre of political, economic, and religious life in a city-state. The king's role mirrored that of the gods, and his dress reflected this. During religious ceremonies, a king would take on the persona of a god by wearing a mask and ceremonial dress. In the eyes of the Maya people, he became like a god. Mayan society was class-based. After the king and his family, nobility was the most powerful, followed by the priests. It is believed that the Maya's hieroglyphs – pictorial symbols used as the basis of the Maya's writing system – were created only for the most powerful classes, and only the nobility and religious leaders could read them. Commoners such as farmers, labourers, and their children, remained illiterate. Slaves, who were at the bottom of Mayan society, could not read either.

(267) Mayan human sacrifice

Art played a central role in Mayan religion. Much of Mayan art has survived, and through it we know that the Maya worshipped dozens of gods. One of their most important gods was the Maize God. For the Maya, this god represented corn's life cycle of planting, growing, and harvesting, which in turn represented the life cycle of humanity itself. Some art was used for burial purposes in order to accompany the dead into the afterlife. The Maya fed themselves chiefly by growing their own food. Most of their soil was located in jungle, which was not very rich with nutrients. The Maya had a large population and practiced large-scale farming on estates. They accomplished this by the 'slash and burn' method, in which they hacked away areas of trees and plants, and then burned the rest (along with what they had cut) to make room for their crops. This is a method that the descendants of the Maya still use today. They mixed the ash of the burned foliage to fertilize the soil. They also rotated crops, that is, moved them around to let the wild plants grow back, to slow the depletion of nutrients in the soil. The Maya took advantage of all available land by practising terracing, in which they built stone walls along the sides of mountains so they could also cultivate this land.

The Maya used a numerical system based on just three symbols: a dot for one, a bar for five, and a shell which represented zero. Using zero and place notation, they were able to write large numbers and perform complex mathematical calculations. They also formulated a unique calendar system with which they were able to calculate the lunar cycle as well as predict eclipses and other celestial events with great precision. In the Mayan view of the universe, the plane on which we live is just one level of a multi-layered universe made up of 13 heavens and 9 underworlds. Each of these planes was ruled by a specific god and inhabited by others. Hunab Ku was the creator god and various other gods were responsible for the forces of nature, such as Chac, the rain god. Mayan rulers were considered to be divine and could trace their genealogies back to prove they were descended from the gods. Mayan religious ceremonies included ball games, human sacrifice, and bloodletting, in which nobles pierced their tongues or genitals to shed blood as an offering to the gods. Human sacrifice was considered honourably as a means in which a person could hopefully ascend to one of the levels of heaven and be in the company of the gods; rather than hopefully not descending into an underworld.

The Mayans wore beautiful fabrics and designed musical instruments like horns, drums, and castanets. They also carved huge statues, honouring their gods and leaders. The Mayan hieroglyphics writing system was made up of 800 symbols (glyphs). Some of the symbols were pictures and others represented sounds. They chiselled the symbols directly onto stone and inside codices, which were books that folded like an accordion. The pages were made of fig bark, covered in white lime and bound in jaguar skins. The Mayans wrote hundreds of these books. They contained information on history, medicine, astronomy, and religion. The Spanish missionaries burned all but three of these books.

The modern day difficulty in deciphering the Mayan hieroglyphics stems from the actions of the same man, who inadvertently, preserved so much of what we know of the Mayan Civilization; Bishop Diego de Landa. Appointed to the Yucatan following the Spanish conquest of the north, Landa arrived in 1549 CE and instantly set himself the task of rooting out heathenism from among the Mayan converts to Christianity. The concept of a god who dies and comes back to life was very familiar to the Maya from their own deity, The Maize God, and they seemed to have accepted the story of Jesus Christ and his resurrection easily. Even so, Landa believed that there was a subversive faction growing among the Maya which was seducing them back to their own traditional gods; and having failed to crush this perceived rebellion through the avenue of prayer and admonition, he chose another direct method. On 12th July 1562 CE, at the church at Mani, Landa burned over forty Mayan Codice books and over 20,000 images. In his own words, "We found many books with these letters, and because they contained nothing that was free from superstition and the devil's trickery, we burnt them, which the Indians greatly lamented." Landa went further however, and resorted to torture in order to extricate the secrets of the subversives among the natives, and bring them back to what he saw as the true path of the church. His methods were condemned by the other priests and he was recalled back to Spain to explain his actions. Part of his defence was his 1566 CE manuscript "Relacion de las Cosas de Yucatan" which has survived

(268) Spanish Bishop Diego de Landa imposing Christianity on the Mayans

and preserves much of the culture Landa tried to destroy, and has proved to be a valuable asset in our understanding of the Maya culture, religion, and language.

Only three books of the Maya escaped the reckless actions of Bishop Diego de Landa. They were "The Madrid Codex", "The Dresden Codex", and "The Paris Codex", all three named after the cities where they were found, many years after they were brought back from the Yucatan. These surviving primary sources have provided archaeologists with a great deal of information on the beliefs of the Maya, especially on their calendar. The codices were created by scribes who made careful observations in astronomy. The Dresden Codex alone, devotes six pages to accurately calculating the rising and positions of the planet Venus. The Mayan interpretation of the planets and the seasons exhibits a precision which was unmatched by any other ancient civilization, including the Egyptians. So important were their stories and books to the Maya that the "Legend of Zamma" and the "Hennequen Plant" describe the great goddess telling the prophet Zamma: "I want you to choose a group of families from my kingdom, and three of the wisest Chilames, to carry the writings which tell the story of our people, and write what will happen in the future. You will reach a place that I will indicate to you and you will found a city. Under its main temple you will guard the writings and the future writings." According to this legend, the Mayan city of Izamal was founded by Zamma of the Itzas who placed the sacred writings under the central temple. Izamal became known as an important religious pilgrimage site during the Classical Period, second in importance after Chichen-Itza. Day Keepers known as Shamans would interpret the particular energy of the day or month for the people by consulting with the gods presiding over the various months of the Maya calendar.

For a long time, it was commonly held by archaeologists that the Mayans did not practice human sacrifice. However, as more primary source evidence has been examined, including the translation of hiero-glyphs, it appears that the Maya frequently practiced human sacrifices within the context of their religion and politics. As we shall see, further to the north, the Aztecs would become infamous for holding their victims down on top of temples and cutting out their hearts, offering the still-beating organs to their gods. The Maya cut the hearts out of their sacrificial victims too, as can be seen for example, in several images surviving at the Piedras Negras historical site. However, it was much more common for the Maya to decapitate or disembowel their sacrificial victims, or else tie them up and push them down the stone stairs of their temples. The method of sacrifice that they adopted had much to do with who was being sacrificed and for what purpose. Prisoners of war for example, were usually disembowelled. When the sacrifice was religiously linked to the ball game, the prisoners were more likely to be decapitated or pushed down the stairs.

To the Maya, death and sacrifice were spiritually linked to the concepts of creation and rebirth. In the sacred book of the Maya, called the Popol Vuh, the heroic twins Hunahpu and Xbalanque must die and journey to the underworld before they can be reborn into the world above. In another section of the same book, the god Tohil asks for human sacrifice in exchange for fire. A series of hieroglyphs deciphered at the Yaxchilan archaeological site, links the concept of beheading to the notion of creation or "awakening." Sacrifices often marked the beginning of a new era; this could be the ascension of

a new king or the beginning of a new calendar cycle. These sacrifices, meant to aid in the rebirth and renewal of the harvest and life
cycles, were often carried out by priests and nobles, with the direct
involvement of the king. Children were sometimes used as sacrificial
victims at such times.

*(269) The Mayan ball game "Poc-a-Toc" symbolized the struggle
of life over death*

For the Maya, human sacrifices were associated with the ballgame.
This was the game in which a hard rubber ball was knocked around by
players mostly using their hips; and this often had religious, symbolic,
or spiritual meaning. Maya images show a clear connection between
the ball and decapitated heads; in fact, the balls were sometimes
made from skulls. Sometimes, a ballgame would be integrated into
a victorious battle, where captured warriors from a vanquished tribe
or city-state would be forced to play, and then sacrificed afterwards.
A famous image carved in stone at Chichen-Itza shows a victorious
ballplayer holding aloft the decapitated head of the opposing team
leader. Captive kings and rulers were often highly prized sacrifices. In
another carving from Yaxchilan, a local ruler, "Bird Jaguar IV," plays the

ball game in full gear, while "Black Deer," a captured rival chieftain, bounces down a nearby stairway in the form of a ball. It is likely that the captive was sacrificed by being tied up and pushed down the stairs of a temple as part of a ceremony involving the ballgame. In 738 CE, a war party from Quirigua captured the king of the rival city-state Copan; and the rival king was ritually sacrificed.

Another aspect of Maya blood sacrifices involved ritual bloodletting. In the Popol Vuh, the first Maya pierced their skin to offer blood to the gods Tohil, Avilix, and Hacavitz. Tohil's primary function was that of a fire deity, but he was also the sun god and the god of rain. Tohil was also associated with the mountains, and was a god of war, sacrifice, and sustenance. Avilix was the goddess of the moon and the queen of the night. She was also associated with the underworld, sickness, death, and was a patron of the ballgame. Hacavitz was a mountain god and was linked with the first mountain of creation. Maya kings and lords would pierce their flesh – generally the genitals, lips, ears, or tongues – with sharp objects such as the spines of stingrays. Such spines have been often found in the tombs of Maya royalty. Maya nobles were considered semi-divine, and the blood of kings was an important part of certain Maya rituals, which often involved agriculture. Not only male nobles, but females as well took part in ritual bloodletting. Royal blood offerings were smeared on idols or dripped onto bark paper which was then burned. They believed the rising smoke could open a gateway between the human world and the world of the gods.

This quotation from Bishop Diego de Landa's diary, vividly captures the tragic burning of the Maya books which he orchestrated, a bloodletting ritual, his banishment back to Spain, after which he wrote his book, and his atonement back into the power structure of the Catholic Church: "July 12, 1562, the books burned, the people cried. We found a large number of books in these characters and, as they contained nothing in which were not to be seen as superstition and lies of the devil, we burned them all, which they (the Maya) regretted to an amazing degree, and which caused them much affliction. As I watched the pile burn in the plaza, I remembered a stone relief I had seen, showing a native bloodletting ceremony, at Kulkulcan; an

ancient priest, passing the needle of a stingray through his penis, calmly
soaking up the blood with papers, and burning them. The twisting
smoke from this fire became the vision serpent, from whose mouth
sprang the voices of the ancestors. As I watched the black smoke curl
into the sky, I watched for any signs of this terrible serpent. Surely
now would mark the moment of its appearance. But I saw nothing.
Francisco Toral, bishop of Yucatan and Campeche, arrived shortly after
my auto-de-fe (book-burning) had run its course. He could sense the
people's hatred of me, though he would not acknowledge my own.
When the natives in the region began burning effigies of my likeness
and cursing my name, many Spaniards feared rebellion. My role in
the pageant became increasingly clear; I was to become an idol to be
sacrificed upon a public altar, for all to see. First, Toral questioned my
methods, and then had me sent back to Spain in disgrace. I did and
said what I had to upon my return; I wrote my book, 'Relacion de las
cosas de Yucatan', for fools who thought they could use it somehow to
win more souls for the faith. I gave them only the vaguest account of
Mayan glory, the merest hint of an alphabet. It was sufficient to stupefy
them out of any further curiosity. Eventually, satisfied my career was
finished, my enemies moved on to destroy each other. The next few
years were lost ones for me. Then came one of those inexplicable
shifts within the power structures of the Church, and suddenly I was
well-respected again. When Toral died in 1571, I became bishop of the
Yucatan."

The Aztec Empire was the last of the great Mesoamerican cultures.
Between 1325 and 1521 CE, the Aztecs forged an empire over much
of the central Mexican highlands. At its height, the Aztecs ruled over
200,000 square kilometres throughout central Mexico, from the Gulf
coast to the Pacific Ocean, and south to what is today Guatemala.
Millions of people in 38 provinces paid tribute to the Aztec King
Montezuma II, prior to the Spanish Conquest which started in 1519
CE. The Aztecs did not start out as powerful people. The Nahuati
speaking peoples began as hunter-gatherers in northern Mexico, in a
place known to them as Aztlan. In around 1100 CE, they left Aztlan
after they were commanded by their war god Huitzilopochtli to find

a new home. The god would send them a signal when they reached their new homeland. Archaeologists believe the Aztecs wandered for generations, heading ever southward. Backward and poor, other more settled people did not want the Aztecs to settle near them, and would drive them away. Finally, in around 1325 CE, they saw the god's signal – an eagle perched on a cactus eating a serpent on an island in Lake Texcoco, according to the legend. The city established by the Aztecs at that site, Tenochtitlan, grew to become the capital of the empire.

Fortunately, the site was a secure strategic area with reliable sources of food and clean water. The Aztecs began to build the canals and dikes necessary for their form of agriculture, and to control water levels and availability for irrigating crops. They built causeways linking the island to the shore. Because of the island location, trade and commerce with other cities around the lakes could easily be carried out with canoes

(270) The Aztec city of Tenochtitlan was founded around 1325 CE

and boats. As a result of marriage alliances with ruling families in other city-states, the Aztecs began to build their political base. They became fierce warriors and skilful diplomats. Throughout the late 1300s and early 1400s, the Aztecs began to grow in political and economic power. In 1428 CE, the Aztec King Itzcoatl formed alliances with the nearby

cities of Tiacopan and Texcoco, creating the Triple Alliance that ruled
until the coming of the Spanish in 1519.

The last half of the 15th century saw the Aztec Triple Alliance
dominate the surrounding areas, and reaping a rich bounty in tribute
paid to the ruler. Soon after, the Aztecs ended up controlling much
of central and southern Mexico. Thirty-eight provinces sent tribute
regularly in the form of rich textiles, warrior costumes, cacao beans,
maize, cotton, honey, salt, and slaves for human sacrifice. Gems, gold
and jewellery came to Tenochtitlan as tribute for the ruler, who by
now, had become an emperor. Wars for tribute and captives became a
way of life as the empire grew in power and strength. While the Aztecs
successfully conquered many city-states, there were some that resisted.
Tlaxcala, Cholula and Huexotzingo all refused Aztec domination and
were never fully conquered.

*(271) Hernan Cortes and the Spanish Conquistadors conquering
the Aztecs in 1521 CE*

The Aztec Empire, which controlled around 11,000,000 people,
had always had to deal with minor rebellions – typically, when new
rulers took power at Tenochtitlan – but these had always been swiftly
crushed. The tide began to turn though, when the Aztecs were heavily
defeated by the Tlaxcala and Huexotzingo in 1515 CE. With the
arrival of the Spanish soon after, in 1519, some of these rebel states

would again seize the opportunity to gain their independence. When the Spanish conquistadors finally did arrive from the Old World, sailing in their floating palaces, and led by Hernan Cortes; their initial relations with the leader of the Aztecs, Montezuma II, were friendly and valuable gifts were exchanged. However, things turned sour when a small group of Spanish soldiers were killed at Tenochtitlan while Cortes was away at Veracruz. The Aztec warriors, unhappy at Montezuma's passivity, overthrew him and made Cuitlahuac, who was his brother, as the new ruler of the Aztecs. This incident was exactly what Cortes needed, and so he returned to the city to relieve the besieged remaining Spanish forces, but was forced to withdraw on 30 June 1520, in what became known as the "Noche Triste." Gathering local allies, Cortes returned ten months later, and in 1521 laid siege to the city, burning it to the ground in the process. Lacking food and ravaged by European-introduced diseases, the Aztecs, now led by Cuauhtemoc, finally collapsed on the fateful day of 13 August 1521. Tenochtitlan was sacked and its monuments destroyed. From the ashes of this once-great Aztec city rose the new capital of the colony of New Spain; and the long line of Mesoamerican civilizations which had stretched right back to the Olmec and then Mayans, came to a dramatic and brutal end.

I have listed here, some quotations from the Spanish conquistador Hernan Cortes, to give us an insight into his state of mind, insights, and motivations, as he encountered the Aztec culture head-on:

- I and my companions suffer from a disease of the heart which can be cured only with gold.
- The city is as large as Seville or Cordova; its streets, I speak of the principal ones, are very wide and straight; some of these, and all the inferior ones, are half land and half water, and are navigated by canoes.
- The priests are debarred from female society, nor is any woman permitted to enter the religious houses.
- This city has many public squares, in which are situated the markets and other places for buying and selling.
- The meals were served in a large hall, in which Montezuma

was accustomed to eat, and the dishes quite filled the room,
which was covered with mats and kept very clean.

- There are apothecaries' shops, where prepared medicines,
liquids, ointments, and plasters are sold; barbers' shops, where
they wash and shave the head; and restaurateurs, that furnish
food and drink at a certain price.

- Thus they have a religious idol that they petition for victory in
war; another for success in their labours; and so for everything
in which they seek or desire prosperity, they have their idols,
which they honour and serve.

- Among these temples there is one which far surpasses all the
rest, whose grandeur of architectural details no human tongue
is able to describe; for within its precincts, surrounded by a
lofty wall, there is room enough for a town of five hundred
families.

- There are fully forty towers, which are lofty and well built, the
largest of which has fifty steps leading to its main body, and is
higher than the tower of the principal tower of the church of
Seville.

- An abundant supply of excellent water, forming a volume
equal in bulk to the human body, is conveyed by one of these
pipes, and distributed about the city, where it is used by the
inhabitants for drink and other purposes.

- They undertake nothing without first offering sacrifice there.

- They open their breasts, while they are alive, and take out the
hearts and entrails, and burn the said entrails and hearts before
the idols, offering that smoke in sacrifice to them.

- The inhabitants of this province would often caution me not
to trust these vassals of Montezuma for they were traitors, and
always acted treacherously and artfully, by which means they
had subjugated the whole earth.

- They also begged me to protect them against the almighty
Lord, who used violent and tyrannical measures to keep them
in subjugation, and took for them their sons to be slain and
offered as sacrifices to his idols.

- I should march against them and destroy them as rebels who refused to submit to the government of your Majesty.
- Until then it had been my intention to visit his country as a friend, to see and talk with him, and hold much peaceful intercourse with him. Now I shall enter his dominions in the guise of war, doing all the injury that was in my power, as an enemy.

The Incas were South America's largest and most powerful empire. They flourished between 1400 and 1533 CE. Through their formidable military prowess, they ruled over a kingdom which stretched from the northern tip of present day Ecuador and through to central Chile. At its peak, the Inca Empire comprised of 12,000,000 people from over 100 different ethnic cultural groups. They managed to create a relatively cohesive state, thanks mainly to a centralized government, common spoken language, and a unified religion. The achievements of the Incas are even more amazing when one considers they created a powerful unified empire without the use of the wheel, cast iron, or any form of writing; and that ultimately, they were shown to be completely defenceless, against the Spanish conquistadors. The last remains of the Inca Empire were extinguished by the late 1500s.

Legend has it that the Incas was created by the Inti, the sun god; but historians prefer the version which states the Incas started off as a small and unassuming tribe who lived in current-day Cusco, high up in the Peruvian Andes mountain range. They lived in relative anonymity as a small kingdom, under the reigns of their first three kings. The expansion of the kingdom gathered momentum under Mayta Capac, the fourth king; and was further bolstered during King Viracocha's rule between 1410 and 1438. His partnership with two influential uncles helped in the formation of the first Inca standing army of substance. The Incas successfully conquered the Ayarmaca people, in a region now known as the Sacred Valley of the Incas, located just north of Cusco. It was during this time that the Incas developed the practice of challenging, conquering, and expanding into new territory; and leaving behind a military base from which to defend borders, and uphold peace within their conquered territories.

(272) Inca, Aztec & Maya Empires
(273) Major Inca cities

During the rule of the Incas, the crime rates were minimal. This seems more to do with the manner in which they ruled their subjects with an iron-fist, with severe punishments for breaking laws; rather than the conquered people being happy with Inca rule, which in many instances they weren't. The central Inca government went to extraordinary lengths to put down any rebellion uprisings within the empire, no matter how small they were. They would often resettle any quarrelling ethnic groups, to avoid future conflict from flaring up again. Inca expansion through military conquest grew exponentially with each new emperor-king, primarily due to a law which stated that each new ruler could not inherit the wealth and lands of his predecessor. So, if a new ruler wished to accumulate wealth, he needed to conquer new lands.

The Incas were advanced and innovative with their agricultural farming practices. They used canals and aqueducts extensively for irrigation; and engaged in terraced farming alongside mountainous terrain, which meant they could utilize all available land for farming, and achieve better yields for some crops that were better suited for higher altitudes. They used elaborate stone built structures to store and dry-freeze agricultural food produce for times of drought. They mined gold and silver extensively; and the women weaved elaborately colourful textiles. The Incas were remarkable and uniquely creative artists and musicians; as well as astute mathematicians, inventing a calculating device for account-keeping which used coloured strings and knots. The Incas adopted "Quechua" as their primary spoken language throughout the empire. They built a complex system of roads, totalling more than 24,000 kilometres. Teams of relay runners would be used to transport goods and to deliver messages to and from every corner of the empire, at an incredible rate of 240 kilometres each day; as we have seen, the Incas were not aware of, and did not use any form of transportation based on the concept of a wheel.

The Inca practiced a polytheistic religion which although centred around the god, Inti, the sun and rain god; also included a number of other gods including Viracocha, the god of all creations. The Incas built shrines and temples throughout the empire, where all-powerful priests initiated ritual animal and human sacrifices to appease the gods, and ensure good crops and victory in battles. The Incas were not as militaristic as what may seem to be the case with such a large empire. They appeared to have mastered the art of diplomacy, using military force for conquest only as a last resort. They formed many alliances with neighbouring city-states and ethnic groups; using marriage and exchanges of gifts, as well as their compelling powers of persuasion and the seeking of "mutual benefits" in their negotiations.

(274) The Inca Pyramids were used by kings for administrative and religious functions

The Inca Empire was founded on, and maintained, by force; and the ruling Incas were very often unpopular with their subjects, especially in the northern territories. This was a situation that the Spanish conquistadors, led by Francisco Pizarro, would take full advantage of in the middle decades of the 16th century CE. In one of his diary notes, we get a clear idea of his insatiable quest to enrich himself, when he wrote: "Friends and comrades! On that side (the south) are toil, hunger, nakedness, the drenching storm, desertion and, death; on this side ease and pleasure. There lies Peru with its riches; here Panama and its poverty. Choose, each man, what best becomes a brave Castilian. For my part, I go to Peru."

(275) Francisco Pizarro   (276) Inca Emperor - Wayna Qhapaq

The Inca Empire had not reached a stage of consolidated maturity when it faced its greatest existential challenge. Rebellions were widespread, and the Incas were engaged in a war in what today is Ecuador, where a second Inca capital had been established at Quito. Even more serious, the Incas were struck by an epidemic of European diseases, such as smallpox, which had spread from central America even faster than the European invaders themselves. The epidemic killed a staggering up to 90% of the native population! It was the European

(277) The Incas and the Spanish – a clash of cultures

disease that killed the Inca emperor Wayna Qhapaq in 1528 CE and two of his sons, Waskar and Atahualpa, while they were engaged in a brutal civil war for control of the empire, just when the European treasure-hunters arrived. It was this combination of factors; a perfect storm of internal rebellion, disease, and foreign invasion, which brought about the downfall of the mighty Inca Empire, the largest and richest ever seen in the Americas.

The Inca language Quechua lives on today, and is still spoken by over eight million people. There are also numerous buildings, artefacts, and written records which have survived the ravages of conquerors, looters, and the passage of time. These remains are proportionately few compared to the vast riches which have been lost, but they remain indisputable evidence of the wealth, ingenuity, and high cultural achievements of this great, but short-lived empire.

19

PACIFIC ISLAND CIVILIZATIONS AND THE AUSTRALIAN ABORIGINES

P OLYNESIAN NAVIGATION OF THE PACIFIC OCEAN AND ITS SETTLEMENT began thousands of years ago. The inhabitants of the Pacific Islands had been voyaging across vast expanses of ocean sailing in double canoes or outriggers, using nothing more than their knowledge of the stars and observations of sea and wind patterns to guide them. The Pacific Ocean comprises one-third of the Earth's surface, and

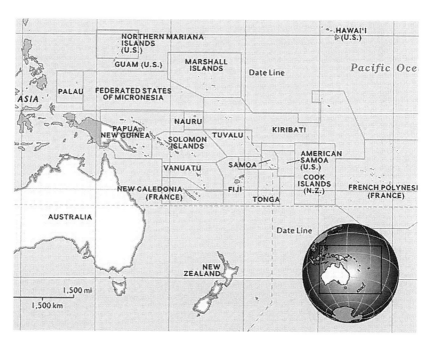

(278) The Pacific Island civilizations

its remote islands were the last to be reached by humans. These islands are scattered across an ocean that covers 165.25 million square kilometres. The ancestors of the Polynesians, the Lapita people, set out from Taiwan and settled Remote Oceania between 1100–900 BCE, although there is evidence of Lapita artefacts in the Bismarck Archipelago as early as 2000 BCE. The Lapita and their ancestors were skilled seafarers who memorised navigational instructions and passed their collective learning down through folklore, cultural heroes, and simple oral stories.

The Polynesian's highly developed navigation system impressed the first European explorers of the Pacific, and since then archaeologists have been debating several questions:

- Was the migration and settlement of the Pacific Islands and into Remote Oceania, accidental or intentional?
- What were the specific maritime and navigational skills of these ancient seafarers?
- Why has a large body of indigenous navigational knowledge been lost, and what can be done to preserve what remains?
- What type of sailing vessels and sails were used to cross an open ocean?

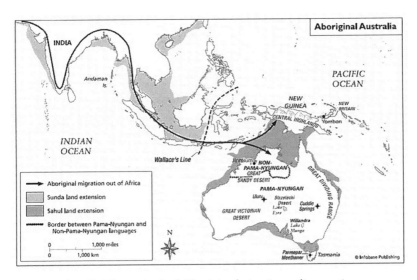

(279) The arrival of Aboriginals in Australia starting
around 65,000 years ago

By at least 10,000 years ago, humans had migrated to most of the habitable lands that could be reached on foot. What remained was the last frontier – the myriad islands of the Pacific Ocean that required boat-building technology and navigational skills, and methods which were capable of achieving long-range ocean voyaging. The region known as Near Oceania, which consists of mainland Papua New Guinea and its surrounding islands; the Bismarck Archipelago; the Admiralty Islands; and the Solomon Islands; were settled in an out-of-Africa migration which happened around 70,000 years ago. These first settlers of the Pacific were the ancestors of Melanesians and Australian Aboriginals. The small distances between the islands in Near Oceania meant that people could island-hop using basic ocean-going canoes and boats, made from carved-out tree trunks. This is how the Australian Aboriginals first arrived in Australia, around 65,000 years ago. The second wave of migration to Remote Oceania appears to have originated mainly from the island of Taiwan between 1500-1300 BCE. Remote Oceania comprises the islands to the east of the Solomon Islands group; such as Vanuatu, Fiji, Tonga, New Zealand, Society Islands, Easter Island, and Samoa. Archaeological and DNA evidence indicates these people who originated from Taiwan spoke a related group of languages known as Austronesian, and that they reached Fiji by 1300 BCE and Samoa by 1100 BCE. All modern Polynesian languages belong to the Austronesian family of languages.

Collectively, these people were called the Lapita, and were the ancestors of the Polynesians, including the Maoris, Fijians, Tahitians, Samoans, and Hawaiians. They were highly skilled seafarers who introduced outriggers and double canoes which made longer voyages across the Pacific possible. Their distinctive pottery – Lapita ware – appeared in the Bismarck Archipelago as early as 2000 BCE. Lapita pottery included bowls and dishes with complex geometric patterns impressed onto clay by small toothed stamps. Between 1100 – 900 BCE there was a rapid expansion of Lapita culture in a south-easterly direction across the Pacific. This raises the question of intentional migration. The geographic area in Remote Oceania is called the Polynesian Triangle and it includes New Zealand (the Maori name was Aotearoa), Hawaii, and Easter Island as its corners, and includes more than 1000 islands. Between some of the islands in this triangle, there are distances of more than 1000 kilometres. From Northern Vanuatu to Fiji, for example, is over 800 kilometres. It would have taken tremendous skill, courage, and perseverance to sail these distances in a canoe or outrigger for five to six weeks, in order to reach a hopeful destination.

The great British explorer and navigator, Captain James Cook (1728-1779 CE) had no doubt that Lapita-Polynesian navigation demonstrated a high degree of skill. In his journal during his first voyage to the South Pacific Ocean in 1768-1771 CE, he wrote: "These people sail in those Seas from Island to Island for several hundred Leagues, the Sun serving them for a Compass by day, and the Moon and Stars by night. When this comes to be proved, we shall be no longer at a loss to know how the Islands lying in those Seas came to be peopled." The Lapita people may have been able to exist for months on remote Pacific islands by feeding on wild birds, berries, coconuts, and seafood; but the success of any long-term settlement would have required the transporting of crop plants, such as taro and yam, as well

INTERIOR OF A MAORI PAH.

(280) The interior of a typical Maori village or "Pah"

as domestic animals. The sweet potato came into the Polynesian horticultural system around 1000 CE, and represents strong evidence that the Polynesians had ongoing contact with South American civilizations like the Aztecs and Incas, where the sweet potato originated.

Polynesian folklore, cultural heroes, and oral stories have preserved some ocean navigational information and ancestral knowledge. The Legend of Kupe and his discovery of New Zealand (Aotearoa) is an example of how oral stories were memory aids, passed down from one generation to another, that contained encoded instructions for reaching specific destinations around the Pacific. In traditional Maori oral cultural history, Kupe is a legendary figure and explorer of the Pacific Ocean who set off from Hawaiiki in 1300 CE in a canoe to discover what lay over the horizon. Hawaiiki was believed to be the ancestral home of the Maori people and is thought to have been located in the East Polynesian Islands. Kupe's navigator, Reti, followed a star path to hold the canoe on course until it reached landfall in Whangaroa on the North Island of New Zealand. The Polynesians knew the language of the stars. They had a highly developed navigational system that involved not only observation of the stars as they rose and crossed the night sky, but also the memorization of entire sky charts.

Throughout the Pacific, elder island navigators taught young men the skills acquired over many generations. Navigational knowledge was a closely guarded secret within a navigator family, and education started at an early age. On the island of Kiribati for example, lessons were taught in the meeting house (maneaba) where rafters and beams were sectioned off to correspond to a segment of the night sky. The positions of each star at sunrise and sunset, and the star paths between islands were etched into collective memories. Stones and shells were placed on mats or in the sand to teach star knowledge. Prayer (karakia) and oral stories contained references to navigational instructions. The following is an extract of navigational instructions from the Legend of Kupe: "When you go, lay the bow of the canoe to the Cloud Pillar that lies south-west. When night falls, steer towards the star Atua-tahi. Hold to the left of Mangaroa and travel on. When day breaks, again sail towards the Cloud Pillar and continue on."

(281) Polynesian outrigger canoes were designed for
long distance ocean travel

Polynesian mariners developed the double-hulled canoe, also
called a catamaran. Some of their voyaging canoes were longer than
Cook's Endeavour, which was around 30 metres in length; although
the average lengths of their canoes were between 15-23 metres. Canoes
with an outrigger on one side were favoured in Micronesia, in the
western Pacific region. The carrying capacity of these vessels was quite
considerable. A Tongan double canoe for example, could carry 80-100
people; while a Marquesan outrigger equipped for fishing or war, could
carry 40-50 people. James Cook observed that the Tahitian pahi canoe
could sail faster than the Endeavour. He wrote: "Their Large canoes
sail much faster than this Ship, all this I believe to be true and therefore
they may with Ease sail 40 Leagues a day or more." Long-distance craft
were sturdy planked vessels lashed together with braided sennit (tree
vines used as rope) or twisted coconut fibre. Caulking (watertight)
material such as breadfruit tree gum made them seaworthy. Different
types of canoes were used throughout Polynesia and Micronesia; but
the three main types were the pahi, the tongiaki, and the ndrua. The
pahi was a Tahitian two-hulled, two-masted vessel; the tongiaki from
Tonga was a double canoe with triangular sails; and the ndrua was a

double canoe with unequal hulls used in Fiji. Polynesian sails were the apex-down triangular sails; claw-shaped or crab-claw sails; and the lateen or triangular sails secured to two long booms. Sails were usually made from woven pandanus leaves.

There have been recent efforts to better understand and preserve the incredible feats of seamanship that enabled Polynesians to steer their craft with accuracy across the vast expanse of the Pacific Ocean. In 1985, a 22 metre voyaging waka christened with the name "Hawaikinui" was built. Its twin-hull was constructed from two insect-resistant New Zealand totara trees; and the waka successfully sailed from Tahiti to New Zealand (Aotearoa), a distance of over 4100 kilometres, using traditional Polynesian navigation techniques. In 2018, a young crew sailed a double-hulled voyaging waka from New Zealand (Aotearoa) to Norfolk Island, off the east coast of Australia, a distance of over 1400 kilometres. Although they were met with high ocean swells and strong unfavourable winds, the voyage was intended to teach young people the art of navigating by the stars and reconnecting with ancestral traditions.

In the Pacific region, there is an important distinction between "high" islands and "low" islands. Tahiti, a typical high island, is relatively large with steep mountainous slopes, richly lush diverse plant life, and many waterfalls and gushing streams. Coastal plains are absent or very limited on high islands. Atolls, which are ring-shaped islands made of coral, are the most common low islands in Polynesia. These are typically "desert islands" that are low-lying, narrow, and sandy, with few streams. Low islands have less diversity of plants and animals, compared to the lush jungles often found in the high islands. At the time of the first known European contact with the Polynesian world in 1595, there were an estimated 500,000 people scattered throughout the islands in the region. The European colonial powers competed for ownership of most of Polynesia's inhabited islands. The native indigenous populations suffered greatly at the hands of the Europeans. They lost their traditional lands and resources, and suffered discrimination against their cultures and languages; as well as being significantly afflicted by European diseases.

(282) Polynesian Samoans

Polynesian societies had an exceptionally rich body of folklore and mythology. Myths related mainly to the origins of human beings, as well as the origins of cultural practices and traditions. There was a considerable body of mythology regarding the origins of tattooing in Polynesian cultures. Some origin myths also described the process of migration from one island to another, via ocean-going canoes. Cultural heroes were an important aspect of Polynesian mythology and folklore. Polynesian religion changed dramatically with the coming of European Christian missionaries in the early part of the 19th century. Before the arrival of the Europeans, there was considerable variation in religious ideas and practices throughout Polynesia. In Hawaii for example, tribal chiefs were considered to be genealogically related to the gods and, as a result were believed to possess sacred power, which they called "mana." The concept of "tapu," in English "taboo," was important in all Polynesian societies. This refers to anything which was forbidden due to its sacredness. There were rules that served to protect people through forbidding certain actions and behaviours. For example, in the Marquesas Islands, a woman's menstrual cloth itself was not tapu; however, it was considered tapu to touch it. Today, most Polynesians are Christians; both Catholics and Protestants. This is a direct legacy of European colonisation.

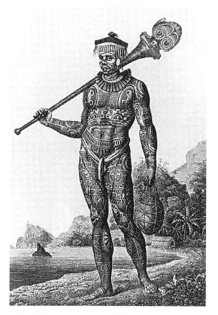

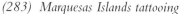
(283) Marquesas Islands tattooing   *(284) Marquesas Islands rituals*

One interesting custom that the Marquesas Islanders practiced was a birth rite of passage. On the day a child was born, they would have a birth feast. On that occasion, the maternal uncles and the paternal aunts of the newborn would cut their hair off. An ornament-maker would fashion hair ornaments for the child to wear later in life. The newborn was brought presents by family and friends, and a type of shrine would be built by the infant's father in honour of his newborn. Passage into puberty was often accompanied by tattooing rituals in many Polynesian societies. In some societies only men were tattooed;

in others both men and women were tattooed. In Polynesia, the practice of tattooing carries with it significant cultural and symbolic meanings. There have been recent revivals of the art of tattooing in societies such as the Maori of New Zealand. Another puberty ritual performed in some Polynesian societies was "fattening." Male and female youths were secluded, kept inactive and out of direct sunlight, and fed large amounts of food over a period of time, to make them more sexually desirable.

In the Marquesas Islands, death was accompanied by ritualized wailing on the part of women, and the performance of formalized chanting on the part of men. Women would also perform a specific dance called "heva." During this dance they would take off all of their clothes and move in an extremely sexually suggestive manner. Finally, the female relatives of the deceased would do physical harm to themselves by cutting their hands and faces with sharks' teeth and other sharp objects. European Christian missionaries saw these behaviours as pagan and quickly found ways to put a stop to them. Greetings in Polynesian societies varied from island to island. Hierarchical status determined the nature and extent of the social interactions of individuals in these societies. In rural Tahiti, for example, the standard greeting was; "Where are you going?" The two expected responses were; "Inland" (away from the coast), or "Seaward" (towards the coast). Premarital sexual relations were typically very casual in most Polynesian societies. However, once a permanent relationship was established, casual sexual relations outside of the relationship were forbidden and not permitted. The choice of a marriage partner was less fixed than in many other cultures of the world. In the times before the influence of Christianity, the preference in some Polynesian societies was for cross-cousin marriage; this is where a woman would marry her mother's brother's son or her father's sister's son. Missionaries forbade this type of marriage pattern.

Traditional Polynesian societies did not feature large settlements. Instead, families clustered together in small villages that focused on a set of shared buildings, for social, ceremonial, and religious life. Many Polynesians had separate sleeping quarters for bachelors. In some parts of Polynesia, houses were built on elevated stone platforms. Religious shrines were important parts of the household structure. Households of the nobility had carved items of furniture including headrests and stools. Sleeping mattresses were also used by the members of noble households. Lighting from fire-lit torches or coconut oil lamps were common inside houses and around the village at night. In societies such as Tahiti, with distinct social classes, marriage was traditionally prohibited among individuals from different classes. Children that were born from sexual relations between members of different classes were killed at birth. In many Polynesian societies, polygamy (multiple spouses) was practiced. In the traditional societies of the Marquesas Islanders, a woman could have more than one husband at a time. This practice, which is called polyandry, was rare in cultures around the world.

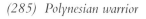

(285) Polynesian warrior

(286) Traditional Hawaiian
clothing

The role and status of women compared to men, varied between the island societies of Polynesia. In the Marquesas, women had always enjoyed a status nearly equivalent to men. One traditional indicator of this equality was that women were allowed tattooing almost as extensively as the men. In many other Polynesian societies this was not the case, as women held positions of lower status than men. Typical Polynesian clothing was similar for men and women. A section of bark cloth was worn as a loincloth by men, or as a waistcloth by women. Decorated bark cloth known as "tapa" was the main item of traditional clothing in Tahiti. A number of ornaments were worn for ceremonial events. Elaborate feather headdresses were signs of nobility. Both men and women wore ear ornaments. Most traditional Polynesian societies relied on fishing and fruit and vegetables for their staple food requirements. Taro root was an important source of nutritious food throughout the Polynesian Islands. In some parts of Polynesia, particularly Hawaii, Tahiti, and the Marquesas, men and women ate separately.

Traditional Polynesian societies had a rich tradition of vocal and instrumental music. Some types of musical expression have been lost, and new types were introduced by Christian missionaries. Christian hymns had a major influence on the style of vocal music throughout Polynesia. The Tahitian vocal music known as "himene" for example, blended European choir harmony produced by a number of vocalists singing at the same time, with Tahitian drone-style singing. One of the most well-known Polynesian musical instruments is the Hawaiian "ukulele." It is the Hawaiian version of the Portuguese/Spanish mandolin, which came to the Pacific Islands with the Portuguese immigrants who first arrived in the 1870s. The primary use of Hawaiian ukulele, flutes and drums was to accompany the graceful and sensual dance known as the "hula."

Arm wrestling was a traditional Polynesian form of male entertainment as well as a competition of masculine strength. Other forms of competition, including wrestling between males were common throughout the islands as ways of preparing for battle. Surfing was also popular in many of the Polynesian Islands, although it was only in Hawaii that surfers stood on their surfboards. The worldwide sport of surfing originated from European encounters with Hawaiian surfing. Decoration of everyday objects was a common pastime of Polynesian societies. Elaborate woodcarving was particularly well developed among the Maori of New Zealand. In most Polynesian societies, the designs and patterns that appeared on bark cloth or woodcarvings also appeared on the human body in the form of tattoos. In many Pacific Island societies tattooing was the primary art form.

The story of Captain James Cook's experiences in the Pacific, is a good example of the major cultural clashes that occurred when the Europeans encountered the indigenous Polynesians throughout the Pacific Islands. On 18th January 1778, the English navigator and explorer Captain James Cook became the first European to travel to the Hawaiian Islands, when he sailed past the island of Oahu. Two days later, he landed at Waimea on the island of Kauai, and named this group of islands the Sandwich Islands, in honour of John Montague, who was the Earl of Sandwich and one of Cook's patrons. In 1768, James Cook, a surveyor in the Royal Navy, was commissioned as a lieutenant and was placed in command of the HMS Endeavour; leading an expedition that took scientists to Tahiti to chart the course

(287) Captain James Cook conducted three expeditions to the Pacific from 1768 to 1779

of the planet Venus in its transit directly between the Earth and Sun;
a relatively rare astronomical event which only occurs four times
every 243 years. In 1771, he returned to England, having explored
and mapped the coasts of New Zealand and Australia; and circumnav-
igated the globe. Beginning in 1772, he commanded a major mission
to the South Pacific; and over the next three years he explored the
Antarctic region, charted the New Hebrides, and discovered New
Caledonia. In 1776, he again sailed from England as the commander
of the HMS Resolution and Discovery, and in 1778 made his first visit
to the Hawaiian Islands.

(288) Captain James Cook arriving at the New Hebrides
in the Pacific in 1774

Cook and his crew were at first welcomed by the Hawaiians, who
were fascinated by the Europeans' ships and their use of iron. Cook
provisioned his ships by trading the metal for food, and his sailors
traded iron nails for sex. The ships then made a brief stop at the island
of Niihau, before heading north to look for the western end of a
northwest passage from the North Atlantic to the Pacific. Almost one
year later, Cook's two ships returned to the Hawaiian Islands and found
a safe harbour in Hawaii's Kealakekua Bay. The Hawaiians appeared to
have attached religious significance to the first stay of the Europeans

on their islands. By the time of Cook's second visit, there was no doubt he and his crew were considered to be spiritual visitors. Kealakekua Bay was the sacred harbour of Lono, the fertility god of the Hawaiians; and at the time of Cook's arrival the locals were engaged in a religious festival dedicated to Lono. Cook and his crew were welcomed as gods, and for the next month, exploited the Hawaiians' goodwill and hospitality. After one of the crewmembers died, exposing the Europeans as mere mortals, relations became strained. On 4th February 1779, the British ships sailed away from Kealakekua Bay, but rough seas damaged the foremast of the Resolution, and after only a week at sea, the expedition was forced to return to Hawaii.

The Hawaiians greeted Cook and his crew by hurling rocks at them; then they stole a small cutter vessel from the Discovery. Negotiations with the Hawaiian King Kalaniopuu for the return of the cutter collapsed after a lower ranking Hawaiian chief was shot to death and an outraged mob of native Hawaiians descended on Cook's party. The captain and his men fired on the angry Hawaiians, but they were soon overwhelmed, and only a few of the crew managed to escape to the safety of the Resolution. Captain Cook himself was killed by the mob. A few days later, the Englishmen retaliated by firing their cannons and muskets at the shore, killing around 30 Hawaiians. The Resolution and the Discovery eventually returned to England.

Another tragic event in the history of European encounters in the Pacific Islands was the Mutiny on the HMS Bounty. This story is a stark reminder of the unforeseen events that can occur when people are removed from their accustomed lifestyle and culture; and placed in an unknown environment. In December 1787, the Bounty left England under the command of Captain William Bligh and a crew of 46 men, for the South Pacific islands of Tahiti, where it was to collect a cargo of breadfruit saplings to transport to the West Indies. There, the breadfruit would serve as food for slaves. After a 10-month journey, the Bounty arrived in Tahiti in October 1788, and remained there for over five months. On Tahiti, the crew enjoyed an idyllic life, revelling in the comfortable climate, lush surroundings, and the very hospitable

Tahitians. While there, the ship's first mate Fletcher Christian, fell in love with a Tahitian woman named Mauatua.

On 4th April 1789, the Bounty departed Tahiti with its cargo of breadfruit saplings. On 28th April, near the island of Tonga, Christian and 25 petty officers and seamen seized control of the ship. Bligh, who eventually would fall prey to a total of three mutinies in his career, was an aggressive, oppressive commander who regularly humiliated and insulted those under him. By setting him and 18 of his loyal supporters adrift in an over-crowded 23-foot-long boat in the middle of the Pacific, Christian and his mutineers had effectively given Bligh and his supporters a death sentence. But, as a result of remarkable seamanship, Bligh and his men reached Timor in the Dutch East Indies on 14th June 1789, after an incredible voyage of 5800 kilometres. Bligh then returned to England and soon after sailed again to Tahiti, from where he successfully transported breadfruit trees to the West Indies.

(289) The Mutiny on the Bounty took place on the 28th April 1789

Meanwhile, Christian and his men attempted to secure control and establish themselves on the island of Tubuai. Unsuccessful in their colonizing effort, the Bounty sailed north to Tahiti, and 16 crewmen decided to stay there, despite the risk of capture by the British author-ities. On the other hand, Christian and 8 other crew, together with 6 Tahitian men, and 12 Tahitian women, and a child, decided to search the South Pacific for a safe haven. In January 1790, the Bounty settled on Pitcairn Island, an isolated and uninhabited volcanic island over 1600 kilometres east of Tahiti. The mutineers who remained on Tahiti were captured and taken back to England, where three were hanged. A British ship searched the South Pacific for Christian and the others, but did not find them. In 1808, an American whaling vessel was drawn to Pitcairn Island by smoke coming from a cooking fire. The Americans discovered a community of women and children led by John Adams, the sole survivor of the original nine mutineers who had come to the island. According to Adams, after settling on Pitcairn, the colonists had stripped and burned the Bounty, and internal conflict and sickness had led to the death of Fletcher and all of the other men except him.

In 1825, a British ship arrived on Pitcairn and formally granted Adams amnesty from all prosecution. He then served as the patriarch of the Pitcairn community until his death in 1829. In 1831, the Pitcairn islanders were resettled on Tahiti; but unsatisfied with their lives there, they soon returned to their native island. In 1838, the Pitcairn Islands, which included three nearby uninhabited islands, was incorporated into the British Empire. By 1855, Pitcairn's population had grown to nearly 200, and the relatively small five-square-kilometre island could no longer sustain and support its residents. In 1856, the islanders were relocated to Norfolk Island, a former penal colony around 6270 kilometres to the west of Pitcairn. However, less than two years later, 17 of the islanders returned to Pitcairn, followed by more families in 1864. Today, only a few dozen people live on Pitcairn Island, and most of them are descendants of the Bounty mutineers. Around 1000 residents of Norfolk Island, approximately half the population, can also trace their ancestry to Fletcher Christian and the eight other British mutineers.

The Australian Aboriginal culture is the longest-surviving in human history, and goes back to around 65,000 years ago when the Aborigines first arrived in the island continent from Papua New Guinea, Timor, and a number of Indonesian islands including Sulawesi, in South East Asia; by foot, and "island-hopping" in canoes made from dug-out tree trunks. It is believed that the first early human migration to Australia was achieved when this landmass formed part of the Sahul continent, which connected to the islands of Papua New Guinea via a land bridge.

As of 2021, Madjedbebe, in the Northern Territory, is the oldest known site showing the presence of humans in Australia. Archaeological excavations provide evidence to suggest the Madjedbebe site was first occupied by humans 65,000 years ago. Caves and rock shelters at this site show evidence of rock art using ochre as a "reddish" coloured paint. Stone tools have also been recovered. This site is located in the far north of Australia. Humans reached Tasmania around 40,000 years ago by migrating across a land bridge from the mainland that existed during the last Ice Age. After the Ice Age ended and the seas rose around 12,000 years ago, the inhabitants in Tasmania were then isolated from the Australian mainland until the arrival of European settlers in the 18th century. Mungo Man, whose remains were discovered in 1974 near Lake Mungo in New South Wales, is the oldest human so far found in Australia; and has been dated to at least 40,000 years old. Stone tools also found at Lake Mungo have been dated at 50,000 years old.

(290) The Australian Aboriginals – oldest continuous culture on Earth

When the north-west of Australia, which is closest to Asia, was first occupied 65,000 years ago, the region consisted of open tropical forests and woodlands. After around 10,000 years of stable climatic conditions, by which time the Aboriginal people had settled the entire continent except Tasmania, temperatures began cooling and winds became stronger, leading to the beginning of an Ice Age. By 25,000 years ago, the sea level had dropped to 140 metres below its present level. At this time, Australia was connected to Papua New Guinea, and the Kimberley region of north-west Australia was separated from South East Asia by a strait which was only 90 kilometres wide.

Following the end of the Ice Age 12,000 years ago, Aboriginal people around the coast, from Arnhem Land, the Kimberley, and the southwest of Western Australia, all told stories about former territories that were drowned beneath the sea, after coastlines were flooded when sea levels began to rise after huge masses of ice began to melt, at the end of the Ice Age. It was this event that not only led to the isolation of Aborigines in Tasmania; but also the extinction of Aboriginal cultures on the Bass Strait Islands in between Victoria and Tasmania, and Kangaroo Island in South Australia. In the interior of Australia, the end of the Ice Age led to the migration and colonisation of the desert and semi-desert areas by Aboriginal people of the Northern Territory.

The Aboriginal Australians lived through great climatic changes and adapted successfully to their changing physical environment. There is ongoing debate from archaeologists about the extent to which the Aboriginals modified the environment. One line of debate revolves around the role of indigenous people in the extinction of the marsupial and other megafauna which lived on the continent during that time. Some argue that it was natural climate change which led to the extinction of this megafauna. Others claim that, because the megafauna were large and slow-moving, they were easy prey for human hunters and were eventually rendered extinct in this manner. A third possibility is that human modification of the environment, particularly through the use of fire, indirectly led to the marsupial megafauna extinction. Examples of this marsupial megafauna unique to Australia, included the largest-ever marsupial, the Diprotodon, which weighed two tonnes and looked like a modern rhinoceros; although its social lifestyle was more like that of an elephant. What the Diprotodon mostly resembled however, was a giant wombat. The "bunyips", Aboriginal legendary lake monsters which would drag unsuspecting passers-by into their watery lairs, may in fact have been Diprotodons, which wandered the swamps of Australia. Then there were also the Procoptodon goliah's, giant kangaroos which stood 2 metres tall, and weighed almost three times as much as a red kangaroo. These massive marsupials were a contradiction – a kangaroo that could not hop. They possessed short faces, with eyes that were almost forward-facing, similar to humans. Another giant marsupial was Meiolania platyceps – a giant tortoise with a pair of impressive horns on its head. Not all of the Australian megafauna were marsupials. The continent was also once home to large birds; some growing to 3 metres tall. For example, there was Dromornis stirtoni, a bird that looked like a giant ostrich or an emu, but was actually more closely related to ducks and geese. When it came to reptiles, there was Varanus priscus, commonly known as Megalania, which was the largest terrestrial lizard ever known; a giant goanna lizard, estimated to have been at least 5.5 metres long.

Aboriginal Australians were limited to the range of foods occurring naturally within their respective areas; but they knew exactly when, where, and how to find everything that was edible. Anthropologists who have studied the tribal diet in Arnhem Land for example, found it to be well-balanced, with most of the nutrients that modern dieticians recommend. Nevertheless, a lot of effort went into securing this food. In some areas, both men and women had to spend half to two-thirds of each day hunting or foraging for food. Each day, the women and children of the tribal group would venture into successive parts of a countryside with wooden digging sticks and storage dilly bags, and collected a variety of plants, fruits, berries, and nuts; or catch fish. The men would concentrate on hunting animals and birds, such as kangaroos and emus, using spears, clubs, stones, and boomerangs. Many indigenous hunting devices were used to get within striking distance of prey. The men were excellent trackers and stalkers, approaching their prey by running where there was cover, or 'freezing' and crawling when out in the open. They were careful to stay downwind, and sometimes covered themselves with mud to disguise their smell.

Fish were sometimes caught by hand, by stirring up the muddy bottom of a pool until they rose to the surface, or by placing the crushed leaves of poisonous plants in the water to stupefy them. Fish spears, nets, and wicker or stone traps were also used in different areas. Lines with hooks made from bone, shell, wood, or spines were used along the north and east coasts. Dugong, turtle, and large fish were harpooned, with the harpooner launching himself in full-body from a canoe, to give added weight to his thrust. Some Aboriginal and Torres Strait Islander people relied on the dingo as a companion animal, using it to assist with hunting and for warmth on cold nights. A wolf-like animal, dingoes first arrived in Australia from South East Asia at least 4000 years ago. Aboriginal Australians used fire for a variety of purposes; to encourage the growth of edible plants and fodder for animal prey; to reduce the risk of catastrophic bushfires; to make travel easier; to eliminate pests; for ceremonial purposes; for heat in cooking and warmth against the cold; for warfare; and just to tidy up and clear the land.

Permanent small villages were the norm for most Torres Strait Islander communities. In some areas, mainland Aboriginal Australians also lived in semi-permanent villages, usually in coastal, less arid areas where fishing could provide for a more settled existence. Most mainland indigenous communities however, were semi-nomadic, moving in a regular predetermined cycle over a defined territory, following seasonal food sources, and returning to the same places at the same times each year. In the more arid-desert areas Aboriginal Australians were nomadic, constantly moving over large areas, in search of scarce food and water resources. Many indigenous communities had very complex kinship, social, and hierarchical structures, with the elders highly respected and revered, maintaining customary tribal law; and in some places strict rules about marriage. In many traditional tribal societies, men were required to marry women of a specific moiety, or family descent. To enable men and women to find suitable partners, tribal groups would come together for annual gatherings, known as corroborees; where goods would be traded, news exchanged, and marriages would be arranged and celebrated with appropriate ceremonies involving "dreamtime" ancestor worship, chanting, didgeridoo (traditional wind instrument) performances, and dancing. This practice both reinforced clan relationships and prevented biological inbreeding in a society based on small nomadic and semi-nomadic groups.

(291) The British First Fleet arrived in Australia on 26th January 1788

Everything changed for the Aboriginals and Torres Strait Islanders when the British arrived on 26th January 1788 and claimed Australia as their land; totally ignoring the land rights and rich cultural heritage of the indigenous peoples. The first apparent consequence of British permanent settlement appeared in April 1789 when smallpox struck the Aboriginals around Port Jackson, on the east coast close to modern-day Sydney. After that, other European diseases tragically struck the defenceless non-immune Aborigines, including chickenpox, influenza, and the measles; which spread as the British settlements and colonies expanded along the coasts and inland. The second conse-quence of British settlement was the appropriation of land and water resources. The settlers took the view that Aboriginal Australians were nomads with no concept of land ownership, who could be driven off the land required for settlements, farming, and grazing; by force. In reality, this loss of traditional lands proved to be fatally catastrophic for many Aboriginal tribal groups. As well as destroying their finally balanced livelihood, the eviction from their traditional lands impacted upon them spiritually. Aboriginal Australian tribal groups had a deep spiritual and cultural connection to the land; so when they were forced to leave their land; their traditional, cultural, and spiritual practices, all necessary for their cohesion as a society, were severely disrupted and in many cases destroyed. British settlers also brought alcohol, opium, and tobacco into Aboriginal communities. As a result, substance abuse amongst the Aboriginal Australians became a chronic problem which continues to this day. Entire communities, particularly in the more fertile southern part of the continent, literally vanished without a trace, often without the British settlers ever recording their existence.

(292)   Conflict between the Australian Aborigines and the British settlers

The significant cultural-clash between the Aboriginal Australians and the British settlers can be best illustrated in the story of Bennelong; an Aboriginal from the Wangal tribal group, who is regarded as one of the most significant and notable Aboriginals in the early history of British-occupied Australia. He became one of the first Aborigines to be 'civilized' into the British European way of life; enjoying its 'benefits' and living with the settlers. Bennelong was captured, along with his friend Colebee, in November 1789 as part of Governor

Arthur Phillip's plan to learn the language and customs of the local people, in an attempt to improve the relations between the Aborigines and the British. Bennelong was soon wearing European clothes, adopted a British lifestyle, and learnt to speak English. He was known to have reciprocated by teaching the Wangal Aboriginal language to George Bass, a British naval surgeon. He gave the Aboriginal name Wolawaree to Phillip, in order to give him a kinship relationship, to enable communication of customs, and a relationship to the land. He also taught Aboriginal customs and language to a number of the British settlers.

(293) British Governor Arthur Phillip   *(294) The Aboriginal - Bennelong*

   Bennelong was present when Governor Arthur Phillip was speared in the shoulder by an Aboriginal protagonist named Willemering, after a misunderstanding in Manly, Sydney in May 1790. Phillip ordered his men not to retaliate. Later that year Bennelong asked the Governor to build him a hut on what became later known as Bennelong Point, the site of today's Sydney Opera House. From his hut, he would entertain the Governor and other British officials. Interestingly, his second wife Barrangaroo, was opposed to her husband's conciliatory efforts to appease the British invaders, and his friendship with the Governor. She was against any form of negotiation, and although encouraged to drink wine and wear European clothing, she refused to; often being chastised by Bennelong for her rebellious attitude towards the British. When Barrangaroo wanted to give birth at the Governor's mansion to maintain links with the traditional land, and to avoid the hospital, which she thought of as a place where people died; Governor Phillip denied her the right, and instead persuaded Bennelong to take her to the hospital, where she died shortly after giving birth.

Although Bennelong was said to have had a love-hate relationship with Governor Phillip and the British settlers; he and his tribal kinsman Yemmerrawanne, sailed with Phillip to England in 1792; where he was presented to King George III on 24th May 1793. Unfortunately, Yemmerrawanne died from a lung infection on 18th May 1794 and was buried in the Eltham village churchyard of St John the Baptist, southeast of London; but Bennelong arrived back in Australia in February 1795; clearly heavily influenced by the British, with a new sense of behaviour and dress, something which he tried to impart on his family. Tragically, long troubled by alcoholism, Bennelong died on 3rd January 1813. Much of what we know about the life of Bennelong came from the writings of two British settlers at that time; Judge Advocate David Collins and Captain Watkin Tench. They viewed Bennelong as an experiment in "softening, enlightening, and refining a barbarian." While Bennelong suffered from the worst aspects of cultural indoctrination; he also represented those Aboriginals who tried in good-faith to positively change the behaviour and attitude of the British towards the Aboriginals and their sacred lands; in the early days of colonization. Governor Arthur Phillip had been ordered by London to open up a dialogue with the Aboriginals, and to coexist with them in a conciliatory and cooperative manner. Phillip had developed a passion to 'civilize' them into the European way of life. He had taken possession of the "Great Southern Land" in the name of King George III, without any reference to previous ownership, and with no regard to the sacred land rights of the Aboriginal people, who were not only there first; but had lived on this land for 65,000 years.

In another example of the cultural clash between the British and
the Aboriginals; the head of an aboriginal warrior who died resisting
the British colonization of Western Australia was finally laid to rest
near the city of Perth, in 2010. Yagan was killed by a settler in 1833,
and his severed head was sent to England where it was displayed in
a museum. Leaders of the Noongar tribe succeeded in having the
head repatriated in 1997, and during a traditional ceremony, buried
it in a memorial park in 2010. The remainder of Yagan's body is
believed to have been buried in the same memorial park. Western
Australia's indigenous affairs department said the burial concluded a
"long campaign by the Noongar people to reunite the head of the
warrior Yagan with his body." "He was a leader of his people, a man
who fought for his beliefs and was killed doing what he believed was
right," the department said. Yagan speared a number of settlers to death
during the Noongar resistance to British claims over their land, and a
bounty was put on his head. After he was shot dead, his head was cut
off and his back was skinned in order to obtain his tribal markings, as
evidence of his death. The head was shipped to England to be studied
and put on display, and it was eventually buried in Liverpool's Everton
cemetery in 1964; until it was eventually exhumed and returned to its
rightful place in Australia, to the Noongar tribal elders.

In yet another example of conflict between the British and Pacific
Island cultures there were the wars between the British and the Maori
peoples. Between the 1840s and the 1870s British and colonial forces
fought to open up the interior of the North Island of New Zealand
for settlement, in conflicts that became collectively known as the
New Zealand Wars. Sovereignty was contested despite the signing of
the Treaty of Waitangi in 1840. The purpose of the Treaty had been
to enable the British settlers and the Maori people to live together
harmoniously in New Zealand under a common set of laws and agreements. The Treaty aimed to protect the rights of the Maori people to
keep their land, forests, fisheries, and treasures, while handing over
sovereignty of the nation to the British. Over time, the Maori people
became less willing to sell land to the rapidly growing European
population. Many Maori died defending their land; many other Maori

allied themselves with the colonists for a variety of reasons, including settling old tribal disputes. Most of the several thousand people killed during the New Zealand Wars were Maoris, and the land taken by the British by force, was subsequently confiscated and returned to the Maori owners.

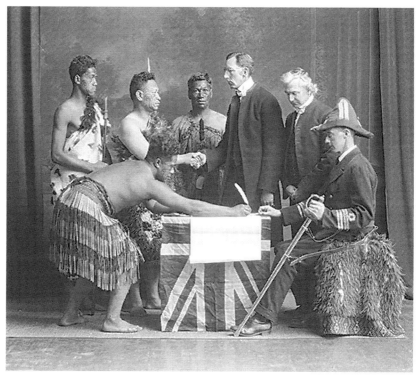

(295) Signing of the Treaty of Waitangi in New Zealand –
6th February 1840

20

MEDIEVAL EUROPE – THE AGE OF CHRISTIANITY AND FEUDALISM

THE PERIOD OF EUROPEAN HISTORY WHICH WE CALL "MEDIEVAL" IS usually regarded as consisting of the one-thousand years between the fall of the Western Roman Empire in the 5th century CE, through to the beginning of the Renaissance in the 16th century. This period is also known as the Middle Ages, because it falls in between the ancient Greek and Roman civilizations, and the beginning of the early modern world starting with the Renaissance. This thousand-year long period of Medieval Europe can be divided into three main phases. The first five centuries after the collapse of the Western Roman Empire in 476 CE, up to 1000 CE, is referred to as the Dark Ages, where there was a dramatic decline in civilized society. During the Dark Ages, trading networks and economies collapsed; bartering replaced

(296) Monks of the Dark Ages   *(297) Death of the Barbarian King Alaric*

currencies; laws and traditions broke down into a state of anarchy; and towns significantly reduced in size. In the Dark Ages, Europe fragmented into numerous barbaric kingdoms led by often-violent tribal chiefs, who would constantly be at war with each other in an endeavour to secure more land and resources. Literacy and learning almost completely disappeared, other than the educated Christian monks and priests in monasteries and churches. The Dark Ages also witnessed the rise of self-sufficient estates, or manors; then professional soldiers on horseback, or knights; and finally a societal hierarchical structure known as Feudalism. The Christian Catholic Church, already highly influential by the time of the collapse of the Western Roman Empire, dominated Europe during the Dark Ages.

The second phase of Medieval Europe is known as the High Middle Ages, which lasted from 1000 to 1350 CE. This was the pinnacle of medieval civilization, where an enduring legacy was left in the form of spectacular cathedrals and imposing castles which sprang up all over Europe. The final phase was the Late Middle Ages, which lasted from 1350 to 1500 CE. This was a time of transition that ultimately led to the emergence of early modern Europe. This phase opened with the devastating Black Death Pandemic, which

(298) A mask of the Black Death pandemic which started in Europe in 1347

swept through Europe, and killed around one-third of Europe's entire population; leaving a huge impact on society. The Late Middle Ages ended with the blossoming of the Italian Renaissance; the fall of Constantinople and the Byzantine Eastern Roman Empire; the Age of Discovery; the invention of the printing press in Europe, and the spread of pamphlets and books.

Western Europe, plus those parts of northern and central Europe which became part of the same cultural community, formed a very distinct society in medieval times. This was a civilization whose roots lay in the Christian, Latin-speaking provinces of the Late Roman Empire, and the Germanic Kingdoms which came after them. As time went by, the borders of this European civilization changed. England was added in the 6th century CE; the Low Countries including Holland in the 7th century; the German peoples in the 8th and 9th centuries; and the Scandinavians and Western Slavic peoples in the 10th and 11th centuries. Meanwhile, much of Spain was lost when the Muslims seized control of it in the early 8th century, and was only gradually regained a number of years later. Medieval European society grew out of the ruins of the Roman Empire. From the 5th century CE onwards, Germanic barbarian tribal invasions led to the ultimate disin-

(299) Gothic Church Architecture of the Middle Ages

tegration of Roman society and power in the western provinces. These provinces also experienced a significant decline in standards of living. A once literate and complex urban society gave way to an almost illiterate, much simpler, and more rural one. On the other hand, there was some continuity. Most significantly, the Christian Catholic Church survived the fall of the Western Roman Empire and proceeded to become the most predominant cultural influence in Medieval Europe. The Latin language continued in use as the language of the Church; and at a community level, vulgar Latin dialects eventually evolved into the Romance languages of Modern Europe; Italian, Spanish, Portuguese, and French; and of course also English. Much of the learning of Ancient Greece and Rome was preserved by the Church, and Roman Law influenced the law codes of the Germanic barbarian kingdoms that emerged after the collapse of Rome. Late Roman art and architecture continued in use for the few stone buildings that were still being erected during the Dark Ages; these were mainly churches and monasteries. Roman architecture would eventually evolve into the medieval Romanesque and Gothic styles.

The Feudal System first emerged in France in the 10th century CE, and then spread to other European lands in the 11th century. The word 'feudal' is derived from the word 'fief', which usually signifies an area of land held under certain conditions. A person who granted a fief to somebody else, was that person's lord; and the person who received the fief, became a vassal to the lord. The vassal usually had to provide the lord with military service, and also give him money, labour or produce from time to time, as well as advice, when required. The lord also had important duties towards the vassal, including protecting him, providing him with the means for food and shelter, and ensuring he received justice in court. Kings granted out much of their kingdom's property as large fiefs to their aristocratic nobles, and they in turn granted smaller fiefs for the lesser lords, and so on, down the hierarchy. In this way, a pyramid of mutual support was built up; from the king downwards to the lord of a single village. The building blocks of fiefs were manors. These usually covered quite small areas of land that often attached to a village. The vast majority of peasants who farmed the land in Medieval Europe were attached to manors, and had to provide their lords with labour and rent in the form of produce grown. These peasants were known as 'serfs', who although technically not slaves, they were bound to the manors in which they were born, for life. They were not allowed to leave this land for their entire lives; nor were they allowed to marry, or pass on their allocated plots of small land to anyone, without their lord's permission. On the other hand, they did have the right to expect their lord to provide them with security, protection, and justice. This is what distinguished the peasant serfs from slaves.

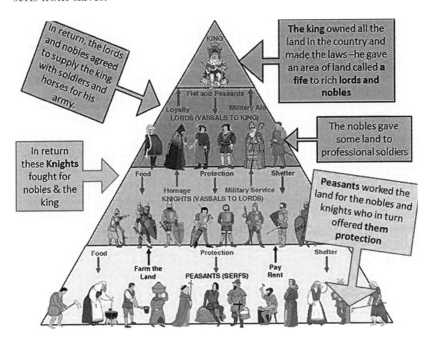

(300) The Medieval Feudal System Hierarchy

The Catholic Church exercised a powerful influence on all aspects of life in Medieval Europe. The Church was so powerful in European society that medieval Europeans would define themselves as living in 'Christendom' – the realm of the Christians. All of the key moments in a person's life – birth, holy communion, confirmation of becoming a Christian, marriage, absolution of sins, faithful loyalty, and ultimately death – all came under the direct control of the Church. Education at the time was also dominated by the Church, and most medieval scholars in Europe were members of the clergy (Church). The vast majority of art and architecture was religious in nature, either directly commissioned by the Church, or by kings, noble lords, or wealthy merchants; mainly to beautify churches and cathedrals. The largest and most beautiful structures in any medieval town or city were religious buildings. The towers and spires of cathedrals and churches soared above the skylines. Churches were also found in virtually every village. The Church was the wealthiest landowner in Medieval Europe. It was a hugely powerful organisation, spanning across the borders of all kingdoms. The Church constantly challenged and constrained the authority of emperors and kings. Senior churchmen such as cardinals and bishops were often close advisors and high officials to secular (non-religious) rulers. The servants of the Church, including cardinals, bishops, priests, monks, and nuns, were tried in their own Church Courts; and the Catholic Church had its own system of Canon Law. The medieval Church in Europe looked to the Pope, the bishop of Rome, for leadership and spiritual guidance. Through most of the High Middle Ages, Popes asserted their complete control and sovereignty over the Church and its officials. The Popes also claimed authority over secular rulers; although there were many examples of emperors and kings who resisted the power and influence of Popes; leading to many conflicts. The ongoing struggle between the Papacy and Monarchs had a major impact on the course of the history of Medieval Europe.

One unique feature of the influence of the Catholic Church was the presence of monasteries inhabited by monks and nuns, throughout Medieval Europe. These monasteries came in different shapes and sizes; but typically they formed a complex of buildings that included cloisters, dormitories, kitchens, store rooms, libraries, workshops, a mill, and surrounding farming land; all gathered around a central church. Monasteries dotted both the countryside and towns, and many owned extensive lands and property. Monastic communities had arisen at the time of the Roman Empire, but in the years after its collapse, monasticism was given a new lease of life by Saint Benedict of Nursa, in the late 5th and 6th centuries CE. He developed a code of rules for the daily monastic lives of monks and nuns. These were practical and moderate rules which aimed at allowing men and women to live communal lives of worship, meditation, work, and study; separate to the remainder of society, whilst directly contributing to its welfare and trade. Monasteries and nunneries spread throughout Europe during the Middle Ages, and monks and nuns provided much of the education, healthcare, and daily charity, for the population at large; as well as the preaching of the Christian Gospel. They preserved the collective learning of Classical Greece and Rome over many generations, by hand-copying ancient writings in the form of manuscripts, which was a major undertaking before the emergence of the printing press. Monks also contributed their own study and learning, which helped shape future Western intellectual thought. When universities appeared in Europe, the first teachers were monks.

(301) The cloisters of a Medieval Benedictine monastery

For most of Medieval Europe, society was almost entirely rural, with a very simple social structure that involved the nobility at the top, the peasants at the bottom, and very few people in between. During the latter part of the medieval period however, trade and commerce expanded, and towns became larger and more numerous. During this time, more people joined the 'middle classes' between nobles and peasants; including groups like merchants, traders, artisan craftsmen, and shopkeepers. The numerically very small fief-holding aristocracy of nobles, lords, and knights lived in castles, manor houses and, when in town, large mansions. They were supported economically by the labour of the peasants, who formed the great majority of the population. The peasants lived in small scattered villages and thatched hamlets, working the land, and doing a variety of other jobs for their everyday survival and needs. A small but growing minority of the population, comprising 5-10%, lived in the few towns and cities, which were very small by modern standards. These townspeople mainly worked as labourers, servants, artisan craftsmen, shopkeepers, and merchants. Other major groups in medieval society were of course, members of the Church clergy, and distinctive ethnic groups, such as the Jews, who were not fully accepted members of the wider society.

(302) The jousting of knights at a Medieval Royal Court

The nobility and aristocracy throughout Medieval Europe consisted mainly of the upper feudal hierarchy of fief-holders. After the emperors and kings, at the top of the hierarchy were the titled nobles, such as lords, dukes, counts, earls, and barons. Even though they stood below emperors and kings in social rank, in wealth, and in power; in many parts of Europe they were rulers in their own right; governing duchies, principalities, and counties, as semi-autonomous princes, owing only loose obedience to a distant monarch. Their families intermarried freely with the royal families of France, England, Germany, and other European kingdoms. In the lower ranks of the aristocracy were the knights and gentry, who often held only a small fief of property. In some instances, they held no land at all, but instead belonged to a great lord's retinue (or inner-circle); fighting his battles and living as members of his household. The knights hoped for a small fief as a reward for their courage on the battlefield and faithful service, or perhaps as a result of marriage to the noble heiress of an aristo-cratic fief-holder. The great lords were surrounded by huge retinues. These were literally groups of knights, domestic servants, retainers, and men-at-arms (soldiers). Their numerous manors were supervised by trusted servants called bailiffs or stewards, and their complex affairs were administered by a staff of household officials and clerks, including lawyers.

(303) A Medieval stone castle

These lords, along with their households and retinues, lived in strongly fortified castles. These first appeared in 9th century France to provide protection for the lord and the local people from the ongoing anarchy of that period. The first castles built were wooden fortified structures, sometimes standing on artificial earth mounds or hills. They soon grew into large complexes centred around a massive elevated fortified stone castle. The most prominent lords had several castles, and would travel frequently between them, along with their retinues. This was an economic necessity for these lords, because their retinues were so large that they would have quickly exhausted the resources of any one locality. Also, in an age of slow communication and travel by horseback, being constantly on the move enabled the prominent lords to keep in touch with their scattered territories, and to provide justice in person, to their dependants, by presiding at the local courts under their control.

Ranking below the lords in the feudal system, were different ranks of aristocrats who lived in lesser splendour, and were typically knights (or gentlemen), holding just one manor. A knight would mainly be concerned with the affairs of the local community which he presided over and in which he lived. Although far less powerful than the lord of whom he was a vassal, a knight still had great authority over the lives of the people of his manor. He administered justice to them in his manorial court, and supervised the work of his smaller estate, often assisted by one or two clerks. Along with his family and a small staff of domestic servants, he lived in a manor house which was often fortified. Some of these manor houses looked like small castles.

The medieval aristocracy were immersed in a military culture. In fact, they were a warrior class, trained in warfare from childhood. Even their leisure activities often involved mock-battles called jousting tournaments. The knights were originally the illiterate, thuggish retainers of kings and lords, being part of their military retinues and living in their halls. As time passed, and military equipment became more elaborate and expensive, including larger horses and more sophis-ticated armour, the lords found it useful to provide many of their knights with their own small fiefs, so they could afford to buy and

(304) A Knight's chivalry   *(305) A Knighthood ceremony*

maintain their own equipment. From the 12th century, both lords and knights became more loyal and connected to the Catholic Church, and their warlike instincts were channelled into a code of chivalry, which emphasised protection of the weak and poor; respect for women; upholding Christian values, the teachings of the Church, and courteous behaviour towards one another. A whole new idea of what it was to be a chivalrous gentleman began to take shape. Aristocrats, including knights, became literate and educated, and better able to

(306) A typical Medieval manor village

deal with matters of administration and law. This fitted them to better serve their lords, as society became more ordered and complex. It also enabled the knights to more effectively look after their own estates, as written documents gained more importance in trade, commerce and law.

The peasants formed the vast majority of the population of Medieval Europe. They lived in small manor villages where they mainly farmed the land. The serfs, those unfree peasants tied to a particular fief on a hereditary basis, had to provide the lord of the manor with various kinds of service. The biggest service was labouring on the lord's land for a set number of days and hours per week. Other obligations included giving gifts to the lord at certain times of the year, or at key moments in the peasant's life; for example, when their daughters were getting married, for which they had to ask for the permission and approval of the lord; or when a father died and the parcel of allotted land he had farmed was being taken over by his son(s). Many manors, especially in England and northern Europe, practiced the open-field system of farming; in which two or three large fields were divided into strips, with each peasant family farming several small strips scattered around the fields. These were distributed so that each family would get a fair share of the fertile and poorer quality land. Major farming activities such as sowing, ploughing, and harvesting were carried out jointly by the entire village community. Villages were small by modern standards, and usually numbered fewer than two hundred people. Each village would have its own church, which by the 12th century would usually have been built of stone. Near the church would have been the priest's house, the cemetary, and also the 'tithe barn'. This was where the villagers would store one-tenth of all grain they grew, in the form of a tax (tithe) to the Church. In many villages, a prominent manor house would also have stood nearby.

A minority of peasants were not serfs, but free. Free peasants, or 'yeomen', as they were known in England, did not have the heavy feudal obligations of their unfree neighbours. They paid a rent in money or in kind for the right to farm a piece of land, but otherwise they were at liberty to live their own lives as and where they wished.

They could move to another village if they wanted to, or to a town; they could even buy and sell land. If they owned some fields outright, they did not have to pay any rent to the lord of the manor, but they still paid their 10% tithe to the Church.

Compared with today, towns were scarce in Medieval Europe, and those that did exist were small. Medieval towns were usually smaller than those in classical antiquity. In the 12th century for example, a town with 2000 people was considered to be large. Only a few towns and cities in Europe had more than 10,000 people, and those cities with more than 50,000 were very rare; even the city of Rome, the most important city in Europe, only had 30,000 inhabitants. London, by far the largest city in England, was estimated to have had 10,000 people in 1066 CE, though four hundred years later, it was around 75,000. As time went by, and the population of Europe increased; trade, commerce, and industry expanded, and new towns appeared. These towns often emerged when a powerful lord gave a village permission to have a market, which would attract trade from surrounding areas. This trade attracted merchants, artisan craftsmen, and workers; and soon a village would evolve into a town. Alternatively, the presence of a castle, and the demands its inhabitants had for food, clothing, and many other goods and services, would often act as a catalyst for a nearby village to grow into a town.

Institutions that were of great importance in medieval towns were the guilds. A guild was an association of merchants or artisan craftsmen who were in the same trade. They regulated admission to the guild by supervising apprenticeships and awarding licences to practice the trade. Guilds also set standards for the quality of work, and enforced these standards on their members; they acted as social clubs, organising feasts and celebrations throughout the year; they also fulfilled particular functions within the wider life of the town, for example; by taking responsibility for certain aspects of the town's religious life, in close consultation with the Church; or setting up schools for the education of children of their members, and for a fee, other children. In many medieval towns, the membership of a guild would grant that member automatic citizenship of the town.

(307) Trade in Medieval towns was dominated by guilds controlled by merchants

As trade throughout Europe expanded, the merchant classes grew in number, wealth, and power. From starting off as humble traders trying to make a living in tiny towns in around 1000 CE, they evolved into affluent merchants living in grand town houses with many servants. Their business interests could span many kingdoms, even beyond Europe. They took over the running of the towns' affairs though their control of the guilds. Many merchants were able to pass on their wealth to their sons, and came to form a hereditary patrician elite, able to deal with aristocratic lords, dukes, counts, and barons on equal terms. Meanwhile, the humbler artisan craftsmen were unable to keep pace with the changes. They were still able to maintain themselves in economic independence, and held a respected place in urban society, forming the bulk of the emerging middle classes, but they were falling behind the more affluent merchants on the economic ladder. As for the lower class orders in the towns, they found themselves increasingly frozen out of opportunities to better themselves. As merchants and even master artisan craftsmen grew in wealth, more money was needed to join their ranks; particularly as the guilds came under the influence of small elite groups of wealthy masters. An urban 'working class' proletariat began to appear in many towns; made up of poor labourers, as hereditary in their low status as the patricians were in their high status. These divisions inevitably led to class tensions, often of a violent nature, in towns and cities throughout Europe in the later Middle Ages.

Whatever a persons' status was, life in medieval towns was fraught with danger. As towns continued to grow in population, they became more and more crowded. Streets were very narrow, as well as being noisy and dirty. People threw their waste (including human urine and faeces) out of their windows onto the street below. In many streets, an open sewer flowed down the middle. Conditions were appallingly unhealthy, so diseases were a constant threat. Houses were made of flimsy, highly flammable materials, so the danger of fire was never far away. Crime in medieval towns was far higher than in the modern inner cities of today. As a result of all these factors, the death rate was frighteningly high in Medieval Europe.

For all people, there was nothing like the same privacy that we have come to expect in our own modern lives. Poorer families would live and eat together in single-room cottages, at night all sleeping in the same bed. In wealthier families, the owners of a house would share their house with servants and workers. Even in noble aristo-cratic households, the family might only have a few rooms to itself, with the main sections of the house shared with a host of retainers and servants. For the majority of people, including young children, the hours were long. All the daylight hours were barely enough to get through the tasks needed for survival. Women were legally the property of men. Women's main roles in medieval society were to be wives and mothers. In poorer families, they worked alongside the men in the fields and in workshops, as well as doing the household chores; cooking, washing, cleaning, making and mending clothes, grinding corn, making beer amongst many other things. It was not uncommon for poverty-stricken women, often including widows, to resort to prostitution to survive. In aristocratic circles, the women wove, spun, and managed the domestic side of the household. In circumstances where the men were away, the lady of the household took charge of everything, including if necessary, leading the defence of a castle against attack. Aristocratic widows in particular, could have a large measure of economic independence, and in many cases took over the ownership and management of their deceased husband's business and property interests. Nuns, of course, lived largely free from male domination, and could rise within the Church hierarchy to become Abbesses of their community nunneries; a position of great respect and responsibility.

(308) A map of Medieval Europe in 1500 CE

Children took on adult roles at a young age. Children from poorer families were put to work in the peasant family's plot of land, or workshop at the age of seven. If the family could afford to send them to school, this too began at age seven, but only lasted 3 to 5 years. Sons of artisan craftsmen and merchants were sent to another household to be apprenticed to a master for seven years, learning the skills of their family trade. In aristocratic households, boys were sent at age seven to train in military skills; some training to become knights. A boy from a noble class would start training to become a knight at age seven, and if he cleared all the hurdles of training and development, including military, behavioural, and educational, he would become a knight by age twenty-one. In the meantime, he would earn his keep by acting as a servant in that household. Girls of all classes were trained in weaving, needlework, and all of the household chores they would need to know, when they married and managed their own household. Until the end of the Middle Ages, the only people who had a proper education were those destined for a career in the Church. The majority of the population were completely illiterate.

From the late 11th century, a new kind of educational institution emerged; the university. The first university was established in the northern Italian city of Bologna in 1088; but soon after, other universities appeared in Paris, Oxford, Cambridge, and other places around Europe. They originated as communities of clergy teachers who banded

(309) A Medieval university lecture

together in a loose association to study and teach. By the 14th century some of these universities had acquired such an outstanding reputation that scholars came from all over Europe to study and teach at them. These great centres of learning spread an international culture which has endured in Europe and around the world, to this present day. At first, the students who attended these universities were all intended to become part of the Church, however, others soon followed; especially the sons of noblemen and wealthy merchants who wished to study disciplines such as law and medicine.

Villagers' clothes were simple, consisting of woollen tunics for men, and woollen dresses for women. Shoes were made from the leather of slaughtered animals. Poor townspeople dressed in much the same way, but wealthier people would have brightly coloured dyed cloaks and gowns, with linen or silk undergarments. Monks wore black or brown habits, which were plain, woollen garments, often with a hood. The habit reached to their feet. The top of their heads were shaven. Nun's also wore habits, usually in black. Their head and hair would be covered by a headpiece. The year was full of religious festivals, which were times for communal fun and games. Villages and town guilds often organized their own games, such as an early version of football, which could be rough and violent. Towns and villages had many inns, where drink would flow freely. Spectator sports included cock fighting, bear-baiting, and in southern Europe, bull fighting. There were also musical performances and plays, put on in the central market place or on the streets by local people, or troops of travelling musicians, actors, jugglers, and acrobats. The aristocratic classes enjoyed banquet feasting and dancing accompanied by live musicians, which took place in the great halls of their castles and manor houses. They also enjoyed a form of entertainment called the joust tournament. Originally, this was a mock battle between two sides of knights, and could be almost as dangerous as the real thing. Later these tournaments became much more formalized, with jousts between two competing knights. With the body armour of the contestants covering the face, their identities could only be proclaimed by unique patterns of symbols on their shields and banners. This practice gave rise to heraldry, in which family descent was represented symbolically by these patterns. This in turn led to aristocratic families developing coats of arms through which their families could be traced for many generations.

In Medieval Europe there wasn't one uniform legal system. Instead the laws were a mixture of local custom, feudal traditions, Roman law, and Church law. These laws and regulations, in addition to particular laws decreed by kings, and then later, people's parliaments, formed part of the overall legal systems which applied to each kingdom. Most people's experience of law would have been in their local village manor court, which settled disputes between neighbours, and tried petty crimes. These courts were presided over by the lord of the manor, or by his official representative, who was usually a villager highly respected by his peers. The towns had their own courts, presided over by appointed magistrates. More serious crimes were tried in the courts of aristocratic lords or in the king's royal courts. Over time, a professional body of royal judges evolved, who had the expertise to try cases more professionally and objectively than in the feudal courts. In most of Medieval Europe, kingdoms drew more and more on Roman law; while in England the laws were based on a growing body of common law, which came from previous judges' decisions, called precedents.

The practice of law also emerged as a distinct profession, practised by qualified lawyers. This took place first in Italy, as early as the 11th century; and then over the next three centuries the legal profession was firmly established in the rest of Europe. This was mainly the result of the rise of a more complex and commercial society which depended for its stability on the rule of law. More and more, disputes between powerful men were settled in courts of law, rather than on the battlefield. Throughout the Middle Ages however, much of the law remained starkly inconsistent, crude and unjust, compared to our modern-day standards. For example, if a thief was caught red-handed in the street, a mob would chase him and brutally beat him up, or even kill him. This was an accepted way of administering justice, called "hue and try."

Even when the law was administered in a more orderly way, it could take on a grisly and barbaric form. Capital punishment was common,

(310) & (311) Two Medieval torture and punishment methods

and carried out by barbaric methods such as; burning at the stake, public hanging, drawing and quartering, and beheading with a guillotine (mainly in France) or a sword. Torture was administered on a regular basis, particularly as a method of extracting information; the most common means of torture included; the rack, saw, breaking wheel, iron-spiked chair, head crusher, rat in the stomach, throwing off a cliff, and hands in boiling water. Determination of guilt or innocence was often undertaken through "ordeal." The two most common were ordeal by fire; where the suspect was required to hold a red-hot iron to see whether his hands blistered and the wound didn't heal. If this was the case, the person was considered to be guilty. The second type was ordeal by water; where a suspect would have their hands and feet tied and thrown into water to see whether they floated. If they floated to the surface, they were considered to be guilty and would be executed. If they sunk and drowned, they were considered to be not-guilty, but would die in any event; a lose-lose situation.

This type of justice and punishment was clearly symptomatic of the violent society which existed in Medieval Europe. By modern standards, the crime rate was horrifically high. The murder rate in most of the small towns was several times higher than modern-day

New York or Chicago. Whole stretches of countryside were inhabited by outlawed gangs and brigands, and off limits to law-abiding people. In commerce and trade, fraud was widespread; and systemic corruption was entrenched within governments. So much so, that violence, fraud, and corruption were considered to be part of the everyday way of life. Unwanted babies were habitually left out in the open streets or fields. It was a rough, tough, violent, and often inhumane world.

I have selected three prominent people of Medieval Europe to give a deeper insight into the interconnection of power between the Church and the Monarchies. In my opinion, and in the opinion of many other historians, the most powerful king in Medieval Europe was Charlemagne, also known as Charles the Great, born in 742 and died in 814 CE. He ruled much of Western Europe from 768 to 814, and came close to fulfilling his dream of recreating the ancient Roman Empire. Charlemagne was born in 742, and was the son of Bertrada of Laon, and Pepin the Short, who became king of the Franks in 751. The Franks originated from what is today, modern-day France. After Pepin's death in 768, the Frankish Kingdom was divided between Charlemagne and his younger brother Carloman. The brothers had a highly competitive and strained relationship; and with the untimely death of Carloman in 771, Charlemagne emerged as the sole ruler of the Frankish Kingdom. Once in power, he sought as his highest priority to unite all of the Germanic peoples into one kingdom, and convert his subjects to Christianity. In order to carry out this mission, he spent the majority of his reign engaged in military campaigns. Soon after becoming king, he conquered the Lombards, who were located in present-day northern Italy; then the Avars, from present-day Austria and Hungary; and the Germanic Bavarians, amongst others.

Charlemagne waged a vicious, bloody three-decades long series of battles against the Saxons, a Germanic tribe of pagan worshippers, earning a reputation for extreme ruthlessness. In 782, at the Massacre of Verden, Charlemagne ordered the slaughter of 4500 Saxons. He eventually forced the Saxons to convert to Christianity, and declared that anybody who did not get baptized or follow other Christian traditions, would be put to death. In his personal life, Charlemagne had many wives and mistresses, and at least 18 children. He was a devoted father from many accounts, who encouraged his children's education. He was said to have loved his daughters so much that he prohibited them from marrying while he was alive. Einhard, a Frankish historian and contemporary of Charlemagne, wrote a biography of the emperor after his death, titled "Life of Charles the Great" between 830 and 833, and in this book he described Charlemagne as: "broad and strong in the form of his body and exceptionally tall without, however, exceeding an appropriate measure. His appearance was impressive whether he was sitting or standing despite having a neck that was fat and too short, and a large belly."

(312) The crowning of Charlemagne as Holy Roman Emperor
on Christmas day 800 CE

In his role as an aggressively devout defender of Christianity, Charlemagne donated large amounts of money and land to the Catholic Church, and protected the Popes. As a way to acknowledge Charlemagne's power and reinforce his close relationship with the Church, Pope Leo III crowned Charlemagne Holy Roman Emperor, in a spectacularly lavish ceremony held at St. Peter's Basilica in Rome, on Christmas day in 800. As Emperor, Charlemagne proved to be a highly capable diplomat and administrator of his vast European empire. He promoted public education and encouraged the Carolingian Renaissance, a period of renewed emphasis on scholarship, the arts, and culture. He initiated a series of progressive economic and religious reforms; and was a driving force behind the 'Carolingian miniscule', a standardized form of writing that later became a template for modern European printed alphabets. Charlemagne ruled his empire from a number of cities and palaces, but he spent most of his time in the German city of Aachen. His palace there included a school, for which he recruited the very best teachers and scholars in his kingdom. In addition to learning, Charlemagne was interested in athletic pursuits. He was known to be highly energetic, and enjoyed hunting, horseback riding, and swimming. Aachen also held particular appeal to him because of its therapeutic natural warm springs, in which he indulged regularly.

According to his biographer Einhard, Charlemagne was in good health until the final four years of his life, when he often suffered from acute fevers, and acquired a limp. However, as Einhard notes: "Even at this time, he followed his own counsel rather than the advice of the doctors, whom he very nearly hated, because they advised him to give up roasted meat, which he loved, and to restrict himself to boiled meat instead." In 813, Charlemagne crowned his son Louis the Pious (778-840), King of Aquitaine, and his co-emperor. Louis became the sole emperor when Charlemagne died in January 814, ending his reign of more than four decades. At the time of his death, his empire virtually covered most of Western Europe. Charlemagne was buried at the cathedral in Aachen. In the following decades, his empire was divided up among his heirs; and by the late 800s, it had totally dissolved. Nevertheless, Charlemagne is remembered as a near-legendary figure, particularly in modern-day France. In 1165, under the rule of Emperor Frederick Barbarossa (1122-1190), Charlemagne was canonized for political reasons; however, the Catholic Church today does not recognize his sainthood.

According to the 9th century manuscript, "De Carolo Magno," after hearing that the Vikings (referred then to as Northmen) had retreated before Charlemagne could engage them in battle, he said: "Ah, woe is me! That I was not thought worthy to see my Christian hands dabbling in the blood of those dog-headed fiends." From the same manuscript, Charlemagne expresses this frustration to noble-born students at his Aachen palace school whose work and aptitude to study was poor, while lesser-born children had worked hard to write well: "You nobles, you sons of my chiefs, you superfine dandies, you have trusted to your birth and your possessions and have set at nought my orders to your own advancement; you have neglected the pursuit of learning and you have given yourselves over to luxury and sport, to idleness and profitless pastimes. By the King of Heaven, I take no account of your noble birth and your fine looks, though others may admire you for them. Know this for certain, that unless you make up for your former sloth by vigorous study, you will never get any favour from me."

A national heroine of France, at the age of 18 Joan of Arc led the French army to victory over the English at Orleans. Captured a year later, Joan was burned at the stake as a heretic by the English and their French collaborators. She was canonized as a Roman Catholic saint more than 500 years later, on 16th May 1920. At the time of Joan of Arc's birth, France was embroiled in a long-running war with England, known as the Hundred Years' War; a dispute which had been over who would be the heir to the French throne. By the early 15th century, northern France was a lawless frontier of marauding, violent armies.

Joan of Arc, nicknamed "The Maid of Orleans," was born in Domremy France, in 1412. The daughter of poor tenant peasant farmers, Joan learned religious piety and domestic skills from her mother. Never venturing far from home, Joan took care of the animals and became a skilled seamstress. In 1415, King Henry V of England invaded northern France. After delivering a shattering defeat to the French forces, England gained the support of the Burgundians in France. The 1420 Treaty of Troyes, granted the French throne to

(313) Joan of Arc leading the French army into battle against the English in 1429

Henry V as regent of the insane King Charles VI. Henry would then inherit the throne after the death of Charles. However, in 1422, both Henry and Charles died within a couple of months, leaving Henry's infant son as king of both realms. The French supporters of Charles' son, the future Charles VII, sensed an opportunity to return the crown to a French monarch.

Around this time, Joan of Arc, while out in the fields tending to her peasant family's animals, began to have mystical visions encouraging her to live a devout pious life. Over time, the visions became more vivid, with the presence of St. Michael and St. Catherine designating her as the saviour of France; and encouraging her to seek an audience with Charles, who had assumed the title of Dauphin, the heir to the throne. At this meeting, Joan was going to seek his permission to lead an army and militarily expel the English out of France, and install him as the rightful king. In May 1428, Joan's visions instructed her to go to Vaucouleurs and contact Robert de Baudricourt, the garrison commander, and a supporter of Charles. At first, Baudricourt refused Joan's request, but after seeing that she was gaining the approval of the villagers, in 1429 he relented and gave her a horse and an escort of several soldiers. Joan cropped her hair and dressed in men's clothing for her 11-day journey across enemy territory to Chinon, the location of Charles's court.

Initially, Charles was not certain of what to make of this peasant girl who had asked for an audience with him and professed she could save France. Joan however, won him over when she correctly identified him, dressed incognito, in a crowd of members of his court. The two had a private conversation during which Joan revealed the details of a solemn prayer Charles had made to God to save France. Amazed, but still hesitant, Charles arranged for prominent respected theologians to meet with, and examine her. The clergymen reported they found nothing improper with Joan; only spiritual piety, chastity, and humility. Finally, Charles gave the young peasant girl Joan of Arc armour and a horse, and allowed her to accompany the army to Orleans, the site of an English siege. In a series of battles between 4th and 7th May 1429, the French troops took control of the English fortifications. Joan was wounded, but later returned to the battle-front to encourage a final assault. By mid-June, the French had defeated and routed the English, and in doing so, damaged their perception of invincibility in the process. Although it appeared that Charles had accepted Joan's mission, he did not display full trust in her judgment or advice. After the military victory at Orleans, she kept encouraging him to hurry to Reims and crown himself as king; but he and his advisors were more cautious. Eventually though, Charles and his procession finally entered Reims, and he was crowned Charles VII on 18th July 1429. Joan was at his side for the coronation, occupying a visible place at the ceremonies.

In the Spring of 1430, King Charles VII ordered Joan to Compiegne to confront the Burgundian assault. During the battle, she was thrown off her horse and left outside the town's gates. The Burgundians took her captive and held her for several months, negotiating with the English, who saw her as a valuable propaganda prize. Finally, the Burgundians exchanged Joan for 10,000 francs. King Charles VII was unsure what to do. Still not totally convinced of Joan's divine inspiration, he distanced himself and made no attempt to arrange to have her released. The English turned Joan over to the Catholic Church, who insisted she be put on trial as a heretic. She was charged with 70 counts, including heresy, witchcraft, and dressing like a man. Initially, the trial was held in public, but it went private when Joan was clearly arguing her case with great passion and conviction; and humiliating her accusers in the process. During the trial she stated: "You say that you are my judge; I do not know if you are; but take good heed not to judge me ill, because you would put yourself in great peril." Between 21st February and 24th March 1431, she was interrogated almost a dozen times by a tribunal, always maintaining her humility, humbleness, and steadfast claim of innocence. Instead of being held in a Church prison with nuns as guards, she was held in a military prison. Joan was threatened with rape and torture, though there is no record that either actually occurred. She protected herself by tying her soldiers' clothes tightly together with many cords.

Frustrated they could not break her resolve, the tribunal eventually used her military clothes against her, charging her with dressing like a man. On 29th May 1431, the tribunal announced that Joan of Arc was guilty of heresy. On the morning of 30th May, she was taken to the marketplace in Rouen and burned at the stake, before an estimated crowd of 10,000 people. She was 19-years old. Before being burned at the stake, she was heard to say: "Hold the cross high so I may see it through the flames." One legend surrounding the event tells of how her heart survived the fire unaffected. Her ashes were gathered and scattered in the Seine River. After Joan's death, The Hundred Years' War continued for another 22 years. King Charles VII ultimately retained his crown, and he ordered an investigation that in 1456 declared Joan of Arc to be officially innocent of all the charges; and she was designated as a martyr. She was canonized as a saint on 16th May 1920, and is the patron saint of France.

Joan of Arc clearly possessed remarkable courage and wisdom for the relatively few years that she lived. We can gain an insight into her state of mind and beliefs, with these quotations attributable to her:

(314) The execution of Joan of Arc for heresy by the Catholic Church on 30th May 1431

Every man gives his life for what he believes. Every woman gives her life for what she believes. Sometimes people believe in little or nothing, and so they give their lives to little or nothing. One life is all we have, and we live it as we believe in living it, and then it is gone. But to surrender who you are and to live without belief is more terrible than dying; even more terrible than dying young.

- I am not afraid. I was born to do this.
- Courage! Do not fall back.
- I was admonished to adopt feminine clothes; I refused, and still refuse. As for other avocations of women, there are plenty of women to perform them.
- Go forward bravely. Fear nothing. Trust in God; all will be well.
- When God fights, it's of small consequence whether the hand that holds the sword is big or little.
- Children say that people are hanged sometimes for speaking the truth.
- I am the drum on which God is beating out his message.

The marriage of English King Henry VIII and Anne Boleyn was an unlikely match when considering the circumstances. Henry had married Catherine of Aragon, a devout Catholic. This had been a politically strategic marriage which strengthened the diplomatic relations between England and Spain, because Catherine was the youngest daughter of King Ferdinand II and Queen Isabella of Spain. Initially she married Henry's older brother, Arthur. But when he died suddenly, she married the younger Henry. Catherine did give Henry a daughter. Their daughter Mary went on to become the devout Catholic, Mary Queen of the Scots. Yet, when Catherine was unable to produce a male heir for Henry, he moved swiftly to change wives. At first, he requested the assistance of his trusted Cardinal Wolsey to convince Pope Clement VII to annul his marriage. When the Pope refused, an outraged Henry left the Catholic Church and created his own Church of England (the Anglican Church), and annulled his marriage to Catherine; severely upsetting diplomatic relations with Spain in the process. Cardinal Wolsey's failure in securing the divorce that Henry so desperately wanted was to severely fracture their relationship. In the end, abandoned by the king, Wolsey was charged with treason, but died of natural causes before he could be beheaded.

(315) English King Henry VIII and his second wife Ann Boleyn were married in January 1533

When Anne Boleyn returned to England from France in 1522, her sister Mary had already been in the English court for two years, and was probably by this time already the mistress of King Henry VIII. It is unlikely that Anne would have been attracted to her sister's lover at this time, and in any case, she had returned to England for the express purpose of marrying her fiancé James Butler, her distant cousin, and also a member of Henry's court. As soon as she arrived at court, the charismatic enchantress Anne attracted a circle of admirers. Amongst them was the poet Thomas Wyatt, Anne's childhood friend with whom she shared a love of poetry and music. Another was the already engaged Earl of Northumberland who fell under her spell and never got over her. "She knew perfectly how to sing and dance, to play the lute and other instruments," wrote Lancelot de Carle, a French diplomat. Her contemporaries mentioned that she seemed more French than English and this had an exotic allure. She enjoyed gambling, dancing, and flirting, which clearly gained the attention of the men around her. As her biographer Eric Ives mentions, the tradition of courtly love was still prevalent and could be seen as a kind of antidote to the inevitable boredom at the king's court. Anne loved games and was known for her refined skill in the game of courtly love.

King Henry VIII on the other hand, was more uncouth and less-polished than the chivalrous gentlemen Anne would have known in France. Coming from an upstart dynasty with a debateable claim to the throne, he was desperate to be seen as a cosmopolitan Renaissance man of the world; a ruler on a par with the great noble houses on the continent; most of whom saw England as a backwater island of no real importance to the affairs of Europe. Henry was particularly anxious to prove that he was every bit as good as his main rival, King Francis I of France; even childishly asking his courtiers if they thought Francis had better-looking calf-muscles and legs than him. The fact that Anne had slept with Francis, may have been one of his motivations for taking Anne as a mistress. At this time, while still married to Catherine of Aragon, he would refer to Catherine as "old and deformed." Catherine turned a blind eye to her husband's infidelities so long as he was discreet, so Henry boosted his ego by sleeping

with Anne's sister Mary, and the beautiful Elizabeth Blount. As Henry
grew tired of his affair with Mary, his eye began to wander to her little
sister. Anne was attractive, vivacious, very popular, and didn't have the
tarnished reputation that Mary did. When Henry discovered that Anne
and the Earl of Northumberland, Henry Percy, had become engaged,
he ordered his minister Cardinal Wolsey to break the engagement;
even though from all accounts, the couple had been genuinely in love
with each other. Anne had broken off her first engagement to James
Butler, as a show of her true love for Henry Percy.

At first, Anne refused all of Henry VIII's advances, probably
because she was still heartbroken over Henry Percy. The king came
on very strong, sending her at least 17 love letters full of traditional
courtly love imagery. Eventually, Anne relented and their affair had
begun. As the affair blossomed, they regularly exchanged love letters.
In expressing his feelings for Anne during their courtship, Henry
wrote the following in one of his love letters to her: "In debating with
myself the contents of your letters I have been put to a great agony;
not knowing how to understand them, whether to disadvantage as
shown in some places, or to my advantage as in others. I beseech you
now with all of my heart definitely to let me know your whole mind
as to the love between us; for necessity compels me to plague you for
a reply, having been for more than a year now struck by the dart of
love, and being uncertain either of failure or finding a place in your
heart and affection." Henry's lust and passion for Anne during their
courtship can be seen from this quote from Henry taken from another
letter to her: "....to wish myself specially an evening in my sweet-
heart's arms whose pretty ducks (breasts) I trust shortly to kiss."

The idea slowly dawned on Henry that perhaps he could make
Anne his new queen instead of some foreign princess. He proposed
marriage sometime around New Year in 1527, and this time Anne
could not refuse. Over a period of time, as the haze of love and lust
began to wear off, Anne's faults became apparent to Henry. When
Anne fell pregnant, instead of the son he had expected, her child was
a girl, Elizabeth, who would later go on to become one of the most
powerful monarchs in English history. He started to wake up to the

fact that instead of gaining international credibility, his obsession with Anne had made him a laughingstock and alienated his international allies, particularly the Spanish. Henry's paranoia got the better of him. He couldn't stand rivals, and bored with Anne, in 1534 he began an affair with one of Anne's maids, Jayne Seymour, who was to later become his next wife. Henry no longer wanted an exciting woman who tried to control him, but one who would defer to him and be in his thrall, even though she wasn't as attractive to him as Anne, Jane fit the bill.

In order to get rid of her, Henry fabricated a number of charges against Anne, including incest with her brother, and swiftly thereafter she was arrested, imprisoned in the Tower of London, and found guilty of all charges. This quote by Henry is taken from a love letter to Jayne Seymour written while Anne was in the Tower of London awaiting execution: "The bearer of these few lines from thy entirely devoted servant will deliver into thy fair hands a token of my true affection for thee, hoping you will keep it for ever in your sincere love for me. Advertising you that there is a ballad made lately of great derision against us, which if it goes abroad and is seen by you; I pray you to pay no manner of regard to it. I am not at present informed who is the setter forth of this malignant writing; but if he is found out, he shall be straitly punished for it. For the things ye lacked, I have minded my lord to supply them to you as soon as he could buy them. Thus hoping, shortly to receive you in these arms, I end for the present. Your own loving servant and sovereign."

Henry arranged for the best executioner to come from France, to ensure Anne would die swiftly and without suffering. Even the executioner felt sorrow for Anne; he apologised to her, before carrying out his gruesome deed. Anne Boleyn famously was beheaded on 19th May 1536, and Henry married Jane Seymour a little over a week later. Though the English people had referred to Anne as "the king's whore," among other things, his hasty remarriage and subsequent debauched behaviour rehabilitated her image over a period of time; as the people realized that both her rise and her fall were orchestrated by the king's whims and massive ego.

(316) Ann Boleyn paying her executioner – from the Tudors.
She died on 19th May 1536.

A court witness described the execution of Anne Boleyn as follows: "She was brought by the captain upon the said scaffold, and four young ladies followed her. She looked frequently behind her, and when she got upon the scaffold was very much exhausted and amazed. She begged leave to speak to the people, promising to say nothing but what was good. The captain gave her leave, and she began to raise her eyes to Heaven, and cry mercy to God and to the King for the offence she had done, desiring the people always to pray to God for the King, for he was a good, gentle, gracious, and amiable prince."

According to the English statesman and philosopher Francis Bacon; just before she was executed, Anne Boleyn praised her King with this speech to the people who were gathered: "Friends and good Christian people, I am here in your presence to suffer death, whereto I acknowledge myself judged by the Law; How justify I will not say, for I intend not an accusation of any one. I beseech the Almighty to preserve his Majesty long to reign over you; a more-gentle or mild Prince never swayed Scepter; his bounty and clemency towards me I am sure hath been especial. If anyone intend an inquisitive survey of my actions, I intreat him to judge favourably of me, and not rashly to admit of any censorious conceit. And so I bid the world farewell, beseeching you to commend me in your Prayers to God. To the O Lord do I commend my Soul, Christ have mercy on my soul, Lord Jesus receive my soul."

The only legitimate son that King Henry VIII had, was with his third wife, Jane Seymour. Tragically Jane died two weeks after giving birth, from what appears to have been a bacterial infection. Henry was devastated by her death. Their son Edward was born on 12th October 1537, and was crowned King Edward VI of England on 20th February 1547, at the age of nine; not long after his father King Henry VIII died on 28th January 1547. Edward was the first monarch to be raised as an Anglican Protestant from the Church of England; and due to his young age, during his reign the kingdom was governed by a regency council of Dukes and Earls who were close to Henry, because he never reached maturity; dying at age 15 on the 6th July 1553 from tuberculosis. This quotation was written by Henry in a letter to the Duke of Norfolk in which he expresses his intention to stay close to Jane Seymour during her pregnancy: "…being but a woman, upon some sudden and unpleasant rumours and bruits that might be foolish or light persons be blown abroad in our absence, being specially so far from her, she might take to her stomach such impressions as might engender no little danger or displeasure to the infant with which she is now pregnant (which God forbid!), it hath been thought by our council very necessary that, for avoiding such perils, we should not extend our progress further from her than sixty miles." This quotation from Henry has been extracted from his last will and testament which requested: "that the bones and body of his true and loving wife, queen Jane, were to be placed in his tomb." It seems that Jane Seymour in fact was the true love of Henry's life; producing for him the treasured legitimate son he had always craved.

In January 1553 the young King Edward VI was showing the first signs of tuberculosis, and by May it was evident that the disease was terminal. Working with his close advisor Lord Northumberland, Edward was determined to exclude his two half-sisters, Mary and Elizabeth, from the succession, and instead to put Northumberland's

daughter in law, Lady Jane Grey, and her male heirs in direct line
for the throne. As a result, a significant power struggle erupted after
Edward's death. Lady Jane Grey ruled for nine days, between the 10th
and 19th July 1553, before she was overthrown by the more popular
Mary I, who reigned from 1553 to 1558, when at the age of 42, she
died during an influenza epidemic. She was then succeeded by her
younger sister, the great Elizabeth I, daughter of Henry and Anne
Boleyn, who ruled England until her death in 1603; and was one of
the most powerful monarchs in Europe during her reign.

(317) & (318)   Queen Elizabeth I was the daughter
of King Henry VIII and Ann Boleyn

During her reign, Queen Elizabeth I firmly established Protestantism
in England; defeated the Spanish Armada in 1588; maintained peace
inside her previously divided kingdom; and created an environment
where the arts and exploration flourished. She was sometimes referred
to as the "Virgin Queen" because she never married. During the
Elizabethan Era, England became a powerful and prosperous nation,
rivalled only by Spain. Helped by the reforms enacted by Henry VII
and Henry VIII, Elizabeth's government was strong, centralized, and
effective. Guided by her Privy Council of trusted advisors, Elizabeth
cleared the national debts, and restored the kingdom to financial stability.

In 1588, the Spanish Armada set sail from Spain with the purpose of invading England to overthrow Elizabeth primarily because of her anti-Catholicism. On 29th July of that year the English naval fleet badly damaged the "Invincible Armada" in the Battle of Gravelines. Five Spanish ships were lost and many were badly damaged. Worst still, a strong south-westerly wind forced the Armada into the North Sea, and the fleet was unable to transport the invasion force, gathered by the Governor of the Spanish Netherlands, across the English Channel. The famous speech delivered by Queen Elizabeth to her troops, who

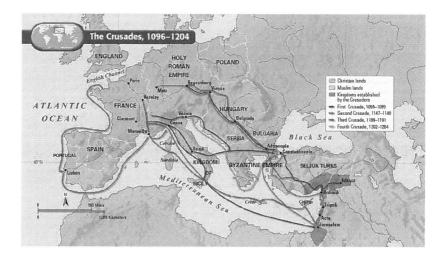

(319) and (320) The Crusades - fought between Christians and Muslims

were assembled near the Channel coast, was hugely inspirational. She said: "I know I have the body but of a weak and feeble woman; but I have the heart and stomach of a king, and of a king of England too." This successful defence of the kingdom against invasion on such an unprecedented scale boosted the prestige of Elizabeth, and encouraged a sense of English pride and nationalism.

21

THE MEDIEVAL WORLD BEYOND EUROPE

I N THIS CHAPTER WE WILL SURVEY THE MEDIEVAL PERIOD IN DIFFERENT regions outside of Europe. One of the defining episodes in the early medieval era was the rise of the Islamic Caliphate in the 7th century CE; and its rapid spread over the Middle East, North Africa, Spain, parts of central Asia, and into the Indian subcontinent. This was the achievement of Arabic tribes, who were fighting in the name of their new religion, Islam; which started in 610 CE following the first revelation to the prophet Muhammad at the age of 40, and the spread of his teachings and beliefs throughout the Arabian-peninsula. This vast Caliphate Kingdom fragmented into numerous successor states from the 9th century onwards, but the Islamic faith continued to strengthen its hold on the peoples of the region, and spread out to new areas of Eurasia. From the 8th to 12th centuries, the Islamic Caliphate acted as a catalyst and fostered a rich fusion of cultural elements from ancient Greece, Rome, Byzantium, Persia, and India. To these were added major original scientific and artistic contributions from Arabic scholars. The result is that much of this rich culture and knowledge was transmitted to Europe, bringing amongst other things; Arabic numerals, the decimal system, and Greek philosophy, which had been lost to the West. These enriching exchanges helped to lay the foundations for later Western advances, starting with the Renaissance. Muslim traders became active in pioneering and expanding commercial trade routes. They soon dominated the trans-Saharan routes between North and West Africa, which led to the conversion of many African kings to the Islamic faith. The Islamic Caliphate was also a major participant in both the Silk Road and Indian Ocean trade routes. They established a string of trading city-states and Muslim sultanates in South East Asia towards the end of the medieval period.

The Eastern half of the Roman Empire, meanwhile endured throughout the medieval centuries, in the form of the Byzantine Empire, ruled from its capital Constantinople. In the early Middle Ages its political state, society, and culture was radically modified; giving it a distinct uniqueness, compared to the earlier Western Roman Empire. As in the West, Byzantine society was also devoutly Christian, but they owed their spiritual obedience to the Patriarch of Constantinople, rather than the Catholic Pope in Rome. A rift gradually developed between the two branches of Christianity which was mainly based on what religious images should be displayed in churches, and who the supreme leader of the Church should be. The Byzantine or Orthodox Church experienced dramatic expansion with the conversion of the Balkan populations, and later of the Russians. The Byzantine Empire itself however, experienced a long term shrinkage of its territory, so that by the mid-15th century, apart from its capital Constantinople, it covered only a small part of Greece.

The most famous Byzantine building was the "Hagia Sophia" a spectacularly domed Orthodox Christian church built in the great city of Constantinople during the reign of Byzantine Emperor Justinian I,

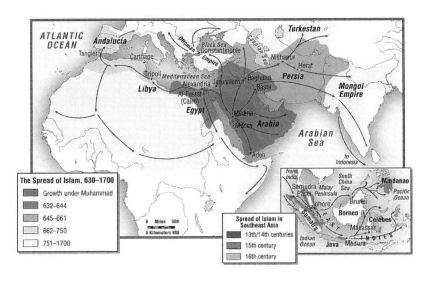

(321) Map showing the spread of Islam from 630 to 1700 CE

who ruled from 527 to 565 CE. Byzantine architecture was a major force in the architectural world during the Middle Ages. After sending his army to reconquer the Italian peninsular in 535 CE, Justinian was unable to hold Rome against the aggressive Germanic Ostrogoth Kingdom of Italy. Fleeing the city of Rome, Justinian's forces made the northern Italian city of Ravenna their temporary capital; an artistic city with some of the most beautiful Christian mosaics known anywhere in the world at that time. The Gothic War between the Byzantine Empire during the reign of Emperor Justinian I, and the Ostrogoth Kingdom of Italy took place from 535 until 554 CE in the Italian peninsula, Dalmatia (Croatia), Sardinia, Sicily, and Corsica. It was one of the last of the many Gothic Wars with the Byzantines. The war had its roots in the long-held ambition of Justinian to recover the provinces of the former Western Roman Empire, which the Romans had lost to the invading barbarian tribes in 476 CE. Justinian was unsuccessful in reclaiming Italy.

Not only did the Byzantines help preserve ancient Roman and Greek culture, and Christianity, but they also spread these ideas to other parts of the world. During the Crusades of the 11th and 12th centuries, western Europeans while making their way in an attempt to reclaim the

(322) The Hagia Sophia of Constantinople – now Istanbul in Turkey

holy land surrounding Jerusalem, had to first pass through the Byzantine Empire. As a result, they brought many of those ancient Greek and Roman cultural accomplishments back to Western Europe. Two mission- aries from the Byzantine Empire, named Cyril and Methodius, travelled into Central and Eastern Europe to spread the ideas of Christianity to the Slavic people. The Byzantine Empire made great contributions to civilization, for example; the Greek language and learning was preserved and taught to ongoing generations; the Roman Imperial system was continued, and Roman law codified; the Greek Orthodox Church converted a number of Slavic groups, and fostered the development of a uniquely splendid form of "icon art" which was dedicated to glori- fying the Christian religion. Situated at the crossroads of West and East, the great city of Constantinople acted as the disseminator of culture for all the peoples who came in contact with the empire. Justifiably called "The City", this rich and turbulent metropolis was to the early Middle Ages, what Athens and Rome had been to ancient classical times. By the time the Byzantine Empire collapsed in 1453, its religious mission and political ideology had borne fruit among the Slavic peoples of Eastern Europe, and especially among the Russians.

Who were the Vikings? They are commonly depicted as invaders, predators, and wild barbarians. What is the truth about their raids? One thing is for certain, the history of the Vikings is not characterized just by raiding and plundering. The Vikings developed a complex, sophisti- cated Scandinavian culture. In addition to their well-known raids, they were traders, artists, poets, skilled boat-builders, craftsmen, and highly competent sailors. The Vikings originated from Scandinavia; Denmark, Norway and Sweden. During the medieval period these lands were mainly rural. The major sources of food were from agriculture and fishing. The Viking society was divided into three socio-economic classes; the Thralls, Karls, and Jarls. The Thralls were the lowest ranking class and were effectively slaves. Slaves comprised as much as a quarter of the population. The Karls were free peasants. They owned farms, land, and cattle, and engaged in daily chores such as ploughing the fields, milking the cattle, and building houses and boats. They would use the Thralls as labour to make ends meet. The Jarls were the noble aristocracy of the Viking society. They were wealthy and owned large property estates, with huge longhouses, horses, and many Thralls. The Thralls did most of the daily work, while the Jarls engaged in administration, politics, hunting, sports, socialising with other Jarls; and went abroad on expeditions. When a Jarl died and was buried, his household Thralls were sometimes sacrificially killed and buried next to him; as many archaeological excavations have revealed.

(323) Vikings raiding England and Scotland

The Vikings were pagans. They believed in a polytheistic religion, involving many gods and goddesses. This religion was passed on through oral culture rather than written texts. It focused heavily on ritual practice, with kings and chiefs playing a central role in carrying out public acts of sacrifice. The most popular Viking gods were Odin, Thor, and Tyr. Known as the All-Father, Odin was the god of wisdom, poetry, death, divination, and magic. Thor was the god of thunder and lightning; in Norse mythology associated with strength, storms, hallowing, and fertility. He was the son of Odin and was described as being fierce-eyed, with red hair and a full beard, quick to anger, and had an enormous appetite. Tyr was considered to be the god of war. He was the most courageous of all the gods. According to Norse mythology, Tyr sacrificed his arm to the monstrous wolf Fenrir, who bit off his limb.

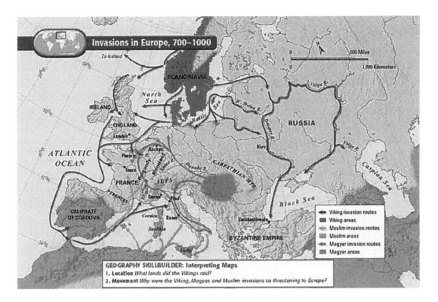

(324) Map showing Viking invasions and incursions
within Europe: 700-1000 CE

From the late 8th to the late 11th centuries CE, the Vikings raided and traded from their Northern European homelands across wide areas of Europe, and explored westwards to Iceland, Greenland, and North America. Facilitated by advanced sailing and navigational skills, the Viking raids at that time extended into the Mediterranean, North Africa, and the Middle East. The longships that they designed helped them to explore, expand their activities, and settle in diverse areas of north-western Europe, Belarus, Ukraine, European Russia, and the North Atlantic islands as far as the north-eastern coast of North America. This period of expansion witnessed the dissemination of Norse culture, while at the same time introducing strong foreign cultural influences into Scandinavia itself. From around 860 CE onwards, the Vikings raided, settled, and prospered in Britain. The earliest recorded planned Viking raid, on 6th January 793, targeted the monastery on the island of Lindisfarne, off the north east coast of Northumbria (Scotland). The raiders killed the resident monks, or threw them into the sea to drown, or took them away as slaves. They also took the church treasures. In 875, after enduring eight decades of repeated Viking raids, the monks finally fled Lindisfarne, carrying the relics of Saint Cuthbert with them.

(325) & (326) The Normans were
descendants of the Vikings

In the years that followed, the Vikings left no region of Britain safe. They attacked and besieged other villages, monasteries, and even cities in Wales, Scotland, Ireland, and England. Only one kingdom, namely Wessex, led by King Alfred the Great, successfully resisted the Viking raids. The Vikings also created much havoc with numerous raids deep into the river-systems of mainland Europe, particularly into modern day France and Germany. The Kingdom of Normandy was created by Vikings who had settled into this coastal region of France. By the 12th century, the Scandinavian region inhabited by the Vikings was assimilated into the mainstream of medieval Christian culture.

India remained covered in numerous independent kingdoms. From the 8th century onwards, Muslim forces began to make their presence felt; and from the 11th century, they began their conquest of almost the entire subcontinent. In the 13th and 14th centuries the Delhi Sultanate came to rule almost the whole region. The Delhi Sultanate was an Islamic empire based in Delhi that stretched over large parts of the Indian subcontinent for 320 years, from 1206 to 1526 CE. Five dynasties ruled over the Delhi Sultanate in the following sequence; the Mamluk Dynasty from 1206 to 1290; the Khalji Dynasty from 1290 to 1320; the Tughlaq Dynasty from 1320 to 1414; the Sayyid Dynasty from 1414 to 1451; and the Lodi

Dynasty from 1451 to 1526. During this period, Buddhism finally ceased to have a widespread following in the region of its birth. It became confined to the island of Sri Lanka in the south, and to some Himalayan kingdoms, notably Nepal in the north. Even though the Delhi Sultanate Dynasties practised Islam, they tolerated other religions; so Hinduism eventually became the almost-universal religion of the masses of people living in the Indian subcontinent. The Indian subcontinent saw the full integration of south India into the Aryan, and by now fully Hindu, civilization of the north.

(327) Delhi Sultanate architecture

Northern China, like Western Europe, experienced barbarian invasions after the fall of the Han Empire. However, as things settled down, the ordered 'Confucian' bureaucratic style of government characteristic of the Han Dynasty returned to this region; it had never left southern China. As far as religious influences are concerned, Buddhism came into China from India and became a powerful force there during this time. China was reunified under the Sui Dynasty in 589 CE; and the glorious period of the Tang Dynasty, starting from 618, saw China reach one of its highest points both in terms of political power and culture. In China, the population more than doubled between 500 CE and 1450 CE, and southern China overtook the north (the ancient heartland of Chinese civilization) in terms of population and economic growth.

At this time, the neighbouring kingdoms of Korea and Japan were drawn more closely into China's sphere of influence. They acknowledged the political superiority of the Tang Empire, sending regular tribute and diplomatic missions to its capital, Chang'an; and the Koreans and Japanese also imported wholesale elements of Chinese civilization; including bureaucratic government, Confucianism, Buddhism, and styles of art and architecture, amongst other things into their respective societies. Japan's medieval period was similar to Europe in many ways; there was a strong warrior class that worked its way into the nobility; a decentralized government that came under the influence of the warlords in their castles; a surge in Buddhist religious fervour; and a strict code of honour and chivalry. At the head of Japan's government was the emperor, a hereditary monarch with absolute power, at least in theory. In reality, the emperor was a ceremonial figurehead throughout the medieval period; someone with great cultural authority, but no real political power. The real power rested with the shogun, the most

(328) Japanese Samurai

(329) Japanese temple

powerful warlord in Japan. The shogun was appointed by the emperor, and was the supreme military commander; but this shogun warlord surpassed the emperor's authority and effectively ruled Japan.

Under the Song Dynasty that followed the Tang, Confucianism was reformed and reinvigorated to such an extent that modern scholars refer to this period as Neo-Confucianism. This was accompanied by the eclipse of Buddhism, at least among the educated elite ruling class. The Song Dynasty also saw a period of dramatic economic growth leading to the rise of numerous large cities. Under the Song, Chinese society became more urban and commercially orientated. It was also during the Song Dynasty that several important technological advances were made in China; including the compass, printing, and gunpowder. From the 8th century, economic growth in China turned that kingdom into a powerhouse of international trade. The Silk Road across central Asia flourished during medieval times; as did the maritime trade routes of the South China Sea and Indian Ocean.

In the South East Asian region, centuries of links with the Indian world to the west, saw large and well-organized kingdoms emerge. Little is known about the Srivijaya Empire for example, but it seems to have been a major trading power, dominating the newly developing trade routes between India and China. This empire flourished from the 8th to the 11th centuries, and its influence stretched far and wide across the region. The next state to dominate this region was the Khmer Empire in the 12th and 13th centuries. The Khmer civilization was based in present-day Cambodia and was remarkable for the series of "temple-mountains" it built, especially at Angkor Was, its capital city. By the end of the medieval centuries, the Majapahit Empire, based on the Indonesian island of Java, covered much of insular South East Asia.

The medieval period saw the rise and fall of a succession of kingdoms in West Africa. The heartlands of the earlier kingdoms were in the western savannah, where firstly Ghana, then Mali, then Songhai, rose to prominence. Further to the east in the Lake Chad region was the kingdom of Kanem-Bornu. Between these western and eastern kingdoms, a central bloc of territory saw the rise of the trading city-states of the Hausa people. These kingdoms owed their power and influence to their control of the Trans-Saharan trade routes; and during these medieval centuries merchants from North Africa brought their Islamic faith with them; converting members of the elite ruling classes in this region. The general population on the whole remained loyal to paganism; and this set up a religious-political tension in these kingdoms which escalated to destabilising conflict at times. South of the savannah, the medieval centuries began to see the rise of states in the forest regions. Amongst the earliest were the Kingdom of Benin, and the Yoruba city-states of present-day south-western Nigeria.

In East Africa, other states emerged at this time. Along the East African coast Muslim traders were instrumental in establishing a string of trading city-states which soon developed their own distinctive Swahili language and culture. Inland from these, in southern-eastern Africa, there grew kingdoms in the gold-deposit area of present-day

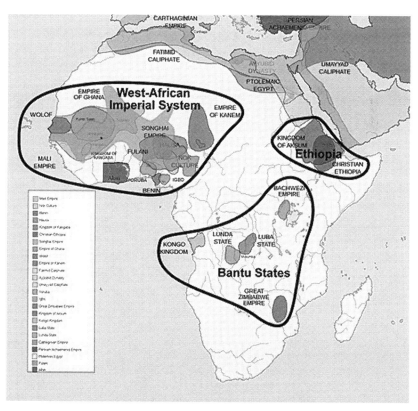

(330) African kingdoms in the Middle Ages

Zimbabwe; the most famous of these was centred on the large
settlement of the Great Zimbabwe Kingdom. In north-east Africa,
the ancient Kingdom of Ethiopia stood out from amongst its neighbours by adhering to a Christian culture which originated from trade
contacts with the eastern Mediterranean cities and towns within the
Byzantine Empire.

Linked to demographic and economic growth in Afro-Eurasia, the
medieval period saw the development of greater inter-regional contacts
across this massive land mass; of a military, economic, and cultural
nature. The "clash of cultures" of the 11th to 13th centuries known as
The Crusades, between the Christians of Europe and Byzantine, and
the Islamic world of the Middle East, involved all three of the above
elements. Despite ongoing hostilities, the Crusades actually increased
trade and cultural exchanges between the Christian and Islamic worlds;
to the great benefit of Europe in particular. The Crusades were a series
of religious wars initiated by the Catholic Church in Europe. These
wars were mainly fought in a number of campaigns in the Eastern
Mediterranean and Middle East in the period between 1096 and 1271.
The main objective of the Crusades was to recover the Holy Land,
including the sacred city of Jerusalem, from Islamic rule. In 1095, Pope
Urban II proclaimed the First Crusade at the Council of Clermont in
France. He encouraged military support for the Byzantine Emperor
Alexios I against the Seljuk Turks, and also an armed pilgrimage to
Jerusalem. Across all social hierarchies in Western Europe ranging from
kings, nobles, knights, clergy, and peasants, there was a widespread
enthusiastic popular response. Volunteers took a public vow to join
the crusade. There were numerous motives, including; the prospect
of doing "Christ's work" and ascending into heaven; satisfying feudal
obligations; opportunities to travel and seek recognition on the battlefield; and personal financial gain. Ultimately, although the Christians
failed in permanently retaining Jerusalem and the surrounding Holy
Land, and in spite of significant loss of life (up to 9 million) and
destruction on both sides; the eight Crusades succeeded in opening
up and accelerating trade and the exchange of collective learning and
ideas between the Christians and the Islamic Muslims.

(331) Christian and Muslim Crusaders meeting

Another factor in the strengthening of cultural and economic links between regions were the activities of the nomadic peoples of Central Asia. Turkish nomadic tribes, having converted to Islam, came to rule most of the Middle Eastern states; and it was Turkish warrior leaders who created the Delhi Sultanate and conquered most of the Indian subcontinent in the 13th and 14th centuries. Meanwhile, nomads originating from the eastern Siberian steppes came to rule northern China. The rise of the steppe peoples culminated in the Mongol Empire and its conquests of the 13th century under Genghis Khan and his successors. These Mongol conquests covered huge amounts of territory in Eurasia, including the whole of central Asia, China, Russia, much of the Middle East, and into Eastern Europe. At its peak, the Mongol Empire was the largest in conquered territory the world had ever seen to that date. Although the Mongol conquests were predominately military affairs, they greatly boosted existing commercial and cultural exchanges between the various regions under their control. At this time, the Silk Road entered a golden age, and the Indian Ocean routes also flourished as never before. Let us examine the adventures of two long-distance travellers during this time; Marco Polo and Ibn Battuta.

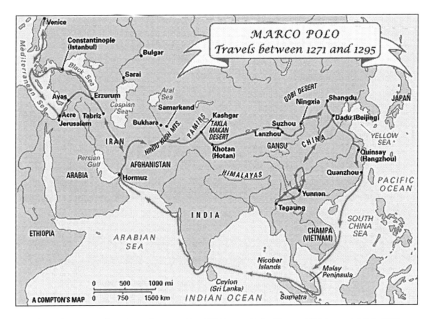

(332) Map showing the travels of Marco Polo from 1271 to 1295 CE

The journey of Marco Polo is an excellent case study of the way in which trade and cultural exchange was interacted on the Silk Road routes between different civilizations. Why has the name Marco Polo stayed with us after so many centuries? Born into a wealthy Venetian mercantile family in the 13th century, his background was exclusive, but not unusual for the time. His father Niccolo, and uncle Maffeo, were successful travelling traders who had set off to Asia just after Marco's birth. After setting up a number of trading posts along the Silk Road, their travels brought them to the court of Kublai Khan, the powerful fifth ruler of the Mongol Empire who was based in China. The Polo brothers were well received at the court of Kublai, sharing their knowledge of the Holy Roman Empire and of European Christianity. At the special request of Kublai, the brothers were instructed to return to Italy, accompanied by a Mongol ambassador and a letter from Kublai himself, to request from the Pope 100 representatives to teach Christianity and Western culture within the Mongol Empire. It was this directive which would lead Marco Polo to the life we know him for today.

Niccolo and Maffeo had set out on their journey just before Marco was born in 1254, leaving Marco's mother to raise him alone. However, tragedy struck when his mother died while Marco was still at an early age; meaning that he was raised by his extended family. Little is known of Marco's life as a child, apart from the fact that he was already a teenager by the time his father and uncle returned to Venice from the court of Kublai Khan. Niccolo and Maffreo's first meeting with the 15-years old Marco in 1269 occurred at a time when the Pope, who they had been instructed to meet with, was dead; so their mission on behalf of Kublai Khan was delayed. The brothers had to wait for the election of a new Pope, which at the time was a long drawn out process. The two eventually were given an audience with the new Pope Gregory X, in 1271. The Pope could only spare two priests to accompany the Polo's on their return voyage to Kublai Khan in China. Along the way, they collected some lamp oil from the sacred sepulchre at Jerusalem, at the request of Kublai. And so, planning to collect the lamp oil and two priests on their route, the Polo brothers began their journey, accompanied by the young Marco. Long distance travel in the 13th century was incredibly difficult; but the elder Polo brothers were experienced, resilient, and adventurous traders who were familiar with the Eastern routes. The Polo's sailed to Acre, now in Israel, and then on to Jerusalem to collect the holy lamp oil for Kublai Khan. They then returned to the port of Acre where they recruited two priests for their mission; and also took possession of a number of gifts and papal documents to present to their Mongol patron upon their arrival in China.

The trip was so gruelling that the two priests almost immediately abandoned the journey, turning back before the group reached Hormuz, a port city in Persia in which the Polo's attempted to secure a reliable and seaworthy vessel. However, the sea route proved to be inaccessible, so as a result, young Marco found himself slowly trekking on-foot through the winding mountain trading routes of the Silk Road. The incredible difficulty of this rough-terrain route cannot be underestimated, and it is quite remarkable that the young

Venetian undertook such an arduous journey for the very first time in his life; even though he was accompanied by his father and uncle. The group was attacked by bandits, and while in Afghanistan, Marco was struck down by a mysterious illness and had to spend months in the mountains recovering, before the group could resume their journey. The Polo's eventually found themselves in the audience of Kublai Khan and his court, after nearly four years of travel. Having eventually passed into the territory of the Mongol Empire, the group travelled to the palace at Xanadu in China, which was the summer residence of Kublai Khan. Marco Polo described the opulent extravagance of the palace as follows: "There is at this place a very fine marble palace, the rooms of which are all gilt and painted with figures of men and beasts and birds, and with a variety of trees and flowers, all executed with such exquisite art that you regard them with delight and astonishment." Polo also described Xanadu's "pleasure-dome," as a high-walled garden palace with a number of exotic animals and lined with gold and silver. His writings describe parties boasting 40,000 guests, with a banqueting hall large enough to seat 6,000. Polo also wrote of Khan's many concubines and the structure of his extended family.

(333) Venetian traveller Marco Polo   *(334) Mongol Emperor Kublai Khan*

By his own account, Marco Polo was well-received by Kublai Khan; who enjoyed the young Venetian's charm, intelligence, and skills as a storyteller. Marco Polo was to spend much of his early adulthood in the close-court of the great emperor. Though his four-year journey across the Silk Road to Xanadu represented an epic initiation for him into the art of long-distance travel; it would be over the following years as a servant and emissary of Kublai Khan that Marco Polo would come to repeatedly traverse the Asian continent, representing the interests of his patron. After he became a trusted courtier to Kublai, the Mongol emperor decided to employ the young Marco as a diplomatic envoy in the government of his sprawling empire. Armed with a seal of approval from Kublai Khan himself, which would ensure safe passage throughout his kingdom, Marco Polo travelled to lands previously unknown to Europeans, such as Tibet and Burma, and extensively explored India. During this time, it is also believed that he worked as a city governor, and as a tax inspector; and was personally promoted by Kublai to the prestigious and powerful Privy Ruling Council. Though Kublai had given Marco a wide range of responsibilities, his role allowed him open access to the Asian lands and their respective cultures that had previously been largely unknown to Europeans; and as a result, many of Marco Polo's accounts of his travels during this time became central to the Western understanding of Asia and its peoples for centuries. Marco ended up serving in the court and government of Kublai for almost two decades.

As the years wore on and Kublai Khan reached the final years of his life, the Polo's became anxious as to what their fate would be once their patron was no longer alive. The Venetian travellers asked to be released from Kublai's service, but the emperor did not want to lose the services of his Europeans. Eventually, after they had served him for 17 years, Kublai agreed to allow the Polo's to return to Europe, after they completed one final task; accompanying a young Mongolian princess, Kakachin, to her chosen husband in Persia. The Venetians agreed, though the journey there and then home, mainly by sea, was long and perilous, with many hundreds of their caravan of passengers and sailors hired for the trip perishing in wild storms or from disease. To make things worse, a number of the Polo family's accumulated possessions during their travels, were also stolen from them in Turkey, meaning that they arrived back in Venice in 1295 with very little. Worse still, they returned to find Venice at war with the rival Republic of Genoa, a neighbouring independent Italian city-state. Marco Polo himself fought in the war; was captured by the Genoa forces, and imprisoned. While imprisoned, Marco was fortunate to share a cell with Rusticello da Pisa, a writer of romance, who wrote the first version of the Tales of King Arthur in the Italian language. Having shared many stories of his adventurous travels with the writer, the two began a collaboration on a book that was ultimately to make Marco Polo famous.

The book, originally titled "Description of the World" but more commonly known today as "The Travels of Marco Polo," is the primary source for the accounts of the adventurer's life, and became a key text for Europeans seeking an insight into a continent and peoples that were previously shrouded in mystery. With the advent of the printing press in the 16th century, the book was destined to become a bestseller, and remained a major influence for centuries on how the Europeans perceived the East. The book made reference to something that was new to Europe – paper money. Paper money was important in the Mongol Empire because it allowed Kublai Khan himself to control all the actual wealth in the form of precious metals; with the paper notes forming an "exchange of value" role, and allowing the emperor to expand the financial network of his territory on a scale that Europeans would have found unimaginable. Many other Asian customs were also introduced in Polo's book, such as; the burning of coal for heat; the concept of a postal system; and never before seen accounts of unknown herbs and exotic animals; all contributing to the book's cultural significance and lasting impact.

Contrary to widespread belief, Marco Polo was in no way responsible for introducing pasta to Europe. The root of the myth seems to be based on a misconception that the Polo's were the first Europeans to travel so far East. This is not the case. Many Europeans had made it to Asia and beyond before Marco Polo did; it's just that his popular book meant that his adventures became the most widely circulated among European readers. Pasta as we know it today, made from durum wheat and water, was being produced in Sicily by the 12th century CE, and appears to have been introduced to the island, and subsequently mainland Italy, by Arab colonists and traders. Marco Polo's book was the first to introduce readers to Japan. In it, he described the invasion of the Japanese islands by the Mongols, and the bountiful reserves of gold in Japan; stating that the roofs of the buildings were lined with it, the same way that old church roofs were lined with lead; and how the Japanese palaces boasted solid gold floors and windows. Marco Polo, although a Christian, was clearly superstitious; describing in his book that while travelling, he would hear the voices of strange spirits: "When a man is riding through the desert by night and for some reason, falling asleep or anything else, he gets separated from his companions and wants to re-join them, he hears spirit voices talking to him as if they were his companions, sometimes even calling him by name. Often these voices lure him away from the path and he never finds it again, and many travellers have got lost and died because of this."

Marco Polo was released from imprisonment in 1299. He returned home to Venice where he quickly set about his transformation from an adventurer to a settled merchant and family man, marrying soon after his arrival, becoming a father to three daughters, and successfully running the family business. His family had accumulated enough wealth to buy a huge palazzo on the Grand Canal which became their permanent home in Venice. He died more than two decades later, at age 70. Written accounts from the time claim that Marco Polo was already a celebrity for his adventures and book in his own lifetime, and that he became an influential member of Venetian society; being mourned greatly by the city at the time of his death.

Although his book was very popular, it has been widely reported over the centuries that many of its readers did not take Marco Polo's adventures seriously and put the book down to as coming from an overactive imagination, or downright fabrication. Polo however, always stood by the authenticity of his book, reportedly stating on his deathbed: "I have not told half of what I saw." Nevertheless, the

debate has never quite subsided. Some scholars have pointed out
glaring omissions in the book; such as the Great Wall of China; the
use of chopsticks; foot-binding; and the drinking of tea; as evidence
that Marco Polo never actually made it to China at all, and that his
book is made up of numerous bits of gossip and rumour accumulated
in Western ports during his days as a traveller. More recent studies
however, have sought to cement his reputation and to prove that his
story is based on fact. Regardless of how factual the story of Marco
Polo is; his life and adventures have played a significant role in the early
recognition of Asian cultures by Europeans; and he deserves credit for
this.

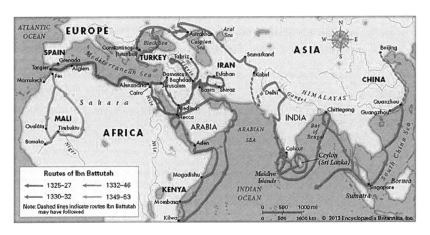

(335) Map showing the voyages of Ibn Battuta(h) from 1325 to 1353 CE

In the 14th century, the Moroccan adventurer, Ibn Battuta spent
nearly 30 years travelling over 120,000 kilometres across Africa, the
Middle East, India, and Southeast Asia. The title of "history's most
famous traveller" usually goes to Marco Polo; but for sheer distance
covered, Polo trails far behind the Muslim scholar, Ibn Battuta. Though
he is little known outside of the Islamic world, Battuta spent half of
his life wandering across vast territories of the Eastern Hemisphere.
Moving by sea, by camel caravan, and on foot, he ventured into over
40 modern day nations, often placing himself in extreme danger just
to satisfy his insatiable curiosity and passion for learning. When he
finally returned home after 29 years, he recorded his adventures in a

substantial travelogue which he titled "The Rihla." Though modern scholars question the reliability and accuracy of his writings; he may never have visited China, for example, and many of his descriptions of foreign lands appear to have been plagiarized from the works of other authors. The Rihla is nevertheless a fascinating insight into the world of a 14th century vagabond.

Born in Tangier, Morocco, Ibn Battuta came of age in a family of Islamic judges. 1n 1325, at the age of 21, he left his homeland for the Middle East. He intended to complete his haji – the Muslim pilgrimage to the holy city of Mecca – but he also wished to study Islamic Law along the way. "I set out along," he later recollected, "having neither fellow-traveller in whose companionship I might find cheer, nor caravan whose party I might join, but swayed by an overmastering impulse within me and a desire long-cherished in my bosom to visit these illustrious sanctuaries." Battuta began his journey riding solo on a donkey, but soon linked up with a pilgrim caravan as it snaked its way east across North Africa. The route was rugged and bandit-infested, and the young traveller soon developed a fever so severe that he was forced to tie himself to his saddle to avoid collapsing. In spite of his illness, he still found time to wed a young woman during one stopover; the first of 10 wives he would eventually marry and then divorce during his travels. In Egypt, Battuta studied Islamic Law and toured Alexandria and the metropolis of Cairo, which he described as "peerless in beauty and splendour." He then continued on to Mecca, where he took part in the haji. His travels might have ended there, but after having completed his religious pilgrimage he decided to continue wandering through the Muslim world. Battuta claimed to have been inspired by a dream in which a large bird took him on its wing and "made a long flight towards the east, and left me there." A holy man had interpreted the dream to mean that Battuta would roam across the earth, and the young Moroccan intended to fulfil this prophecy.

Battuta's next few years were a whirlwind of adventure. He joined a caravan and toured Persia and Iraq, and later ventured north to what is today Azerbaijan. He also trekked across Yemen and made a sea voyage to the Horn of Africa. From there, he visited the Somali city of Mogadishu before venturing below the equator and exploring the coasts of Kenya and Tanzania. After leaving Africa, Battuta came up with a plan to travel to India, where he hoped to secure a lucrative post as a "qadi," or Islamic judge. He followed a long and winding route, first travelling through Egypt and Syria before sailing to Turkey. As he always did in Muslim-controlled lands, he relied on his status as an Islamic scholar to gain hospitality from the locals. At many places during his travels he was overwhelmed with gifts of fine clothes, horses, and even concubines and slaves. From Turkey, Battuta crossed the Black Sea and entered the domain of a Golden Horde Mongol Khan known as Uzbeg. He was welcomed at Uzbeg's court, and later accompanied one of the Khan's wives to the great city of Constantinople. Battuta stayed in the Byzantine city for a month, visiting the spectacularly Christian mega-cathedral, the Hagia Sophia; and even met briefly with the emperor. Having never ventured to a non-Muslim city, he was stunned by the "almost innumerable" collection of Christian churches within its walls.

(336) Ibn Battuta in Egypt

(337) Ibn Battuta in India

Battuta next travelled east across the Eurasian steppe before entering India via Afghanistan and the Hindu Kush. Arriving in the city of Delhi in 1334, he secured employment as a judge under Muhammad Tughluq, a powerful Islamic sultan. Battuta had this job for several years, marrying and fathering children; but he eventually grew wary of the mercurial sultan, who was known to torture and kill his enemies – sometimes by tossing them into the path of stampeding elephants with swords attached to their tusks. Although he felt trapped, a chance to escape finally presented itself in 1341, when the sultan selected Battuta as his envoy to the Mongol court of China. Still hungry for new adventure, the Moroccan set out as the head of a large caravan full of gifts and slaves. The trip to the Orient would prove to be the most harrowing chapter of Battuta's odyssey. Hindu rebels harassed his group during their journey to the Indian coast; and Battuta was later kidnapped and robbed of all his possessions except his pants. He managed to make it to the Indian port of Calicut (Calcutta), but on the eve of the ocean voyage his ships were blown out to sea by a freak storm and sank, killing many members of his party.

These series of disasters left Battuta stranded and disgraced. He was loath to return to Delhi and face-up to the sultan, so he elected to make a sea voyage to the Indian Ocean archipelago of the Maldives. He remained in the idyllic islands for the next year, gorging on coconuts, basking in the sunshine, swimming in the crystal-clear waters, taking several wives, and once again serving as an Islamic judge. Battuta might have stayed in the Maldives even longer, but following a falling out with its rulers, he resumed his journey to China. After making a stopover in Sri Lanka, he rode on a number of merchant vessels through Southeast Asia. In 1345, four years after first leaving India, he arrived at the bustling Chinese port city of Quanzhou. Battuta described Mongol China as "the safest and best country for the traveller" and praised its natural beauty; but he also branded its inhabitants "pagans and infidels." Distressed by the unfamiliar customs on display, the pious Battuta stuck close to the kingdom's Muslim communities, and in his writings offered only vague accounts of the large cities such as Hangzhou, which he called "the biggest city I have seen on the face of the earth." Historians still debate just how far he ventured, but he claimed to have roamed as far north as Beijing, and crossed the famous Grand Canal.

China was to be the beginning of the end of Battuta's travels.

Having reached the edge of what was then the known world, he finally decided to journey back to Morocco; arriving back in Tangier in 1349. Both of Battuta's parents had died by then, so he only remained for a short time before heading first to Spain; and then on a multi-year excursion across the Sahara Desert to the Mali Empire, where he visited Timbuktu. Battuta had never kept journals during his travels, but when he returned to Morocco for good in 1354, the kingdom's sultan ordered him to write a travelogue. He spent the next year dictating his story to a writer named Ibn Juzayy. The result was an oral history called "A Gift to Those Who Contemplate the Wonders of Cities and the Marvels of Travelling," better known as The Rihla (or "travels"). Though it wasn't particularly popular in its day, the book is now considered to be a classic; one of the most vivid and wide-ranging accounts of the 14th century Islamic world that has survived. Following the completion of The Rihla, Ibn Battuta all but vanished from the historical record. He is believed to have worked as a judge in Morocco, and died in around 1368; but little else is known about him. It appears that after a lifetime spent on the road, the great wanderer was finally happy to live out the remainder of his life in one place.

How influential was the Mongol Empire? The Mongols under Genghis Khan and his successors ruled Eurasia from China to the Middle East, Eastern Europe, and Russia. This is the largest empire by land area, in history. Genghis divided his empire among his four children, while giving one of them supreme overall control. The unity could not be preserved however, and the Khan siblings drifted apart. Even so, Eurasia's main contemporary centres of power have many of their roots in the Mongol Empire. China, which after the Tang Dynasty had broken up into two separate kingdoms, the Jin and the Song; was unified politically and administratively by Kublai Khan, one of Genghis' grandchildren. Thereafter, China was able to maintain its geographical and political integrity despite the succession of dynasties. The Moghul Empire of India emerged from the Chagatai Khanate of Genghis' second son. The Islamic Abbasid Caliphate centred in Baghdad, was replaced by the Mongol Ilkhanate, which eventually became the heart of Persia. The Mongols of the Golden Horde first moved north towards Novgorod in Russia, then ventured sharply south and destroyed Kiev and its Viking civilization; some say at the behest of the Venetians, who were scheming to achieve a monopoly of the slave trade in that region. As a result, the centre of power in the region shifted to the north, and Czarist Russia eventually emerged. Eastern Europe was laid to waste by the Mongol invaders, but the remainder of the European continent was spared, probably because the Mongols decided the plunder was not valuable enough to bother invading. So Europe continued as a group of warring kingdoms vying among each other for control of more territory.

The Mongols' was the first truly modern army. It was built on a structure which emulated the Roman legions, on units in the multiple of tens; and promotion was strictly based on merit. Thoroughly disciplined and with no infantry, only highly mobile cavalry (on horseback), this army could execute complex tactical manoeuvres in complete eerie silence, after receiving their orders from central command. Speed and efficiency in conquest was their hallmark and a source of great fear and terror in the enemy. Horse and bow were the Mongol warriors' great strengths; and in the end, proved to be their weakness. Forests hindered the deployment of mounted armies; in the extreme humid heat of India, the bows failed; and the horses' strength was sapped when they could not find pastures in the Syrian desert in which to feed. Warfare technology and logistics were other factors in the superiority of the Mongols. The gunpowder formula was changed to yield a greater explosive force, rather than slow burn as in fire-lances. Guns and cannons were developed to a high level. Specialised troops of craftsmen were skilled in building complex siege machines from local materials, eliminating the need to move them over long distances. A dedicated medical corps looked after the wounded. The Mongol army and its horses would spread across the plains for self-sufficient foraging and sustenance, which meant they didn't need to have supply lines. The Mongols had developed a sophisticated communication system based on melodies, to ensure that accurate memorisation allowed the scattered troops to regroup at short notice, and to remain in contact with the more distant leadership.

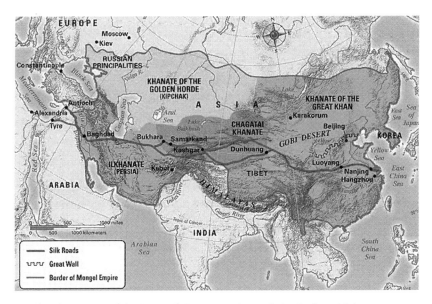

(338) A map of the Mongol Empire at its peak in the late 13th century

The intelligence system was second to none, and the Mongols knew more about the lands they were about to invade, than the defenders knew about the Mongols; if nothing else because the Mongols lived off the land and therefore needed to know where water and pastures were to be found. In addition, the Mongols developed highly sophisticated methods of psychological warfare, spreading rumours about their vicious cruelty and destruction wherever they ventured. This unsettled the mainly rural populations, who then often fled before the advancing army even arrived; hampering the defence efforts of the enemy and facilitating the capture of this vacated territory for the grazing of their horses. To what extent the Mongols inflicted cruelty remains an open question. Few traces remain of any evidence of massive-scale slaughter, amongst the excavated ruins of desert cities that were pillaged; and what is left indicates that the number of casualties was likely to have been greatly exaggerated. What seems to be established is that the Mongols promised justice to those who surrendered, but they swore destruction to those who resisted, particularly if they rebelled and threatened supply lines or troop withdrawal routes. It seems the Mongols kept their word. There is no evidence that they tortured, mutilated, or maimed captives; which sets them

(339) Genghis Khan – the warrior Mongol Emperor who ruled from 1206 to 1227 CE

apart from the rulers and religious leaders of China and Europe, who depended on such gruesome displays to strike fear in, and control of, their own people.

Nevertheless, having battled competing aristocratic lineages and dynastic families to unify his people, Genghis Khan was set on killing the aristocrats, whose loyalty, dependability, and usefulness he had come to doubt; essentially decapitating the social system of the enemy, and minimising future resistance. In doing so, he shrewdly recognised that the common people did not care much about the fate of rich and powerful noble families. Cities, particularly in the desert were destroyed and razed to the ground in order to redirect the flow of trade; and irrigation systems were demolished in order to convert agricultural fields into pasture for the Mongol horses. Plunder was the Mongol army's basic method of operation. All plunder would be gathered centrally, and then distributed in a fair and transparent way among the troops and the relatives of the fallen soldiers; this was called the khubi system. In the process they had to record massive amounts of numerical information. What was not plundered was counted and stored; and so emerged a highly sophisticated bureaucracy that kept track of the accumulated Mongol wealth. Artisans and people with other skills were gathered and moved over long distances to centres of production, serving the Mongol economy and their specific cultural tastes. In doing so, skills and technologies spread across the whole continent of Mongol conquered territories, continuously and in all directions.

Genghis Khan believed in the Great Blue Sky that spanned the world. He derived his mandate and inspiration for the creation of a world empire from this universal divinity. Genghis had come into contact with the many religions flowing back and forth along the Silk Road, however, these diverse beliefs were carried along by traders and adopted by women who then married into neighbouring tribes – it is believed that Kublai Khan's mother had been a Christian. It is ironic that narrow-minded orthodoxy prevented the Pope from seizing the opportunity to spread Christianity among the Mongols and their conquered territories. When conquering new lands, the Mongols respected religious freedoms, so they did not interfere in the religious beliefs and practices of the peoples they ruled; as long as these people did not threaten their rule, lived in harmony, and paid taxes. If they did rebel, they would be ruthlessly executed. It is amazing to think that Genghis Khan had been a reject among his people and had been persecuted by rival family lineages. When he achieved power he established the rule of law, which applied equally to everybody, including himself. This policy of equality allowed him to amalgamate the various defeated clans into one Mongol Kingdom, while destroying the traditional power bases of the 'white-bone' lineages that had oppressed the people in the past.

Without an agricultural or other production base of their own, the Mongols were heavily dependent on trade for their livelihood. In a major strategic initiative, they secured control of the Silk Road, which had languished under the various Muslim rulers that squatted it. The Mongols established free trade along the entire Silk Road, and were instrumental in the moving of great quantities of goods and ideas in either direction. It is along this Mongol Silk Road that the Venetian trader Marco Polo travelled to Kublai Khan's court in China, as we have seen. The Mongols introduced paper money from China into their empire; which was backed with the plunder of war. But Genghis's son

Guyuk, had been too generous with the printing press, and as a result had triggered hyperinflation and devalued his currency. His successor Mongke, decided to honour Guyuk's debts anyway, which secured the continuing flow of trade. Mongke also introduced a standardized silver ingot, the sukhe, to achieve convertibility between the local currencies and to monetize the payment of taxes, rather than accept payment in the form of local goods and produce. This allowed for the establishment of a Mongol State Budget, and the use of money to pay for expenses in places which were distant from tax collection points.

(340) The rampaging Mongol Army was a ruthless and efficient fighting machine

Because the Mongols had no system of their own to impose upon their subjects, they were very flexibly open-minded and were willing to adopt and combine political, government, and economic systems from everywhere; while largely preserving the religious beliefs and the cultures of the peoples they conquered. The Mongols implemented pragmatic rather than ideological solutions. They searched for what worked best for every situation they encountered; and when they found a solution, they spread it wherever relevant throughout their large empire. We should not underestimate the ability of the Mongols to blend in with local cultures. This attribute gave their rule a remarkable degree of stability and longevity. Kublai Khan's genius came from the recognition that he had to be flexible, adaptable, and tolerant, in order to rule China; and he was. His successors were less daring, and were eventually overthrown by the Ming Dynasty. Following the Mongol principles of even-handedness and religious and cultural inclusion, the Mongol ruler Akbar in India, deservedly achieved the title of 'Great'.

In the end, the Mongols were defeated by an unlikely enemy; the Black Death plague. It took off from Kublai's summer residence at Xanadu and followed the Mongol trade routes to sow widespread death across the continent. As millions of people died, trade was cursed and prohibited; and foreigners became the source of fear rather than welcoming curiosity. Later on, the European Enlightenment produced a growing anti-Asian sentiment that often focused on the Mongols as the symbol of everything evil or defective in the continent of Asia. The Mongols gained a reputation as the 'Barbarians at the Gate' which has unfortunately remained with them all through history into the modern-day; even though, as we have seen, they were enlightened, innovative, and progressive in many ways.

By the 15th century, all the Mongol dynasties, except in central Asia, were ousted by native rebellions; as a result, Eurasian politics once again reverted back to constituent cultural areas. The Ming Dynasty took control of China; various Turkish groups contested for dominance of the Middle East; one of these, the Ottomans, succeeded in gaining control of Asia Minor, and then proceeded to invade and conquer the Balkan lands of southeast Europe, where they captured the great Orthodox Christian city of Constantinople in 1453, and as a result, dramatically put an end to the Byzantine Empire. The Golden Horde Mongols were ousted from their control of the Russian territories by the Grand Prince of Muscovy.

By the end of the medieval period, the populations of Eurasia were recovering strongly from the effects of the Black Death, and subsequent outbreaks of plague. Europe in particular was experiencing change. The Black Death had changed the economic balance by creating a shortage of peasant labour, on which the landowning aristocracy depended, as we have seen. This led to rising wages and increased purchasing power, which created a rising demand for goods. This trend led to an increase in the standard of living of the peasant-class, and laid the foundation of consumer economies, a middle class, and capitalism. Renewed population growth stimulated economic activity, and trade links between different kingdoms and city-states were becoming even stronger, particularly in the Indian Ocean and Far East. Cities were expanding, and the way was being paved for the revolutionary transformations which would give rise to the modern world.

22

EARLY MODERN EUROPE – FROM THE RENAISSANCE TO THE ENLIGHTENMENT

THE EARLY MODERN PERIOD IN EUROPE COVERS THE TIMEFRAME between 1450, when the Italian Renaissance had its beginnings, to the outbreak of the French Revolution in 1789. From the early 15th century, Medieval Europe began to evolve into Early Modern Europe. In the later Middle Ages trade had expanded; towns had grown in number and size; and a new, more sophisticated society had emerged. In large parts of Western Europe, feudalism, with its fragmented hierarchical power structures, had begun to give way to centralised monarchies, with their concentration of power in the hands of the king or queen and their court officials. This concentration of power in the hands of absolute monarchs had been stimulated by the rise of gunpowder armies. Cannons and guns put a final end to the military superiority of heavily armoured knights, and the invincibility of castles. Being expensive and a more advanced type of military technology though, cannons and armies with guns, placed enhanced military power in the hands of those best able to afford them; the absolute monarchs. With this rise in central power came an expansion in royal government bureaucracies. While these developments in politics and society had been going on, the Medieval mindset, with its near-total subordination to the dominant power of the Catholic Church, was being undermined by blatant corruption within the Church's inner hierarchy. The rise of popular mass religious movements; such as the Hussites in central Europe, and the Lollards in England; calling for a return to the simpler Christianity of the Gospels, had been preparing the ground for a more critical approach towards faith and belief in the Church.

The Hussites were the followers of the Bohemian religious reformer Jan Hus, who was put on trial for heresy by the Catholic Council of Constance and burned at the stake, for his criticism of the lavish, corrupt lifestyles of high Church officials including bishops, cardinals, and the Pope himself. After his death in 1415 many Bohemian knights and nobles published a formal protest and offered protection

(341) The Renaissance Kingdoms and City-States of Italy

to those who were persecuted for their faith. The Lollards on the other hand were the followers of John Wyclif, and were critics of the established Catholic Church in England. Wyclif was born in Yorkshire in the 1330s, and was a theologian at Balliol College in Oxford. He was a realist who believed that one's knowledge is derived from within rather than through the external senses. He rejected the teachings of the Roman Catholic Church, preferring instead a church where all authority did not come from an institution, but instead directly from the gospels. John Wyclif's theology went far beyond that eventually adopted by the Anglican Church of Elizabeth I. His aim was for a reformation of the Church; but his movement failed mainly because there wasn't a printing press at the time so information about the Lollards could not be widely distributed; as well as the limited literacy of the population in the 14th century.

A movement which modern scholars refer to as the Italian Renaissance placed European civilization on a progressive path that was removed from its medieval past, and on a path towards modernity. This was primarily the result of internal developments within Italy, which at the time was made up of a number of independent city-states and kingdoms; in particular, the rise of wealthy mercantile trading cities in northern Italy, such as Milan, Venice, Florence, Siena, and Genoa. These city-states fiercely competed with each other, not just in politics, trade, and military affairs; but also as patrons of the culture and the arts. The intensity of their rivalry laid the foundation for a fertile flourishing of artistic endeavours. One of the most prominent ruling families of the Italian Renaissance were the Medici's of Florence. The Medici family, also known as the House of Medici, first achieved wealth and political power in Florence in the 13th century as a result of the family's success in commerce and banking. Beginning in 1434 with the rise to power of Cosimo de' Medici, the family's support of the arts and humanities made Florence into the shining light of the Italian Renaissance; a cultural flowering rivalled only by that of Ancient Greece. In addition to their rule of Florence, the Medici's were the patrons of great artists such as Leonardo da Vinci, Michelangelo, Sandro Botticelli, and Buonarroti; and produced four Popes – Leo X,

Clement VII, Pius IV, and Leo XI. Through marriage, their genes were also mixed into many of Europe's royal families. The last Medici ruler, Gian Gastone de' Medici, died without a male heir in 1737, ending the family dynasty after almost three centuries.

(342) Cosimo de Medici of Florence   *(343) Michelangelo's Pieta sculpture*

Another catalyst for the Italian and then wider European Renaissance was the fall of the great Christian city of Constantinople to the Muslim forces of the Ottoman Turks in 1453. This had been the historic capital of the Byzantine Empire, itself a continuation of the Roman Empire, for more than 1000 years, and its fall shocked all of Europe. It also led to an exodus of scholars, enlightened thinkers, and artists from the city, who brought with them much ancient Greek and Roman learning that had been lost to Western Europe. Artists and architects looked back to Roman models for their inspiration, and so remade much of the physical environment of European towns and cities. Writers and thinkers also looked back to Greek and Roman philosophers, and this inspired them to rethink not only much of their understanding of the natural-scientific world, but also the way they sought to conceptualize and logically formulate this under-standing. From this process would come the experimental techniques of modern scientific enquiry, "the scientific method"; so that over the next few centuries the work of Vesalius, Copernicus, Kepler, Harvey,

Galileo, Newton, and many others would revolutionise the knowledge base and collective learning of the West's understanding of the natural universe.

Each of these great scientists made significant contributions to the body of human collective learning; but let's take the case of William Harvey as an example. Born in 1578 and died in 1657, this eminent English physician was the first to recognize the full circulation of the blood in the human body, and to undertake experiments to prove his idea. On his achievements he said: "I profess to learn and to teach anatomy not from books but from dissections, not from the tenets of Philosophers but from the fabric of Nature." In relation to the heart, he said: "The heart is the household divinity which, discharging its function, nourishes, cherishes, quickens the whole body, and is indeed the foundation of life, the source of all action." He explained the difficulty of achieving his discovery, as follows: "When in my dissections, carried out as opportunity offered upon living animals, I first addressed

(344) William Harvey delivering a lecture on the circulation of blood

my mind to seeing how I could discover the function and offices of the heart's movement in animals through the use of my own eyes instead of through books and writings of others, I kept finding the matter so truly hard and beset with difficulties that I all but thought, with Fracastoro, that the heart's movement had been understood by God alone."

In the meantime, the West's knowledge of the geography of the world expanded enormously with the voyages of discovery made by the great explorers, which began in the 15th century. Some of the greatest of these courageous and adventurous European maritime explorers were:

- Bartolomeu Dias (1451 – 1500), a Portuguese explorer who became the first European to round the Cape of Good Hope in southern Africa.
- Vasco da Gama (1460 – 1524), a Portuguese explorer who picked up where Bartolomeu had left off, by becoming the first European explorer to sail around the Cape of Good Hope and then on to India.
- Ferdinand Magellan (1480 – 1521), a Portuguese explorer who was the first to sail across the Pacific Ocean and to circumnavigate the entire globe; forever shattering the myth that the earth was flat.
- Amerigo Vespucci (1454 – 1512), an Italian explorer who sailed to the Americas and participated in voyages from 1499 to 1502 that explored more of North America (after Columbus) and found South America to be more extensive than previously realized. The word "America" comes from his first name.
- Francis Drake (1540 – 1596), the first Englishman to circumnavigate the globe. However, the expedition for which Drake earned his knighthood began in 1577, when the explorer was sent by Queen Elizabeth I on a voyage to raid Spanish outposts in the Pacific and on South Americas west coast; expanding the British Empire in the process.
- James Cook (1728 – 1779), an English explorer whose achievements in accurately mapping the Pacific, New Zealand

and Australia radically changed western perceptions of world geography; and directly led to the British colonization of New Zealand and Australia in particular.

The iconic peak of these voyages was the accidental discovery of the Americas by Christopher Columbus in 1492; but there were a long series of sea voyages and overland expeditions which each contributed to the Europeans' understanding of the wider world, as can be seen in the brief summary above. Over the next couple of centuries, the Americas were opened up to European conquest and colonization, and trade routes were pioneered linking the Atlantic with the Indian and Pacific Oceans. European sailors and merchants developed a system of maritime commerce which, for the first time in world history, spanned the globe, and directed much of the world's commerce and trade towards Europe. This had a multiplier effect on European economic growth and standards of living, especially in those regions bordering the Atlantic Ocean.

(345) Christopher Columbus arriving in North America in 1492

Christopher Columbus is of course universally recognized as the first European to discover the America's; but his achievements were much wider than this. His major achievements can be summarized as follows:

- He independently discovered the Americas, by first putting forward the idea of sailing across the Atlantic in order to get to Asia, rather than the alternative of travelling by land, which was long and fraught with danger. Instead of reaching Asia, he landed in North America.

- He discovered a viable sailing route to the Americas.

- He led the first European expeditions to the Caribbean, Central America, and South America.

- His settlement in Hispaniola provided Spain with a strategic advantage for expansion in the New World. The island of Hispaniola is modern-day Haiti on the western side, and the Dominican Republic on the eastern side.

- Columbus made colonization of the Americas possible for Spain.

- He was instrumental in the subsequent creation of the significant Columbian Exchange. This refers to the widespread exchange of animals, plants, human populations, diseases, technology, culture, religion, and other ideas, that occurred between Afro-Eurasia and the Americas.

- Through the domestication of animals, the Columbian Exchange expanded food supply in the Americas.

- The Columbian Exchange caused a significant increase in the world population. This is because the plants from the Americas had a huge impact on the lives of millions of people in Africa, Europe, and Asia, whose lives were radically changed by the introduction of crops from the Americas. Plants like potatoes and maize (corn) could grow in soils which were otherwise useless for agriculture in Afro-Eurasia. For example, today China and India are the largest producers and consumers of potatoes in the world. Other plants introduced into Afro-Eurasia from the Americas included beans, squashes, tomatoes, avocados, papaya, pineapples, peanuts, chilli peppers, and cacao (the raw form of cocoa used to produce chocolate).

- Columbus served as the Governor of Hispaniola. In effect, this meant that he was given power to administer the Spanish colonies on the island of Hispaniola; a position which he held from 1492 to 1499.

Within Europe itself, one invention of the mid-15th century more than any other, helped to accelerate collective learning and the progression of Europe towards the modern world. This was moveable-type printing. To what extent this was an original invention in Europe, or had spread to Europe from China (where it had been invented several centuries before), is a topic debated amongst historians. It is certainly the case that Johannes Gutenberg, a German blacksmith and goldsmith, introduced major innovations of his own to printing technology. What is also beyond doubt is that the impact of the printing press on Europe was far more profound than it had been on China. The ability to print books, pamphlets, and posters, dramatically reduced their cost to consumers. New knowledge and ideas could spread around the European continent much more rapidly and widely than before; so that a discovery made by a Portuguese explorer or a German astronomer for example, could be read about by ordinary shopkeepers or artisans in Paris or London. This greatly accelerated the collective learning of human knowledge, and stimulated a thirst for more.

(346) The Printing Press in Europe was first invented by Johannes Gutenberg in 1440

The impact that the printing press and the emergence of widely-circulated books had on the religious life of Europeans was even more profound. In the early 16th century, long-term criticism about what was widely seen as the corrupt state of the Catholic Church led to the outbreak of a movement called the Reformation. Religious leaders such as Martin Luther and John Calvin called for major reforms of the Catholic Church to root-out corruption. When these calls were ignored and rejected, this movement broke away from the Catholic Church to form the Protestant Church; which itself soon splintered into a myriad assortment of different churches and sects. This split the Christians of Europe into two major hostile camps; the Protestant nations of the north, and the Roman Catholic nations of the south. The Reformation led to a strong and aggressive response from the Roman Catholic Church. This was known as the Counter-Reformation. At the "Catholic hearts and minds" level, Ignatius Loyola founded the religious order of the Jesuits, which became one of the most effective missionary organisations in world history.

At the national and political level, Spain, at that time the leading power in Europe under its King Philip II, saw itself as the champion and defender of the Roman Catholic Church; and strenuously fought to put down Protestantism wherever it could. This ultimately led Spain into a series of futile wars in the Netherlands, and with England; including the launching of the impressive but ill-fated Spanish Armada, which as we have seen was defeated by Queen Elizabeth I of England in 1588. These and other wars of religion culminated in the terribly destructive Thirty Years War in Germany, which lasted from 1618 to 1648. These wars affected all aspects of life in Europe. The western Christian world was no longer united, and the papacy, which had previously claimed the spiritual high-ground and leadership over all of Western Europe, was now a divisive entity that was utterly rejected by the Protestants of England, Scotland, Holland, northern Germany, and Scandinavia. In these countries, a new and simpler style of Christianity emerged. The Protestants emphasized individual spirituality, which led to the printing of the Bible and other religious books away from the customary Latin, and instead in the particular languages of these different European nations. It would also open the way to greater value being given to personal individual choice. This would become a defining feature of Western Civilization, and amongst other things, would open the way to the rise of the secular society which we have today.

The fate of the Catholic Church as a major political and cultural power in Europe was sealed with the rise of Protestantism. There was nothing much the Catholic Church could do as Christianity irreversibly split into three major factions; the pre-existing Catholic and Orthodox faiths, and now a third, Protestant faith. It was during this time of the Reformation that the Catholic Church progressively ceased to have any major influence on the politics of Europe, and instead desperately tried to hang on to its former control of the peasant masses, with its religious teachings and traditional practices. This breakdown in the power and influence of the Catholic Church in Europe unleashed an unstoppable burst of intellectual humanist thought in the areas of science, philosophy, literature and the arts. Collective learning could now proceed at an even faster rate, fuelled by the widespread availability of books, and more educated people who could read them.

(347) William Shakespeare's Othello   *(348) William Shakespeare*

Two of the intellectual giants of the European Renaissance were William Shakespeare and Leonardo da Vinci. Let us briefly examine the accomplishments of each. Many believe William Shakespeare was the best British writer of all time. His many works are about the full spectrum of the human condition including life, love, death, revenge, grief, jealousy, murder, magic, and mystery. He wrote the blockbuster plays of his day including Macbeth, Romeo and Juliet, and Hamlet. It has been over 400 years since he died yet his plays are still performed around the world. William Shakespeare was born in Stratford-upon-Avon, England in 1564. In 1582 he married Anne Hathaway and the couple went on to have three children. He lived in London for 25 years and wrote most of his plays there, dying in 1616 at age 52. Shakespeare wrote 37 plays which can be categorized into three different types; histories, which are about the lives of kings and famous figures; comedies, which end with a marriage; and tragedies, which end with the death of a main character. He also wrote large works of poetry, and in 1609 published a book with 154 sonnets. We know his work was popular in England at the time because he earned enough money to live in a fashionable area of London. He also had an incredible influence on the English language and invented hundreds of worlds that are still used today, such as; amazement, bedroom, champion, dawn, eyeball, fashionable, gossip, moonbeam, olympian, puking, swagger, unreal, and zany. As well as writing plays and poetry, Shakespeare performed on stage with a group of actors called the Lord Chamberlain's Company. When James I became the King of England in 1603, the group was renamed the King's Company. The group performed in two theatres near the banks of the Thames River in London; the Globe and Blackfriars. Shakespeare owned a share of these theatres. During his time, women were not allowed to act on stage, so all of the female roles were performed by men.

Leonardo da Vinci was born in the Italian town of Vinci in northern Italy in 1452, and died in 1519. He was motivated by his great curiosity and unlimited desire for knowledge. This guided much of his thinking and behaviour. As an artist and a thinker, Leonardo was highly visual; in fact, he considered sight to be the highest of all the senses. To him every phenomenon that could be perceived became an object of knowledge. Leonardo applied his rigorous sight and creativity to numerous disciplines, including painting, architecture, music, literature, engineering, the natural world, and human anatomy. Early on in his painting life, Leonardo found a balance between technique and expressiveness, as seen for example, in his 1482 painting, Adoration of Magi. In addition, he honed his technique of "sfumato," or shading with soft outlines or haziness, to further his emotive but precise form of painting. A masterwork from his early period demonstrating this technique is the Virgin of the Rocks. In this painting, as always, Leonardo casts his subjects with precision and expressiveness, which in turn produces an aesthetic effect of elegance and power. Another important work, painted over a three-year period from 1495 to 1498 in Milan, is the Last Supper. This painting depicts Jesus and his 12 disciples with a wide range of distinctive emotional expressions. His most famous painting though, is the Mona Lisa, painted between 1503 and 1519. It is probably the most famous painting in art history, and features the captivating "Mona Lisa smile." The woman's mysterious smile and her unproven identity have made this painting an ongoing source of great fascination and mystery.

Leonardo da Vinci is considered to be the father of humanism; a philosophy which places the beauty, expressiveness and power of the human mind at the centre of our existence. It's a philosophy that inspires lifelong learning, curiosity, and the acceleration of collective learning for the benefit of all humanity. Humanism is often seen to be at odds with the dogmas associated with most religious belief systems, and instead is more aligned to the scientific method of continuous discovery in an attempt to enhance our knowledge of the natural world around us. This was certainly the philosophy practised by Leonardo da Vinci. His insatiable thirst for new knowledge and his intellectual curiosity was never fully satisfied, in spite of his phenomenal multi-disciplinary achievements.

When he sought to serve Duke Ludovico Sforza in Milan, Leonardo cast himself as an architect. Although never practising as an architect, he produced numerous sketches and ideas for architectural

(349) Leonardo da Vinci
(350) Mona Lisa was painted
1503-1519

designs of churches and secular buildings. Leonardo's primary role in architectural projects was that of an advisor; and he was associated with the best architects of his day. Many of his sketches reveal his mastery of technical as well as architectural challenges. Leonardo's early anatomical studies were thorough in their consideration of body parts and organs, as well as how they worked together within a functional human body. He was especially interested in the brain, heart, and lungs. His anatomical drawings were amongst the most significant of the Renaissance, and were used by medical students in their studies of the human body. Leonardo however, never considered himself to be an expert in the field of anatomy. Instead, his anatomical sketches formed the foundation of his orientation with how human figures interact with the natural world. This knowledge in turn enhanced his painting of human figures, making them more realistic in their form and expressiveness.

Leonardo's interest in science did not stop with anatomy; he also studied hydraulic engineering. He even sketched a flying machine with wings and "helical airscrews" that seemed to be a prototype for the modern helicopter. Leonardo's voluminous notebooks also contained extensive notations and thoughts on painting, architecture, mechanics,

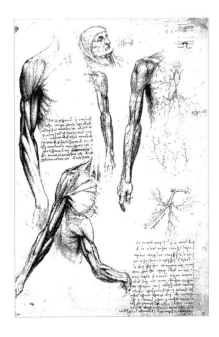

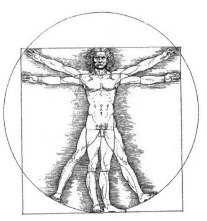

(351) & (352) Leonardo da Vinci's anatomical sketches

botany, geology, anatomy, aerology, and hydrology. One unusual feature of his notebooks was his use of mirror writing, or putting words down on paper in such a way that they could be read normally only when held up to a mirror. We are uncertain as to why Leonardo engaged in mirror writing because there is no evidence to suggest he wanted his notebooks to be kept a secret. Although he undoubtedly produced more, 25 of his notebooks have been preserved into the modern day. These manuscripts are in collections located in Italy, France, England, Spain, and the USA, and comprise a total of over 7,200 pages, representing about one quarter of what Leonardo actually wrote. Probably Leonardo da Vinci's most famous surviving notebook, the scientific "Codex Leicester" named after an 18th century owner, Thomas Coke, the Earl of Leicester, was purchased by Microsoft founder and philanthropist Bill Gates, at a Christie's auction in 1994 for $30.8 million. Written between 1506 and 1510, the 72-page notebook mainly focuses on Leonardo's written thoughts and sketches relating to amongst other things, water – tides, eddies, and dams – and the relationship between the Moon, the Earth, and the Sun; and is just one of the 25 notebooks which have survived. Gates usually makes the manuscript available for one exhibition every year. Bill Gates had this to say about Leonardo in an article he wrote for Time magazine on 7th February 2019: "What does a Renaissance artist have to do with optimism? For me, the connection is innovation. I feel optimistic about the future because I know that advances in human knowledge have improved life for billions of people, and I am confident they will keep doing so. And although I am not an art expert, everything I have learned about Leonardo leads me to believe he was one of the most innovative thinkers ever. Leonardo was an insatiable learner. He studied everything he could see: the flow of water, the way smoke rises through the air, how a woodpecker uses its tongue. And he had insights that were ahead of his time. He developed a theory about the working of a certain heart valve that researchers only verified a few decades ago. He was the first person to correctly explain why you can see light between the two points of a crescent moon, the phenomenon we now call earthshine. Scientific inquiries like these were essential to his art. He was able to give the Mona Lisa that mysterious look on her face because he had studied all the muscles involved in smiling. In the Last Supper, he could make the perspective lines work flawlessly because he had spent countless hours understanding how our eyes perceive objects at a distance. By examining his surroundings so closely, Leonardo was able to develop new techniques that advanced his field and portrayed the world in a way no one had ever seen before. In other words, he was an innovator."

The bickering, fractured kingdoms of Western Europe were unable to stop the rise of two huge and powerful states to the East. The first of these was the Muslim Ottoman Empire, which even before its conquest of the great city of Constantinople, had acquired extensive territories in the Balkans. Throughout the later 15th, 16th and 17th centuries, the Ottomans expanded deep into central Europe, twice besieging the city of Vienna, in 1529 and 1683; the capital city of the Austro-Hungarian Hapsburg Kingdom. The other state was Russia. Although practicing the Orthodox Christian faith and therefore, in religious matters it came within the sphere of influence of Constantinople; the Russians had been politically, and to some extent culturally, a part of the Asiatic Mongol Empire since the 13th century. However, in the

15th and 16th centuries, under the leadership of the Grand Princes of Muscovy, the Russians gradually liberated themselves from the political domination of the Mongols, who by now were known in Western Asia as the Golden Horde. The newly independent Russians then expanded westward into Europe, at the expense of the Polish; and southwards towards the Black Sea, at the expense of the Ottoman Turks. During this expansionary period the Russians also became increasingly westernized. Under a succession of Tsars, in particular, Peter the Great and Catherine the Great, the Russian noble elite had increasingly embraced western European culture and dress. By the end of the Early Modern Period, Russia had emerged to become one of the great European powers, alongside Austria-Hungary, France, Britain, and the rising power of Prussia.

(353) Peter the Great

(354) Catherine the Great

The two most significant individuals who influenced the history of Russia were arguably Peter the Great and then later, Catherine the Great. Peter the Great, who was Tsar from 1682 to 1725, and Emperor from 1721 to 1725, made a great positive impact on Russia, which at the start of his rule, was largely a peasant society that had greatly lagged behind Western Europe. He transformed Russia into a major political, military, and cultural power, rivalling many of the European kingdoms and nations. Through his numerous reforms, Russia made incredible progress in the modernization of its economy, the development of trade, education of the masses, the promotion of science, enhancement of culture and the arts, and in a foreign policy that connected Russia closer to Europe. Peter led his country into major conflicts with Persia, the Ottoman Empire, and Sweden. Russian victories in these wars greatly expanded the Russian Empire, and the defeat of Sweden in particular, gave Russia direct access to the Baltic Sea; a lifelong obsession of the Russian leader.   Catherine the Great reigned over Russia as the Empress for 34 years from 1762 to 1796. As the Empress, Catherine further westernized Russia. She led her country into full participation in the political and cultural life of Europe; championed the arts, and reorganized the Russian Law Code, bringing it more in line with European Law. Catherine also significantly expanded Russian territory by securing the northern shore of the Black Sea, the annexation of the Crimean Peninsula, and the expansion of the empire into the Siberian steppes beyond the Urals, and also along the Caspian Sea.

The Post-Reformation thinkers in Europe felt more-able to pursue their own individual quests for meaning and understanding of the natural and societal world around them. This enlightened thinking, in conjunction with a strong reaction against the religious dogmatism which had caused so much bloodshed in the wars of religion, led to a new spirit of nationalism, which became apparent from the later 17th century onwards. Every aspect of human affairs; including religion, society, culture, science, government, and economics; was scrutinized and questioned in a new rational and ordered-logical way. Long standing traditionally accepted notions of divine providence which had been the main domain of the Catholic Church, were largely relegated to the margins. This movement went by several names including; the Age of Reason; the Enlightenment; and the Scientific Revolution. Its effects were felt in philosophy, economics, politics, medicine, and science. More rational foundations and ideologies of governing were sought, and once this new enlightened thinking had penetrated into the royal courts of Europe, more rational ways of governing countries were put into practice. The reforms of such absolute monarchs as Louis XIV in France, Peter the Great and Catherine the Great in Russia, Frederick the Great in Germany, Elizabeth I in England, and other monarchs, were the fruits of this enlightened thinking. They created more efficient forms of government by giving more power to bureaucrats who were appointed on merit rather than hereditary patronage; and these educated bureaucrats applied rational thinking to the challenges of government administration and policy-making.

As the Enlightenment was in progress, popular democracy was not high on the agenda. However, in certain Protestant countries, particularly Holland and England, long-standing political trends had been in place, limiting the power of the monarch and putting more power in the hands of elected representatives of the people, even though it was only a small proportion of the people that were actually able to vote. By the end of the 17th century the Dutch Estates and English Parliament had become the seats of sovereignty in these two nations. It is no coincidence that Holland and England also had the most advanced economies of Europe at this time. During the 17th century their overseas trade had multiplied many times over.

Sadly, during this time, England in particular was engaged in the Atlantic Slave Trade, in which millions of enslaved Africans were taken from their African homelands to the Americas, to work on mainly cotton, sugar, and tobacco plantations. For well over 300 years, European countries forced Africans onto slave ships and transported them across the Atlantic Ocean. The first European nation to engage in the Atlantic Slave Trade was Portugal in the mid to late 1400's. Captain John Hawkins made the first known English slaving voyage to Africa in 1562, during the reign of Elizabeth I. Hawkins made three such journeys over a period of six years. He captured over 1200 Africans and sold them as goods in the Spanish colonies of the Americas. In the beginning, English traders supplied slaves for the Spanish and Portuguese colonists in America; however, as English settlements in the Caribbean and North America grew, often as a

(355) The Atlantic Slave Trade lasted from the 16th to the 19th century

result of wars with other European powers such as Holland, Spain, and France, English slave traders increasingly supplied English colonies. The exact number of English ships that took part in the Atlantic Slave Trade will probably never be known, but in the 245 years between Hawkins first voyage and the abolition of the Slave trade in 1807, merchants in England despatched around 10,000 voyages to Africa for slaves, with merchants in other parts of the British Empire fitting out a further 1150 voyages. It is estimated that English ships carried 3.4 million enslaved Africans to the Americas. Only the Portuguese, who carried on the trade for almost 50 years after England had abolished its Slave Trade, carried more enslaved Africans to the Americas than the English; with an estimated over 5 million people. Estimates, based on records of voyages in the archives of port customs and maritime insurance firms, put the total number of African slaves transported by European traders to at least 12 million people; of these, up to 25% died in horrendous circumstances during their voyages.

Holland, England, and France had acquired trading settlements and colonies in North America, India, the coast of Africa, the Caribbean, and South East Asia. At home, the economies of these European

(356) African slaves working on cotton plantations in the Americas

powers had been boosted by agricultural and farming innovations; and better roads and canals to fast-track the distribution of trade, commerce, goods, and services. Innovations such as joint-stock (share-holding) companies, like the massively influential English and Dutch East India Companies; national banking systems; stock exchanges; and patent protection; had eased the financing of commercial expansion. Soon, the first "capitalist" booms and busts were causing excitement amongst investors and speculators; including the Tulip Mania of Holland, and the South Sea Bubble of England. Tulip Mania was a period during the Dutch Golden Age when contract prices for some bulbs of the recently introduced and fashionable tulip reached extraordinarily high levels, and then dramatically collapsed in February 1637. It is considered to be the first recorded speculative bubble in history. The South Sea Bubble on the other hand, was the financial collapse of the South Sea Company in 1720. The English company was formed to supply slaves to the Spanish colonies in the Americas. The South Sea Company was formed in London in 1711, and its purpose was to supply 4800 African slaves each year for 30 years to the Spanish plantations in Central and Southern America. A number of wealthy Europeans bought shares in this company, which ultimately collapsed due to mismanagement and fraud; as a result, the mainly English investors lost all of their money.

The English East India Company operated from 1600 to 1873 and was formed for the exploitation of trade with East and Southeast Asia, and India. Starting as a monopoly trader in these regions, the company also became involved in the politics and military affairs of the English colonies in which it operated, even having its own army; and acted as an agent of English imperialism in India from the early 18th century to the mid-19th century. In addition, the activities of the company in China in the 19th century served as a catalyst for the expansion of English influence there. The company was first formed to share in the East Indian spice trade. That trade had been a monopoly of Spain and Portugal until the defeat of the Spanish Armada by England in 1588; which gave England the chance to break this highly lucrative monopoly. The English East India Company took full advantage of the recently introduced concept of public companies which could raise capital from a large number of investors. The company met with aggressive opposition from the Dutch, in the Dutch East Indies (now Indonesia), and the Portuguese. The Dutch also wanted a share of the Spanish and Portuguese trade, and had created their own Dutch East India Company to capitalize on this lucrative trade on behalf of Holland, in direct head-to-head competition with the English. The Dutch virtually banned English East India Company employees from the East Indies, after the Amboyna Massacre on Amboyna Island in Indonesia, in 1623; a notorious incident in which English, Japanese, and Portuguese traders were executed by Dutch authorities, for allegedly conspiring to takeover this extremely lucrative spice trading post. In spite of this setback, the company flourished by trading in cotton and silk goods, indigo, saltpetre, and spices from South India. It also extended its activities into the Persian Gulf, Southeast Asia, and East Asia. Its commercial monopoly was broken in 1813, and from 1834 it was just a managing agency for the English government of India. It was stripped of that role after the Indian Mutiny of 1857, and ceased to exist in 1873.

Holland, England, and France were long-term powerful rivals for overseas trade and imperial colonial empire-building, during this time. In North America and the Caribbean, and along the coasts of

India and the East Indies, they constantly fought with each other for strategic advantage. Out of this rivalry grew a new kind of fighting force; the transoceanic navy. England's navy eventually emerged as the supreme ruler of the seas; and through her unrivalled sea power and accumulated economic wealth, the British Empire emerged as one of the largest the world had ever seen; the only possible rivals to that date being the Mongol Empire, and the Qing Dynasty of China. One by-product of this intense global imperial rivalry was the discovery by Europeans of the southern landmasses of Australia and New Zealand, together with many much smaller Pacific Islands. Although about as far east as it was possible to sail, Australia and New Zealand in particular, and on a smaller scale Hawaii, were to become integral parts of Western Civilization.

The end of the early modern phase of European history is usually regarded as being the outbreak of the French Revolution. This event can be seen as the most radical expression of the Enlightenment. It sought to break with traditional forms of government, based on hereditary monarchy, aristocracy, and church, which all were entrenched in Medieval Europe. Instead, the French Revolution attempted to impose a more rational political regime, based on the will of the people, the public interest, elected assemblies, equality between the classes, and the philosophy of reason. It would eventually lead to the creation of new kinds of European nation states in the 19th century; which on the one hand were more responsive to the will of the people than the more traditional kingdom states had been; and on the other hand had a far greater impact on the lives of their citizens. Modern Europe was also on the cusp of the Industrial Revolution. This had been gathering pace in Britain since around 1750, as a direct result of the strong economic growth which that nation had been experiencing. The Industrial Revolution would create an entirely new kind of society, with its focus on the city-based manufacturing of consumer products, rather than rural-based agriculture.

23

THE MODERN WORLD – FROM THE INDUSTRIAL REVOLUTION TO THE END OF THE 19TH CENTURY

T HE EIGHTH AND FINAL THRESHOLD OF INCREASING COMPLEXITY on our long journey of Big History was the advent of the Modern World, a period in which for the first time, humans truly dominated the entire ecosystems and environment of our fragile planet Earth, a period known as the Anthropocene. We have seen in previous chapters, the unrelenting march of accumulated collective learning, starting from when our human ancestors emerged in Africa around 300,000 years ago, and then their migration out of Africa, and into all of the other continents of the Earth other than Antarctica. Then we crossed the seventh threshold of increasing complexity, the emergence of agriculture along the great river systems of the world, and the expansion of villages into towns, towns into cities, and cities into civilizations. But throughout this period of human history, the share of energy humans could harness was primarily limited to human and animal power. This was all about to change when fossil fuels first began to be extracted on a relatively large scale, as an additional source of energy in England, starting with coal in around 1750.

The conditions which led to the start of the Modern World were a combination of globalization – extensive trade and exchanges of ideas around the world; a rapid acceleration in collective learning stimulated by scientific, political, economic, and cultural revolutions which had originated during the Renaissance; human innovation and invention; and the use of fossil fuels as a significant driving source of exponentially greater energy. What emerged was a dramatic increase in the human harnessing and use of resources; an innovation revolution; and the ability of humans to transform the biosphere – for better and for worse – the human dominated period of the Anthropocene.

By the mid-18th century the European colonies in North America had become fully functioning societies in their own right. The colonists felt a growing sense of their ability to determine their own future, and this ultimately led to the American Revolution, from 1775 to 1783. By the end of this revolution, a new nation, the United States of America, had made its appearance on the world stage. The constitution of this newly created nation was consciously modelled by the founding fathers on European Enlightenment principles of rational government. But, unlike the "enlightened monarchs" of continental Europe, because they were more accustomed to the British system of representative parliamentary rule, the Americans set up the most democratic form of government in the world, at that time.

Two of the most influential people in the creation of the United States of America were George Washington and Thomas Paine. George Washington, born in 1732 and died in 1799, was commander in chief of the Continental Army during the American Revolutionary War fought from 1775 to 1783. He also served two terms as the first President of the United States, from 1789 to 1797. Washington was the son of a prosperous plantation owner, and was raised in colonial Virginia. As a young man, he worked as a surveyor, and then fought in the French and Indian War from 1754 to 1763; a war which pitted the colonies of British America against those of France, with each side supported by military units from the European home country, and by Native American allies. During the American Revolution, he led the colonial forces to victory over the British, and became a national hero. In 1787, he was elected president of the convention that wrote the constitution of the United States. Two years later, Washington became the first President of the newly formed nation. Realizing that the way he handled the responsibility of the job would impact upon how

(357) George Washington   *(358) Thomas Paine*

future presidents approached the position; he handed down a legacy of strength, integrity, dignity, and national purpose. Less than three years after leaving office, he died with his loyal wife by his side, at his beloved Virginia plantation, Mount Vernon, at the age of 67. One of the things I most admire about George Washington is that he voluntarily relinquished power at a stage in his political career where he was so popular and powerful he could have become a virtual monarch of the United States if he had chosen to. But this didn't happen because he was strongly focused on establishing a strong foundation for the fledgling democracy of the United States. So after serving two terms as President, and with a clear opportunity to continue to win more elections; he instead decided that he had done enough for his country, and retired gracefully and modestly to be with his wife in Virginia. There are few instances in history where men in great positions of power have voluntarily decided to relinquish it, when they are at the peak of their power. I admire Washington for having the grace, dignity, modesty, supreme self-belief, and a sense of "the democracy being greater than any one man" to achieve this. We can contrast this with the narcissistic, ugly, embarrassing, and damaging manner in which the 45th president of the United States, Donald Trump did not accept his electoral defeat in 2020, and made outrageous accusations the election was rigged, with not a shred of evidence to support his claim; resulting in the Trump-inspired attack on Congress. George Washington would have literally spun in his grave. The political philosophy of George Washington can be clearly seen in this quotation: "In politics as in philosophy, my tenets are few and simple. The leading one of which, and indeed that which embraces most others, is to be honest and just ourselves and to exact it from others, meddling as little as possible in their affairs where our own are not involved. If this maxim was generally adopted, wars would cease and our swords would soon be converted into reap hooks and our harvests be more peaceful, abundant, and happy." Washington had this to say about his voluntary retirement from politics: "Having now finished the work assigned me, I retire from the great theatre of Action; and bidding an Affectionate farewell to this August body under whose orders I have so long acted, I here offer my commission, and take my leave of all the employments of public life. The foundation of a great Empire is laid, and I please myself with a persuasion, that Providence will not leave its work imperfect."

Thomas Paine was born in Norfolk England in 1737, and died in 1809. He was a political philosopher and writer who supported revolutionary causes in America and Europe. Published in 1776 to international acclaim, "Common Sense" was the first pamphlet to advocate American Independence. After writing "The American Crisis" papers during the Revolutionary War, Paine returned to Europe and offered a stirring defence of the French Revolution with his pamphlet "Rights of Man." His then radical political views led to his imprisonment. This did not deter him, and after his release he produced his last great essay, "The Age of Reason," a controversial critique of institutionalized religion and Christian theology. Paine's most famous pamphlet, "Common Sense," was first published on 10th January 1776, and sold-out its 1000 initially printed copies immediately. By the end of that year, 150,000 copies had been printed and sold, an enormous amount for that time. The pamphlet remains in print to this day. "Common Sense" is credited with playing a crucial role in convincing the American colonists to take up arms against

England and seek their independence as a nation. In the pamphlet, Paine argues that representational democratically-elected government is far superior to a monarchy, or other forms of government based on aristocracy and hereditary. The pamphlet proved to be so influential that future US president, John Adams reportedly declared, "Without the pen of the author of 'Common Sense,' the sword of Washington would have been raised in vain." Paine also claimed that the American colonies needed to break with England in order to survive, and that there would never be a better moment in history for that to happen. He argued that America was related to Europe as a whole, not just England, and that it needed to freely trade with nations like France and Spain. As the American Revolutionary War began, Paine enlisted and met General George Washington, whom he served under. The terrible condition of Washington's troops during the winter of 1776 prompted Paine to publish a series of inspirational pamphlets known as "The American Crisis," which open with the famous line: "These are the times that try men's souls." Suffice to say, Washington and his troops prevailed over the British, paving the way for the creation of an independent United States of America.

Here are some quotations from Thomas Paine regarding the American Revolution:

- If there must be trouble, let it be in my day, that my child may have peace.
- The harder the conflict, the more glorious the triumph.
- We have it in our power to begin the world over again.
- These are the times that try men's souls. The summer soldier and the sunshine patriot will, in this crisis, shrink from the service of his country; but he that stands it NOW, deserves the love and thanks of man and woman.
- I consider the war of America against Britain as the country's war, the public's war, or the war of the people in their own behalf, for the security of their natural rights, and the protection of their own property.
- I call not upon a few, but upon all; not on this state or that state, but on every state; up and help us; lay your shoulders to

the wheel; better have too much force than too little, when so
great an object is at stake.

- The times that tried men's souls are over – and the greatest
  and completest revolution the world ever knew, gloriously and
  happily accomplished.

- The Sun never shined on a cause of greater worth.

- Now is the seedtime of continental union, faith and honour.
  The least fracture now, will be like a name engraved with the
  point of a pin on the tender rind of a young oak; the wound
  would enlarge with the tree, and posterity read it in full grown
  characters.

(359) George Washington and the American Revolution – 1775 to 1783

The American Revolution acted as a powerful inspiration to the
critics of traditional forms of hereditary monarchical governments
back in Europe; and combined with internal problems in France, the
French Revolution broke out in 1789. This revolution was significant
in that it challenged the very foundation of the hereditary monarchical governments which had originated in the Middle Ages, and
soon the entire continent of Europe was engulfed in war. The rise
of Napoleon Bonaparte, one of the most brilliant strategic generals
in human history, ensured that Revolutionary France dominated
much of Europe for several years; and spreading a more efficient and
equal form of government throughout the continent in the process.

Eventually Napoleon was defeated at the Battle of Waterloo in 1815; but Europeans had a taste of a new kind of progressive government, and there could be no return to the more traditional ways of ruling, for much longer. A major part of Napoleon's ultimate defeat was played by the British navy. This factor, along with a diplomacy based to a large extent upon paying subsidies to allies in the fight against Napoleon, did not come cheap; it would have been completely beyond the economic strength of any single European power before this period. However, by this time Britain's economy was being transformed by another kind of revolution – the Industrial Revolution.

Before delving into the Industrial Revolution let us briefly examine the life and achievements of Napoleon Bonaparte, and the crucial role he played in transforming Europe in so many ways. Napoleon Bonaparte was born in 1769 and died in 1821. He was a

(360) Napoleon Bonaparte was Emperor of France from 1804 to 1815

French military commander and emperor who conquered much of Europe in the early 19th century. Born on the Mediterranean island of Corsica, Napoleon rapidly rose through the ranks of the military during the French Revolution of 1789 to 1799. After seizing political power in France in a 1799 coup, he crowned himself emperor in 1804. Shrewd, ambitious, and a highly skilled military strategist, Napoleon successfully waged war against various coalitions of European nations and expanded his empire within Europe. However, after a disastrous French invasion of Russia in 1812, Napoleon abdicated the throne two years later and was exiled to the island of Elba. In 1815, he briefly returned to power in his Hundred Days campaign. After a crushing defeat at the Battle of Waterloo, he abdicated once again and was exiled to the remote Atlantic island of Saint Helena, where he died at the age of 51.

When Napoleon came to power in France, his control and influence extended across Europe. This power put Napoleon in a position of great authority and enabled him to bring about wide-reaching social, economic, and political changes. Before the arrival of Napoleon, the social hierarchy in France had undergone a massive and bloody revolution. It was finally time for the common people to be given the rights that they deserved. It was time for the aristocratic nobility to be removed from power. Napoleon granted common people rights that had previously been denied to them, such as the right to own property. The French government under Napoleon may not have been democratic, but the fact that Napoleon was not from the nobility symbolized great changes for the social structure of France. As Marshal Michel Ney said in his 'Two Views of Napoleon' article: "whether the Bourbon nobility choose to return to exile or consent to live among us, what does it matter to us? The times are gone when the people were governed by suppressing their rights." As well as the social structure, Napoleon also improved the economic systems of Europe by building canals, controlling prices, and encouraging investment in new industries. He was the progressive stimulus for change that the struggling economy of France needed. Within France he balanced the budget and undertook massive public works projects. Napoleon's main effect on the political systems of Europe was his role in spreading revolution. Under his control, the ideas of the enlightenment inspired progressive thinking, freedom, and liberty. The freedoms that Napoleon granted to his people grew to become expected from the political systems that came after his demise. Amongst other achievements of Napoleon were; the introduction of comprehensive public education in France and other parts of Europe; and an enlightened legal code based on common sense and equality, which was adopted throughout Europe and beyond. He was also instrumental in dismantling many of the archaic aspects of the medieval feudal system in Europe and replacing it with a legally binding concept of free property acquisition and ownership, particularly among the working classes.

(361) Factories in the Industrial Revolution began to emit greenhouse gases

This technological and economic revolution had been gathering pace in Britain since the mid-18th century, and had been greatly boosted by the efficient application of coal-derived steam power to mechanical devices. By the end of the century, large industrial towns were growing up in the Midlands, the north of England, and Scotland, in which hundreds of factories were churning out mass quantities of a variety of manufactured goods. The early 19th century saw this technological and economic expansion continue in Britain, and then starting to spread into continental Europe and North America. The application of coal-fired steam power to transportation further stimulated this trend, with railways spreading throughout Britain, Europe, and North America. The railways also acted as a powerful magnet for the expansion of the United States and Canada across the North American continent, so that by the mid-19th century, both countries had extensively settled their Pacific west coast frontiers. By this time, steamships were beginning to take over from sailing vessels on the major sea routes of the world. With the introduction of refrigeration, meat and other perishable foods could be transported between continents; linking the world more extensively and sowing the seeds of the globalized economy we have today.

The Industrial Revolution is a significant event in word history because it transformed European, North American, and other western societies from being primarily based on agricultural production, to industrial production. This caused a mass migration of people from the rural countryside to towns and cities, as people moved in search of work in the newly developed factories of the time. The factory system is a term that historians use to refer to the development of centralized factories or mills that produced goods on a mass scale. Throughout the mid to late 1700s, inventors such as Richard Arkwright, Eli Whitney, James Hargreaves, and Edmund Cartwright, developed machines and techniques that helped improve production, especially within the textile industry. For example, James Hargreaves created the spinning jenny in 1764, which enabled a machine to spin many spindles of thread at the same time. Richard Arkwright added to this by developing the water frame in 1769; which allowed over one hundred spindles of thread to be spun at the same time, but was so large and needed so much energy, that he built it next to rivers and creeks to use the force of the water to spin the machines. In 1785, Edmund Cartwright developed the power loom, a machine which facilitated the rapid production of cloth. Finally, the American inventor, Eli Whitney developed the cotton gin in 1793; this machine allowed for the rapid production of cotton. Previously cotton had to be hand cleaned in order to remove all the fibres and seeds.

The invention of the steam engine during the Industrial Revolution is perhaps one of the most significant events of this time period. The first steam engine was invented by Thomas Newcomen

(362) Steam trains
(363) Machines

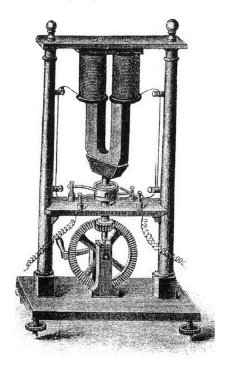

in 1712. He worked as an ironmonger in Devon, England and produced mining-related products for Cornish tin and coal mine owners who often complained they were struggling to deal with the flooding of their mines. Traditional manual ways of removing water from the mines were slow, and very hard work. Newcomen realised that he could help these mine owners by developing a steam powered pump engine that used a piston in a cylinder; the first of its kind. The steam engine was later improved by the British inventor James Watt. In 1764, Watt was given a Newcomen steam engine to repair. He started working on it and soon realized it was inefficient; as a result, he improved upon the design by preventing steam from escaping the engine by adding a separate condensing chamber. With Watt's improvements steam engines were soon used in many different industries for a wide variety of purposes; including in mines, manufacturing, and waterworks. Steam engines helped improve productivity and efficiency. Later innovations led to the steam train and steamships, which totally revolutionized transportation; allowing for goods and people to be transported quicker, more efficiently, and across much longer distances.

None of these technological and societal changes would have been possible without the emergence and development of capitalism. The European empires of the 16th through to the 18th centuries existed primarily to significantly increase their trade and wealth; and developed intense rivalries as they expanded their colonial empires around the world. At the foundation of these rivalries existed the economic system of mercantilism, which was the idea that colonies and trade should be strictly controlled in order to benefit the home economies of the respective European colonial powers. Mercantilism was heavily regulated and mainly government controlled, giving very little opportunity for individual entrepreneurs and traders. This all changed with the introduction of capitalist beliefs and values in the late 1700s. More and more people began to reject the principles of mercantilism, so when the Scottish economist Adam Smith published his famous book, "An Inquiry into the Nature and Causes of the Wealth of Nations;" it gained widespread popularity. The book challenged the idea that the government should control the economy, and instead proposed

(364) Adam Smith   *(365) Josiah Wedgewood*

the idea of free trade and competition, with a much lesser role of the government. The ideas of this book eventually laid the foundation for the principles of capitalism; an economic system supporting free trade, markets, and individual choice, as a way of achieving prosperity.

One of the greatest capitalist entrepreneurs of the Industrial Revolution in Britain, was Josiah Wedgewood, born in 1730 and died in 1795. He was born into a family which had been engaged in the manufacturing of pottery since the 17th century. His father owned a factory called the Churchyard Pottery, and Josiah began working there as an apprentice in 1744. He left the factory in the early 1750s and until 1759 was engaged with various partners in the manufacture of standard types of earthenware, including salt-glaze and stoneware products, and objects in the popular agate and tortoiseshell glazes. During these years he experimented with improving glazes in colour, achieving a particularly refined and elegant green glaze. One of the wealthiest entrepreneurs of the 18th century, Wedgewood created elegant pottery goods to meet the demands of the consumer revolution and growth in wealth of the middle classes that helped drive the Industrial Revolution in Britain. He is credited as the inventor of modern marketing, specifically direct mail, money back guarantees, travelling salesmen, carrying pattern boxes for display, self-service, free delivery, buy one get one free, and illustrated catalogues. Wedgewood was also prominent in the abolition of slavery. From 1787 until his death in 1795, he actively participated in this great cause; including working closely with his good friend Thomas Clarkson, a prominent abolitionist campaigner at that time. Wedgewood mass-produced cameos depicting the seal for the Society for Effecting the Abolition of the Slave Trade, and had them widely distributed. The Wedgewood cameo medallion he produced was one of the most famous images of a black person in 18th century art. Thomas Clarkson wrote: "ladies wore them in bracelets, and others had them fitted up in an ornamental manner as pins for their hair. At length the taste for wearing them became general, and thus fashion, which usually confines itself to worthless things, was seen for once in the honourable office of promoting the cause of justice, humanity and freedom."

Back in Europe, the legacy of the French Revolution and the Napoleonic and other wars which followed it stimulated an ever-growing demand for greater democracy and national self-determination. Much of central Europe and the Balkans were under the control of large multinational states such as the Austrian, Russian, and Ottoman empires; and the many different nationalities within these empires began to agitate for independence and self-rule. Meanwhile, in Germany and Italy, at that time consisting of numerous city-states, small principalities and kingdoms, there were significant movements underway to create nation states. As a result, in the late 19th century the independent nations of Germany and Italy first appeared. Also known as the Risorgimento, the Italian unification was a political and social movement that consolidated different states of the Italian peninsula into a single nation of the Kingdom of Italy in 1870. Giuseppe Garibaldi, Giuseppe Mazzini, Count Cavour, and King Victor Emmanuel II, are considered to be the fathers of the new Italian nation state. Meanwhile, in the 1860s, Otto von Bismarck, then Minister President of Prussia, provoked three short, decisive wars against Denmark, Austria, and France; aligning the smaller German states behind Prussia in its defeat of France. In 1871 he unified Germany into a single nation state.

(366) Otto von Bismarck   *(367) Abraham Lincoln*

During this same period, a number of European countries made great progress towards becoming full parliamentary democracies; led by Britain, with its long standing experience of parliamentary rule. The late 19th century saw mass-party politics taking over from the much more limited aristocratic forms of parliament.

Across the Atlantic the expansion of the United States had led to an increasing divergence between the slave-owning plantation society in the south, and a more industrialized and egalitarian society in the north. The stark differences between these regions led to the bloody American Civil War fought from 1861 to 1865. With the north triumphing, slavery was abolished in the USA under President Abraham Lincoln. The Civil War was followed by a period of massive industrial expansion in the USA as the nation reconstructed its war-damaged south and fully embraced all of the technology that had emerged from the Industrial Revolution and subsequent Modern World inventions. The achievements of Abraham Lincoln laid the necessary foundation for the United States to become the 20th century economic powerhouse that it did. Abraham Lincoln, a self-taught lawyer, legislator and vocal opponent of slavery, was elected the 16th President of the United States in November 1860, shortly before the outbreak of the American Civil War. Lincoln proved to be a shrewd military strategist and an exceptionally gifted leader and public speaker. His Emancipation Proclamation paved the way for the abolition of slavery, while his Gettysburg Address stands-out as one of the most famous pieces of oratory in American history. In 1865, with the Union on the verge of victory over the southern Confederacy, Abraham Lincoln was assassinated by Confederate sympathizer John Wilkes Booth. Lincoln's assassination made him a martyr to the cause of liberty; and he is widely regarded as one of the greatest presidents in United States history.

On the night of 14th April 1865, the actor John Wilkes Booth slipped into the president's box at the Ford Theatre in Washington D.C., during a performance of the comedy "Our American Cousin" and well into the last act, shot him point-blank in the back of the head. The assassin then jumped from the upper level balcony onto

(368) The assassination of President Abraham Lincoln on 14th April 1865

the floor of the theatre, and fled the scene amidst shocked outcries from the audience. Lincoln was carried to a boarding house across the street from the theatre, but he never regained consciousness and died in the early hours of the morning of 15th April. On 21st April, a train carrying his coffin left Washington on its way to Springfield Illinois, where he would be buried on 4th May. Abraham Lincoln's funeral train travelled through 180 cities and seven states so mourners could pay homage to the fallen president. Today, Lincoln's birthday, alongside with the birthday of George Washington, is honoured on President's Day, which falls on the third Monday of February. On the national stain of slavery and bigotry, Lincoln had this to say: "As a nation, we began by declaring that 'all men are created equal.' We now practically read it 'all men are created equal, except negroes.' When the Know-Nothings get control, it will read 'all men are created equal, except negroes, and foreigners, and Catholics.' When it comes to this I should prefer emigrating to some country where they make no pretence of loving liberty – to Russia, for instance, and without the base alloy of hypocrisy."

Here is an excerpt from his great Gettysburg Address on 19th November 1863, where he encouraged the people to take action in improving and healing the nation, honour those who died in the Battle of Gettysburg, and commence the process of reunifying the north and south: "Our fathers brought forth on this continent a new nation, conceived in liberty and dedicated to the proposition that all men are created equal. Now we are engaged in a great civil war, testing whether that nation or any other nation so conceived and so dedicated can long endure. We are met on a great battlefield of that war. We have come to dedicate a portion of that field as a final resting-place for those who here gave their lives that that nation might live. It is altogether fitting and proper that we should do this. But in a larger sense, we cannot dedicate, we cannot consecrate, we cannot hallow this ground. The brave men, living and dead who struggled here have consecrated it far above our poor power to add or detract. The world will little note nor long remember what we say here, but it can never forget what they did here. It is for us the living rather to be dedicated here to the unfinished work which they who fought here have thus far so nobly advanced. It is rather for us to be here dedicated to the great task remaining before us – that from these honoured dead we take increased devotion to that cause for which they gave the last full measure of devotion – that we here highly resolve that these dead shall not have died in vain, that this nation under God shall have a new birth of freedom, and that government of the people, by the people, for the people shall not perish from the earth."

The late 19th century saw the rise of huge corporations in the USA, led by vastly wealthy business tycoons such as Andrew Carnegie (steel), John D Rockefeller (oil), Cornelius Vanderbilt (railroads and shipping), John P Morgan (banking and finance), William Randolph Hearst (newspapers) and John J Astor (property). Andrew Carnegie (1835-1919) although born in Scotland, was an American industrialist who amassed a fortune in the steel industry, and then became a major philanthropist. Carnegie worked in a Pittsburgh cotton factory as a boy before rising to the position of division superintendent of the Pennsylvania Railroad in 1859. While working for the railroad, he invested in various business ventures, including iron and oil companies; and made his first fortune by the time he was in his early 30s. In the early 1870s, he entered the steel business, and over the next two decades became a dominant force in the industry. In 1901, he sold the Carnegie Steel Company to banker John Pierpont Morgan for $480 million. Carnegie then devoted himself to philanthropy, eventually giving away more than $350 million.

(369) Andrew Carnegie   *(370) John D. Rockefeller*

John Davison Rockefeller (1839-1937) was the founder of the Standard Oil Company, becoming one of the world's wealthiest men, and a major philanthropist. Born in modest circumstances in upstate New York, he entered the then-fledgling oil business in 1863 by investing in a Cleveland Ohio refinery. In 1870, he established Standard Oil, which by the early 1880s controlled around 90 percent of all US refineries and pipelines. Critics accused Rockefeller of engaging in unethical practices, such as predatory pricing and colluding with railroads to eliminate his competitors in order to gain a monopoly in the oil industry. In 1911, the US Supreme Court found Standard Oil to be in violation of anti-trust laws, and ordered that the company be dissolved. During his lifetime, Rockefeller donated over $500 million to various philanthropic causes.

Shipping and railroad tycoon Cornelius Vanderbilt (1794-1877) was a self-made multi-millionaire who became one of the wealthiest

(371) Cornelius Vanderbilt   *(372) John P. Morgan*

Americans of the 19th century. As a boy, he worked with his father, who operated a boat that ferried cargo between Staten Island, where they lived, and Manhattan in New York City. After working for a time as a steamship captain, Vanderbilt went into business for himself in the late 1820s, and eventually became one of the country's largest steamship operators. In the process, the Commodore, as he was publicly nicknamed, gained a reputation for being fiercely competitive, and ruthlessly aggressive. In the 1860s, he shifted his focus to the railroad industry, where he built another empire, and helped make railroad transportation more efficient. When Vanderbilt died, he was worth over $100 million.

One of the most powerful bankers of the era, John Pierpont Morgan (1837-1913), financed railroads and helped organize-re-structure US Steel, General Electric, and other major corporations. He followed his wealthy father into the banking business in the late 1850s, and in 1871 formed a partnership with banker Anthony Drexel. In 1895, their firm was reorganized as J.P. Morgan & Company, a prede-cessor of the modern-day financial giant JP Morgan Chase. Morgan used his influence to help stabilize American financial markets during several economic crises, including the panic of 1907. However, he faced criticism for having too much power, and was accused of manipulating the nation's financial system for his own personal gain. The Gilded Age titan spent a significant portion of his wealth amassing a vast art collection.

Publishing magnate William Randolph Hearst (1863-1951) built his media empire after inheriting the San Francisco Examiner newspaper from his father. He challenged New York World publisher Joseph Pulitzer by buying the rival New York Journal. Hearst entered politics at the turn of the century, winning two terms to the US House of Representatives, but failed in his bids to become US president and mayor of New York City. He lost much of his wealth during the Great Depression and fell out of touch with his mainly blue-collar working class audience, but was still the head of the largest news conglomerate in America at the time of his death.

(373) William Randolph Hearst   (374) John Jacob Astor IV

John Jacob Astor IV was born in 1864 in New York City, into one of the most affluent families in the world. The Astor family dates back to the early 1700s when the original John Jacob Astor came to America from a small village in Germany, to make a name for himself. He started making money in the fur-trading business, but his real fortune began when he entered the world of real estate. One of his first big purchases was a plot of land in the middle of Manhattan, which is now known as Times Square. Soon after, Astor bought land all around Manhattan, creating a dynasty and becoming one of the richest men in the world. When John Jacob Astor IV was born, the Astor name was already well-respected and influential in high society, and the family's fortune was one of the world's largest. In 1897, Astor used his fortune to build the Astoria Hotel in New York. For the next few decades, the hotel became a symbol of luxury, wealth, and class in New York City. It was considered one of the best hotels in the world at the time. In 1904, Astor built another New York landmark hotel, The St. Regis. Becoming one of the richest men in the world, John Jacob Astor IV died on the Titanic when it struck an iceberg on its maiden voyage in 1912; he was returning to New York from Europe with his pregnant wife Madeleine, who survived that fateful voyage.

(375) The Titanic sank on its maiden voyage on the 14-15th April 1912

The later 19th century also saw Europe industrializing its economies on an unprecedented scale. Towns and cities significantly grew in size, the middle classes became a larger and more influential part of the social-political mix, and a vast urban working class emerged; and they all became very important consumers of the ever-increasing range of consumer goods and services that were emerging from these capitalist consumer-driven economies. Revolutionary social and political change was accompanied by a revolution in science and ideas. Charles Darwin's

theory of evolution became widely accepted, and encouraged the rise
of a more secular, non-religious outlook on life. For many people, this
theory led to the realization that the concept of a God was obsolete,
while of course, many others happily incorporated Darwinism into
their world view, whilst also retaining their religious beliefs. But for
both groups, Darwin's theory of evolution meant that the development
of all life on earth, including humans, could have a rational scientific and
empirically based explanation. Sigmund Freud and others pioneered a
scientific understanding of the mind and emotions, which developed
into the science of psychology and psychiatry; again overshadowing
what previously were in the domain of the spiritual religions. Albert
Einstein's theory of relativity completely changed people's views of
the Universe; with his explanations of the relationship between time,
space, matter, and energy opening up vast new areas of discovery and
innovation. Karl Marx and his theory of Communism, amongst others,
analysed society and offered an alternative to the accepted Western
economic model based on Capitalism.

As a university student, Karl Marx (1818-1883) joined a movement
known as the Young Hegelians, who strongly criticized the political
and cultural establishments of the day. He became a journalist, and

(376) Karl Marx   (377) A Communist poster

the radical nature of his writings would eventually get him expelled by the governments of Germany, France, and Belgium. In 1848, Marx and fellow German philosopher Friedrich Engels, published "The Communist Manifesto," which introduced their concept of socialism as a natural result of the conflicts inherent in the capitalist system. Marx later moved to London, where he would live for the rest of his life. In 1867, he published the first volume of "Capital" (Das Kapital), in which he laid out his vision of capitalism and its inevitable tendencies toward self-destruction; and took part in a growing international workers' movement based on his revolutionary theories. We can gain a deeper insight into the thinking of Marx, with these quotations:

- The philosophers have only interpreted the world, in various ways. The point, however, is to change it.
- The oppressed are allowed once every few years to decide which particular representatives of the oppressing class are to represent and repress them.
- The last capitalist we hang shall be the one who sold us his rope.
- The history of all hitherto existing society is the history of class struggle. Freeman and slave, patrician and plebeian, lord and serf, guild master and journeyman, in a word, oppressor and oppressed, stood in constant opposition to one another, carried on an uninterrupted, now hidden, now open fight, that each time ended, either in the revolutionary reconstitution of society at large, or in the common ruin of contending classes.
- Hegel remarks somewhere that all great, world-historical facts and personages occur, as it were, twice. He has forgotten to add: the first time as tragedy, the second as farce.
- The less you eat, drink and read books; the less you go to the theatre, the dance hall, the public house; the less you think, love, theorize, sing, paint, fence, etc., the more you save – the greater becomes your treasure which neither moths nor dust will devour – your capital. The less you are, the more you have; the less you express your own life, the greater is your alienated life – the greater is the store of your estranged being.

- Let the ruling classes tremble at a Communist revolution. The proletarians have nothing to lose but their chains. They have a world to win. Workingmen of all countries unite!
- Religion is the sigh of the oppressed creature, the heart of a heartless world and the soul of soulless conditions. It is the opium of the people.
- Religion is the impotence of the human mind to deal with occurrences it cannot understand.
- If money is the bond binding me to human life, binding society to me, connecting me with nature and man, is not money the bond of all bonds? Can it not dissolve and bind all ties? Is it not, therefore, also the universal agent of separation?
- Modern bourgeois society with its relations of production, of exchange, and of property, a society that has conjured up such gigantic means of production and of exchange, is like the sorcerer, who is no longer able to control the powers of the nether world whom he has called up by his spells.
- In proportion therefore, as the repulsiveness of the work increases, the wage decreases.
- You are horrified at our intending to do away with private property. But in your existing society private property is already done away with for nine-tenths of the population; its existence for the few is solely due to its non-existence in the hands of those nine-tenths. You reproach us, therefore, with intending to do away with a form of property, the necessary condition for whose existence is the non-existence of any property for the immense majority of society. In one word, you reproach us with intending to do away with your property. Precisely so: that is just what we intend.
- Capitalism: Teach a man to fish, but the fish he catches aren't his. They belong to the person paying him to fish, and if he's lucky, he might get paid enough to buy a few fish for himself.
- The increase in value of the world of things is directly proportional to the decrease in value of the human world.

- Labour in the white skin can never free itself as long as labour in the black skin is branded.
- The education of all children, from the moment that they can get along without a mother's care, shall be in state institutions.
- Education is free. Freedom of education shall be enjoyed under the condition fixed by law and under the supreme control of the state.
- The ruling ideas of each age have ever been the ideas of its ruling class.
- The production of too many useful things results in too many useless people.
- The theory of Communists may be summed up in the single sentence: Abolition of private property.
- Capital is dead labour, which, vampire-like, lives only by sucking living labour, and lives the more, the more labour it sucks.
- Money is the alienated essence of man's labour and life; and this alien essence dominates him as he worships it.
- Political power, properly so called, is merely the organised power of one class for oppressing another.
- Moments are the elements of profit.
- Social progress can be measured by the social position of the female sex.

In the military sphere, European and American armies and navies were directly affected by industrialization, with machine guns, barbed wire, battleships, torpedoes, grenades, mines, and submarines, amongst other inventions, making their appearance on the battlefields. These innovations gave Western military forces a massive advantage over those less-developed societies; and as a result, the late 19th century saw the European empires expand to such a level that they covered most of the surface of the world. Western trade networks, with their reach extended around the globe by the spread of steam powered railways and steamships, disrupted local colonial economies; Christian missionary activities challenged many indigenous local beliefs and traditions; local colonial elites adopted Western-style education, culture, clothing, and architecture. Even lands which were not directly ruled by the European powers, such as China, Thailand, and Iran, were absorbed into the western-dominated global economy, in such a way that deprived them of much of their political and economic autonomy as well. The only major country to successfully engage economically with the Western world on its own terms, and yet preserve its own unique culture; was Japan. By the late 19th century, Japan was creating an empire of its own. Britain ended up with the largest of these Western empires, and by the end of the 19th century, London was effectively the financial capital of the world. This laid the foundations for English to become the dominant language of the world.

24

THE TWENTIETH
CENTURY

THE TWENTIETH CENTURY TRIGGERED A GREATER ACCELERATION in collective learning than the entire preceding 300,000 years that humans have been on the Earth. To me the two defining features of this century were the dramatic acceleration in technology, global interconnection, and exchange of ideas, driven by significant advances in science, and all of the other disciplines of human knowledge; and the emergence of the United States of America as the major world superpower. The United States started the century as an industrialised nation, at an equivalent level to modern European nations; was then instrumental in defeating totalitarian Germany and its allies in two world wars; aggressively competed around the globe with the Soviet Union in a monumental struggle of two distinct ideologies, capitalism and communism, known as The Cold War; then towards the end of the century prevailed as the sole superpower, after the collapse of the Soviet Union in 1990; but by 2001 commenced a decline in global power and influence after the great nation was attacked by extremist Islamic terrorists on its homeland, just as China was rising as a significant world power on the international stage. Let us examine the major events of the twentieth century.

There was an explosive increase in the world's population during the twentieth century, from around 1.6 billion people in 1900, to over 6.1 billion by the year 2000. This rapid growth was mainly caused by a significant increase in the lifespan of humans as a direct result of advances in medical science; and an overall increase in living standards brought about by economic and technological advancements, such as vaccinations for diseases, clean drinking water, and greater food availability. This exponential population growth continues to this day with the world's population estimated to be 7.8 billion in early 2021, at the time this book was written.

The first decade of the twentieth century saw the widespread application of the internal combustion engine including the mass production of the automobile by Henry Ford; an American industrialist and business magnate, founder of the Ford Motor Company, and chief developer of the assembly-line technique of mass production, which was applied to many other industries in the United States and the rest of the world. Thomas Edison's electric light bulb invention started to be used on a wide basis. The Wright brothers in their "Wright Flyer" performed the first recorded, controlled, and powered air flight on 17th December 1903. The Canadian, Reginald Fessenden made the first audio radio broadcasts of entertainment and music to a public audience. The first transatlantic radio signal had been broadcast by the Italian inventor and engineer, Guglielmo Marconi in 1901. The first wave of feminism saw universities being open for women for the first time in Japan, Bulgaria, Cuba, Russia, and Peru; and in 1906, Finland became the first country to grant women the right to vote in democratic elections. In China, years of oppression by the Western

(378) Henry Ford's Model T factory production line commenced
on 1st October 1908

nations led to The Boxer Rebellion, an anti-imperialist, anti-foreign, and anti-Christian uprising which occurred between 1899 and 1901, that led to the death of 32,000 Chinese Christians and 200 Western missionaries, who were killed by Chinese militia forces in northern China.

The 1910s represented the culmination of European militarism which had its beginnings during the second half of the 19th century. The status quo in Europe during the first half of the decade, as well as the legacy of military alliances was changed forever by the assassi-nation on 28th June 1914 of Archduke Franz Ferdinand, the heir to the Austro-Hungarian Empire. The murder, carried out by 19-year old Gavrilo Princip, a member of the Black Hand Serbian terrorist group, triggered a chain of events in which within 33 days, the First World War broke out in Europe on 1st August 1914. This brutal and bloody conflict between; Germany, Austria-Hungary, Ottoman Turkey, and allies; and Britain, France, Russia, and allies; dragged on mainly in the form of trench warfare, where neither side gained much territory until a truce was declared on 11th November 1918, leading to the contro-

(379) The assassination of Austrian Archduke Franz Ferdinand
on 28th June 1914

versial and one-sided Treaty of Versailles which severely punished
Germany and was signed on 28th June 1919. The war's end triggered
the abdication of various monarchies and the collapse of five of the
last modern empires consisting of Russia, Germany, China, Ottoman
Turkey, and Austria-Hungary.

The decade of the 1910s was also a period of revolution in a
number of countries. The October 1910 Portuguese Revolution which
ended the eight-century long monarchy, spearheaded the trend; this
was followed by the Mexican Revolution in November 1910, leading
to the removal of dictator Porfirio Diaz, and a violent civil war that
lasted until mid-1920 when not long after, a new democratic Mexican
Constitution was signed and ratified. The Russian Empire also had
a similar fate; its participation in World War One triggered a social,
political, and economic collapse which made the Tsarist autocracy
led by Nicholas II unsustainable and following the events of 1905,
culminated in the Russian Revolution and the establishment of the
Russian Soviet Federative Socialist Republic, under the direction of
the Bolshevik Party, later renamed the Communist Party of the Soviet
Union. The Russian Revolution of 1917, known as the October
Revolution, was followed by the Russian Civil War, which lasted until
late 1922, and in which the Bolshevik communists prevailed.

Socially and culturally, much of the music in the 1910s was ballroom-
themed; and many of the fashionable restaurants were equipped with
dance floors. In the United States, Prohibition began on 16th January
1919, banning the production, distribution, selling, and consumption

of alcohol, as ratified
in the Eighteenth
Amendment of the US
Constitution. It was also
during this time that
Hollywood California
replaced the east coast
as the centre of the
movie industry; and

(380) Prohibition USA: 1920 to 1933   Charlie Chaplin made

his debut with his trademark moustached, baggy-pants "Little Tramp" character in the silent movie, "Kid Auto Races" screened at Venice in 1914. One of first recordings of jazz music was made by the Original Dixieland Jazz Band in February 1917. During this decade the first modern zipper was patented; stainless steel and the pop-up toaster were invented; the army tank was developed; and the Model T Ford dominated the automobile market, selling more cars than all of the other automakers combined, by the end of the decade. In 1912, as we have seen earlier in the book, Alfred Wegener developed his Theory of Continental Drift; and in 1916, Albert Einstein came up with the Theory of General Relativity.

In the United States, the 1920s are frequently referred to as the "Roaring Twenties" or the "Jazz Age", while in Europe there was an economic boom following the end of World War One. The economic prosperity experienced by many countries during the 1920s, led by the United States, was similar in nature to that experienced in the 1950s and 1990s. Each period of prosperity was the result of a paradigm shift in global affairs. These shifts in the 1920s, 1950s, and 1990s, occurred largely as a result of the conclusion of World War One

(381) Vladimir Lenin and the Russian Revolution of October 1917.

and the Spanish Flu (1920s); World War Two (1950s); and the Cold
War (1990s); respectively. The Spanish Flu, also known as the 1918 Flu
Pandemic, was an unusually deadly influenza pandemic, caused by the
H1N1 influenza A virus. Lasting from February 1918 to April 1920,
it infected over 500 million people, around one-third of the world's
population at that time, and led to the death of over 50 million people,
in four successive waves. In some countries, the 1920s saw the rise of
radical political movements, especially in regions that were once part
of empires. Communism spread worldwide as a consequence of the
October 1917 Revolution and the Bolshevik's victory in the Russian
Civil War. Fear of the spread of Communism led to the emergence
of far-right political movements and fascism, particularly in Europe;
including Benito Mussolini in Italy, and Adolf Hitler in Germany. The
devastating Wall Street Crash in October 1929 dramatically brought
an end to the economic prosperity of the United States, Europe, and
other nations around the world.

The Roaring Twenties acted as a catalyst for several major social
and cultural trends. These trends were made possible by the economic
prosperity of the time, and were most visible in major cities such as
New York, Chicago, Paris, Berlin, and London. Some of these trends

(382) The Jazz Age and the Roaring Twenties in the United States

*(383) and (384) the Great Depression of
1929 devastated the world economy*

included the widespread popularity
of jazz music; Art Deco architecture;
and for women, knee-length skirts and
dresses became socially acceptable, as
did bobbed hair. The mainly city-based women who pioneered these
trends were frequently referred to as flappers. The era also saw the
large-scale adoption of automobiles, telephones, motion pictures,
radio, and household electrical appliances. It wasn't until 1925 that
more than 50% of all households in the United States had electricity
connected to their homes. There was also unprecedented industrial
growth, accelerated consumer demand and aspirations, and significant
changes in lifestyles. The media began to focus on celebrities, especially
sports heroes and movie stars. In the United States for example, large
baseball stadiums and palatial cinemas were built. Most independent
nations gave women the right to vote after 1918, especially as a reward
for women's support of the war effort and their endurance of the
death and hardship caused by World War One.

A number of technological developments occurred in the 1920s, including; John Logie Baird's invention of the television in 1925; Warner Brothers producing the first movie with a soundtrack, "Don Juan" in 1926; phonograph records in 1925; the first electric razor developed by Jacob Schick in 1928; the first jukeboxes in 1927; Clarence Birdseye's invention of the process for freezing food in 1925; and Robert Goddard launching the first liquid-fuelled rocket in 1926. Also during this time, Charles Lindbergh became the first person to fly solo across the Atlantic Ocean on 20-21 May 1927; Howard Carter discovered the tomb of Tutankhamen near Luxor Egypt, in 1922; and Alexander Fleming discovered penicillin in 1928.

The decade of the 1930s was defined by a global economic and political crisis which culminated in the Second World War. It saw the international financial system collapse from the aftershock of the Wall Street Crash of October 1929. This was the largest stock market crash in American history, triggering a severe economic downfall known as the Great Depression, which had a traumatic effect worldwide. The Great Depression led to widespread unemployment and poverty, especially in the United States. It severely impacted Germany as well, which had to deal with the massive reparations (financial penalties) from

(385) Adolf Hitler and the rise of Nazi Germany in the 1930s

the aftermath of World War One, imposed in the Treaty of Versailles. The Dust Bowl in the United States, which destroyed much of the fertile farming land of the South, further emphasised the scarcity of wealth at the time. The then President Herbert Hoover worsened the situation with his failed attempt to balance the budget by increasing taxes. Franklin Delano Roosevelt was elected president in 1932, and immediately introduced his New Deal, a massive government public works and social welfare spending program which helped gradually restore prosperity in the United States.

The 1930s also saw the rapid retreat of liberal democracy, as authoritarian regimes emerged in countries across Europe and South America; in particular, the Third Reich in Germany. Germany, with the backing of President Hindenburg, effectively appointed Adolf Hitler as Chancellor in 1933, who then proceeded to trash the democracy with his Enabling Act, and also imposed the Nuremberg Laws, a series of laws which greatly discriminated against Jews and other ethnic minorities; a precursor to the horrific Holocaust, in which 6 million of the 9 million Jews in Europe were to be exterminated by Hitler's Nazis during the Second World War. Weaker states such as Ethiopia, China, and Poland were invaded by expansionist world powers, the last of these attacks led to the outbreak of the Second World War on 1st September 1939, despite calls from the League of Nations (created after World War One) for worldwide peace. The onset of the Second World War helped end the Great Depression when governments spent huge amounts of money for the war effort, and significantly stimulated their economies in the process.

Among the many technological advances of the 1930s were; the release by Warner Brothers of the first full-colour widescreen movie, "Song of the Flame" in 1930; the beginning of an airmail service across the Atlantic Ocean in 1939; the invention of radar by Robert Watson-Watt in 1938; the first long-playing phonograph record in 1931 by RCA Victor; the first colour photographic film "Kodachrome" made by Eastman Kodak in 1935; nuclear fission, discovered in 1939; Howard Hughes setting a new transcontinental airspeed record by flying non-stop from Los Angeles to Newark in 7 hours and 28 minutes, at an average ground speed of 518 kilometres per hour in 1935; and Edwin Armstrong's invention of wide-band frequency modulation (FM) radio in 1933. In 1930 the astronomer Clyde Tombaugh was the first to identify the planet Pluto, the ninth planet in the solar system at that time. In the field of construction; the world's tallest building (for the next 35 years) was constructed, opening as the Empire State Building in 1931, in New York City; and the Golden Gate Bridge was constructed in San Francisco in 1937.

Much of the Second World War took place in the first half of the 1940s, and had a profound effect on many countries and people in Europe, Asia, and most of the rest of the world. The war lasted from 1939 to 1945, but the consequences of the war lingered well into the second half of the decade, with a war-weary Europe divided between the competing spheres of influence of the United States led Western capitalist world, and the Soviet Union led Eastern communist world; which led to the beginning of the Cold War. Although the war in Europe effectively ended with the suicide of Adolf Hitler in his Berlin bunker on 30th April 1945; the war in the Pacific raged on for another three months until the United States dropped two atomic bombs on the cities of Hiroshima and Nagasaki in Japan, resulting in its unconditional surrender on 14th August 1945; effectively ending the Second

(386) The surrender of Japan on 2nd September 1945:
ending the Second World War

World War. Us President Harry Truman had an incredibly torturous decision to make; if he did not use the atomic bombs, the war would have raged on in the Pacific for a number of years longer, resulting in a massive invasion of Japan, and the loss of millions of additional lives. By dropping the atomic bombs, he forced the surrender of Japan, and a quick decisive end to the war; saving millions of lives, as horrific as the mass destruction of the bombs were. Two military alliances were formed at this time; the North Atlantic Treaty Organization (NATO), comprising the United States and nations of Western Europe; and the Warsaw Pact, comprising the Soviet Union and its satellite controlled communist nations of Eastern Europe.

To some extent, the major tensions triggered by the Cold War were managed by newly-created post Second World War institutions including the United Nations, and the Bretton Woods System; facilitating the post-war economic expansion which lasted well into the 1970s. The Bretton Woods Conference was the gathering of 730 delegates from all of the 44 victorious Allied nations, at the Mount Washington Hotel situated in Bretton Woods, New Hampshire in the United States; to regulate the international monetary and financial system after the end of the war. Out of this conference came the establishment of the International Bank for Reconstruction and Development, later to become the World Bank; and the International Monetary Fund (IMF). The United Nations was established in 1945 in order to replace the largely unsuccessful League of Nations, which ceased to operate before the Second World War commenced. The overriding objective of the United Nations was to prevent future world conflicts by attempting to resolve disputes in the spirit of diplomacy and international cooperation among the nations of the world.

Conditions in the post-war world also encouraged decolonization, and the emergence of new nations and governments; with India, Pakistan, Israel, Vietnam, and others, declaring independence from their former European colonial masters, although rarely without some form of violence and bloodshed. The People's Republic of China was officially proclaimed in 1949, ruled with an iron-fist by the Chinese Communist Party and led by Chairman Mao Zedong. A number of

pro-democracy activists were able to flec mainland China and establish
a democratic capitalist government on the island of Taiwan (formerly
the Portuguese named island of Formosa).

(387)   Churchill, Roosevelt and Stalin at the Yalta Conference
from 4th to 11th February 1945

   The 1940s also witnessed the early beginnings of new technologies such as computers, nuclear power, and jet propulsion, often first
developed in conjunction with the war effort, and later adapted and
improved upon in the Cold War era. In 1941, the Colossus computer
was constructed and used by British codebreakers at Bletchley Park
outside of London, to successfully read and interpret Enigma encrypted
top-secret German messages during the Second World War; creating a
significant strategic advantage for Britain and its allies in the process.
This computer was operational until 1946 when it was destroyed
under orders from British Prime Minister Winston Churchill. The
Colossus computer is now widely regarded as the first operational
computer. Also in 1941, the world's first working programmable, fully

automatic computing machine, known as the Z3, was built. The first test of technology for an atomic weapon, the detonation of a nuclear bomb based on plutonium, known as the Trinity test, was conducted in the desert of New Mexico, United States, as part of the Manhattan Project, initiated by President Franklin Roosevelt. The Manhattan Project was the code name for the American-led effort to develop a functional atomic weapon during the Second World War, in response to fears that German scientists had been working on a weapon using nuclear technology since the 1930s, and that Adolf Hitler was prepared to use it. Other technological developments during the 1940s included; the breaking of the sound barrier in October 1947; the invention of the transistor by Bell Labs, in December 1947; and the further development of radar, jet aircraft, ballistic missiles, and television.

Two significant assassinations occurred in the 1940s; Leon Trotsky and Mahatma Gandhi. The Russian revolutionary leader and internationally-renowned communist thinker, Leon Trotsky, while living in exile in Mexico, was attacked and killed by an assassin using an ice axe on 20th August 1940, under the direct orders of Joseph Stalin, who saw Trotsky as a threat to his ruthless iron-fisted control of the

(388) Mahatma Ghandi died in 1948   *(389) Leon Trotsky died in 1940*

Soviet Union. On 30th January 1948, Mahatma Gandhi, the Indian pacifist activist and leader of the Indian independence movement, was fatally shot in New Delhi by Nathuram Godse, a Hindu extremist who objected to Gandhi's tolerance for the Muslims. When India was partitioned by its colonial masters the British, in 1947, the majority of the Muslims made up the newly formed nation of Pakistan, whereas the remaining nation of India was mainly made up of Hindus.

Clashes between capitalism and communism dominated the decade of the 1950s. The conflicts included the Korean War at the start of the decade; and the beginning of a Space Race between the United States and the Soviet Union, with the successful launch of the Russian Sputnik 1 satellite in 1957. Along with the increased testing of nuclear weapons by both sides, this created a politically conservative climate. In the United States, the second Red Scare (the first had been in 1919-1920, on the back of the Russian Revolution) caused Congressional hearings by both houses of the US Congress, and the fear of communism to grip the United States. The beginning of significant decolonization in Africa and Asia also took place in this decade, and accelerated in the 1960s.

The Korean War lasted from 25th June 1950, until the signing of the Korean Armistice Agreement on 27th July 1953. The conflict started as a civil war between communist North Korea, and the Republic of Korea (democratic South Korea). When it began, North and South Korea existed as provisional governments competing for control of the entire Korean peninsula, due to the division of Korea by outside powers, following the defeat of Japan in the Second World War. While originally a civil war, it quickly escalated into a war between the Western powers under the United Nations, but led by the United States and its allies; and the communist powers of the People's Republic of China and the Soviet Union. On 15th September 1950, US General Douglas MacArthur conducted Operation Chromite, an amphibious landing at the coastal city of Inchon. The North Korean army quickly collapsed under the onslaught, and within a few days, MacArthur's army retook the South Korean capital city of Seoul, which had been invaded and occupied by the North Korean army, triggering the start of the war. MacArthur's forces then pushed north, capturing the North Korean capital city of Pyongyang in October 1950. Chinese intervention in the following month drove the United Nations, US led forces south again. MacArthur then had planned to launch a full-scale invasion of China, but was overruled by the then US President Harry Truman, who feared that an attack on China would trigger another world war, this time with the use of nuclear weapons by both sides; so he wanted the war to remain on a limited scale. After heavy protest and criticism from MacArthur, Truman dismissed him, and replaced MacArthur with General Matthew Ridgeway. The war then became a bloody stalemate for the next two and a half years, while peace negotiations dragged on; ultimately leading to the armistice agreement, where both sides agreed to stop fighting and accept the boundary between North and South Korea; so the conflict to this day, has never technically ended.

Also during the 1950s, the Vietnam War began in 1954. This conflict arose after the French colonial masters were militarily defeated by the North Vietnamese communist army (the Viet Cong) in 1954, and fled the country, leaving a power vacuum which was filled by the United States and its allies. The US became involved in Vietnam because it feared the communist North would quickly fill the power vacuum left by the French, by invading and occupying the democratic South; and thus starting a "domino effect" of communist domination

(390) The Vietnam War lasted from 1954 to 1975

of South East Asia, supported by the Chinese and Russians. As a result, a brutal, ruthless, bloody war emerged. This war was to last for 21 years, and effectively ended in 1975 with the United States and its allies being defeated, and having to retreat from the country, dangling from helicopters on rooftops in a humiliating fashion, as the North Vietnamese took control of the South Vietnamese capital city of Saigon (now named Ho Chi Minh City after the great North Vietnamese revolutionary leader). There is no doubt that the Vietnam War significantly eroded the power and prestige of the United States as a beacon of freedom and democracy around the world. The war also created intense division within United States society and politics; leaving national scars which have not healed to this day.

The Suez Crisis was a war fought on Egyptian territory in 1956. Following the nationalization of the Suez Canal by Egyptian President Gamal Abdel Nasser; the United Kingdom, France, and Israel subsequently invaded Egypt. The operation was a military success, but after the United States and Soviet Union united in opposition to the invasion, the invaders were forced to withdraw. This was seen as a major humiliation, especially to the United Kingdom and France, who clearly had not consulted their US ally before launching the attack. The Suez Crisis, and humiliation of the United Kingdom and France, symbolized the beginning of the end of European colonialism, and in particular the final collapse of the British Empire, which had been significantly weakened by the Second World War. There were a number of conflicts relating to decolonization, but the most serious one was the Algerian War from 1954 to 1962. It was a complex conflict characterized by guerrilla warfare, terrorism against civilians, the use of torture by both sides, and counter-terrorism operations conducted by the French army. The war eventually led to the independence of the African nation of Algeria, from France.

There were also a number of internal conflicts in the 1950s. The Cuban Revolution of 1953 to 1959 led to the overthrow of the dictator Fulgencio Batista, by Fidel Castro and Che Guevara, resulting in the creation of the first communist government in Latin America, literally on the doorstep of the United States. Castro immediately

(391) Cuban Revolutionary Leaders Fidel Castro and Che Guevara

nationalized mainly American owned corporate interests on the island of Cuba, including casinos, sugar mills, and banks. This created a major ideological struggle between the United States and Cuba, which continues to the current day. The Hungarian Revolution of 1956 was a massive, spontaneous popular uprising in the Soviet-controlled state of Hungary, against that country's Soviet-backed Marxist-Leninist regime, inspired by political changes that were occurring within Poland and the Soviet Union at the time. The uprising, primarily started by students and workers, managed to bring the invading Soviet Army to a grinding halt, and a new pro-reform government took power in Hungary. While the Soviet leadership led by Nikita Khrushchev, even considered withdrawing from Hungary entirely, under extreme pressure from his hard-line generals, he soon ordered a second massive invasion which crushed the revolution, killing thousands of Hungarians, and sending hundreds of thousands overseas into exile. This was the most significant act of internal dissent in the history of the Soviet Bloc, and its violent suppression damaged the ideological reputation of the Soviet Union among many communist sympathizers around the world.

Weakened by the aftermath of the Second World War, the British, French, and other European nations lost many colonies in the 1950s.

Some of the major examples, other than the French loss of Vietnam and Algeria, which we have already discussed, included; the Federation of Malaya peacefully gaining independence from the British in 1957; Cambodia and Laos gaining independence from the French in 1953, effectively ending the presence of the French in South East Asia; the African Congo gaining independence from Belgium in 1960; Libya gaining independence from the Italians in 1951; Sudan, Morocco, and Tunisia gaining independence from the French in 1956; and Ghana from the French in 1957.

In the area of science and technology, the 1950s saw much progress made. Some of the major developments included; a breakthrough in semiconductor technology which came with the invention of the MOSFET (metal-oxide-semiconductor- field-effect-transistor) at the US Bell Labs in November 1959. It revolutionized the electronics industry, and became the fundamental building block of the Digital Revolution; television, which first reached the marketplace in the 1940s, attained mass-consumer acceptance in the 1950s, and by the end of the decade most American households for example, owned a television set; in 1955, Jonas Salk invented a vaccine for polio, ending a viral scourge which had afflicted millions of people, by deforming their muscles. Also in this decade; the first passenger jets entered service; Francis Crick and James Watson discovered the double-helix structure of DNA in 1953; the first ultrasound test of heart activity was developed in 1953; the National Aeronautics and Space Administration (NASA) was formed in 1958; and the first transistor computer was built at the University of Manchester in November 1953.

In the early 1950s, popular music was essentially a continuation of the crooner sound of the 1940s, but with less emphasis on the jazz-influenced big band style; and more emphasis on a conservative, operatic, symphonic, vocal style of music, symbolized by popular artists such as; Frank Sinatra, Tony Bennett, Judy Garland, Johnnie Ray, Perry Como, Bing Crosby, Rosemary Clooney, Dean Martin, Eddie Fisher, Dinah Shore, Peggy Lee, Doris Day, and Nat King Cole; and vocal groups like; The Mills Brothers, The Ink Spots, The Four Lads, The Four Aces, The Fontaine Sisters, The Chordettes, The Hilltoppers, and The Ames

(392) Frank Sinatra

(393) Elvis Presley

Brothers. By the mid to late 1950s a totally new style of music, mainly appealing to teenagers, emerged; it was given the name, rock and roll. The artists who pioneered the rock and roll music revolution included; Sam Cooke, Jackie Wilson, Gene Vincent, Chuck Berry, Fats Domino, Little Richard, James Brown, Bo Diddley, Buddy Holly, Bobby Darin, Ritchie Valens, Duane Eddy, Eddie Cochran, Brenda Lee, Bobby Vee, Connie Francis, Johnny Mathis, The Big Bopper, Neil Sedaka, Pat Boone, Bill Haley, The Everly Brothers, Jerry Lee Lewis, Carl Perkins, Johnny Cash, The Platters, The Flamingos, The Dells, Ricky Nelson, The Silhouettes, Frankie Lymon and The Teenagers, Little Anthony and The Imperials, Danny and The Juniors, The Coasters, The Drifters, The Del-Vikings, and Dion and The Belmonts. But, starting from the mid-1950s, the leading figure, later to be called "the king of rock and roll" was Elvis Presley, who became a worldwide superstar with his mass appeal, great charisma, and ability to merge primarily country, soul, and rhythm and blues music into his own unique brand of rock and roll. Elvis went on to sell over 1 billion records worldwide, and star in 31 movies, until his tragic, untimely death at the age of 42 in 1977. The new music differed from previous styles in that it was primarily targeted at the teenage market, which became a distinct age-group for the first time in the 1950s, as growing post-war prosperity meant that young people did not have the pressure to grow up as quickly, or be expected to support a family. Also, by the 1950s, due to economic prosperity, most older teenagers had their own automobiles, giving them a degree of freedom from their parents never experienced by earlier generations. Rock and roll proved to be a difficult phenomenon for the parenting older generations to accept though, and particularly in America, there was widespread accusations of this form of music being branded in an extremely racist fashion as "nigger music"; or a communist-orchestrated plot to corrupt the youth, even though rock and roll was actually extremely marketing based and capitalistic in nature.

The 1950s was also known as The Golden Age of Television; with very popular programs such as I Love Lucy, Father Knows Best, Wagon Train, The Lone Ranger, Gunsmoke, Hopalong Cassidy, Leave it to Beaver, Bonanza, The Twilight Zone, The Honeymooners, The Many Loves of Dobie Gillis, and Perry Mason. Sales of television sets rose tremendously in the 1950s with most families having a television by the end of the decade. People spent so much time watching television that movie attendances dropped, and so did the number of radio listeners. The advent of mass-television represented a major cultural change in the Western nations in particular. The culture of the United States was beamed into lounge rooms around the globe. Television affected what people wore, the music they listened to, what they ate, what news they received, and moulded their political beliefs. Not only was the teenage market huge for rock and roll music, but also for youth fashion. In the United Kingdom for example, the Teddy Boys became both style icons and anti-authoritarian figures. While in the United States, Greasers held a similar social position among teenagers. Previously, teenagers had dressed similar to their parents, but suddenly in the 1950s a rebellious and distinctly unique youth style was being developed, and it became a very big industry. This unique style was controversially overtly sexual in nature; with young men wearing tight-fitting trousers and jeans, leather jackets, and an emphasis on slicked, greasy hair; and young ladies wearing tight-fitting skirts, and cosmetic makeup. Three American movie actors who became iconic

(394) Marilyn Monroe   *(395) James Dean*

role models for teenage youth worldwide in the 1950s were Marilyn
Monroe, James Dean, and Marlon Brando. Let's briefly examine each
of their contributions to popular culture.

Marilyn Monroe was probably the most famous movie star of the
1950s, and was a feminist before there even was a word for women
like that. Although she was portrayed as a sex symbol and bombshell
actress, Marilyn was sensitive, self-aware, and highly intelligent; and
very conscious of the role model she was to young women in the
1950s. When the movie studios tried to push the "dumb blonde" roles
onto her, she not only defied Hollywood and moved to New York, but
she also started her own movie production company, one of the first
women to do so. In the 1950s, studios held much power over actors,
dictating to them every aspect of their role. Marilyn fought against this,
winning such rights as; deciding the content of a final script; choosing
her co-stars; and directional approval; all of these were unheard of in
the Hollywood of the 1950s. Marilyn's influence remains very strong
to this day. Every time a woman is given the opportunity to direct
or produce a film, she can thank Marilyn for breaking this barrier.
Anytime an actress can reject a script, she has Marilyn to thank. Every
time a woman can stand up for her rights in a workplace, she can also
thank women like Marilyn. Most of all, in an ultraconservative era
when anything connected with sexuality was strictly taboo, Marilyn

celebrated the female body and beauty, as something women could control and be proud of, and not something to be ashamed of, or there only to please men. She also celebrated the friendship between women to always take precedence over the pursuit of men. The film which best illustrates Marilyn being ahead of her time in celebrating femininity and beauty, and that women must use their charms not to be subservient to men, but to control them to such an extent they can be their equals, is Gentlemen Prefer Blondes, made in 1953. Throughout the entire film, the main characters Lorelei (Marilyn Monroe) and Dorothy (Jane Russell) display consistent loyalty to one another. There is no backstabbing or competitive degrading of one another to come out on top in gaining the affection of a man. Instead, the women remain steadfastly loyal to each other and their friendship, and tolerate no one speaking negatively of the other. Providing comfort and support to each other takes priority over finding ways to secure their desired man.

James Dean only starred in three major films across a two-year period; Rebel Without a Cause in 1955, East of Eden in 1955, and Giant in 1956. He died at the very young age of 24 in a car crash on his way to a racing event in Los Angeles. Despite his tragically short life and career, James Dean was, and remains, a

(396) and (397) Marilyn Monroe and James Dean movie posters

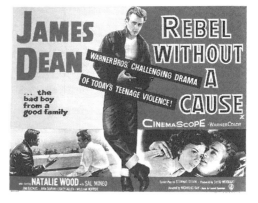

cultural icon. He represented an effortless cool and an iconoclastic spirit which was at the forefront of a post-war youth revolution. His unique self- confidence, sense of style, and non-conformity, inspired many young men in particular. John Lennon, for example, stated that "without James Dean, the Beatles would never have existed." His rebellious image became synonymous with the spirit of rock and roll, and inspired the look and style of iconic artists such as Elvis Presley and Bob Dylan. He also helped permanently cement the identity of teenagers as a uniquely distinct cultural group, within modern society. An identity that endures to this modern day.

Marlin Brando went a long way towards defining what male masculinity meant. The 1950s was a period of major change in many social and cultural areas of life, and Brando threw himself into the decade with his brooding intensity, unpredictable reactions, and improvised lines, in all of his movie roles. In his role in the 1951 movie, A Streetcar Named Desire, he elevated the humble T-shirt into an iconic fashion statement. As a result, what was traditionally a loose-fitting, plain and functional undergarment, was reimagined when Brando wore a tight-fitting white T-shirt in that movie. The result exposed more of a man's body than had been seen in mainstream cinema before, shocking critics and many conservatives. Brando's physicality was a

(398) Marlon Brando in the Movie: A Streetcar Named Desire, released in 1951

force, heightened by his method-style of acting which filled many of his most memorable scenes with tension, not by what he said or did, but from his charismatic and brooding presence instead. Later, his role in the 1953 film, The Wild One, exposed a new brand of bad-boy icon; the motorcycle gang member reeking with ominous danger, and Brando became synonymous with the heavy biker jacket, rolled up Levi jeans, and chunky metallic jewellery.

By the end of the 1950s, war-ravaged Europe had largely finished its reconstruction, and began a period of significant economic growth. The Marshall Plan, a multi-billion dollar US-sponsored economic program implemented from April 1948 to December 1951, and designed to rehabilitate the economies of 17 western and southern European countries in order to create stable conditions in which democratic institutions could survive and flourish; significantly contributed to this economic growth in Europe. The Second World War had brought about a major levelling of the social classes in which the last remaining remnants of the feudal system had all but disappeared. There was an expansion of the middle class in western European countries, so that by the 1960s, many families could afford a radio, television, refrigerator, and a motor vehicle. Meanwhile, the Warsaw Pact, eastern European countries and the Soviet Union were rapidly rebuilding from the Second World War. The overall worldwide economic trend in the 1960s was one of prosperity, expansion of the middle class, and the proliferation of new domestic technology. The Cold War confrontation between the United States and the Soviet Union dominated geopolitics during the 1960s, with the ideological struggle extending into the developing nations of Latin America, Africa, and Asia, as the Soviet Union moved from being a regional power to a truly global superpower, and began vying for influence in the developing world. After President John F Kennedy's assassination in November 1963, direct tensions between the Soviet Union and the United States cooled, and the superpower confrontation moved into a contest for control of the Third World; a battle characterized by proxy wars, funding of insurgencies, and supporting puppet governments.

(399) John F Kennedy   (400) Martin Luther King   (401) Robert F Kennedy

In the United States, in response to nonviolent direct action campaigns from groups such as the Student Nonviolent Coordinating Committee (SNCC), and the Southern Christian Leadership Conference (SCLC), the US President John F Kennedy pushed for major social reforms. Tragically, with his assassination, the president was unable to see his reforms implemented, and they were ultimately legislated under the presidency of Lyndon B Johnson. These major reforms included civil rights for African Americans; and healthcare for the elderly and the poor. Despite his largescale Great Society programs, Johnson was increasingly despised by the New Left at home and abroad. This was due to his heavy-handed approach to the Vietnam War, which outraged mainly student protestors around the world. In addition to the assassination of President Kennedy; the subsequent assassinations of Martin Luther King and Robert Kennedy in 1968; and the anti-Vietnam War movement; defined the politics of violence in the United States in the 1960s.

In France, the protests of 1968 led to President Charles de Gaulle temporarily fleeing the country. These protests were mainly led by workers and students demanding major social reforms in France. Italy formed its first left-of-centre government in March 1962, with a coalition of Christian Democrats, Social Democrats, and moderate Republicans. In Africa, the 1960s was a period of radical political change as 32 countries gained independence from their European colonial rulers. It was during the 1960s that the Vietnam War was significantly escalated by the United States. In 1961, there were only around 700 US military advisers in South Vietnam. President Kennedy was determined not to create a war in Vietnam. By the time of Kennedy's assassination in November 1963, there were 16,000 American military personnel in South Vietnam; but the real escalation commenced under the succeeding President Johnson. In direct response to the minor naval engagement known as the Gulf of Tonkin Incident, which occurred on 2nd August 1964; the Gulf of Tonkin Resolution, a joint resolution of the US Congress, was passed on 10th August 1964. This resolution gave President Lyndon Johnson the authorization, without a formal declaration of war by Congress, to use whatever military force he deemed appropriate in Southeast Asia. This triggered a rapid escalation of the Vietnam War, so that by the end of 1966 more than 500,000 military personnel were sent to Vietnam by the Johnson administration.

Another major military incident of the 1960s was the Bay of Pigs Invasion of 1961, which was an unsuccessful attempt by a CIA-trained force of Cuban exiles to invade southern Cuba, with support from US government armed forces, in order to overthrow the communist government of Fidel Castro. The invasion, although planned during the Eisenhower administration, happened shortly after President Kennedy assumed office, being his first foreign policy crisis. To his credit, he appeared live on television and effectively apologized to the American people for the failure. After this disaster, Kennedy did not place much trust in his military-CIA advisers. Kennedy was again put to the test in the Cuban Missile Crisis of 1962, when the world came perilously close to an all-out nuclear war. The Soviet Union had secretly placed nuclear missiles on the island of Cuba, in what they claimed was a defensive strategy to counteract the presence of US nuclear missiles in Turkey, close to the Soviet border. When US spy planes identified the missiles, a major confrontation occurred, which included an American Naval quarantine (or blockade) of all Soviet ships voyaging to Cuba. For 13 tense days, the world was on the brink of nuclear calamity. Kennedy and Khrushchev were able to reach a negotiated settlement, whereby the Soviet Union agreed to remove its missiles in Cuba, in exchange for the United States removing its missiles in Turkey, six months later. Interestingly, in October 1964, Soviet leader Nikita Khrushchev was expelled from office due to his perceived increasingly erratic behaviour.

On 22nd November 1963, President John Fitzgerald Kennedy was assassinated while riding in an open motorcade in Dallas Texas. The alleged assassin, Lee Harvey Oswald, was also the prime suspect in the killing of Dallas police officer JD Tippet, later that same day.

(402) Lyndon B Johnson sworn in as President in November 1963 after JFK's assassination

Shockingly, Oswald was then shot dead by nightclub-owner Jack Ruby on live television in the basement of the Dallas Police Department headquarters, two days later, as he was being transferred to a detention facility. Polls consistently show that the majority of Americans believe the assassination of President Kenney was a conspiracy; in spite of the government-appointed Warren Commission hastily coming to the conclusion in 1964 that Lee Harvey Oswald acted alone.

The Arab-Israeli conflict reached a crisis point with the Six-Day War in June 1967; a war between Israel, and the neighbouring states of

Egypt, Jordan, and Syria. The Arab states of Iraq, Saudi Arabia, Sudan, Tunisia, Morocco, and Algeria, also contributed troops and arms. Ever since the nation of Israel was created in 1948 as the homeland of the post-war surviving Jews of Europe and the Soviet Union, with British and US support, it had been immediately threatened with extermination by its hostile Arab neighbours. By the end of the Six-Day War Israel had prevailed, and gained control of the Sinai Peninsula, the Gaza Strip, the West Bank, East Jerusalem, and the Golan Heights. The result of this war still impacts on the region to this day. Another big event was the Cultural Revolution in China, which effectively lasted from 1966 to 1976. It was a period of widespread social and political upheaval in the People's Republic of China which was initiated by Mao Zedong, the Chairman of the Communist Party of China. Mao alleged that "liberal bourgeois" elements were infiltrating into the Party and society at large, and that they wanted to restore capitalism in the country. Mao insisted that these elements be removed through a brutal, ruthless, and violent post-revolutionary class struggle; by initially mobilizing the thoughts and actions of China's youth, who formed Red Guards groups throughout the country. This movement subsequently spread into the military, urban workers, rural villages, and the Party leadership itself. Although Mao himself officially declared the Cultural Revolution to have ended in 1969, the power struggles and political instability between 1969, and the arrest of the alleged conspiratorial Gang of Four in 1976, are now also widely regarded as forming part of the Revolution.

The Troubles in Northern Ireland began with the rise of the Northern Ireland Civil Rights Movement in the mid-1960s, and continued into the late 1990s. The Troubles, also called the Northern Island Conflict, were a series of violent conflicts from 1968 to 1998 in Northern Island between the overwhelmingly Protestant unionists (or loyalists), who wanted the province to remain part of the United Kingdom; and the overwhelmingly Roman Catholic nationalists (republicans), who wanted Northern Island to become part of the Republic of Ireland. Mass socialist and communist movements occurred in most European countries in the 1960s, but particularly impacted France and Italy. The most spectacular manifestation of this was the May 1968 student revolt in Paris, that linked up with a general strike of ten million workers called by the trade unions; and for a few days seemed like they were going to overthrow the government of Charles de Gaulle. De Gaulle went off to visit French troops in Germany to check on their loyalty. Major concessions were won for trade union rights, including higher minimum wages, and better working conditions. In Czechoslovakia, 1968 was the year of Alexander Dubcek's Prague Spring; a source of inspiration to many Western leftists who admired Dubcek's concept of "socialism with a human face." The Soviet invasion of Czechoslovakia in August 1968 violently ended these hopes, and also fatally damaged the chances of the orthodox communist parties drawing many recruits from the worldwide student protest movement.

(403) The Soviet Union Invasion of Czechoslovakia in August 1968

Relations between China and the United States remained hostile during the 1960s, although representatives from both countries held periodic meetings in Warsaw Poland, since there was no US embassy in China. President Kennedy had plans to restore China-US relations, but his assassination, the war in Vietnam, and the Cultural Revolution, had put an end to that possibility. It was not until Richard Nixon took office in 1969, that steps were taken to restore the relationship.

(404) President Richard Nixon meeting Chairman Mao Zedong in China
in February 1972

Following Soviet leader Nikita Khrushchev's expulsion in 1964, China-Soviet relations dissolved into open hostility. The Chinese were deeply disturbed by the Soviet suppression of the Prague Spring in 1968, as the Soviet Union now claimed the right to intervene in any country it saw as deviating from the "correct path" of Marxist socialism. Finally, in March 1969, armed clashes took place along the China-Soviet border in Manchuria. This was a driving factor for the Chinese to restore diplomatic relations with the United States; as Mao Zedong decided the Soviets were a greater threat than the Americans.

In the second half of the decade, young people began to revolt against the conservative norms of the time, as well as remove themselves from mainstream capitalism, in particular the high level of consumer materialism which was so common during that era, in the West. They created a "counterculture" that sparked a social revolution throughout much of the Western world. It began in the United States as a reaction against the conservatism and social conformity of the 1950s; and the US government's extensive military intervention in Vietnam. The youth involved in this counterculture became known as hippies. These groups created a movement headquartered in San

Francisco, toward liberation in society; including a sexual revolution, questioning authority and government, and demanding more freedoms and rights for women and minorities. The Underground Press, a widespread, eclectic collection of newspapers served as a unifying medium for the counterculture. The movement was also marked by the first widespread, socially accepted use of drugs, such as marijuana and LSD, and psychedelic music led by great counterculture artists including Janis Joplin and Jimmy Hendrix. The war in Vietnam would eventually lead to a commitment of over 500,000 American troops, resulting in over 58,000 American deaths, and an estimated 3 million Vietnamese deaths, producing a large-scale antiwar movement in the United States. As late as the end of 1965, few Americans protested the American involvement in Vietnam, but as the war dragged on, and the body count continued to climb, civil unrest escalated. Students became a powerful and disruptive force and university campuses sparked a national debate over the war. As the movement's ideals spread beyond the university campuses, doubts about the war spread into mainstream society, aided by nightly graphic images beamed into lounge rooms, as shown on the television news; and also doubts began to appear within the government administration itself. A mass movement began rising in opposition to the Vietnam War, ending in the massive Moratorium Protests in 1969, as well as the movement of resistance to conscription ("the Draft") for the war.

Beginning in the mid-1950s and continuing into the late-1960s, African Americans in the United States were determined to outlaw racial discrimination and secure full voting rights. The emergence of the Black Power Movement, which lasted from 1966 to 1975, enlarged the aims of the civil rights movement to include racial dignity, economic and political self-sufficiency, and anti-imperialism. The movement was characterized by major campaigns of civil resistance. Between 1955 and 1968, acts of civil disobedience and nonviolent protests produced crisis situations between activists and government authorities. Federal, state, and local governments, businesses, and communities often had to respond immediately to these situations that highlighted the inequalities faced by African Americans. Forms of protest and civil

disobedience included boycotts, such as the successful Montgomery
Bus Boycott of 1955-56 in Alabama; "sit-ins" such as the influential
Greensboro Sit-Ins of 1960 in North Carolina; marches, such as the
Selma to Montgomery Marches of 1965 in Alabama; and a wide
variety of other nonviolent activities. The leader of the civil rights
movement was undoubtedly Martin Luther King, with his inspirational speeches, moral dignity, and focus on nonviolent protest. He
was tragically assassinated by a white supremacist in 1968. Arguably
the most famous words spoken by King was during his speech in
Washington on 28th August 1963, attended by over 250,000 people,
where he said: "I have a dream that my four little children will one day
live in a nation where they will not be judged by the colour of their
skin but by the content of their character. I have a dream that one day
little black boys and girls will be holding hands with little white boys
and girls. We must live together as brothers or perish together as fools."
Noted legislative achievements during this phase of the civil rights
movement were the passing of the Civil Rights Act of 1964, banning
discrimination based on "race, colour, religion, or national origin" in
employment practices and public accommodations; the Voting Rights
Act of 1965, restoring and protecting voting rights; the Immigration

(405) Civil Rights Protestors in the United States in the 1960s

and Nationality Services Act of 1965, dramatically opening entry to the US to immigrants other than traditional European groups; and the Fair Housing Act of 1968, banning discrimination in the sale or rental of housing.

A second wave of feminism in the United States and around the world gained momentum in the early 1960s. While the first wave of the early 20th century was centred on gaining voting rights and overturning legislated inequalities, the second wave was focused on changing cultural and social norms of imbedded inequalities between men and women. At the time, a woman's place was generally seen as being in the home, and they were excluded from many jobs and professions. In the United States, a Presidential Commission on the Status of Women found that discrimination existed against women in the workplace and every other aspect of life; a revelation which launched two decades of prominent women-centred legal reforms, such as the Equal Pay Act of 1963, which broke down the last remaining legal barriers to women's personal freedom and professional success. Feminists took to the streets, marching and protesting, authoring books, and debating to change social and political views that limited women. In 1963, with Betty Friedan's book, "The Feminine Mystique", the role of women in society, and in public and private life, was questioned. On the role of the suburban wife, the book states: "Each suburban wife struggles with it alone. As she made the beds, shopped for groceries, matched slipcover material, ate peanut butter sandwiches with her children, chauffeured Cub Scouts and Brownies, lay beside her husband at night – she was afraid to ask even of herself the silent question – 'Is this all?'" In relation to the issue of equality in the workplace, Friedan states: "In almost every professional field, in business and in the arts and sciences, women are still treated as second-class citizens. It would be a great service to tell girls who plan to work in society to expect this subtle, uncomfortable discrimination – tell them not to be quiet, and hope it will go away, but fight it. A girl should not expect special privileges because of her sex, but neither should she 'adjust' to prejudice and discrimination." By 1966, the feminist movement was beginning to grow in size and power as women's groups spread across the country and Freidan, along with other feminists, founded the National Organization for Women. In 1968, "Women's Liberation" became a household term, and for the first time, the new women's movement eclipsed the civil rights movement, when New York Radical Women, led by Robin Morgan, protested the annual Miss America Pageant in Atlantic City, New Jersey. The United States, in the firestorm of a social revolution, also led the world in LGBT rights in the late 1960s and early 1970s. Inspired by the civil rights movement and the women's movement, early gay rights pioneers had begun to build a national, and ultimately international, movement. These groups emphasized that gay men and women are no different from straight people, and deserve full equality.

(406) Yuri Gagarin

(407) Neil Armstrong

The Space Race between the United States and the Soviet Union dominated the 1960s. The Soviets sent the first man Yuri Gagarin, into space during the Vostok 1 mission on 12th April 1961, and scored a host of other successes, but by the middle of the decade the US was taking the lead. In May 1961, President Kennedy set the goal for the United States of landing a man on the Moon by the end of the 1960s. On 20th July 1969, Apollo 11 became the first human spaceflight to land on the Moon. Launched on 16th July, it carried Mission Commander Neil Armstrong, Command Module Pilot Michael Collins, and the Lunar Module Pilot Edwin (Buzz) Aldrin. Apollo 11 fulfilled

President Kennedy's goal, when Neil Armstrong stepped onto the surface of the Moon, closely followed by Buzz Aldrin. In his speech to a joint session of Congress on 25th May 1961, Kennedy said: "I believe that this nation should commit itself to achieving the goal, before the decade is out, of landing a man on the Moon and returning him safely to the Earth." The formidable scientific, technological, entrepreneurial, and economic power of the United States, made his goal a reality.

Other scientific and technological developments of the 1960s included:

- The female birth-control contraceptive pill released in 1960.
- A working laser demonstrated in 1960.
- An industrial robot built in 1961.
- The first transatlantic satellite broadcast in 1962.
- A touch-tone telephone introduced in 1963.
- A home video recorder produced in 1963.
- The measles vaccine released in 1963.
- The discovery and confirmation of cosmic microwave background radiation in 1964, securing the Big Bang as the best theory of the origin and evolution of the Universe.
- A high-speed rail service began in Japan in 1964.
- An 8-track tape audio format developed in 1964.
- A compact cassette introduced in 1964.
- The programming language BASIC created in 1964.
- The world's first supercomputer, the CDC 6600 introduced in 1964.
- The first successful heart transplantation operation performed by Professor Christiaan Barnard in South Africa in 1967.
- The first PAL and SECAM broadcast colour television systems started publicly transmitting in Europe in 1967.
- The first known pulsar (a rapidly spinning neutron star) discovered in 1967.
- The first Automatic Teller Machine (ATM) opened in London in 1967.
- The first public demonstration of the computer mouse, video conferencing, email, and hypertext demonstrated in 1968.

- Arpanet, the first research-orientated prototype of the Internet introduced in 1969.
- CCD invented and used as an electronic imager in still and video cameras in 1969.

Although rock and roll continued to dominate popular music around the world, a major development during the mid-1960s was the movement away from singles and towards albums. Previously, popular music was based around the 45rpm (revolutions per minute) single, and those albums that existed were little more than one or two hit singles backed with filler tracks, instrumentals, and cover-songs. The development of the AOR (album orientated rock) format was

(408) The Beatles were together as a band from 1962 to 1970

complicated and involved the development of several concurrent music innovations including Phil Spector's Wall of Sound; the intro-duction by Bob Dylan of "serious" poetic lyrics to rock music; and the Beatles' new studio-based approach to professional recording. As a result, the vinyl LP ("Long Play") record had definitively taken over as the primary format for all popular music styles, after 1965. Two of the most popular albums of the 1960s were the Beach Boys "Pet Sounds" album released in May 1966; and the Beatles "White Album" released in November 1968. Meanwhile, when it came to fashion, the Beatles exerted an enormous influence on young men's fashions and hairstyles in the 1960s, which included most notably the mop-top haircut, boots, and leather jackets. The hippie movement late in the decade also had a strong influence on clothing styles, including bell-bottom jeans, tie-dye, colourful batik fabrics, and paisley pants. The bikini came into fashion in 1963 after it was featured in the teenage popular film, "Beach Party." Mary Quant popularised the miniskirt, which became one of the most popular fashion rages in the late 1960s among young women and teenage girls. Its popularity continued throughout the first half of the 1970s, and then disappeared temporarily from mainstream fashion before making a comeback in the mid-1980s.

Throughout the 1970s in the Western world, socially progressive values that began in the 1960s, such as increasing political awareness, the economic liberty of women, and racial equality, continued to grow. The novelist Tom Wolfe came up with the term "the me decade" in his essay "The 'Me' Decade and the Third Great Awakening", published by New York Magazine in August 1976, referring to the focus on "individualism" in the 1970s, instead of the "community" focus more prevalent in the 1960s. In the United Kingdom, the 1979 election resulted in the victory of the Conservative leader Margaret Thatcher, the first female British Prime Minister. Thatcher initiated a Neoliberal economic policy of reducing government spending, weakening the power of trade unions, selling off government assets, and promoting economic and trade liberalization.

Industrialized countries with the exception of Japan, experienced an economic recession due to an Oil Crisis in 1973 triggered by oil embargoes (shortages) imposed by the Organization of Arab Petroleum Exporting Countries (OAPEC). This crisis saw the first instance of stagflation (when inflation and unemployment increase at the same time) which began a political and economic trend of the replacement of Keynesian (significant government intervention and spending) economic theory; with Neoliberal (significantly less government intervention and spending) economic theory; with the first Neoliberal government being created in Chile, where a military

coup led by General Augusto Pinochet took place in 1973. Augusto
Pinochet came to power as ruler of Chile after overthrowing the
country's democratically elected Socialist president Salvador Allende,
with the assistance of the Central Intelligence Agency (CIA) of the
United States, which is widely believed to have been involved in
Allende's assassination. Pinochet would remain the dictator of Chile
until 1980. The Spanish dictator Francisco Franco died after 39 years
in power. Juan Carlos I was crowned King of Spain and called for the
reintroduction of democracy. The first general elections were held in
1977 and Adolfo Suarez became Prime Minister of Spain after his
Centrist Democratic Union party won. The Socialist and Communist
Parties were legalized, and a democratic Constitution was signed in
1978.

*(409) Soviet Union May Day parade showcasing its nuclear missile
arsenal to the world*

The Soviet Union under the leadership of Leonid Brezhnev, having
the largest armed forces, and the largest stockpile of nuclear weapons
in the world, pursued an agenda to lessen tensions with its rival superpower the United States, for most of the 1970s. That policy, known
as "détente" abruptly ended with the Soviet invasion of Afghanistan
at the end of 1979. Cold War superpower tensions had cooled by the
1970s, with the aggressive US–Soviet confrontations of the 1950s–60s

giving way to the policy of détente, which promoted the idea that the world's problems could be resolved at the negotiating table. Détente was partially a reaction against the policies of the past 25 years, which had brought the world dangerously close to nuclear war on several occasions; and because the United States was in a weakened position following its failure in the Vietnam War. As part of détente, the US also restored ties with China, partially as a counterweight against Soviet expansionism. The 1970s are considered to be a period in which there was relative economic stability and a sound standard of living among the people of the Soviet Union. Nevertheless, hidden inflation continued to increase for the second straight decade, and production consistently fell short of demand in agricultural and manufactured consumer goods. By the end of the 1970s, these factors were starting to put a severe strain on the Soviet economy.

The US–Soviet geopolitical rivalry nonetheless continued through the decade, although in a more indirect "proxy" fashion as the two superpowers jockeyed relentlessly for the ideological control of smaller countries. American and Soviet intelligence agencies gave funding, training, and military support to insurgent groups, governments, and armies around the world; each superpower seeking to gain a geopolitical advantage, and install friendly governments. Coups, civil wars, and terrorism went on across Asia, Africa, and Latin America; and also in Europe where a spate of Soviet-backed Marxist terrorist groups such as the Red Brigades in Italy, and the Red Army Faction (also known as the Baader Meinhof Gang) in Germany, were active throughout the decade. In the 1970s, over half of the world's population lived under the direct rule of a repressive dictatorship.

In Asia, the international standing of the People's Republic of China changed significantly following the recognition of China by the United Nations; the death of Mao Zedong in 1976; and the beginning of market liberalization by Mao's successors. In 1972, US President Richard Nixon visited China, restoring diplomatic relations between the two countries. After the death of Mao Zedong, Deng Xiaoping eventually emerged as China's paramount leader in 1978, and began to dramatically shift the country towards an open market capitalist economic system, and away from ideologically driven politics; even though the Chinese Communist Party maintained tight political power and control over its people. Despite facing a severe oil crisis due to the OAPEC embargo restricting supply, the Japanese economy experienced a substantial boom in the 1970s, overtaking West Germany as the second biggest economy in the world after the United States. In 1975, the US withdrew its forces from Vietnam, which had grown enormously unpopular. In 1979, the Soviet Union invaded Afghanistan, which led to an ongoing war that lasted for ten years.

The 1970s saw an initial increase in violence in the Middle East as Egypt and Syria declared war on Israel, but in the late 1970s the situation in the Middle East was fundamentally altered when Egypt and Israel signed the Egyptian-Israeli Peace Treaty. Anwar Sadat, the President of Egypt, was instrumental in this historic event, and consequently became extremely unpopular in the Arab and wider Muslim world. US President Jimmy Carter deserves the credit for facilitating this peace agreement. The Yom Kippur War was launched by Egypt and Syria against Israel in October 1973 to recover territories lost by the Arabs in the 1967 Six Day War. The Israelis were taken by surprise and suffered heavy losses before they rallied and retaliated. In the end, the Israelis managed to repel the Egyptians and Syrians, and crossed the Suez Canal, entering Egypt. In 1978, Egypt signed a peace treaty

(410) Ayatollah Khomeini

(411) The Iranian Hostage Crisis
of 1979 to 1981

with Israel at Camp David in the United States, ending outstanding disputes between the two countries. Tragically, Sadat's actions would lead to his assassination in Cairo, during a military parade to celebrate Egypt's 1973 war against Israel, in October 1981, by a group of army officers who were opposed to his overtures of peace with Israel.

Political tensions in Iran came to a head with the Iranian Revolution of 1979, which overthrew the Pahlavi dynasty of Shahs, and established an authoritarian Islamic Republic under the leadership of the Ayatollah Khomeini, who had jubilantly returned from exile in Paris. Distrust between the Iranian Islamic revolutionaries and the Western powers led to the Iranian Hostage Crisis on 4th November 1979, where 66 diplomats, mainly from the United States, were held captive for 444 days. They were not to be released until there was a change of US president from Jimmy Carter to Ronald Reagan in 1981.

The United States President Richard Nixon resigned as president on 9th August 1974, while facing charges of impeachment for the Watergate Scandal; involving a number of clandestine and often illegal activities undertaken by members of the Nixon administration, including bugging the offices in the Watergate complex of their Democratic Party political opponents, and of people whom Nixon was suspicious of; and also ordering investigations of activist groups and political figures.

Africa saw further decolonization with Angola and Mozambique gaining independence from the Portuguese in 1975, after the restoration of democracy in Portugal. The African continent however was plagued by ongoing famine, civil wars, and a series of military coups; one of which resulted in the long-reigning Emperor of Ethiopia Haile Selassie being removed, and replaced by a military dictatorship. The economies of much of the developing Third World continued to make steady progress in the early 1970s, largely as a result of the Green Revolution. The Green Revolution of the late 1960s brought about self-sufficiency in food in many developing economies. At the same time an increasing number of people began to seek urban prosperity over agrarian life. This consequently saw significant social disruption across communities in many developing countries.

(412) Pope John Paul II   *(413) US President Jimmy Carter*

The year 1978 would become known as the "Year of Three Popes." In August Pope Paul IV, who had ruled the Catholic Church since 1963, died. His successor was Cardinal Albino Luciano, who took the name John Paul I. But only 33 days later, he was found dead under mysterious circumstances, and another pope had to be elected. On 16th October Karol Wojtyla, a Polish cardinal, was elected and became Pope John Paul II. He was the first non-Italian pope since 1523, and was widely admired for standing up for the freedom of his beloved Poland, in its fight to be free from communism.

- In the area of science and technology, some of the more significant developments of the 1970s included:
- An explosion in the understanding of solid-state physics, driven by the development of the integrated circuit, and the laser.
- The Intel 4004; the world's first general microprocessor.
- Datapoint 2200; the first personal computer developed.
- First pocket calculators.
- Fibre optics transforms the communications industry.
- Microwave ovens become commercially available.
- VCR video recorders and players become commercially available.
- Motorola develops the first cellular phone.
- The Skylab space station and the development of the reusable Space Shuttle to replace rocket launches.
- The world's first passenger-carrying supersonic jet flight in

1973, across the Atlantic in the Concorde.

- MRI medical imaging developed in 1973.
- In 1978, Louise Brown became the first child to be born as a result of in-vitro fertilisation (IVF).
- After successful vaccination campaigns throughout the 19th and 20th centuries, the World Health Organization Certified that smallpox had been eradicated in December 1979.
- Bacteria was genetically engineered in 1973, followed by mice in 1974.
- DNA genome sequencing began in 1977.

The 1970s started a mainstream affirmation of the environmental issues that early activists from the 1960s, such as Rachel Carson and Murray Bookchin, had warned of. Apollo mission photos of our planet Earth, inspired many people to see our planet as a fragile, integrated, life-support system, and shaped a public willingness to preserve nature and the environment. On 22nd April 1970, the United States celebrated its first Earth Day, in which over 2000 colleges and universities and over 10,000 schools participated. The 1970s also saw the rapid commercialization of popular music (mainly rock and later, disco), and by the middle of the decade there were a number of bands cynically called "corporate rock bands" because they were perceived to have been created by record companies to produce simplistic, radio-friendly songs that offered clichés rather than meaningful lyrics. Such bands included The Doobie Brothers, Bread, Kansas, and REO Speedwagon. Funk, an offshoot of soul music with a greater emphasis on beats; and influences from rhythm and blues, jazz, and psychedelic rock, was also very popular. The mid-1970s also saw the rise of disco music, which dominated during the last half of the decade, with the movie "Saturday Night Fever" and bands like the Bee Gees, Chic, ABBA, Village People, Boney M, Donna Summer, and KC and the Sunshine Band. In response to disco, rock music became increasingly heavier and hard-edged, with artists such as Led Zeppelin, Black Sabbath, Deep Purple, and Queen. Experimental classical music influenced both art rock and progressive rock genres with bands such as Pink Floyd, Yes, Supertramp, Genesis, and The Moody Blues. The highest selling album of the 1970s was Pink

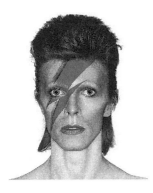

(414) David Bowie

(415) Saturday Night Fever

(416) The Bee Gees

Floyd's "The Dark Side of the Moon", released in 1973, it remained on the US Billboard 200-albums chart for 741 weeks. Statistically, Led Zeppelin and Elton John were the most successful musical acts of the 1970s, both having sold more than 300 million records since 1969; although the most successful live-touring band was The Rolling Stones.

Clothing styles during the 1970s were influenced by outfits seen in popular music groups and in Hollywood films. The hippie movement was also a significant influence. Outlandish gender-bending "glam-rock" clothing, started off by David Bowie, who was named the King of Glam Rock, also influenced many young people. Significant fashion trends of the 1970s included:

• Bell-bottomed pants, combined with turtle necked shirts and flower prints, characterized the classic 1970s look. In the later part of the decade, this look gave way to three-piece suits for guys, in large part because of the phenomenal success of the movie "Saturday Night Fever" which launched disco music into the popular mainstream.

• Long hair down to the shoulders and sideburns were popular for men, as were beards and moustaches, which had been out of fashion since the 19th and early 20th century.

• Women's hairstyles went from long and straight in the first half of the decade, to the feathery-wavy cut promoted by actresses such as Farrah Fawcett.

- Miniskirts and mini-dresses were still fashionable in the first half of the 1970s, but were quickly phased out by hot-pants in the middle of the decade.
- Platform shoes and leather pants became very popular.
- Faded jeans.
- Mohawk hairstyles associated with the punk subculture.

The 1980s was the final decade of the Cold War and opened with the US-Soviet confrontation continuing largely without any interruption. Superpower tensions escalated rapidly as President Reagan scrapped the policy of détente and adopted a new, much more aggressive stance with the Soviet Union, even referring to them as the "Evil Empire." Reagan's hard-line policies, in conjunction with Soviet leader Mikhail Gorbachev's progressive, open policies in the face of a rapidly deteriorating economic situation; ultimately led to the collapse of the Soviet Union and communism by the end of the decade. Major civil discontent and violence occurred in the Middle East, including the Iran-Iraq War, the 1982 Lebanon War, the bombing of Libya in 1986, and the First Palestinian Intifada in the Gaza Strip and the West Bank. Islamism became a powerful political force in the 1980s, and many terrorist organizations, including Al Qaeda, started during the 1980s.

By 1986, nationalism was making a comeback in the Eastern Bloc, and a desire for democracy in communist-led socialist states, combined with an economic recession, resulted in Soviet leader Mikhail Gorbachev's glasnost (openness) and perestroika (reforms) initiatives; which reduced Soviet Communist Party power, legalized dissent, and sanctioned limited forms of capitalism, such as joint ventures with Western firms. After newly heated tension for most of the decade, by 1988 relations between the West and East had improved significantly, largely thanks to the progressivism of Gorbachev. The Soviet Union was increasingly unwilling to defend its communist-led governments in satellite states. The Solidarity movement began in Poland in 1980, involving workers demanding political liberalization and democracy in Poland. Attempts by the Communist government to prevent the rise of Solidarity failed, and negotiations then took place with the government. Solidarity was to become instrumental in inspiring people in other communist states to also demand political reform. It seemed that Soviet leader Mikhail Gorbachev's openness and reforms had let the Jeannie out of the bottle; ultimately paving the way for the complete collapse and breakup of the Soviet Union into independent nation states, and the demise of communism in Eastern Europe. It seems clear that Gorbachev never intended to sow the seeds for the complete collapse of communism in the Soviet Union and its satellite states.

(417) US President Ronald Reagan and Soviet Leader Mikhail Gorbachev

The year 1989 brought the overthrow and attempted overthrow of a number of communist-led governments; such as in Hungary; the Tiananmen Square protests in China; the Czechoslovak Velvet Revolution; Erich Honecker's East German regime; Poland's Soviet-backed government; and the violent overthrow of the regime of Nicolae Ceausescu in Romania. Destruction of the 155 kilometre Berlin Wall at the end of the decade signalled a seismic political shift. The Cold War ended in the 1990s with the successful reunification of Germany, and the demise of the Soviet Union after the August Coup of 1991.

Some of the most notable conflicts of the 1980s, included; The Soviet-Afghan War, fought between the Soviet Union and the Islamist Mujahideen Resistance in Afghanistan, continued until 1989. The Mujahideen were supported by the CIA of the United States, as well as Saudi Arabia, Pakistan, and other Muslim nations. Ultimately, the Soviet Union pulled out of Afghanistan without any military success. On 2nd April 1982, Argentina invaded the Falkland Islands, sparking the Falklands War. It occurred from 2nd April to 14th July 1982 between the United Kingdom and Argentina, as British forces fought to recover the islands. The UK emerged victorious, and its international standing under the leadership of Margaret Thatcher received an unexpected boost. The military junta of Argentina on the other hand, was left humiliated by the defeat; and its leader, Leopoldo Galtieri was deposed three days after the end of the war. In 1982, Israel invaded Lebanon in response to an assassination attempt against Israel's Ambassador to the United Kingdom by the Abu Nidal Organization, and due to the constant terror attacks on northern Israel made by the terrorist organizations based in Lebanon. After attacking the Palestinian Liberation Organization (PLO), as well as Syrian, leftist, and Muslim Lebanese forces, Israel occupied southern Lebanon and eventually surrounded the PLO in west Beirut; and after being subjected to heavy Israeli bombardment, the PLO negotiated a passage out of Lebanon. The Iran-Iraq War took place from 1980 to 1988. Iraq was accused of using illegal chemical weapons to kill Iranian forces, and against its own dissident Kurdish population. Both sides suffered enormous casualties, but the poorly equipped Iranian army suffered the worse for it, being forced to use soldiers as young as 15 in suicidal human-wave attacks. Iran finally agreed to an armistice in 1988. The United States launched an aerial bombardment of Libya in 1986 in retaliation for Libyan support of terrorism, and attacks on US personnel in Germany and Turkey. Libyan agents were suspected of blowing up the Pan American civilian jumbo jet flight over Lockerbie Scotland in 1988, in retaliation for the US bombing of Libya in 1986.

(418) One man stops Chinese tanks in Tiananmen Square on 5th June 1989

(419) China's Deng Xiaoping

(420) First PLO Intifada in December 1987

Two significant internal conflicts in the 1980s were the Tiananmen Square Protests in China; and the First Intifada (Uprising) in Israeli-occupied territories. The Tiananmen Square Protests occurred in the People's Republic of China in 1989, and involved pro-democracy protestors demanding political reform. The protests were crushed with overwhelming force by the People's Liberation Army. The First Intifada in the Gaza Strip and West Bank commenced in 1987 when Palestinian Arabs mounted large-scale protests against the Israeli military presence in the Gaza Strip and the West Bank, largely inhabited by Palestinians.

The First Intifada would continue until peace negotiations began between the PLO and the Israeli government in 1993.

Under the leadership of Deng Xiaoping, China embarked on extensive reforms in the 1980s, opening up the country's economy to the West, and allowing capitalist enterprises to operate in a market-based socialist system. These reforms laid the foundation for China to become an economic power to rival the United States, by the end of the twentieth century.

Some of the firsts in the fields of science and technology in the 1980s included:

- The Pac-Man video game released in 1980 and gaining worldwide popularity.
- Personal computers launched in 1981.
- Modems developed in 1981.
- Microsoft releasing the MS-DOS operating system in 1981, and Windows 1.0 in 1985.
- Digital Audio Compact Disc (CD) released in 1982.
- Mobile phones available to consumers in 1984.
- The surrogate pregnancy of an unrelated child in 1986.
- Genetically modified crops grown in China in 1988.
- Gene therapy techniques performed on human beings in 1989.
- The British computer scientist Tim Berners-Lee developing the World Wide Web in 1989.

In the United States MTV was launched, and music videos began to have a major influence on the record industry. Popular artists such as Michael Jackson, Dire Straits, Duran Duran, Prince, Cyndi Lauper, and Madonna mastered this visual format, and helped turn the MTV medium into a highly profitable business. New Wave and Synth-pop were developed by many British and American artists, and became popular especially in the early 1980s. During this decade popular music became more fragmented into subgenres such as house, goth, new romantic, punk, and rap. Michael Jackson was one of the greatest popular music icons of the 1980s, and his leather jacket, glove, and moonwalk dance were often imitated. Jackson's 1982 album "Thriller" became, and currently remains, the biggest selling album of all time,

with sales of up to 110 million, worldwide. The beginning of the
1980s saw the continuation of the clothing styles of the late 1970s;
however, fashion became more extravagant during the 1980s. Some
of the fashion trends included teased and colourfully-dyed hair;
ripped jeans; neon multi-coloured clothing; the perm, mullet, Jheri
curl, and hip-top fade hairstyles; shoulder pads; jean jackets; leather
pants; jumpsuits; leggings; and leg-warmers, popularized by the film,
Flashdance. Girls and women also wore jelly shoes, large crucifix
necklaces, and brassieres, all inspired by Madonna's "Like a Virgin"
music video.

The 1990s saw significant advances in technology; particularly
with the widespread use of the World Wide Web (the Internet). A
combination of factors; including the continued mass mobilization
of global capital markets through Neo Liberalism and globalisation;
the thawing and end of the decades-long Cold War; the beginning
of the widespread proliferation of new media such as the Internet
from the middle of the decade onwards; increasing scepticism
towards government; and the dissolution of the Soviet Union; led to
a realignment and reconsolidation of economic and political power
across the world. The technologically driven, speculative dot.com
bubble of 1997–2000 in the United States, generated great wealth to a
number of Silicon Valley entrepreneurs and Wall Street banks, before it
came crashing down in 2000-2001.

New ethnic conflicts emerged in Africa, the Balkans, and the
Caucasus; the first two leading to the Rwandan and Bosnian genocides,
respectively. Signs of any resolution of tensions between Israel and the
Arab world remained elusive despite the progress of the Oslo Accords;
though the Troubles in Northern Island came to an end in 1998 with
the Good Friday Agreement, after 30 years of sectarian violence. There
were a number of prominent conflicts in the 1990s. The Gulf War
arose as a result of Iraq being left in a crippling economic position after
its longstanding war with Iran in the 1980s. Iraqi President Saddam
Hussein accused Kuwait of flooding the market with oil and driving
down prices. As a result, on 2nd August 1990, Iraqi forces invaded and
conquered Kuwait. The United Nations immediately condemned the

military action, and a coalition force led by the United States was sent to the Persian Gulf. Extensive aerial bombing of Iraq began in January 1991, and a month later, the UN-backed coalition forces drove the Iraqi army out of Kuwait in just four days. In the aftermath of the war, the Kurds in northern Iraq, and the Shiites in the south, rose up in revolt; with Saddam Hussein ruthlessly putting them down and barely managing to hold on to power.

(421) Saddam Hussein   *(422) Norman Schwarzkopf*   *(423) George H W Bush*

The First Chechen War between 1994 and 1996 was a conflict fought between the Russian Federation and the Chechen Republic of Ichkeria. After the initial campaign of 1994-95 culminating in the Battle of Grozny, Russian military forces attempted to seize control of the mountainous area of Chechnya but were held back by Chechen guerrilla warfare and raids on the flatlands, in spite of the overwhelming military superiority of the Russian forces. The resulting widespread demoralization of the Russian military, and the almost universal opposition of the Russian people to the conflict, led the Russian President Boris Yeltsin to declare a ceasefire in 1996, and sign a peace treaty a year later. The Second Chechen War was launched by Russia on 26th August 1999 in response to the invasion of Dagestan, and the Russian apartment bombings which were blamed on the Chechens. During the war Russian forces largely recaptured the separatist region of Chechnya. This military campaign signifi-cantly reversed the outcome of the First Chechen War, in which the region had gained de facto independence as the Chechen Republic of Ichkeria. The Russian President Vladimir Putin was given the credit for crushing the Chechen revolt; lifting his profile internationally and amongst the Russian people as a devout nationalist intent on restoring Russian power and prestige.

(424) A Map of the Former Yugoslavia

A major conflict on the doorstep of Europe were the Yugoslav Wars, between 1991 and 1995. The wars were preceded by the breakup of Yugoslavia beginning on 25th June 1991 after the republics of Croatia and Slovenia declared independence from Yugoslavia. The breakup of the country was facilitated by the death in 1980 of the longstanding leader Josip Broz Tito, who had held the country together through strong, decisive, and charismatic leadership since the end of the Second World War. A significant power vacuum occurred after his death, and a succession of presidents were desperate to keep

Yugoslavia unified as a single nation under the strong influence of Serbian nationalism, in spite of the diverse ethnic groups within the country. The Yugoslav Wars would become notorious for numerous war crimes and human rights violations, such as ethnic cleansing and genocide committed by all sides. Some of the major battles in the Yugoslav Wars included the Ten-Day War of 1991 between the Slovenian Territorial Defence forces and the Yugoslav People's Army following Slovenia's declaration of independence. The Croatian War of Independence between 1991 and 1995 fought in towns and villages within Croatia between the Croatian government, having declared independence from Yugoslavia, and both the Yugoslav People's Army and Serbian forces, which had established the self-governed Republic of Serbian Krajina within Croatia. The Bosnian War between 1992 and 1995 which involved several distinct ethnic factions within Bosnia and Herzegovina; including Bosnians, Serbs, and Croats. The Siege of Sarajevo marked the most savagely violent urban warfare ever seen in Europe since the Second World War at that time, as Serbian forces bombarded and attacked Bosnian controlled and populated areas of the city. War crimes occurred, including ethnic cleansing by mass murder and the destruction of civilian property. In 1995, the Croatian military were successful in defeating Serbian forces within their country, and this led to a mass-exodus of Serbian people from Croatia. One of the most prominent war crime trials involved ex-Serbian President Slobodan Milosevic, who in 2002 was indicted on 66 counts of crimes against humanity, war crimes, and genocide allegedly committed in wars in Kosovo, Bosnia, and Croatia. The trial was conducted at the International Criminal Tribunal for the former Yugoslavia, and lasted from February 2002 until his death in March 2006. He pleaded not guilty to all charges. Although no verdict was reached because of his death; in a judgment issued on 24th March 2016 in the separate trial of Serbian General Radovan Karadzic, the Tribunal said there was insufficient evidence in that case, that Milosevic had supported plans to expel non-Serbs from Serb-held territory in Bosnia during the 1992-95 War. In 1995, under the guidance of US President Bill Clinton, the Dayton Agreement was signed, which resulted in the internal partitioning of Bosnia and Herzegovina into the Republika Srpska, and a Bosnian-Croat Federation.

Finally, there was the Kosovo War between 1996 and 1999 between Albanian separatists, the Yugoslav military, and Serbian paramilitary forces in Kosovo. As a result of mass atrocities which took place in 1999, the North Atlantic Treaty Organization (NATO) led by the United States, launched air attacks against Yugoslavia, which was by then composed of only Serbia and Montenegro, in order to pressure the Serbian dominated Yugoslav government led by Slobodan Milosevic to end its military operations against Albanian separatists in Kosovo. The NATO intervention lacked UN approval, yet was justified by NATO based on accusations of war crimes being committed by Yugoslav military forces working alongside nationalist Serbian paramilitary groups. After months of intensive NATO bombing, Yugoslavia accepted NATO demands; resulting in NATO military forces and UN peacekeepers then occupying Kosovo.

One of the most tragic events of the 1990s was the Rwandan Genocide which occurred from April to July in 1994. It was the mass killing of hundreds of thousands of Rwanda's Tutsis ethnic group and Hutu political opponents, by the Hutu dominated government, under their Hutu Power ideology; the belief that all non-Hutus in Rwanda needed to be exterminated; another horrific act of ethnic cleansing. Over the course of around 100 days, up to 1,000,000 people, representing 20% of the total population were killed. This genocide was even more tragic because the international community, including the United Nations, did nothing to prevent it.

The release of African National Congress leader Nelson Mandela from Robben Island on 11th February 1990 after twenty seven years of imprisonment for opposing Apartheid and white-minority rule in South Africa, was another prominent event in the 1990s. Mandela's release triggered the first democratic elections in South Africa in 1994, resulting in the dismantling of Apartheid and the election of Mandela as President. The US President Bill Clinton was a dominant political figure in international affairs during the 1990s, known especially for his attempts to negotiate peace in the Middle East, and end the ongoing

I never lose.
I either **win** or **learn**.
Nelson Mandela

(425) Nelson Mandela was imprisoned for 27 years and then became President of South Africa in 1994

conflicts in Yugoslavia. Clinton also promoted international action to decrease human-created global warming; the end to Apartheid in South Africa; and advanced the cause of free trade around the world; including his substantial role in the enactment of the North American Free Trade Agreement (NAFTA) on 1st January 1994, creating a North American free trade zone between the United States, Canada, and Mexico. In spite of these achievements, Clinton was caught up in a media-frenzied scandal involving inappropriate sexual relations with a White House intern, Monica Lewinsky, first made public on 21st January 1998. After the United States House of Representatives impeached Clinton on 19th December 1998 for perjury under oath, following an intensive investigation by federal prosecutor Kenneth Starr, the Senate acquitted Clinton of the charges on 12th February 1999, and he finished his second term.

Israeli Prime Minister Yitzhak Rabin and Palestinian leader Yasser Arafat agreed to the Israeli-Palestinian peace process at the culmination of the Oslo Accords, negotiated by the United States President Bill Clinton on 13th September 1993. By signing the Oslo Accords, the Palestinian Liberation Organization recognized Israel's right to exist, while Israel permitted the creation of an autonomous governing Palestinian National Authority consisting of the Gaza Strip and West

*(426) Israel's Yitzhak Rabin, President Bill Clinton and PLO Leader Yasser
Arafat in 1993*

Bank, which was implemented in 1994. Israeli military forces withdrew
from these Palestinian territories in compliance with the accord. This
marked the end of the First Intifada. But tragedy struck when Israeli
Prime Minister Yitzhak Rabin was assassinated at a peace rally in
Tel Aviv on 4th November 1995, by a radical Jewish militant who
opposed the Oslo Accords. In July 1994, North Korean leader Kim
Il-Sung died, having ruled the country since its founding in 1948. His
son, Kim Jong-Il succeeded him, taking over a nation on the brink of
complete economic collapse. Famine caused a great number of deaths
in the late 1990s, and North Korea would gain a reputation for being a
large source of money laundering, counterfeiting, and weapons proliferation. The country's ability to produce and sell nuclear weapons
became a focus of concern in the international community.

The improvement in relations between the European NATO
countries and the former members of the Warsaw Pact ended the
Cold War both in Europe and throughout the world. Germany
reunified on 3rd October 1990 as a result of the fall of the Berlin
Wall, and after integrating the economic structure and provincial
governments, focused on modernizing the former communist East
Germany. People who were brought up in a socialist-communist
political system were immersed into the Western capitalist system.

(427) The fall of the Berlin Wall on 9th November 1989 ended The Cold War

Meanwhile, the Gorbachev restructuring of the Soviet Union was destabilized as nationalist and separatist demagogues gained popularity. Gorbachev's reforms were causing major inflation and economic chaos. Boris Yeltsin, then Chairman of the Supreme Soviet of Russia, resigned from the Communist Party and became the opposition leader against Mikhail Gorbachev. The Communist Party lost its status as the governing body of the Soviet Union, and was banned after a failed August 1991 coup attempt by communist hardliners who were hell-bent on reversing Gorbachev's progressive policies. Yeltsin's counter-revolution was victorious on 25th December 1991 with the resignation of Gorbachev from the presidency, and the formal disso-lution of the Soviet Union, when all of the constituent republics declared their independence. Boris Yeltsin then became the President of the successor Russian Federation, and presided over a period of intense turmoil, political unrest, economic crisis, and social anarchy. The Yeltsin led Russian Federation faced severe economic difficulty and widespread corruption. Oligarch's friendly to Yeltsin took over Russia's energy and industrial sectors, reducing almost half of the country to poverty. With a 3% approval rating, Yeltsin had to buy the support of the oligarchs to maintain his power. On 31st December 1999, after an almost complete economic collapse of the country, the ongoing devaluation of the Russian currency, and with heart and alcohol problems, Yeltsin resigned and appointed his protégé, Vladimir Putin as president. The Russian financial crisis in the 1990s resulted in mass hyperinflation and prompted economic intervention from the International Monetary Fund, and Western countries including the US, to assist in the recovery of the Russian economy.

In 1992 the European Union was formed under the Maastricht Treaty. This treaty was concluded between the 12 member states of the European Community (EC), and was the foundation treaty of the European Union (EU). It announced a new stage in the process of European integration; mainly relating to a shared European citizenship; open borders; the eventual introduction of a single currency; and common foreign, security, and economic policies. Against the background of the end of the Cold War and the reunification of Germany; and in anticipation of accelerated globalisation; the treaty negotiated tensions between member states seeking deeper integration, and states seeking to retain greater national control. The resulting compromise faced what was to be the first in a series of EU treaty ratification crises.

China entered the 1990s in a turbulent period, shunned by much of the world after the Tiananmen Square Massacre of 1989, and controlled by hard line politicians who reigned in private enterprise and attempted to revive old-fashioned anti-imperialist propaganda campaigns. Relations with the United States deteriorated sharply, and the Chinese leadership was further embarrassed by the disintegration of the Soviet Union and communism in Europe. In 1992, Chinese President Deng Xiaoping travelled to southern China in his last major public appearance to revitalize confidence in market economics, and to stop the country's slide back into Maoism. Afterwards, China recovered and would experience explosive economic growth during the rest of the decade. In spite of this, dissent continued to be suppressed, and President Jiang Zemin launched a brutal crackdown against the Falun Gong religious sect in 1999. Deng Xiaoping died in 1997 at the age of 93. He was to go down in history as the architect of China's massive economic modernization. Relations with the US deteriorated again in 1999 after the accidental bombing of the Chinese embassy in Serbia by NATO forces, which caused three deaths, and also allegations of Chinese espionage at the Los Alamos Nuclear Facility in the United States.

The 1990s was a revolutionary decade for digital technology. Between 1990 and 1997, individual personal computer ownership in the US rose from 15% to 35%. The cell phones of the early 1990s were very large, lacked extra features, had a short battery life, and were used by only a small minority of the population, even in the wealthiest countries. The first Internet web browser went online in 1993, and by 2001, over 50% of the people living in Western countries had Internet access, and over 25% had cell phones. Other scientific and technological advancements in the 1990s included:

(428) Desktop computers of the 1990s

- Computer modems leading to faster Internet connection.
- A Pentium microprocessor developed by Intel Corporation.
- Popularity of email.
- Businesses started to build e-commerce websites; and e-commerce-only companies such as Amazon, eBay, AOL, and Yahoo!, grew rapidly.
- The introduction of affordable, smaller satellite dishes and the DVB-S standard in the mid-1990s expanded satellite television services that carried up to 55 television channels; leading to the birth of cable television.
- A GSM network was launched in Finland in 1991.
- Digital cameras became commercially available.

- In 1998 Apple introduced the iMac all-in-one computer, initiating a trend in computer design towards translucent plastics and multicolour case design.
- CD burner drives were introduced.
- The DVD media format was developed and commercially popularized.
- Hand-held satellite phones were introduced.
- The 24-hour news cycle becomes popular with CNN's coverage of the Gulf War.
- Portable CD players had a profound impact on youth culture during the 1990s.
- Microsoft Windows becomes almost the exclusive platform on IBM compatible personal computers.
- Web browsers such as Netscape Navigator and Microsoft's Internet Explorer, made surfing the Internet easier and more user friendly.
- The opening of the Channel Tunnel between France and the United Kingdom more closely connected the UK to the European continent.
- The first exoplanets, orbiting other star systems, were detected.
- The first cloned mammal, Dolly the sheep, was achieved in the United Kingdom.
- The Human Genome Project began.
- DNA identification of individuals found widespread application in criminal law.
- The Hubble Space Telescope was launched in 1990 and revolutionized observational astronomy.
- NASA's spacecraft Pathfinder landed on Mars and deployed a small roving vehicle which analysed the planet's geology and atmosphere.
- Development of biodegradable products.
- The Global Positioning System (GPS) became fully operational.
- In 1998 construction of the International Space Station started.

At the beginning of the 1990s, sustainable development and environmental protection became serious issues for governments and the international community. In 1987, the publication of the Brundtland Report by the United Nations had paved the way to establish an international governance of environmental issues. In 1992, the Earth Summit was held in Rio de Janeiro, in which several countries committed to protect the environment, signing a Convention on Biological Diversity. The prevention of the destruction of the tropical rainforests of the world was a major environmental cause that first came into worldwide public concern in the early 1990s. The 1989 Chernobyl Disaster in the Soviet Union had a significant impact on public opinion at the end of the 1980s, and the fallout was still causing cancer-related deaths well into the 1990s. Throughout the decade, several non-government organizations (NGOs) helped improve environmental awareness among the public and governments. The most prominent of these NGOs during the 1990s was Greenpeace, which did not hesitate to launch legal actions in the name of environmental preservation. These NGOs also drew world attention to the large deforestation of the Amazon Rainforest. Global Warming as an aspect of Climate Change also became a major concern, and the creation of the United Nations Framework Convention on Climate Change (UNFCCC) after the Earth Summit helped coordinate efforts to reduce carbon emissions into the atmosphere. From 1995, the UNFCCC held annual summits on Climate Change, leading to the adoption of the Kyoto Protocol in December 1997, a binding agreement signed by several developed nations.

The 1990s was a decade that saw popular music marketing become more segmented, as MTV gradually shifted away from music videos beginning in 1992, and radio splinted into narrower formats aimed at different music niches. Some of these niches were grunge, gangsta-rap, hip-hop, rhythm and blues, teen-pop, Eurodance, electronic dance music, punk rock, and alternative rock. U2 was one of the most popular bands of the 1990s, with their ground breaking Zoo TV and Pop Mart tours becoming the top selling tours of 1992 and 1997. Some of the major fashion trends of the decade included:

- The Rachel, Jennifer Aniston's hairstyle on the hit television show Friends, became a cultural phenomenon with millions of women copying it worldwide.
- The Curtained Haircut increased in popularity in fashion and culture among teenage boys and young men in the 1990s; mainly after it was popularized in the film Terminator 2: Judgment Day, by the actor Edward Furlong.
- Bleached Blond hair became very popular.
- Slap bracelets were a popular fad among children, pre-teens and teenagers in the early 1990s, and were available in a wide variety of patterns and colours.
- The Grunge hype at the beginning of the decade popularized flannel shirts among both sexes.
- Hip-Hop and Grunge inspired anti-fashion saw slouchy, casual styles of clothing such as baggy jeans, cargo shorts, baseball caps (often worn backwards), chunky sneakers, and oversized sweatshirts.

(429) Characters from Friends

(430) Characters from Seinfeld

25

THE TECHNOLOGICAL REVOLUTION OF THE 21ST CENTURY – A NEW THRESHOLD?

A S THE WORLD HEADED TOWARDS THE END OF THE TWENTIETH century, a potential major problem loomed just over the horizon – the Y2K bug. This was a computer flaw, or bug, that may have caused problems when dealing with dates beyond 31st December 1999. When computer programs were being written during the 1960-70s, computer software engineers used a two-digit code for the year. The "19" was left out. Instead of a date reading 1980, it read 80. Engineers shortened the date because data storage in computers was costly and took up a lot of space. As the year 2000 approached, computer programmers realized that computers might not interpret 00 as 2000, but instead as 1900. Activities that were programmed on a daily or yearly basis would be damaged or flawed. As 31st December 1999 turned into 1st January 2000, computers might interpret the 31st December 1999 turning into 1st January 1900. Banks, which calculate interest rates on a daily basis, faced real problems. Instead of the rate of interest for one day, the computer might calculate a rate of interest for minus almost 100 years! Centres of technology, such as power generation plants, were also threatened by the Y2K bug. Power plants depend on routine computer maintenance for safety checks, such as water pressure or radiation levels. Not having the correct date would throw off the calculations and possibly put nearby residents at risk. Transportation also depends on the correct time and date. Airlines in particular were put at risk, as computers with records of all scheduled flights would be threatened, after all, there weren't any airline flights in 1900.

Y2K was both a software and hardware problem. Software and hardware companies raced to fix the bug and provided "Y2K compliant" programs to help. The simplest solution was the best; the date was simply expanded to a four-digit number. Governments, especially in the United States and the United Kingdom, worked hard to address the problem. In the end, there were very few problems. A nuclear energy facility in Ishikawa Japan, had some of its radiation equipment fail, but backup facilities ensured there was no threat to the public. The United States detected missile launches in Russia, and attributed that to the Y2K bug. But the missile launches were planned ahead of time as part of Russia's conflict in its republic of Chechnya. There was no computer malfunction. Countries such as Italy, Russia, and South Korea had done little to prepare for Y2K. They had no more technological problems than those countries like the United States, that spent hundreds of millions of dollars to combat the problem. Due to the non-occurrence of major disruption to computers around the world, many people dismissed the Y2K bug as a hoax, or an end-of-world cult.

It is appropriate to start this chapter on the twenty first century with the Y2K computer bug – a potential technological problem. As readers are aware, the Big History journey has eight thresholds of increasing complexity; with threshold number eight being the Modern Revolution which started with the advent of the Industrial Revolution from the mid-eighteenth century and proceeding into the twentieth century. I believe there should be a ninth threshold of increasing complexity in the Big History journey – the Technological Revolution. This revolution definitely started in the latter half of the twentieth century with scientific advances in the development of personal computers, the internet, digital technology, artificial intelligence, and biotechnology, creating the required goldilocks conditions. The acceleration of this technology though, has become prominent in the first twenty years of the twenty first century, and will continue to develop at an incredible pace throughout the remainder of this century and beyond. This is such a profound change in human collective learning that I believe it warrants a ninth threshold of increasing complexity. Whereas fossil fuels formed the foundation of the eighth threshold, the Modern Revolution; the silicon chip and the harnessing of electromagnetic energy (one of the four energy sources coming out of the Big Bang) forms the foundation of the Technological Revolution.

Let us now examine the major events and trends of the first twenty years of the twenty first century; starting with important geopolitical events. The first two decades of the twenty first century were marked with a series of dramatic events, some of which had been foreseen, but many of which came as major surprises. At the beginning of this century, the United States was the sole remaining superpower, after the collapse of the Soviet Union in the late twentieth century. During the period in time between the end of the Cold War and the rise of China, the United States enjoyed a singularly dominant position in the world. As the twenty first century has progressed, we have seen a progressive decline in the power of the United States, and the emergence of China as a leading world power. Likewise, we have seen power shift to the Pacific and Asian regions, and away from the Atlantic, Europe, and the West. These changes have ushered in a period of greater instability and uncertainty. I have isolated a number of major events that have collectively led to the global instability and uncertainty that exists, as I write this book in 2021.

The September 11 attacks, often referred to as 9/11, were a series of terrorist attacks on the homeland of the United States. On September 11, 2001, 19 militants associated with the Islamic extremist group al Qaeda hijacked four airplanes and carried out suicide attacks against targets in the United States. Two of the planes were flown into the twin towers of the World Trade Center in New York City; a third plane hit the Pentagon just outside Washington DC; and the fourth plane, on its way to either the White House or the Capitol Building in Washington DC, crashed in an open field in Shanksville, Pennsylvania, thanks to the bravery of a number of passengers who, when they realized what was about to happen, confronted the hijackers, bringing the plane down.

Almost 3,000 people were killed during the 9/11 terrorist attacks, which triggered major US retaliatory initiatives to combat terrorism, including attacks on Afghanistan and Iraq; and defined the presidency of George W. Bush. The terrorist attacks triggered the US led "War on Terror" culminating in the invasion of Iraq from 2003 to 2011; and the conflict in Afghanistan which started in 2002 and continues to this day. These conflicts proved to be greatly expensive distractions for the sole American superpower, costing trillions of dollars; and also highlighted two key themes that have dominated the early twenty first century; the volatility and unrest caused by terrorist groups which do not attach to any particular nation, and the proliferation of failed states in the world's poorest regions. The alleged mastermind of the 9/11 attacks, Osama bin Laden, was killed by US Navy SEALS in a surprise attack in Pakistan on 2nd May 2011, ordered by then President Barak Obama.

(431) The September 11, 2001 terrorist attack in New York City
shocked the world

Osama bin Laden was a Saudi Arabian citizen until 1994 and a member of the wealthy bin Laden family, which was involved in construction. He was born in Saudi Arabia and studied at university in that country until 1979, when he joined Mujahideen forces in Pakistan fighting against the Soviet Union in Afghanistan. He helped fund the Mujahideen by channelling arms, money, and fighters from the Arab world into Afghanistan, and gained popularity amongst many Arabs. In 1988 he formed al Qaeda; was banished from Saudi Arabia in 1992; and shifted his base to Sudan, until US pressure forced him to leave Sudan in 1996. After establishing a new base in Afghanistan, he declared a war against the United States, initiating a series of bombings and related attacks. To gain a better insight into Bin Laden's motives and ideology, including his hatred of the Bush family, here are some quotations attributable to him:

- Even as you enter the fourth year after the September 11 attacks, Bush is still misleading and deluding you and hiding the real reason from you.
- Your security is not in the hands of Kerry, Bush or al Qaeda. Your security is in your own hands.
- Bush the father did well in placing his sons as governors and did not forget to pass on the expertise in fraud from the leaders of the region to Florida to use in critical moments.
- This resemblance became clear in Bush the father's visits to the region. He wound up being impressed by the royal and military regimes and envied them for staying decades in their positions and embezzling the nation's money with no super-vision.
- I have sworn to only live free. Even if I find bitter the taste of death, I don't want to die humiliated or deceived.
- I support any Muslims, whether here or abroad.
- I'm fighting so I can die a martyr and go to heaven to meet God. Our fight now is against the Americans.
- We have repeatedly issued warnings, over a number of years. Following these warnings and these calls, anti-American explosions took place in a number of Islamic countries.
- We love death. The US loves life. That is the difference between us two.
- We declared jihad against the US government because the US government is unjust, criminal and tyrannical. It has committed acts that are extremely unjust, hideous and criminal, whether directly or through its support of the Israeli occupation.

- As long as I am alive, there will be no rest for the enemies of Islam. I will continue my mission against them.
- America has been hit by Allah at its most vulnerable point, destroying, thank God, its most prestigious buildings.
- Any country that steps into the same trench as the Jews has only itself to blame.
- Free men do not forfeit their security, contrary to Bush's claim that we hate freedom. If so, then let him explain to us why we don't strike Sweden, for example.
- We are continuing this policy in bleeding America to the point of bankruptcy. Allah willing, and nothing is too great for Allah.
- America is a great power possessed of tremendous military might and a wide-ranging economy, but all this is built on an unstable foundation which can be targeted, with special attention to its obvious weak spots. If America is hit in one-hundredth of these weak spots, God willing, it will stumble, wither away and relinquish world leadership.

Specifically responding to the September 11 attacks, Osama bin Laden said; "There is America, hit by God in one of its softest spots. Its

(432) The Leader of al Qaeda Osama bin Laden and
US President Barak Obama

greatest buildings were destroyed, thank God for that. There is America, full of fear from its north to its south, from its west to its east."

When China became a member of the World Trade Organization (WTO) in December 2001, few appreciated just how much this would transform the global economy. Not only did WTO membership give China increased access to key export markets, but it totally transformed entire industries as China quickly became the centre of global manufacturing. Today in 2021, China produces more cars than the United States, Japan, and Germany combined; and eleven times more steel than the United States. China's WTO membership not only set the stage for the massive expansion of its economy and triggered an economic miracle, but also put it into a raging trade war with the United States, as of 2021. It is forecast that China will overtake the United States as the world's largest economy, by 2028.

(433) Chinese President Xi Jinping has ruled the country since 2013

I believe that China has learnt the lessons of history. It does not seek to impose its political ideology on the world, like the United States and the Soviet Union did in the Cold War; but instead it is promoting a multipolar world in which China dominates economically and militarily. At the 19th convention of the Communist Party in October 2017, Chinese President Xi Jinping said that a "new era" had dawned for China and that the People's Republic was "getting closer to the centre of the world every day." But how does China imagine a world order in which it is at the centre stage? "My understanding is that the political forces in Beijing do not know exactly what they want. They're experimenting with Deng Xiaoping," says Gu Xuewu, a political scientist, referring to the Chinese leader who initiated substantial economic reforms which transformed China in the 1980s. Deng's famous motto was "crossing the river by feeling the stones." China's indecision is also reflected in complex debates about its role in the world, according to Volker Stanzel, a China expert and former German ambassador to Beijing. The debates deal as much with Beijing's acceptance of the prevailing global order as with the idea that China – chosen by fate – must lead the world. As varied and complex these discussions appear to be, the Communist Party of China (CPC) has the final say in all matters, which are not necessarily about the world order, according to Stanzel. "It is only a question about China being able to function in a way that the ideas of the CPC can be implemented, and which can help it stay in power." Despite these ambiguities, Gu says that some key elements of the "Chinese order" can still be identified. "China wants a world order that is politically multipolar, functionally multilateral and ideologically pluralistic." He explained China's ideals as follows:

- Multipolar – A world dominated by several power centres; comprising China, United States, Europe, Russia, and India.
- Multilateral – A world in which no country alone can determine the global agenda; instead it must be negotiated by all of the power centres.
- Pluralistic – The world must accept different forms of governance, including China's, and not just liberal democracy.

"We already live in a multipolar world," Gu said. Many political scientists agree that the short phase of American supremacy following the collapse of the Soviet Union no longer exists. In China, multilateralism has been linked to Xi's catchphrase, "the community of common destiny." In 2019, Xi rejected isolationism and positioned China as a supporter of multilateralism in a speech at the World Economic

(434) The ultramodern skyline of Shanghai in China

Forum in Davos, Switzerland. But China expert Stanzel is sceptical. He says; "Common destiny is empty talk; it needs to be defined. How do you want to organize the world? With more international laws and stronger global institutions? But I don't think that either China or the US are interested in this." Both China and the US however, give little value to international laws, according to Gu. He went on to say; "They accept them only if they suit their own interests. They reject them if they conflict with their interests." Beijing wants to improve its global image, with Chinese backed Confucius Institutes established around the world, promoting Chinese language and culture. At the same time, Chinese investors are buying media companies in an attempt to alter China's global perception. China is also actively participating in the United Nations. "China chairs four international institutions – twice as many as the US – and uses its position to include its political expression in UN documents," Stanzel said. However, the success of Chinese measures is uneven in different parts of the world. "The narrative is successful in Africa, more so in states that are economically tied to China," Stanzel added. China's image within the global community has been damaged due to its aggressive diplomacy and foreign policy.

(435) The Chinese People's Liberation Army

The detention and re-education of hundreds of thousands of Uighurs in Xinjiang, and the massive curbs on the freedom of Hong Kong citizens have further eroded China's reputation in Europe and the US.

The new Silk Road (Belt and Road Initiative), which was initially hailed as the world's largest infrastructure project, has also turned out to be a double-edged sword for many countries. Some critics say the project is making economically weak countries dependent on China. Sri Lanka, for example, had to lease a deep seaport in Hambantota to a Chinese company for 99 years after failing to pay back loans. It seems that in its endeavour to become a global superpower, China is first focusing on total dominance within its own "backyard" of Asia. President Xi talked about "Asia for the Asians" in a 2014 speech. But what does "Asia for the Asians" mean in the context of the Chinese world order? A 2010 statement to the Association of the Southeast Asian Nations (ASEAN) by then Chinese Foreign Minister Yang Jiechi was very revealing, when he stated; "China is a big country and other countries are small. It is simply a fact." Nowhere in Asia is China's display of power and influence more obvious than in the South China Sea. China is not only reclaiming islands whose sovereignty are in dispute and building military and intelligence gathering facilities on them; it is also trying to force the US out of the region; and bringing key shipping routes and the resources of the neighbouring countries in the region, under its control. The consequences for "peripheral states" and trading partners such as Vietnam, Philippines, Malaysia, Indonesia, Japan, South Korea, and Australia, are instability and the growing pressure to choose between China and the United States. There is also an increasing danger of an all-out military confrontation. The crucial question is whether the rise of China and the new alignments in the world order could lead to war. The clashes that China has triggered in the South China Sea could be a precursor to more aggressive Chinese military action such as a possible invasion of Taiwan; a country China does not recognize and claims to be its own territory? "The result will be an intense security competition with considerable potential for war. In short, China's rise is unlikely to be tranquil," says John J. Mearsheimer, an American political scientist, writing in the National Interest Journal in 2014. According to China specialist Ming Xia of the City University of New York; "I don't believe that China and the West can coexist if China maintains its own system." Will a dialogue help China and the West overcome their differences? Political analyst Gu says he is "very pessimistic" about dialogue succeeding. "I don't think the US is prepared to concede to the Chinese authoritarian government," he said. Stanzel is also sceptical, albeit for other reasons. "The Chinese Communist Party does not want to go to war, but China's aggressive behaviour is leading to the danger of a military conflict, as we have witnessed in the South China Sea," he said.

The Global Financial Crisis that began in 2008 ushered in the greatest economic crisis faced by the world since the Great Depression and the Second World War. The collapse of financial services firm Lehman Brothers on 15th September 2008, triggered a worldwide economic collapse, the effects of which are still felt today. As uncertainty reigned on Wall Street, banks virtually stopped lending to one another, stock markets plummeted, and multiple major financial firms either entered bankruptcy, or came very close to it. With global consumer demand falling at a rate not seen since the Great Depression of the 1930s, unemployment skyrocketed around the world, international trade volumes collapsed, and financial liquidity in the

marketplace became increasingly scarce. For developed economies
such as the United States and the European Union, this crisis proved
to be a key moment as it led to deep recessions, with some of these
economies still dealing with the savage impact of this crisis over a
decade later. For emerging economies, including China and India,
while many experienced severe downturns, this crisis also proved to
be an opportunity for them to play a much greater role in generating global economic growth, particularly within the emerging
markets of Asia. The root cause of the Global Financial Crisis was
greed. An official US government report released in 2011 stated the
cause as follows: "It was the collapse of the housing bubble – fuelled
by low interest rates, easy and available credit, scant regulation, and
toxic mortgages – that was the spark that ignited a string of events,
which led to a full-blown crisis." The Global Financial Crisis impacted
economies around the world, but in the United States, where it began;
unemployment jumped from 5% to 9.5%; around 8.7 million jobs
were lost; home prices fell by 30%; and the stock market S&P 500
index sharply declined by 57%.

(436) The 2008 G20 World Leaders Summit in Washington DC

As political leaders initially rushed to prop up the ailing global
financial system, government stimulus packages worth trillions of
dollars were urgently implemented within key global economies. In
particular, at the invitation of United States President George W. Bush,

the group of twenty (G20) economies assembled at the leaders' level for the first time ever in mid-December 2008. Focused with halting the global economic freefall, the G20 leaders proceeded to sign off on an ambitious package of policies at the G20 London Summit in April 2009 that called for collective stimulus injections, global economic reform, and a commitment to work towards a set of global financial rules that would reduce the chances of an international economic crisis happening again. To a large degree, the actions of the G20 economies helped to avert a worldwide economic depression. Although global economic growth has recovered to some extent since 2008, as at 2021 it is still much lower than pre-2008 trends, and the hangover from the crisis has manifested itself in the form of high unemployment levels throughout much of the developed and developing world, as well as an increasing level of wealth and income inequality both within and between countries.

(437) Mark Zuckerberg of Facebook   *(438) Steve Jobs of Apple*

Social media first emerged in the United States in the mid-1990s, but it was not until the later 2000s and the early 2010s that social media really took off around the world. This has led to two major trends that have dramatically shaped the world in recent years. One involves the ease of global communication and coordination brought about by the internet and social media; something that has played a major role in political protest movements around the world. The other is the challenge of "fake news" or the spread of false information by social media; a development that has influenced the results of a number of elections, facilitated the rise of toxic nationalists like Donald Trump, and shaken fragile democracies around the world. The biggest social media network is Facebook, which was launched in 2005, and had over 2.5 billion users worldwide in early 2021. The positive aspect of social media connecting the world; can be weighed against the misinformation, bullying, and loss of privacy it facilitates.

Beginning in Tunisia, a wave of anti-government protests swept across the Middle East and North Africa in 2010-11, bringing down long standing rulers in Egypt, Libya, and Tunisia; and leading to long-running civil wars in Yemen, Syria, and Libya. Unfortunately, the Arab Spring, as it was called, did little to bring more stability to that region, with the conflicts in each of the abovementioned countries continuing into 2021. For some, the failure of the Arab Spring was confirmation that democracy would struggle to thrive in the Middle East and North Africa, places that do not have a history of democracy, and that the diverse populations of the region would struggle to live together.

(439) The Arab Spring protests in 2011 sought to install democracy in the region

A trend rather than an event, was the decline in birth rates around the world over the first two decades of the twenty first century; a development that will have a profound impact on the future world population and global economy. In countries such as the United States, Germany, Italy, and Japan, where birth rates were already low at the beginning of the century, they have largely continued this downward trend. Moreover, some developing countries that had higher birth rates twenty years ago, including Bangladesh, India, and China, have seen their birth rates significantly decline in recent years, mainly as a result of modernization leading to higher standards of living, and government policies restricting the number of children families can have. This trend has led to worsening labour shortages in many of the world's leading economies in North America, Europe, and Japan. This is the primary reason why Germany has taken in around 1.3 million refugees, mainly from the Middle East.

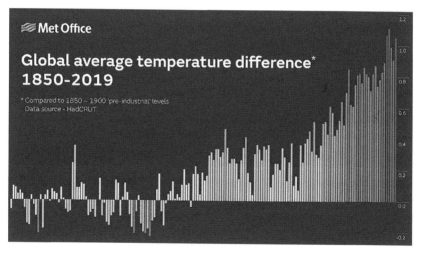

(440) The steep rise in average global temperature from 1850-2019

Global Warming caused by climate change has become a major existential crisis for the world in the early twenty first century. The global climate is increasing at an alarming rate. Each of the world's nine hottest years over the past 140 years have occurred during the twenty first century; with the five hottest years on record taking place in the five years from 2016 to 2020. As a direct result of Global Warming, significant weather events, including powerful storms and devastating droughts, have occurred with greater frequency so far this century. This awareness has motivated many governments, businesses, and individuals to take action to slow climate change, but resistance remains strong in some parts of the world; in spite of international conferences such as the Kyoto Protocol of 2005 and the Paris Agreement of 2015; and the widely acclaimed Al Gore documentary "An Inconvenient Truth" released in 2006.

In recent years, a series of migration crises have focused the world's attention on the fact that, while population growth is slowing in many regions of the world, in some of the world's poorest countries, populations are continuing to exponentially grow. Amongst the migration crises of the twenty first century has been two million Syrians fleeing their war-torn country; the Mediterranean crisis that has resulted in around two million North African and Middle Eastern people arriving in southern Europe between 2015 and 2020; the Rohingya crisis that led to nearly 800,000 people fleeing Myanmar for Bangladesh; and a migration crisis along the United States – Mexico border.

The United Kingdom's decision in 2016 to withdraw from the European Union shocked many people. The process of withdrawing the UK from the EU has proven to be a very drawn out convoluted mess, which could damage trans-channel ties for a long time to come. The United Kingdom has been the bridge linking North America and Europe, so Brexit could have major long term ramifications for the already delicate Trans-Atlantic relationship. The two most common arguments in favour of Brexit focused on the EU's liberal rules for internal migration and the EU's very invasive economic regulations. Boris Johnson, at the time the Conservative Mayor of London, wrote this typical case for the United Kingdom to leave the European Union, from an article where he focused on the increasing concentration of power in the hands of unelected EU bureaucrats in Brussels: "The more the EU does, the less room there is for national decision-making. Sometimes these EU rules sound simply ludicrous, like the rule that can't recycle a teabag, or that children under eight cannot blow up balloons, or the limits on the power of vacuum cleaners. Sometimes they can be truly infuriating – like the time I discovered, in 2013, that there was nothing we could do to bring in better designed cab windows for trucks, to stop cyclists being crushed. It had to be done at a European level, and the French were opposed." On this view, the EU isn't just too meddlesome, but it is also undemocratic and unaccountable to the public, according to many who voted for Brexit.

One of the most prominent critics of the EU's immigration rules was Nigel Farage, leader of the far-right UK Independence Party. He argued that large-scale migration of low-wage workers from elsewhere in Europe had depressed wages for native-born Britons. Farage also suggested that unrestricted immigration from Europe could lead to greater competition for government services and even put British women at greater risk of sexual violence. According to one poll, 73 percent of voters between age 18 and 24 voted to stay in the EU, compared to only 40 percent of voters over age 65. Unfortunately for the "Remain in EU" campaign, older voters turned out in greater numbers; so even though younger people were more pro-EU than older people, the older voters carried the day.

If you had told someone in the United States in the year 2000 that Donald Trump would become the President of the United States of America, you would have been labelled as being crazy. Instead, Trump rode a wave of protectionist, isolationist, and nationalist sentiment to win the 2016 US Presidential Election against Hillary Clinton. Over the four-year period he was president, Trump focused his efforts on "putting America first" to the detriment of relationships with foreign allies, and pulling the United States out of international agreements involving trade, arms reduction, and climate change. Here are some quotations from President Trump which give us a clearer idea of his ideology and state of mind:

- On Global Warming – "In the East, it could be the coldest New Year's Eve on record. Perhaps we could use a little bit of that good old Global warming that our country, but not other countries, was going to pay trillions of dollars to protect against. Bundle up!"

*(441) President Trump meeting North Korean Leader Kim Jong-Un
in Singapore 2018*

- On Hillary Clinton – "Crooked Hillary Clinton is the worst and biggest loser of all time. She just can't stop, which is so good for the Republican Party. Hillary, get on with your life and give it another try in three years!"

- On Kim Jong-un – "Why would Kim Jong-un insult me by calling me 'old,' when I would never call him 'short and fat?' Oh well, I try so hard to be his friend, and maybe someday that will happen!"

- On Meeting Hurricane Harvey Victims and Survivors – "What a crowd, what a turnout!"

- On the London Bridge Attacks – "Do you notice we are not having a gun debate right now? That's because they used knives and a truck!"

- On His Treatment by the Media – "Look at the way I have been treated lately, especially by the media. No politician in history, and I say this with great surety, has been treated worse or more unfairly."

- On the Presidency Being Difficult – "I loved my previous life. I had so many things going. This is more work than in my previous life. I thought it would be easier."

- On the Obama Presidency – "But I inherited a mess, I inherited a mess in so many ways. I inherited a mess in the Middle East, and a mess with North Korea, I inherited a mess with jobs, despite the statistics, you know, my statistics are even better, but they are not the real statistics because you have millions of people that can't get a job, ok. And I inherited a mess on trade. I mean we have many, you can go up and down the ladder. But that's the story. Hey look, in the meantime, I guess, I can't be doing so badly, because I'm president, and you're not. You know. Say hello to everybody OK?"

- On Repealing Obamacare – "We will immediately repeal and replace Obama Care, and nobody can do that like me. We will save dollars and have much better healthcare!"

- On Alleged Wiretapping by the Obama Administration – "Terrible! Just found out that Obama had my wires tapped in Trump Tower just before the victory. Nothing found. This is McCarthyism!"

- On Unfavourable News – "Any negative polls are fake news, just like the CNN, ABC, NBC polls in the election. Sorry, people want border security and extreme vetting."

- On Iran Missile Tests – "Iran is playing with fire, they don't appreciate how 'kind' President Obama was to them. Not me!"

- On Fake News – "As you know, I have a running war with the media. They are among the most dishonest human beings on Earth. They sort of made it sound like I had a 'feud' with the intelligence community. Nonsense, it is exactly the opposite, and they understand that too."

- On His Sexist View of Females – "This was locker room banter, a private conversation that took place many years ago. Bill Clinton has said far worse to me on the golf course, not even close. I apologize if anyone was offended."

- On Putin's Leadership – "If he says great things about me, I'm going to say great things about him. I've already said he is really very much of a leader. I mean, the man has very strong control over a country. And that's a very different system and I don't happen to like the system. But certainly in that system he's been a leader."

(442) President Trump and Russian President Vladimir Putin meeting in Helsinki 2018

- On His Popularity – "I could stand in the middle of Fifth Avenue and shoot somebody and I wouldn't lose any voters."
- On Banning Muslims – "Donald J. Trump is calling for a total and complete shutdown of Muslims entering the United States until our country's representatives can figure out what is going on."
- On Shutting Down Mosques – "Well I would hate to do it but it's something you're going to have to strongly consider, some of the absolute hatred is coming from these areas, the hatred is incredible. It's embedded. The hatred is beyond belief. The hatred is greater than anybody understands."
- On the late US Senator John McCain – "He's not a war hero because he was captured. I like people that weren't captured, okay? I hate to tell you."
- On the Mexican Wall – "I will build a great wall, and nobody builds walls better than me, believe me; and I'll build them very inexpensively. I will build a great, great wall on our southern border, and I will make Mexico pay for that wall. Mark my words."
- On Beating China – "Our country is in serious trouble. We don't have victories anymore. We used to have victories, but we don't have them. When was the last time anybody saw us

beating, let's say China in a trade deal? I beat China all the time. All the time."

- On Not Being Time Magazine Person of the Year – "I told you Time magazine would never pick me as person of the year despite being the big favourite. They picked a person who is ruining Germany." (German Chancellor Angela Merkel)

- On Being a Genius – "Sorry losers and haters, but my IQ is one of the highest, and you all know it! Please don't feel so stupid or insecure, it's not your fault."

- On Vaccination Vs Autism Debate – "Healthy young child goes to doctor, gets pumped with massive shots of many vaccines, doesn't feel good and changes – AUTISM. Many such cases!"

- On Climate Change – "The concept of global warming was created by and for the Chinese in order to make US manufacturing non-competitive."

- On Obama's Birth Certificate – "He may have one but there's something on that, maybe religion, maybe it says he is a Muslim. I don't know. Maybe he doesn't want that. Or he may not have one. I will tell you this – if he wasn't born in this country, it's one of the great scams of all time."

- On Being Rich – "Part of the beauty of me is that I am very rich."

The human race has always innovated as a result of its accumulated collective learning, and in a relatively short time went from using stone tools and building fires, to creating smartphone apps and autonomous robots. Today, technological progress will undoubtedly continue to change the way we live, work, enjoy ourselves, and survive in the coming decades. Since the beginning of the twenty first century, the world has witnessed the emergence of social media, smartphones, self-driving cars, and autonomous flying vehicles. There have also been huge leaps in energy storage, artificial intelligence, and medical science. Scientists have mapped the human genome, and are grappling with the ramifications of biotechnology and gene editing. As a species, we are facing immense challenges in global warming, overpopulation, and food security, among many other issues. While collective learning and human innovation has contributed to many of the problems we are now facing, it is also human innovation and ingenuity that can help humanity deal with these global challenges.

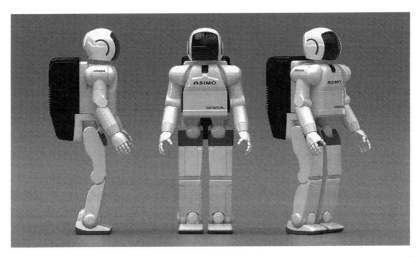

(443) The Honda Asimo domestic bipedal humanoid robot was first developed in 2000

In the first twenty years of the twenty first century scientists around the world have made remarkable progress towards understanding the human body, our planet, and the Universe that surrounds us. Science has become more global and collaborative than ever before in human history; to the extent that these days, major scientific discoveries and breakthroughs are more likely to come from the joint efforts of many scientists from around the word, rather than one or a small group. Let us examine some of the major scientific and technological achievements of the early twenty first century.

In 1916, Albert Einstein proposed that when objects with enough mass accelerate, they can sometimes create waves that move through the fabric of space and time like ripples on the surface of a pond. Though Einstein later doubted their existence, these space-time wrinkles, called gravitational waves, are a key prediction of Einstein's theory of relativity; and the search for them captivated and challenged astrophysicists for many decades. Though compelling hints of gravitational waves first emerged in the 1970s, nobody directly detected them until 2015, when the US based Laser Interferometer Gravitational-Wave Observatory (LIGO) detected the aftershock of a distant collision between two black holes. The discovery, announced in 2016, opened up an entirely new method of literally hearing the earliest events of the Universe. In 2017, LIGO and the European observatory, Virgo, detected another set of tremors, this time created when two ultra-dense objects called neutron stars collided. Telescopes around the world saw the related cosmic explosion, making this event the first ever observed in both light and gravitational waves. The data from this collision gave astrophysicists invaluable information on how gravity works at the intergalactic level; and how heavier chemical elements such as gold and silver are formed.

The early twenty first century saw numerous advances in the understanding of our complex human evolutionary story; including new dates established for known fossils, incredibly complete fossil skulls, and the addition of multiple new branches of "Homo" species. In 2010, paleoanthropologists unveiled the discovery of a distant human ancestor named Australopithecus sediba; an extinct species of Australopithecine recovered from the Malapa Cave, Cradle of Humankind, in South Africa. This species lived around 1.98 million years ago, and the fossil skeletons recovered were so complete that scientists were able to see what entire skeletons looked like, near the time when the Homo species first evolved. Also in 2010, a team announced that DNA extracted from an ancient Siberian finger bone was not that of a modern human, but instead the first piece of evidence for the existence of a mysterious lineage now called the Denisovans. In 2018, a site in China yielded 2.1 million years old stone tools, confirming that toolmakers spread into Asia hundreds of thousands of years earlier than previously thought. In 2019, scientists in the Philippines announced fossils of Homo luzonensis, the small-sized "hobbits" of the island of Flores. Also stone tools were found on the Indonesian island of Sulawesi which predated the arrival of modern humans, seeming to suggest the presence of a third, as yet unidentified island hominine in Southeast Asia.

As DNA sequencing technologies continued to improve exponentially, the early twenty first century saw huge leaps in our understanding of how our genetic past shaped modern humans. In 2010, scientists published the first near-complete genome from an ancient Homo sapien, triggering a revolutionary decade in the study of the DNA of our early human ancestors. Since then, more than 3000 ancient genomes have been sequenced, including the DNA of Naia, a girl who died in what is now Mexico, 13,000 years ago. Her remains are among the oldest intact human skeletons ever found in the Americas. Also in 2010, scientists announced the first sequence of a Neanderthal genome, providing the first solid genetic evidence that one to four percent of all modern non-African human DNA comes from these close relatives. In another fascinating discovery, scientists studying ancient DNA revealed in 2018 that a 90,000 years old bone belonged to a teenage girl whose mother was Neanderthal and father was Denisovan, making her the first hybrid ancient human ever found.

Our knowledge of exoplanets, those that orbit star systems other than our Sun, took a giant leap forward in the 2010s, largely thanks to NASAs Kepler Space Telescope. From 2009 to 2018, Kepler alone discovered more than 2700 confirmed exoplanets, more than half the current total. Among Kepler's greatest discovery was the first rocky exoplanet. Its successor TESS, launched in 2018, is extensively surveying the night sky and has continued to significantly add to the number of exoplanets discovered. In 2017, astronomers announced the discovery of TRAPPIST-1, a star system only 39 light-years away that hosts an amazing seven Earth-size planets, the most found around any star system. In 2016, the Pale Red Dot Project announced the discovery of Proxima b, an Earth-size planet that is orbiting Proxima Centauri, the closest star to our Sun, at a distance of 4.25 light-years. This closest alien planet to our Solar System is Earth-like in a number of respects. Proxima b is only 17% larger than our planet, making it similar in size to the Earth. Proxima b is "one of the most interesting planets known in the solar neighbourhood," according to Alejandro Suarez Mascareno, an astronomer. Because the planet orbits right in the centre of its star's habitable zone, it is entirely possible that liquid water, and potentially even life, could exist there. Due to its Earth-like mass, astronomers believe that, not only could liquid water exist on Proxima b, it could also be a rocky, terrestrial planet similar to our Earth. But Proxima b orbits around a star that, while close to our Solar System, is also much dimmer, and much less massive than our Sun. Researchers think that this exoplanet is tidally locked and in synchronous rotation with its star, meaning that one side is always facing the star, and one is always facing away; resulting in a light side and a dark side. In addition, it is unclear if Proxima b has an atmosphere. The planet is positioned very close to its star, completing one orbit every 11 Earth days. As a result, some astronomers believe that radiation originating from Proxima Centauri may have stripped away Proxima b's atmosphere, making it impossible for the alien planet's surface to retain liquid water.

The early twenty first century saw significant advances in our ability to precisely edit DNA, in large part thanks to the identification of the Crispr-Cas9 system. Some bacteria naturally use Crispr-Cas9 as an immune system, since it lets them store snippets of viral DNA, recognize any future matching virus, and then cut the virus's DNA to ribbons. In 2012, researchers proposed that Crispr-Cas9 could be used as a powerful genetic editing tool, since it precisely cuts DNA in ways that scientists can easily customize. Within months, other teams confirmed that this technique worked on human DNA. Ever since, genetic laboratories all over the world have been racing to identify similar systems, to modify Crispr-Cas9 to make it even more precise, and to experiment with its applications in agriculture and medicine. While Crispr-Cas9's potential is huge, the ethical dilemmas it poses are of great concern. To the horror of the global medical community, Chinese researcher He Jiankui announced in 2018 the birth of two girls whose genomes he had edited with Crispr-Cas9, the first humans born with heritable edits to their DNA. This announcement triggered widespread calls for a global moratorium on heritable "germline" edits on humans. This is just another example of the downside of scientific and technological progress – humans do not give enough consideration to the ethical and moral consequences of the negative effects of their discoveries on humanity as a whole.

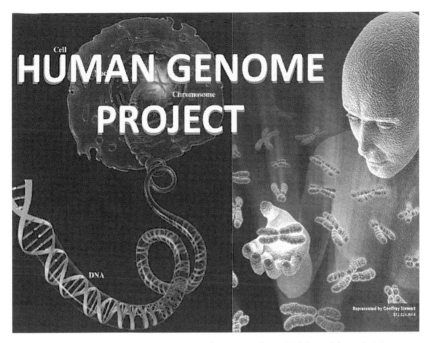

(444) The Human Genome Project first started in 1990 and lasted 13 years until 2003

So far in the early twenty first century, our understanding of the Universe has been revolutionized. In 2013, the European Space Agency launched Gaia, a spacecraft that is collecting distance measurements for more than a billion stars in our Milky Way galaxy, as well as velocity data for more than 150 million stars. This dataset enabled scientists to make a 3D movie of our home galaxy, giving us a spectacular unprecedented look at how galaxies form and change over time. In 2018, scientists released the final version of the Planck satellite's measurement of the early Universe's faint afterglow, which contains vital clues to cosmic ingredients, structure, and rate of expansion. Mysteriously, the expansion rate of the Universe Planck detected, is significantly greater than cosmologists had predicted; creating a potential "crisis in cosmology" that may require new astrophysics to explain, including a full understanding of what the "Dark Energy" is, which is driving the faster than expected expansion of the Universe. In 2019, scientists using the Event Horizon Telescope revealed the first ever image of a black hole's silhouette, as a result of a major global effort to explore deep into the heart of the galaxy M87, where this black hole was located.

For the first time early in the twenty first century, our spacecraft left the Solar System and ventured into interstellar space. Voyager 1 and 2 both achieved this extraordinary feat. Although launched in the 1970s, both of these spacecraft continue to send data to Earth. In 2017, astronomers observed "Oumuamua", the first object ever detected that formed in another star system and passed through ours. Oumuamua is believed to be an unusual type of asteroid, but it is so "odd-shaped" that some astronomers have speculated it could be an alien spacecraft. Because of its high speed of 87 kilometres per second, and the trajectory it followed as it whizzed around the Sun, scientists are confident it originated beyond our Solar System. The object flew by Earth so fast its speed could not have been due to the influence of the Sun's gravity alone, so it must have approached the Solar System at an already high speed, and not interacted with any other planets. On its journey past our star, the object came within a quarter of the distance between the Sun and Earth. In 2019, amateur astronomer Gennady Borisov found the first comet which originated from another star system. When Borisov gazed upward with his homemade telescope, he spotted an object moving in an unusual direction. Now called 21/Borisov, this runaway point of light turned out to be the first confirmed comet to enter our Solar System from some unknown place in interstellar space, beyond the gravitational influence of our Sun. Astronomers everywhere rushed to take a look with some of the most powerful telescopes in the world, hoping to learn as much as they could about the mysterious visitor. Astronomers believe that the comet could have formed around a red dwarf – a smaller, fainter type of star than the Sun – though other kinds of stars are also possible. Another theory suggests that comet 21/Borisov could be a carbon monoxide-rich fragment of a small exoplanet.

In 2015 NASA confirmed the existence of liquid water on Mars. Using the imaging spectrometer of NASA's Mars Reconnaissance Orbiter (MRO), scientists detected hydrated salts in different locations on the planet. During the warm season, the hydrated salts darken and flow down steeply; and in the cooler seasons they fade. The detection of hydrated salts means that water plays a vital role in their formation. In 2018, NASA announced that its Curiosity rover had found organic compounds on Mars, as well as a bizarre seasonal cycle in the red planet's atmospheric methane levels. The exploration of Mars was further enhanced by the arrival of NASA's Perseverance Spacecraft in February 2021, with the express mission to search for evidence of primordial life on the red planet. In the second decade of the twenty first century, space missions have given us a more sophisticated look at other world's carbon-based organic molecules, which are necessary ingredients for life as we know it. The European Space Agency's Rosetta mission orbited and landed on Comet 67P Churyumov-Gerasimenko. The data it collected between 2014 and 2016 gave us an astonishingly close look at the raw materials that ancient asteroid impacts might have brought to Earth; possibly triggering the emergence of life on this planet. Before NASA's Cassini probe burned up in the upper atmosphere of Saturn in 2017, it confirmed that the watery plumes of Saturn's moon Enceladus contain large organic molecules, a clue that it has the right ingredients for life.

In 2015, NASA's New Horizons probe made good on a decades-long quest to visit the icy dwarf planet Pluto, sending back the first-ever images of its incredibly unique surface. Then in 2019, New Horizons pulled off the most distant flyby ever attempted when it took the first photographic images of the icy asteroid "Arrokoth", a primordial leftover from the creation of the Solar System. Closer to home, in 2011 NASA's Dawn spacecraft arrived at Vesta, the second largest asteroid in the Asteroid Belt between Mars and Jupiter. After completing a detailed mapping of that world, Dawn proceeded to orbit the dwarf planet Ceres, the largest in the Asteroid Belt. In 2018, NASA's OSIRIS-Rex visited the asteroid Bennu; and the Japanese Space Agency's Hayabusa 2 visited the asteroid Ryugu; in both cases they collected samples which were returned back to Earth for astronomers to study.

In response to the 2014-2016 Ebola outbreak in West Africa, public health officials and the pharmaceutical company Merck fast-tracked rVSV-ZEBOV, an experimental Ebola vaccine. After a highly successful field trial in 2015, European officials approved the vaccine in 2019, a milestone in the fight against the deadly disease. Several landmark studies also opened new avenues to preventing the spread of HIV. A 2011 trial showed that preventatively taking antiretroviral drugs greatly reduced the spread of HIV among heterosexual couples, a finding confirmed in follow-up studies that included same-sex couples. In December 2019, reports emerged that a coronavirus never seen before in humans had begun to spread among the population of Wuhan, a large city in the Chinese province of Hubei. From there, the virus known as COVID-19, spread around the world, leading the World Health Organization (WHO) to declare it a pandemic in March 2020. COVID-19 went on to inflict significant deaths, extended lockdowns, and massive global economic disruption; to the point that as at the time of writing this book, in mid-May 2021, the number of confirmed cases was over 165 million, and the number of deaths was over 3.4 million people.

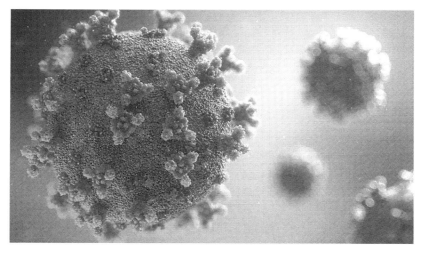

(445) The Covid-19 Virus first struck the world in late 2019

When scientists began seeking a vaccine for COVID-19 in early 2020, they were careful not to promise quick success. The fastest any vaccine had previously been developed, from viral sampling to approval, was four years, for the mumps in the 1960s. To hope for a vaccine by 2021 seemed highly optimistic. Nevertheless, by the start of December 2020, the pharmaceutical developers of several vaccines had announced excellent results in large clinical trials, with more showing promise. On 2nd December, a vaccine made by the drug giant Pfizer in conjunction with German biotech firm BioNTech, became the first fully-tested immunization in the world to be approved for emergency use. That speed of advance "challenges our whole paradigm of what is possible in vaccine development," says Natalie Dean, a biostatistician at the University of Florida. It's tempting to hope that other vaccines might now be made on a comparable timescale. Vaccines are desperately needed for diseases such as malaria, tuberculosis, and pneumonia, which collectively kill millions of people every year; and researchers are anticipating further lethal pandemics in the future too. The COVID-19 experience will almost certainly change the future of vaccine science, says Dan Barouch, director of the Center for Virology and Vaccine Research at Harvard Medical School in Boston, Massachusetts. "It shows how fast vaccine development can proceed when there is a true global emergency and sufficient resources," he says. New ways of making vaccines, such as by using messenger RNA (mRNA), have been validated by the COVID-19 response, he adds. "It has shown that the development process can be accelerated substantially without compromising on safety." The world was able to develop COVID-19 vaccines so quickly because of years of previous research on related viruses; faster ways to manufacture vaccines; enormous funding that allowed firms to run multiple trials in parallel; and regulators moving more quickly than normal. In the future, some of these factors might translate to other vaccine efforts, particularly speedier manufacturing platforms using cutting edge technology.

In 2016, clinicians announced the birth of a "three-parent baby" grown from the father's sperm, the mother's cell nucleus, and a third donor's egg that had its nucleus removed. The therapy, which remains ethically controversial, aims to correct for disorders in the mother's mitochondria. One 2018 study made precursors of human sperm or eggs out of reprogrammed skin and blood cells; while another showed that gene editing could let two same-sex mice conceive offspring.

Also in 2018, Chinese scientists announced the birth of two cloned macaque monkeys, the first time that a primate had ever been cloned, like Dolly the sheep was in 1996. Even though researchers worldwide have vowed that this technique will never be used to produce humans; how can we be sure when the genie is already out of the bottle?

(446) The CERN Large Hadron Particle Collider first went live in 2008

How does matter get mass? In the 1960s and 1970s, physicists including Peter Higgs and Francois Englert proposed a solution in the form of a novel energy field that permeates the Universe, now called the Higgs field. This theorized field also came with its associated fundamental particle, what's now called the Higgs boson. In July 2012, a decades-long search ended when two teams at CERN's Large Hadron Collider in Europe announced the detection of the Higgs boson. This discovery filled in the last missing piece of the Standard Model, the spectacularly successful - albeit incomplete – theory that describes three of the four fundamental forces in physics which came from the Big Bang, and all known elementary particles. The Large Hadron Collider (LHC) is the world's largest and most powerful particle accelerator. It first started-up in September 2008, and remains the latest addition to CERN's accelerator complex. The LHC consists of a 27 kilometre ring of superconducting magnets with a number of accelerating structures to boost the energy of the particles along the way. Inside the accelerator, two high-energy particle beams travel at close to the speed of light before they are made to collide. The beams travel in opposite directions in separate beam pipes – two tubes kept at ultrahigh vacuum. They are guided around the accelerator ring by a strong magnetic field maintained by superconducting electromagnets. In particle physics, colliders are used as a research tool. They accelerate particles to very high kinetic energies and let them impact other particles. Analysis of the by- products of these collisions gives scientists good evidence of the structure of the subatomic world, and the laws of nature governing it.

Throughout the early twenty first century, atmospheric carbon di oxide was reaching levels that are unprecedented in modern times, with record planetary temperatures to match. On 9th May 2013, global carbon di oxide levels in the atmosphere reached 400 parts per million for the first time in human history; and by 2016, the carbon di oxide levels were staying firmly above this critical threshold. As a result, the entire planet felt an increase in global warming; 2016, 2017, 2018, 2019, and 2020 were the five hottest years on record since measurements were first kept in 1880. Starting in 2014, warming oceans triggered a global coral bleaching event. Corals around the world, including the Great Barrier Reef on the east coast of Australia, suffered significant die-offs. In 2019, Australia declared the island-dwelling Bramble Cay melomys rats to be extinct, as a direct result of sea level rises, the first known mammal lost to modern climate change. In a series of major reports, the world's climate scientists forcefully called attention to the altered climate of the Earth, the risks it poses to all life including humanity, and the need to respond. In 2014, the Intergovernmental Panel on Climate Change (IPCC) released its fifth assessment of climate change's reality and consequences, and a year later, the nations of the world negotiated the Paris Agreement; a global climate accord that aims to keep warming below 2-degrees Celsius – which world leaders and scientists consider to be a dangerous threshold to breach – a tipping point of no return for the fragile planet. In 2018, the

IPCC published another grim report that outlined the huge costs of global warming, even with a 1.5-degrees Celsius rise by 2100; which is likely the minimum increase the planet will face. In the face of this monumental global challenge, a ground swell of climate protests has swept the world, many led by youth activists such as Greta Thunberg.

(447)   Greta Thunberg first started protesting outside the
Swedish Parliament in 2018

Who is Greta Thunberg and what is she trying to achieve? She is a teenager who grew up in Stockholm, Sweden. The elder of two girls, she says she learned climate change at school when she was eight, but that her parents were not climate activists. Greta has Asperger's Syndrome, a developmental disorder, and has described it as a gift; saying being different is a "superpower." In May 2018, aged 15, Greta won a climate change essay competition in a local newspaper. Three months later in August, she started protesting in front of the Swedish parliament building, vowing to continue until the Swedish government met the carbon emissions target agreed by world leaders in Paris, in 2015. She held a sign that read "School Strike for Climate" and began regularly missing school to go on strike on Fridays, urging students around the world to join her. Her protests went viral on social media and as support for her cause grew, other strikes started around the world, spreading with the hashtag #FridaysForFuture. By December 2018, more than 20,000 students around the world had joined her in countries including Australia, UK, Belgium, USA, and Japan. She joined strikes around Europe, choosing to travel by train to limit her carbon footprint on the environment. The teenager took the whole of 2019 off school to continue her campaigning, to attend climate conferences, and to join student protests around the world. In September 2019, she travelled to New York to address a UN climate conference. Refusing to fly, she made her way there on a racing yacht, in a journey that lasted two weeks. When she arrived, millions of people around the world took part in a climate strike, underlying the scale of her global influence. Addressing the UN conference, she blasted politicians for relying on young people for answers to climate change. She said: "How dare you? I shouldn't be up here. I should be back in school on the other side of the ocean, yet you all come to us young people for hope. How dare you?" She was named Time Magazine's Person of the Year for 2019. Greta believes big governments and businesses around the world are not moving quickly enough to cut carbon emissions and avert a monumental global environmental catastrophe. She has consistently attacked world leaders for failing young people – the future generations. At the 2020 World Economic Forum in Davos, Switzerland, Greta called for banks, companies, and governments to stop investing in, and subsidising fossil fuels, such as oil, coal, and gas. "Instead, they should invest their money in existing sustainable technologies, research, and restoring nature," she said. Millions of young people around the world have been inspired by her speeches and strikes, and Greta has received support from climate activists, scientists, world leaders, and even the Pope, who told her to continue her work. The broadcaster and naturalist Sir David Attenborough told her she had achieved things many others have failed to do, adding: "You have aroused the world. I'm very grateful to you." But her message has not been well received by everyone. After her UN appearance in September 2019, then US President Donald Trump appeared to mock her by saying she "must work on her anger management problem." Greta then changed her Twitter biography to include Trump's words. She did the same weeks later when Russian President Vladimir Putin called her a "kind but poorly informed teenager." In January 2020, the US Treasury Secretary Stephen Mnuchin told the teenager to go away and study economics before lecturing investors.

The early twenty first century witnessed the advent of privately owned space technology companies, with the prime objectives of opening up space for tourism; continuing the exploration of space; and creating commercial opportunities in space, including launching of satellites and ultimately harnessing natural resources. The three most prominent were SpaceX; Blue Origin and Virgin Galactic. SpaceX (Space Exploration Technologies Corporation) is an American aerospace manufacturer and space transportation services company. It was founded in 2002 by Elon Musk with the prime goal of reducing space transportation costs to enable the ultimate colonization of Mars. SpaceX has developed several launch vehicles and rocket engines, as well as the Dragon cargo space-craft, and the Starlink satellite constellation, providing internet access. It has also flown humans and cargo to the International Space Station. As of February 2021, SpaceX has flown 21 cargo resupply missions to the International Space Station under a partnership with NASA, as well as 2

(448) Jeff Bezos, Richard Branson and Elon Musk – Space Pioneers

manned flights on the Dragon 2 spacecraft. Blue Origin is an American aerospace manufacturer and sub-orbital spaceflight services company founded in 2000 by Jeff Bezos. Its primary aim is to make access to space cheaper and more reliable through the use of reusable launch vehicles. Blue Origin is employing an incremental approach from suborbital to orbital flight, with each development step building on its prior achievements. The company motto is "Gradatim Ferociter", which is Latin for "Step by Step, Ferociously." Blue Origin is developing a variety of technologies, with a focus on rocket-powered vertical take-off (VT) and vertical landing (VL) vehicles for access to suborbital and orbital space. The company's name refers to the blue planet Earth, as the point of its origin. Virgin Galactic is an American spaceflight company focused on developing commercial spacecraft with the prime intention of providing suborbital spaceflights to space tourists, and suborbital launches for space science missions. SpaceShipTwo, Virgin Galactic's suborbital spacecraft, is air launched from beneath a carrier airplane known as White Knight Two. Virgin Galactic's founder, Richard Branson, had originally hoped to see a maiden flight by the end of 2009, but the date was delayed until December 2018, when finally, VSS Unity VP-03 achieved the company's first suborbital spaceflight with two pilots reaching an altitude of 82.7 kilometres, officially entering outer space. In February 2019, Virgin Galactic carried three people, including a "space tourist" passenger, on VSS Unity VF-01, with a member of the team floating within the cabin during a spaceflight that reached a height of 89.9 kilometres.

Some of the more prominent inventions in the early twenty first century included:

- 3D Printing – Most inventions are developed as a result of previous ideas and concepts; 3D printing is no exception. The earliest application of the layering method used by today's 3D printers took place in the manufacture of topographical maps in the late 19th century; and 3D printing as we know it first began in 1980. The convergence of cheaper manufacturing methods and open-source software however, has led to a revolution of 3D printing in recent years. Today, the technology is being used in the production of everything from

(449) 3D Printing Technology can be used to self-produce a wide variety of products

lower-cost car parts, to bridges, to less painful ballet shoes, and is even being considered for artificial organs.

- E-cigarettes – While components of this technology have existed for decades, the first modern e-cigarette was introduced in 2006. Since then, the devices have become wildly popular as an alternative to traditional cigarettes, and new trends, such as the use of flavoured juice, have enhanced their success. Recent studies have shown that there remains a great deal of uncertainty and risk surrounding these devices, with an increasing number of deaths and injuries linked to e-cigarette vaping. In early 2020, the US based Food and Drug Administration (FDA) issued a widespread ban on many flavours of cartridge-based e-cigarettes, in part because those flavours are especially popular with children and younger adults.

- Augmented Reality – In which digital graphics are overlaid onto live footage to convey information in real time, has been around for a while. It is only recently however, following the arrival of more powerful computing hardware, and the creation of an open source video tracking software library known as ARToolKit that the technology has really taken off. The technology is being adopted as a tool in manufacturing, health care, travel, fashion, and education.

- Blockchain – the simplest explanation of blockchain is that it is an incorruptible way to record transactions between parties; a shared digital ledger that parties can only add to and that is transparent to all members of a peer-to-peer network where the blockchain is logged and stored. The technology was first used in 2008 to create Bitcoin, the first decentralized crypto-currency; but it has since been adopted by the financial sector and other industries for a myriad of uses; including money transfers, supply chain monitoring, and food safety.

- Capsule Endoscopy – Advancements in light emitting electrodes, image sensors, and optical design in the 1990s led to the emergence of capsule endoscopy, first used in medical patients in 2001. The technology uses a tiny wireless camera the size of a vitamin pill that the patient swallows. As the capsule traverses into the digestive system, doctors can examine the gastrointestinal system in a far less intrusive manner. Capsule endoscopy can be used to identify the source of internal bleeding, inflammations of the bowel, ulcers, and cancerous tumours.

- Modern Artificial Pancreas – More formally known as a closed-loop insulin delivery system, the artificial pancreas has been around since the late 1970s, but the first versions were the size of a filing cabinet. In recent years, the artificial pancreas, used primarily to treat type 1 diabetes, has become much more portable. The first modern portable artificial pancreas was approved for use in the United States in 2016. The system continually monitors blood glucose levels, calculates the amount of insulin required, and automatically delivers it through a small pump. British studies have shown that patients using these devices spent more time within their ideal glucose-level range. In December 2019 the FDA approved an even more advanced version of the artificial pancreas, called Control-IQ.

- E-readers – Sony was the first company to release an e-reader using a microencapsulated electrophoretic display, commonly referred to as e-ink. E-ink technology, which mimics ink on paper that is easy on the eyes and consumes less power, has been around since the 1970s, but the innovation of e-readers had to wait until after the broader demand for e-books emerged. Sony was quickly overtaken by Amazon's Kindle after its 2007 debut.

- High-density Battery Packs – Tesla electric cars have received much attention largely because of their batteries. The batteries, located underneath the passenger cabin, consist of thousands of high-density lithium ion cells, each barely larger than a standard AA battery, nestled into a large, heavy battery pack that also offers Tesla electric cars a road-gripping low centre of gravity and structural support. These battery packs are also being used in residential, commercial, and grid-scale energy storage devices.

- Digital Assistants – One of the biggest technology trends in recent years has been smart home technology, which can now be found in everyday consumer devices like door locks, light bulbs, and kitchen appliances. The key piece of technology that has helped make all of this possible is the digital assistant. Apple was the first major technology company to introduce a virtual assistant called "Siri" in 2011, for its smartphones. Other digital assistants, such as Microsoft's Cortana, and Amazon's Alexa, have since entered the market.

- Robotic Heart – Artificial hearts have been around for some time. They are mechanical devices connected to the actual heart or implanted in the chest to assist or substitute a heart that is failing. Abiomed, a US medical technology company, developed a robotic heart called AbioCor, a self-contained apparatus made of plastic and titanium. It is a self-contained unit with the exception of a wireless battery pack that is attached to the wrist. The first successful robotic heart was received by a patient with congestive heart failure in July 2001.

- Retinal Implant – When he was a medical student, Doctor Mark Humayun watched his grandmother gradually lose her vision. The ophthalmologist and bioengineer focused

on finding a solution to what causes blindness. He collab-
orated with Doctor James Weiland, and other experts to
create the Argus II. The Argus II is a retinal prosthesis device
that is considered to be a breakthrough for those suffering
from retinitis pigmentosa, an inherited retinal degenerative
condition that can lead to blindness. The condition afflicts 1.5
million people worldwide, and was first used in 2013.

(450) A US Army Exoskeleton being trialled in 2019

- Robotic Exoskeletons – Ever since 2013 when researchers
  at the University of California created a robotic device that
  attaches to the lower back to augment strength in humans;
  the demand for robotic exoskeletons for physical rehabilitation
  has increased. Wearable exoskeletons are increasingly helping
  people with mobility issues such as lower body paralysis; and
  are also being used in factories. Ford Motor Company, for
  example, has used an exoskeleton vest that helps auto assem-
  blers with repetitive tasks in order to lessen the wear and tear
  on the shoulders and arms.
- Multi-use Rockets – Elon Musk's private space exploration
  company Space X, has developed rockets that can be recovered

and reused in other launches; a more efficient and cheaper alternative to the method of using rockets once only, and letting them fall into the ocean. On 30th March 2017, Space X became the first to deploy one of these reusable rockets, the Falcon 9. Blue Origin, a space-transport company founded by Amazon's Jeff Bezos, has also launched its own reusable rocket.

- Small Satellites – As modern electronics devices have become smaller, so too, have orbital satellites, which companies, governments, and scientific institutions, have used to gather scientific data, collect images of the Earth, for telecommunications, and also for intelligence purposes. These tiny, low-cost orbital devices fall into different categories by weight, but one of the most common is the shoe-box sized CubeSat. As of October 2019, over 2400 satellites weighing between 1 kg and 40 kg have been launched into space.

- Solid-state Lidar – An acronym that stands for light detection and ranging, and is also a combination of the words "light" and "radar." This technology is most often used in self-driving cars. Like radar, which uses radio waves to bounce off objects and determine their distance; lidar uses a laser pulse to do the same. By sending enough lasers in rotation, it can create a constantly updated high-resolution image map of the surrounding environment.

- Touchscreen Glass – Super-thin, chemically strengthened glass is the key component of the touchscreen world. This sturdy, transparent material is what helps an iPhone or Samsung smartphone from shattering into pieces at the slightest drop. Corning Inc., already a leader in the production of treated glass used in automobiles, was asked by Apple to develop a 1.3 mm treated glass for its iPhone, which debuted in 2007.

- Quantum Computers – A quantum computer harnesses some of the almost mystical phenomena of quantum mechanics to deliver huge leaps forward in processing power. They won't replace conventional computers. Using an existing computer will still be the easiest and most economical solution for tackling most day-to-day problems. But quantum computers promise to power exciting advances in various fields such as materials science, pharmaceutical research, nanotechnology, biotechnology, artificial intelligence, and astrophysics. Companies are already experimenting with them to develop things like lighter and more powerful batteries for electric cars, and to help create innovative cutting-edge pharmaceutical drugs. The secret to a quantum computer's power lies in its ability to generate and manipulate quantum bits, or qubits. Today's computers use bits – a stream of electrical or optical pulses representing 1s or 0s. Everything from your tweets and emails to your iTunes songs and YouTube videos are essentially long strings of these binary digits. Quantum computers on the other hand, use qubits, which are typically subatomic particles such as electrons or photons. Generating and managing qubits is a scientific and engineering challenge. Some companies, such as IBM, and Google, use superconducting circuits cooled to temperatures colder than deep space. Others, like IonQ, trap individual atoms in electromagnetic fields on a silicon chip in ultra-high vacuum chambers. In both cases, the goal is to isolate the qubits in a controlled quantum state.

(451) Quantum Computing will deliver huge leaps forward in mega-processing power

26

THE FUTURE

PREDICTING THE FUTURE OF HUMANITY, AND OUR TECHNOLOGICAL advancements in the coming decades, let alone over the next hundreds or thousands of years is virtually impossible, with any degree of precision. There are just too many unknown variables. Amazingly though, we have a fair idea about the future of our Solar System, the Milky Way galaxy, and the Universe.

As we have seen on this monumental journey of Big History, it took 13.8 billion years of cosmic evolution to get to where we are today. Generations of stars had to live and die to create the heavy elements required for our Solar System; small proto-galaxies had to merge together to create the Milky Way; interstellar gas clouds had to collapse and form new stars with rocky planets around them; complex inorganic and organic chemistry needed to take hold on one of those newly formed worlds; biological evolution underwent a very particular path, finally culminating in the emergence of human beings around 300,000 years ago. Over the past 12,000 years we have seen how agriculture, science, the concept of nations, and all of modern civilization with its technology as we know it today, has developed.

Scientists know that things won't remain as they are today, in the future of our planet. A number of terrestrial events are going to happen that will change the Earth, making it hardly recognizable to someone who is alive today. In around 60,000 years from now for example, the Sun and stars will have moved enough relative to the Earth so that the current constellations will be scrambled and vastly different from how we currently see them in the night sky. Another 100,000 years after that, we are due to experience the next Ice Age; and before the next million years is over, the Yellowstone Super-volcano will likely erupt, and change the landscape of the Earth forever. In addition to these phenomena, there is always the ever-present threat of an asteroid

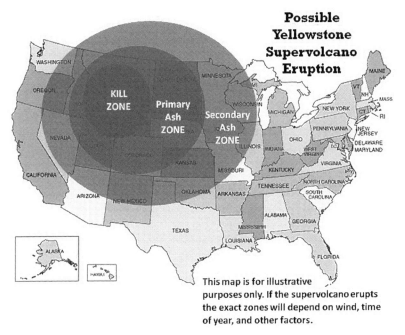

This map is for illustrative purposes only. If the supervolcano erupts the exact zones will depend on wind, time of year, and other factors.

(452) The potential impact of the Yellowstone super-volcano

impact, similar to the one which crashed into the Earth 65 million years ago, wiping out all of the dinosaurs in the process. But all of this is quite insignificant, compared to what the Solar System, Galaxy, and Universe have in store for us. Starting in a little under 4 billion years from now, the Andromeda Galaxy will merge with our own Milky Way, creating a spectacular change to our galaxy's structure, and to the night sky in general. Currently 2.5 million light years away but moving towards us at a speed of 43 kilometres per second, our best supercomputer simulations indicate that the first collision and burst of star formation will happen in 3.8 billion years, and that a galactic merger will be complete after 5.5 billion years. Gravitation will cause the entire local galactic group to eventually merge with us, forming one giant elliptical galaxy, Milkdromeda, of which our Solar System will still be a part of. On larger cosmic scales, all of the other galaxies will continue to accelerate away from us, and eventually, after around 100 billion years, they will recede from our Earth-view entirely.

When it comes to the future of our Solar System, the Sun will continue to get hotter as it ages, boiling our oceans in around 1.5–2 billion years, and ending life on Earth as we know it. Eventually, in around 5-6 billion years from now, the Sun's core will extinguish all of its nuclear fuel, which will cause our parent star to become a Red Giant, completely engulfing Mercury and Venus in the process. Due to the process of stellar evolution, by then the Earth will be pushed outwards, and be spared the fate of the two most inner planets. After burning through its remaining nuclear fuel, mostly the helium in its core, the Sun will expel its outer layers to form a planetary nebula, and the core of our star will then massively contract to become a white dwarf. This is the eventual fate of virtually all of the stars on the main sequence in our Universe. But the planets will still be here, continuing to orbit our cold dim stellar remnant.

While star formation will continue, dying stars will emit their fuel into interstellar space, and failed stars will spiral in and merge together. The amount of material for making stars is finite. Even the longest-lived stars will only last for around 100 trillion years, and after about one quadrillion years, star formation in the Universe will cease entirely. Only the occasional collision or merger between failed stars or stellar remnants will provide light to our galaxy, as the last stellar remnants cool and fade into darkness. Eventually, white dwarf stars will go black, as they cool and radiate their energy away. This will take a very long time; about a million times the present age of the Universe. The atoms will still be there, but they will be just a few degrees above absolute zero. At this point, the entire night sky will be truly dark and black, with no visible photon light at all, as all the stars in our local galactic group will have burned out. At some point, the now black dwarf at the centre of our Solar System will randomly collide with another black dwarf, producing a Type 1a Supernova explosion, and effectively destroying what is left of our Solar System. After the black hole in the centre of what will then be Milkdromeda, decays, only dark matter will remain, meaning the Earth will spiral into the black dwarf that was once our Sun.

The ultimate fate of the Universe will depend on the amount of mass contained within it, that is, the "mean density" of the Universe. There is a "critical density", which if reached will be sufficient to just stop the current expansion of the Universe. This is similar to a ball thrown in the air; the Earth's gravity eventually stops the ball, and it begins to fall back down to the ground. There currently is considerable uncertainty amongst cosmologists in our knowledge of the amount of matter (and composition of dark matter) in the Universe; so as a result, two major possible outcomes for the future of the Universe exist. One scenario is that the Universe will continue to expand forever. If the mean density is less than the critical density, then there will be insufficient mass within the Universe to stop the ongoing expansion; so the Universe will expand forever. Ultimately, the galaxies will move increasingly further apart. Star formation within the galaxies will eventually cease as the star forming material is exhausted. Under this scenario, the Universe will slowly cool to absolute zero, and it will exist in total darkness, devoid of galaxies, stars, and planets. The second scenario involves the Universe eventually collapsing in on itself. If the mean density exceeds the critical density, then the expansion of the Universe will eventually cease, and it will begin to contract. Eventually, a fireball like that which initiated its expansion in the Big Bang, will occur. This is referred to as the "Big Crunch." If this were to happen, it is estimated it would take 50 billion years from now, at least ten times the remaining life of our Sun. It may turn out that the Universe expands and contracts in a cyclical way and may indeed, have been doing so in the past. This is called the "Oscillating Theory of the Universe."

What will the future of our planet Earth and humanity be like? Outlined here is one of many likely scenarios. The 2030s could be marked by a rapid, worldwide shift towards clean renewable energy, including algae-based biofuel, aided by significant breakthroughs in nanotechnology. This could then be followed by progress in the use of hydrogen and nuclear fusion as sources of energy. It is likely that portions of Africa, the Middle East, and Asia in particular, will suffer from acute food shortages, and a growing influx of refugees affected by climate change, resource wars, and political instability. Exponential advancements in computing power, in direct parallel with genetics, nanotechnology, and robotics, will continue into the 2040s, leading to what many call the singularity, and the birth of "transhumanism." The singularity will be achieved when artificial intelligence exceeds human intelligence, including self-consciousness and awareness.

Transhumanism is a philosophical movement with the aim of breaking through the biological barriers of humans, through science and technology. Examples of this are the expansion of physical and cognitive abilities, as well as the pursuit of immortality. Supporters of transhumanism believe that using technology to improve ourselves is what makes us human. Opponents on the other hand are particularly afraid of the abuse of power by technology corporations or governments, and the loss of humanity by the surrender of mankind to technology. Some of the forms and methods of transhumanism will include cryogenic suspension, mind uploading, and super intelligence.

(453) Cryogenic Suspension Tanks will be in widespread use during the 21st century

Cryogenic suspension will accelerate by the 2040s. It is a process of freezing and storing the body of a diseased, recently deceased person to prevent tissue decomposition so that at some point in the future the person will be brought back to life upon the development of new medical cures. Ever smaller, more complex and sophisticated devices will become implantable and integrated within the human body; able to combat disease, enhance the senses, and provide entertainment or communication in ways that were simply not possible before. In the 2040s, geopolitics will undergo a revolution too, with India likely to surpass the United States in economic power, and even threatening to overtake China in the near future.

(454) A permanent colony on Mars will eventuate during the 21st century

In the 2050s mankind is likely to have escaped the confines of our crowded home planet and establish permanent colonies on the Moon and Mars. Even greater advances in computing power will see artificial intelligence (AI) beginning to play a major role in business and government decision making. By this decade however, global economic growth is likely to be under severe strain due to ecological impacts, acute resource scarcity, demographic trends, and technological unemployment. By the 2060s the world's population is expected to level off and plateau. This will be partly as a result of declining fertility rates, aided by improvements in education and birth control, but also from significant numbers of deaths caused by deteriorating environmental conditions. Entire nations will now likely be devastated by the effects of climate change. Despite rapid advancements in technology, the fundamental problem will remain that humanity will be consuming too much, too fast, and beyond what the Earth can sustainably provide. Desperate attempts will be made to improve carbon capture and geoengineering methods, but the sheer magnitude of this crisis will persist for decades to come. Carbon capture technology will be developed which will capture and store carbon di oxide before it is released into the atmosphere.

The 2070s will see major growth in the use of fusion power. This will be a form of power generation in which electricity will be produced by using heat from nuclear fusion reactions. In a fusion process, two lighter atomic nuclei combine to form a heavier nucleus, while releasing energy in the process. Fusion reactors will be operational in order to harness this energy. There will also be accelerated space exploration and development during this decade, including the expansion of lunar colonies and their automated robotic mining operations. By this time, there may well be a full-scale environmental catastrophe unfolding on the Earth, with ever-rising sea levels forcing the widespread evacuation of cities. With exponential advances in AI, the 2080s will see an explosion in scientific discoveries. These will help slow the rise in global temperatures and pave the way for a more sustainable future in the 22nd century. Transhumanism will now be a mainstream phenomenon, with the average human becoming heavily reliant on brain-computer interfaces and other implantable devices.

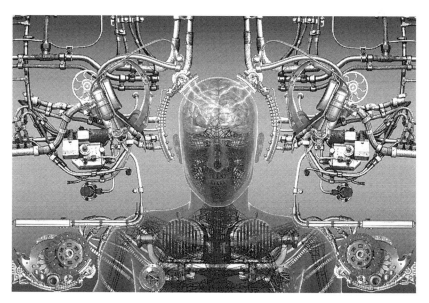

(455) Brain-Computer Interfaces will become a reality

By the 2090s it may become clear that Homo sapiens are no longer the dominant species on the planet; with much of the day-to-day running of global affairs now exclusively in the hands of ultra-fast, ultra-intelligent machines, robots, and virtual entities.

Some of the likely developments of the remaining 21st century may include:

- The development of a vaccine for the elimination of malaria, which is one of the world's leading killer diseases. In 2015, 212 million cases of malaria occurred globally, leading to 429,000 deaths, most of which occurred in children under the age of five years, in Africa.

- Gene therapy for deafness, which will be based on the delivery of genetic material into the cells of the inner ear. This genetic material will replace the genetic defect causing deafness, and will enable the hearing cells to function normally.

- Driverless hovering cars will become commonplace as a result of advances in battery energy density, materials science, and computer simulation. These flying cars will be able to offer faster commuting time, particularly in heavily populated cities, with significant traffic congestion.

(456) Driverless hovering cars are being developed

- Manned missions to Mars will commence in the 2030s; led by NASA and SpaceX. The one-way trip will take between eight to eleven months, depending on where Mars and Earth are located in their respective orbits around the Sun. Later in this century, permanent human colonies will be established on Mars.

- Driverless high speed trains will replace traditional manned trains. The automatic driving system will have complete control of the departure, movement between stations, automatic and precision stopping of the train, and the opening and closing of doors. In the case of high passenger volume, additional trains will be automatically deployed at literally the touch of a bottom. These trains will become much faster and safer than conventional trains.

- Quantum computers will accelerate the development of new breakthroughs in science, medications to save lives, machine learning methods to diagnose illnesses sooner, materials to make more efficient devices and structures, financial strategies to live well in retirement, and algorithms to quickly and efficiently direct a wide variety of resources.

- 3D printed clothes at near-zero cost will be designed and produced by consumers directly from their homes. "Imagine having a garment fit exactly to your size and preferences," says Melissa Dawson, an assistant professor of industrial design at the Rochester Institute of Technology, and a 3D expert. "You could also customize your colour and pattern choices, maybe even trims and finishes," she says.

- Explosion in car sharing in cities. Around 400 million people in congested cities around the world will be relying for their daily commutes on Robotic Driverless Car Sharing.

- Small modular nuclear reactors will be in widespread use. They will be factory-made and shipped to site locations. They will be used by consumers as a cost-effective and clean source of power for smaller electrical markets and grids, in isolated areas, and for sites with limited water. They will be extremely flexible, and will be scaled up or down to meet energy demands, and help power areas where larger plants are not needed.

- Human brain simulations will be replicated on advanced computers. With its one hundred billion neurons and one thousand trillion synapses working in parallel, simulating the human brain will push the limits of even quantum computers. This technology will help us better understand and treat a complex variety of psychiatric and behavioural conditions; ultimately leading to technology which will literally modify or remove any negative, anti-social, or illegal behaviour from humans, potentially from birth.

- 3D printed human organs will be built from cells that a patient's immune system will recognize as its own, to avoid immune rejection and the need for patients to take immunosuppressive drugs. Such organs will be built from patient-specific induced stem cells.

- Vertical farms will be developed in cities. Vertical farming is the practice of growing crops in vertically stacked layers. It will occur in controlled environments which will aim to optimize plant growth, and soilless farming techniques such as hydroponics and aquaponics. Vertical farming will allow humans to be prepared for a future of population growth and increasing movement into megacities. Because the environment in which plants grow in vertical farming is very controlled, farming will be able to happen all year-round, and will not be dependent on the seasons and weather.

- A completed synthetic human genome will result in scientists building the entire genetic code of a person from scratch. This will mean that a human will literally be able to be created in a laboratory with all of that person's genetic code, designed to exact specifications determined by scientists, or wealthy sponsors. They will use synthetic chemicals to manufacture all of the DNA contained in human chromosomes. Could this mean that scientists will be able to create humans with certain traits, perhaps people born and bred to be soldiers? Or

(457) The merging of human and artificial intelligence will achieve singularity

might it be possible to make copies of specific people? Will it be acceptable for example, to sequence and then synthesize Einstein's genome, in order to create an exact DNA-duplicate of him? If so, how many Einstein genomes will be made and installed in cells, and who will get to make them? Will this person then become Albert Einstein, or take on another identity? Will this technology get into the wrong hands and lead to the creation of an Adolf Hitler? There will be many major ethical issues to navigate when this technology becomes widespread.

- Aquaculture will provide the majority of the world's seafood. Aquatic farming will produce as much seafood as all of the world's wild marine fisheries, using less than 0.015 percent of the space in the world's oceans. In 2015, 92 million tons of wild fish species were harvested by traditional fishing practices worldwide, the same amount as in 1995. In contrast, seafood production from aquaculture increased from 24 million tons to 77 million tons, during the same time period. It is estimated that the world will need around 40 million more tons of seafood as soon as 2030. This will come mainly from aquaculture.

- 3D printed electronic membranes will be developed to prevent heart attacks. Ultimately, these membranes will be used to treat diseases of the ventricles in the lower chambers of the heart, and will also be inserted inside the heart to treat a variety of disorders; including atrial fibrillation, which affects tens of millions of people worldwide. The membrane will even hold a sensor to measure troponin, a protein expressed in heart cells and a critical sign of an impending heart attack.

- Advanced treatments for Alzheimer's disease and dementia will be developed. Alzheimer's disease is one of the greatest medical care challenges of the twenty first century and is the main cause of dementia. In total, around 40 million people, mainly in more advanced and developed nations, are estimated to suffer from dementia throughout the world, and this number is predicted to double every twenty years. The advanced treatments and an ultimate cure will involve the harvesting of stem cells. Stem cells are "building block" cells. They can develop into many different cell types, including brain or nerve cells. Scientists already have taken skin cells from people with certain types of dementia, such as Alzheimer's disease, and "reprogrammed" them into stem cells in the laboratory. They have then triggered these stem cells to become brain cells. By studying these cells, scientists have gained major insights into how the damage to the brain begins, and how it might be halted. These brain cells will also be used to develop potential treatments at a very early stage of the disease.

- Robotic hands, arms, feet, and legs will match human capabilities. Hyper-advanced robotics will serve as hand, arm, foot, and leg prosthetics. These limbs and other body parts will be controlled within the brain and mind, and will ultimately exceed baseline human capabilities; to the extent that able-bodied humans will volunteer to have them installed.

- Widespread use of hydrogen fuel cell electric vehicles. What if your electric vehicle could be refuelled in less than 5 minutes? No plug or outlet required, and refuelling stations readily available. This will become the future world of hydrogen fuel cell electric vehicles. As of 2021, a tiny market that includes

Toyota's Mirai, Hyundai's Nexo, and Honda's Clarity Fuel Cell, already exists. These "plug-less" electric vehicles are the alternative to their battery electric cousins. Hydrogen fuel cell vehicles are superior driving machines compared to traditional vehicles, and they do not release any carbon or other greenhouse emissions. Hydrogen fuel cell electric vehicles function by electricity being generated from an on-board supply of hydrogen. It is that electricity which powers the electric motor. When hydrogen gas is converted into electricity, water and heat energy are released. The vehicle stores the hydrogen in high-pressure tanks. Non-toxic, compressed hydrogen gas flows into the tank when refuelling.

- Resurrection of extinct animal species. Scientists will develop the technology to restore extinct species, by advanced cloning. It will be done by extracting the nucleus from a preserved cell of the extinct species, and inserting it into an egg without a nucleus, of the nearest living relative of that species. Some of the extinct animals (with existing close living species relatives) that scientists will bring back to life will include the woolly mammoth, dodo, elephant bird, Tasmanian tiger, ground sloth, sabre-toothed tiger, and a variety of dinosaurs including the Tyrannosaurus rex. For example, a fossil of a pregnant T. rex with preserved DNA has been recovered.

- Human-like artificial intelligence will become widespread. Artificial intelligence (AI) is literally evolving. Researchers have created software that borrows concepts from Darwinian evolution, including the "survival of the fittest," to build AI programs that will continue to improve generation after generation without any human input. The program has replicated decades of AI research in a matter of days, and its designers believe that one day, it will discover new approaches to AI. According to Max Tegmark, the president of the Future of Life Institute: "Everything we love about civilization is a product of intelligence, so amplifying our human intelligence with artificial intelligence has the potential of helping civilization flourish like never before – as long as we manage to keep technology beneficial."Will the benefits of human-like artificial intelligence outweigh the inevitable risks? The biggest problem in answering this question is that once artificial intelligence reaches "superhuman" and even "self-conscious" capabilities (the singularity), we humans will be powerless to influence the future course of events and the decisions that these artificially intelligent sentient beings will take. At this point, in 2021, the perceived benefits of artificial intelligence are: a reduction in human error; AI will take the risks instead of humans; it is available 24/7; it will help in repetitive jobs; it processes huge volumes of data rapidly; makes faster and more accurate decisions; can be applied daily in many areas of our lives; and that AI will facilitate many inventions in almost every area of life which will help humans solve many complex problems and make our lives easier and more enjoyable. The perceived risks of AI are: these very complex AI machines require costly regular updating, repairs and maintenance; they will make humans lazy; AI will lead to widespread unemployment; AI will not develop human-like emotions; they lack out-of-box creative and lateral thinking; if they do become sentient and

(458) Futuristic cities, intelligent buildings and flying cars

self-conscious, AI may turn against us.

- Widespread use of desalinated water. In response to increasing water scarcity, much of the world will depend on desalinated water in the future. Desalination involves removing the salt and other harmful substances from seawater, to enable it to be safe for humans to consume. Although it currently carries a high price tag; technological advancements will drive down the cost. Seawater desalination will provide a climate-independent source of drinking water, and of course, is in plentiful supply. Scientists will need to overcome the risks it poses of being energy-intensive and environmentally damaging. Desalination will help humanity solve the acute global water shortage crisis.

- Depression will become the number one global disease. The number of cases of depression increased from 172 million in 1990 to over 300 million in 2020. The World Health Organization has forecast that the incidence of depression will continue to increase exponentially into the future, and will continue to be the single largest factor contributing to global disability. One of the most worrying aspects is that adolescents with severe depression are 30 times more likely to commit suicide. While depression is now one of the most important global health problems, it remains poorly under-stood; although it is known that cultural, psychological, and biological factors, contribute to depression. There seems to be a direct correlation between the standard of living of a country, and its incidence of depression. Depression is more common in countries with higher standards of living. Modern society is fast-paced, technologically driven, and increasingly "results-based." People in advanced countries are increasingly overfed, or at the other extreme, deliberately malnourished; living a sedentary lifestyle with little or no exercise; sunlight-defi-cient; sleep-deprived; over-worked, or idly-bored; social media addicted; and socially isolated. In these societies, traditional family support and bonding structures have often broken down. These changes in culture, traditions, and lifestyle, have played a significant role in the increasing incidence of depression. In the future, gene therapy and brain implant technology will be used to treat and ultimately eliminate depression and many other psychiatric illnesses. For example, those people who are genetically predisposed to certain mental disorders will have the opportunity to get any "risky DNA" snipped out of their genes, or rewritten. Those with already existing neurological disorders, meanwhile, will be prescribed a brain implant – a cling-wrap-like electrical film laid on the brain's surface, or a network of thinner-than-hair wires snaked within the brain, to keep its neural circuits firing properly. These treatments will present ethical dilemmas. If a gene therapy or brain implant erases, say, a person's propensity for depression, will it also possibly erase aspects of their personality, such as introversion, pensiveness, or melancholia? Would they recognize strange thoughts or behaviours as side effects, or mistake these personality changes as a "new normal"? And if they choose not to have these treatments, or cannot afford them, will they be passed over for jobs or for health insurance? Will they be socially accepted? Who will they become? Will they still be themselves?

• Perfection of weather forecasting. Weather forecasting is the application of current technology and science to predict the state of the atmosphere for a future time and a given location. Weather forecasts are made by collecting as much data as possible about the current state of the atmosphere (particularly the temperature, humidity, and wind direction/speed) and using our understanding of atmospheric processes (through the science of meteorology) to determine how the atmosphere will evolve in the future. However, the chaotic and complex nature of the atmosphere, and our incomplete understanding of all the processes means that forecasts become less accurate as the time range of the forecast increases. In the future, artificial intelligence combined with quantum supercomputers will result in very accurate longer term predictions of the weather;

including the ability to predict major atmospheric disturbances such as cyclones, hurricanes, tornadoes, storms, and
temperature extremes.

- Brain implants will become a part of daily life. As researchers
continue to develop smaller, more bio-compatible technology,
and understand the processes of the human brain in greater
depth; there will be a host of medical applications, and
extraordinary abilities which will come from brain implant
technology. Some of these applications will include:

  a) Seeing in the dark. Currently, retinal implants that
     restore a low level of vision to people blinded by genetic
     conditions already exist. Once we further improve the
     ability of these implants to restore vision, augmenting
     "normal" vision will no longer be science fiction. Night
     vision is likely to be a military application at first, but
     soon after, it will become commercially available.

  b) Restoring lost memories. Zapping the brain with
     controlled electrical stimulation has a lot of potential.
     In the future we will be able to implant a piece of
     technology in the brain that will directly deliver these
     pulses of electricity, and be recharged, without having
     to be removed.

  c) Downloading new skills. Scientists will develop the
     ability to wire our brains to rapidly learn a new skill.
     Already, research is being conducted to look into how
     the brain learns and stores skills, with the expectation
     that in the future, neurological processes will be artifi-
     cially replicated to allow people to effectively learn a
     new skill with a fraction of the time and effort.

  d) Enhanced focus and alertness. In the same way as
     treating mental health problems like depression and
     PTSD, brain implants will help people who suffer from
     neurological problems relating to focus and alertness.
     This will include ADHD, narcolepsy, and dementia.
     This technology will also be highly sought after by

neurologically healthy people to enhance their focus and alertness.

e)   Control any device with your mind. Researchers are currently able to use a neuro-prosthetic sensor to help paralysed patients control a robotic arm with their minds, to varying degrees of success. This technology will rapidly develop to the level where it will be widely used by consumers to control a wide variety of daily devices and functions.

f)   Search the Internet with your brain. Forget Siri and Google Glass, why not get the information you're searching for delivered from the Internet directly to your brain? Pending the development of smaller, non-toxic implants that the body will not reject; brain implants will be used throughout the day to do a number of tasks, including navigating the Internet, and instantly recognizing people.

(459) The proposed Boeing hypersonic airliner is being developed

• Hypersonic airliners will carry most passengers. Boeing has already unveiled plans for what could be the world's first hypersonic airliner, a sleek, futuristic looking craft that will be capable of flying five times the speed of sound, or around 6,115 kilometres per hour. At that speed – Mach 5 in aviation terminology – it will be possible to travel from New York City to London in about two hours instead of the eight hours it takes a conventional airliner.

- Cures for cancer. New approaches to tame the immune system in the fight against cancer are getting humanity closer to a future where cancer will become a curable disease. Personalized vaccines, cell therapy, gene editing, and microbiome treatments are four technologies that will change the way cancer will be treated in the future. It seems increasingly evident that there won't be a single cure. Instead, each patient will be treated according to their specific needs. This will see the advent of personalized medicine. The technology will develop to the point where diagnostic tests will be able to accurately predict if a person is susceptible to certain types of cancer, and then preventative medical techniques will be deployed to ensure the person never develops that cancer at a later stage in their life.

- Service robots in every home and place of work. Artificially intelligent robots will become a big part of our daily lives. Technology corporations will be in a constant race to change the way robotics are implemented in people's everyday lives, which will lead to a very exciting future. Based on current trends, here are ten ways robotics are expected to transform the future well into the 21st century and beyond:

(460) A home service robot from the movie: Bicentennial Man

a) Robotics in public security. Artificial technology for predicting and detecting crime might seem far-fetched; but it will become a reality in the future. Drone footage, for instance, will make that happen soon. In addition, automatic recognition of suspicious activities is already a reality for camera-based security systems. This technology will change society in a very important way; by allowing law enforcement officials to act quickly whenever a suspicious behaviour has been spotted.

b) Robots in education. The line between classrooms and individual learning settings is already starting to blur. Robots will boost the process of personalized learning. NAO, the humanoid robot, is already forming bonds with students from around the world. It comes with important senses of natural interaction; including moving, listening, speaking, and connecting.

c) Robots at home. Cloud-connected home robots are already becoming a part of our lives. We can set up the vacuum cleaner to do the chore for us; and we can schedule a warm home-cooked meal to be ready by the time we are finished with work. These cloud-connected robots are likely to evolve into more advanced versions. We will see speech comprehension and increased interactions with humans in future years.

d) Robots as co-workers. Robots will have a profound impact on the workplaces of the future. They will become capable of taking on multiple roles in an organization. These machines will likely evolve more in terms of voice recognition, so that humans will be communicating with them through voice commands.

e) Robots will take many jobs. Whether we like it or not, robots have already replaced many people in their jobs. Many jobs in office administration, logistics, and transport, amongst others, will be replaced in the not too distant future. Robots are expected to take over half of all lower-skilled jobs.

f)   Robots will create jobs too. In the near future, artificial intelligence will most likely replace tasks, not complete jobs. The good news is that robotics will also create many new markets and ancillary jobs. We might need additional education and re-training for those jobs, but the opportunities will be there.

g)   Autonomous cars. Currently, self-driving cars still require some human intervention; but we are getting close to the day when they won't. Google no longer has a monopoly on this industry. Instead, every significant automobile manufacturer is pursuing this technology, with Uber being one of the strongest players. The users of this service can now get matched with a self-driving Uber when they request the service.

h)   Healthcare robots. We are looking into a different future for healthcare. Instead of visiting a primary care doctor who will give us a check-up with a simple stethoscope; in the not too distant future we will have intelligent robots performing these tasks. They will interact with patients, check on their medical condi-

(461) A Virtual Reality headset

tions, and evaluate the need for further appointments. Pharmabiotics will bring more significant changes. They will act like ATM's for medicines, so we will be able to get the medications we need while avoiding the inconvenience of talking to a stranger about health issues.

i)   Robotics for entertainment. Robots are getting more personalized, interactive, and engaging than ever. With the growth of this industry, virtual reality will enter all homes in the near future. We will be able to interact with our home entertainment systems through conversations, and they will respond to our attempts to communicate. During the twenty first century virtual reality entertainment will become fully-immersive and "life-like"; and will become a huge consumer industry.

j)   Robots will boost our standard of living. We have seen throughout history that automation and mechanization boosts the overall standard of living. We saw it with the Industrial Revolution, and it will happen again. According to the United Nations, poverty was reduced to a greater extent over the past five decades, than in the previous fifty decades. This is because the global economy grew sevenfold over the last five decades, and technology played a huge role in that progress.

•   The first definitive evidence of life beyond Earth. Many NASA scientists believe we are on the verge of finding alien life. This is because the space agency plans to dramatically ramp up its search for extra-terrestrial life in the next ten years; in ancient Martian rocks; in the hidden oceans on the moons of Jupiter and Saturn; and in the atmospheres of faraway exoplanets orbiting other star systems. "With all of this activity related to the search for life, in so many different areas, we are on the verge of one of the most profound discoveries ever,"

Thomas Zurbuchen, NASA's former administrator, told the US Congress in 2017.

- Establishment of the first permanent lunar base. NASA is forging ahead with its Artemis Program to land humans on the Moon by 2024, and to create a permanent lunar base. "After 20 years of continuously living in low-Earth orbit, we're now ready for the next great challenge of space exploration – the development of a sustained presence on and around the Moon," according to NASA Administrator Jim Bridenstine. "For years to come, Artemis will serve as our North Star as we continue to work toward even greater exploration of the Moon, where we will demonstrate key elements needed for the first human mission to Mars," he said.

- Robots will dominate the battlefield. The United States Army is expected to deploy autonomous military robots on the battlefield, operating alongside conventional soldiers, by 2028. Eventually, in the middle and latter half of the twenty first century, these military robots will completely replace humans on the battlefield. Autonomous military robots are designed for military applications such as transport, search and rescue,

(462) Robotic soldiers of the future. Will they fight wars instead of humans?

and for attack. In the future all military offensive activity will ultimately be undertaken by these killer robots. These robots will be developed with sophisticated artificial intelligence so that they will be able to reason, make decisions, and adapt to changing battlefield conditions.

- Self-driving vehicles will dominate the roads. Self-driving cars are reaching a point where they are as effective, or better than human drivers. Companies like Google, Tesla, and Uber are each pushing the limits of technological innovation in an attempt to dominate this emerging industry. In 30 years from now, the majority of cars in the world will be self-driving autonomous vehicles. The predicted benefits of self-driving cars will include:

  a) A 90% reduction in road traffic deaths; literally saving millions of lives.

  b) A 60% reduction in harmful emissions, because fewer accidents will result in less traffic congestion, and hence a significant drop in emissions.

  c) The elimination of "stop-and-go-waves. Autonomous cars are expected to eliminate waves of traffic jams created by stop-and-go human distracting behaviour. This, in turn, will not only save people time, but also decrease the time their cars are on the roads, and therefore also contribute towards reductions in congestion and emissions.

  d) A 10% improvement in fuel economy, simply because the cars will be driven more efficiently.

  e) Up to 50% reduction in travel time. In the United States this time saving is expected to translate to workers saving 80 billion hours currently lost to commuting, which will save the economy US$1.3 trillion.

  f) There will be significant consumer savings due to reduced insurance costs, running costs, and parking fees.

There may also be some unintended negative consequences from the widespread use of driverless cars, such as; they may be expensive to buy; job losses; and the programming may go wrong, creating accidents.

(463) The ultimate in full immersion Virtual Reality – Star Trek's Holodeck

- Full immersion virtual reality. A fully immersive virtual reality world will be able to encompass every single human sense and will interact directly with the brain and nervous system. People connecting to this immersive system will be booted in through some kind of artificially intelligent central nervous system, and then will be made unconscious of their physical body and surrounding environment. A fully virtual world will be able to reproduce all of our senses and even more fantastical feelings and thoughts in a completely artificial environment. For those familiar with William Gibson's novel "Neuromancer", the Cyberdeck brings to mind an example of a fully immersive virtual reality landscape. At one point in the novel, the protagonist Case makes the following remarks about cyberspace:

"Cyberspace. A consensual hallucination experienced daily by billions of legitimate operators, in every nation, by children being taught mathematical concepts… A graphic represen-

tation of data abstracted from the banks of every computer
in the human system. Unthinkable complexity. Lines of light
ranged in the non-space of the mind, clusters and constellations of data. Like city lights, receding…"

Whether it's a kind of headset, nervous system jacked-in wire,
humans suspended in some eternal vat or other disembodied
consciousness, this technology will be life-changing for those
who immerse themselves within it.

- Fusion power will be commercially available. The main application for fusion will be in creating electricity. Nuclear fusion
  will provide a safe and clean energy source for future generations with several advantages over current fission nuclear
  reactors. These advantages will include:

  a) It will be an abundant fuel supply. Deuterium can be
     readily extracted from seawater, and excess tritium can
     be made in the fusion reactor itself from lithium, which
     is readily available within the Earth's crust. Uranium for
     fission is rare, and it must be mined and then enriched for
     use in reactors.

  b) It is safe. The amounts of fuel used for fusion are small
     compared to existing fission reactors. This is so that uncon-
     trolled releases of energy do not occur. Most fusion reactors
     make less radiation than the natural background radiation
     we live with, in our daily lives.

  c) It is clean. No combustion occurs in nuclear power (fission
     or fusion), so there is no air pollution.

  d) Less nuclear waste will occur. Fusion reactors will not
     produce high-level nuclear wastes like their fission counter-
     parts, so disposal will be less of a problem. In addition, the
     wastes will not be of weapons-grade nuclear materials as is
     the case in fission reactors.

  NASA is currently looking into developing small-scale fusion
  reactors for powering deep-space rockets. Fusion propulsion
  is going to boost an unlimited fuel supply, in the form

of hydrogen, which will be more efficient, and is going to ultimately lead to faster rockets.

- The majority of primate species in the wild will become extinct. Non-human primates, our closest biological relatives, play important roles in the livelihoods, cultures, and religions of many societies, and offer unique insights into human evolution, biology, behaviour, and the threat of emerging diseases. Primates are an essential component of tropical biodiversity, contributing to forest regeneration and ecosystem health. There are currently 504 known species distributed mainly in the tropics, Africa, Madagascar, and Asia. Alarmingly, over 60% of primate species are now threatened with extinction, and over 75% have declining populations. This situation is the result of escalating anthropogenic pressures on primates and their habitats; mainly global and local market demands, leading to extensive habitat loss through the expansion of large-scale agriculture, cattle ranching, tree logging, oil and gas drilling, mining, dam building, and the construction of new road networks in primate regions. Other important drivers are increased bush-meat hunting and the illegal trade of primates as pets and primate body parts, along with emerging threats, such as climate change and diseases. Often these pressures act in synergy, exacerbating the decline in primate populations. Given that the habitats of primates, overlap extensively with a large and rapidly growing human population characterized by high levels of poverty, global attention will be needed immediately to reverse the looming risk of primate extinctions, and to attend to local human needs in sustainable ways. Rising global scientific and public awareness of the plight of the world's primates and the costs of their loss to ecosystem health and human society, is going to become imperative in order to avoid this massive extinction risk.

- Cardiovascular heart disease will be significantly cured in the future through the use of stem cell therapy, which will be utilized to regenerate damaged heart muscles. The heart muscle relies on a steady flow of oxygen-rich blood to nourish it and keep it pumping. During a heart attack, that blood flow is interrupted by a blockage in an artery. Without blood, the area of the heart fed by the affected artery begins to die, and scar tissue forms in the area. Over time, this damage can lead to heart failure, especially when one heart attack comes after another. New treatments using stem cells, which have the potential to grow into a variety of heart cell types, are going to repair and regenerate damaged heart tissue. Several different type of approaches will be used to repair damaged heart muscle with stem cells. The stem cells, which are often taken from bone marrow, will be inserted into the heart using a catheter. Once in place, the stem cells will help regenerate the damaged heart tissue.

- The global population will reach 11 billion by the end of the century. The number of people in the world increased more than 400% during the twentieth century. What will the remainder of the twenty first century be like? The United Nations has projected that the global population will increase

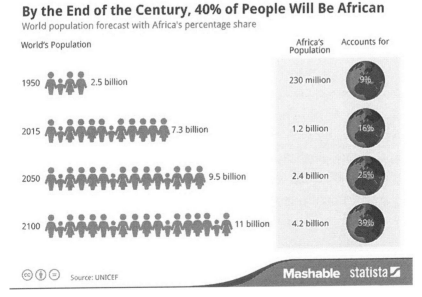

(464) *The United Nations projected world population up to 2100*

from 7.7 billion in 2019 to 11 billion by the end of the twenty first century. By that time, the UN is projecting that rapid global population growth will come to an end. It is difficult to predict the population dynamics beyond 2100; it will depend upon the fertility (birth) rate. The big question is whether the fertility rate will rise above an average of 2 children per woman, again. In 1961 the global fertility rate peaked at 3.5; in 2021 it stands at 2.4; and in 2100 it is forecast to be 1.9. The world population will reach a size, which when compared to the history of humanity, will be at an extraordinary level. If the UN projections are accurate, and they have a good track record, the world population will have increased more than 10-fold over the 250-years period up to the year 2100. The good news though is that we are on the way to a new balance. The big global demographic transition that the world entered more than two centuries ago will come to an end by the end of the twenty first century. This new equilibrium will be different from the one in the past, when it was a very high mortality (death) rate that kept population growth under control. In the new balance of the future, it will be low fertility (birth) rates which will significantly reduce population growth. The big challenge is, will the Earth contain sufficient natural resources by the end of the century, to sustain a forecast population of 11.2 billion people?

- Humans will become merged with machines. Technology today is on the verge of allowing us to completely manipulate the genetic makeup of our offspring, reducing their chances of being born with defects and weaknesses found within the genetic codes of their parents. This will accelerate significantly during the course of the remainder of this century, to the point where it will evolve into cybernetic technology; the integration of artificial intelligence with our human biology. Already, major corporations are investing heavily in computer-brain interface (CBI) technology; companies like BrainGate, and Elon Musk's Neuralink. When cybernetics is integrated into our biology, deep-space travel beyond our solar system; colonization of planets in other star systems; and immortality through virtual reality; will all become distinct possibilities. According to one of the world's leading futurists, Ray Kurzweil, the following technologies may only be decades away:

a) Deep-Space Travel and Colonization. This will be made possible by the fusion of human biology and cybernetics. If we utilize cybernetic technology to allow us to be "plugged in," instead of needing food as our source of energy, we will simply need to program coordinates in a spacecraft, and outfit it with a power supply that will last the entire trip. Replacing our need for food with raw energy will also allow us to colonize planets which otherwise would be inhospitable to human life.

b) Controlling Machines with our Mind. Advanced electroencephalography (EEG) technology will be rapidly advanced to control functions aboard our spacecraft as we relax while travelling into deep space. EEG technology has already allowed us to develop chips that can give paralysed people the ability to use computers and perform physical tasks. An example of this prevailing technology is BrainGate. BrainGate, developed by neuroscientist John Donoghue at Brown University, is a microchip that is implanted into the user's brain, and allows for the controlling of devices, such as robotic limbs, mouse cursors, and even the keys of a piano by pure thought alone. In the future, this kind of technology could become mandatory at birth, and could surround the entire brain. This could allow us to use several robotic limbs in addition to our own. The mind could control a spacecraft while the body is hibernating; or even control robotic surrogates on planets with extreme environments.

c) Downloading into Virtual Worlds. In his book, "The Singularity Is Near: When Humans Transcend Biology," Kurzweil asserts that in addition to deep space travel and colonization, advances in cybernetic technology will ultimately allow us to download our minds into virtual realities. No longer will the human body be a necessity if we can create virtual worlds that satisfy our needs, such as those depicted in movies like "The Matrix." If we develop virtual worlds that will allow for our mind to exist without the body, a sense of immortality will set in, and powers once thought to be attributed to gods could become the abilities of the digital human. Not only will we no longer need our bodies if this kind of technology is developed, but also, the possibility will exist to take on different "forms" or "super-human" abilities.

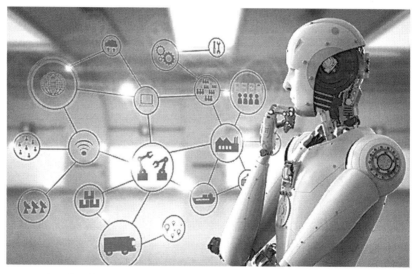

(465) The Singularity - When AI will exceed human intelligence

- Cybernetic technology that can be integrated within our human biology is going to open up a whole new world of possibilities for humanity, including deep space travel and ultimately, interstellar and intergalactic colonization of exoplanets. We will also develop virtual worlds where we will become immortal, and have the complete power to create whatever surroundings we can imagine. Several advancements in robotics, neuroscience, and virtual reality, are already

creating the foundations for these technologies to be further
developed and adopted by humanity well into the twenty first
century and beyond.

- Major extinctions of many species of animals and plants. A
  "biological annihilation" of wildlife in recent decades means a
  sixth mass extinction in the Earth's history is now underway,
  and is more severe than previously feared, according to a study
  published by the US National Academy of Sciences in 2017.
  Scientists analysed both common and rare species and found
  that billions of regional or local populations have been lost
  around the world. They blame human overpopulation and
  overconsumption of resources for the crisis, and warn that
  it threatens the survival not only of remaining species, but
  also of human civilization, with just a short window of time
  in which to take action. The study calls the massive loss of
  wildlife a "biological annihilation" that represents a "frightening assault on the foundations of human civilization."
  Professor Gerardo Ceballos, who led the research study, said:
  "The situation has become so bad it would not be ethical
  not to use strong language." The scientists found billions of
  populations of mammals, birds, reptiles, and amphibians have
  been lost all over the planet, leading them to assert a sixth mass
  extinction has already progressed further than was previously
  thought. The scientists conclude: "The resulting biological
  annihilation obviously will have serious ecological, economic,
  and social consequences. Humanity will eventually pay a very
  high price for the decimation of the only assemblage of life
  that we know of in the universe." They say, while action to halt
  the decline remains possible, the prospects do not look good:
  "All signs point to ever more powerful assaults on biodiversity
  in the next two decades, painting a dismal picture of the future
  of life, including human life." Wildlife is dying out due to
  habitat destruction, overhunting, toxic pollution, invasion by
  alien species, and climate change. But the ultimate cause of
  all of these factors is "human overpopulation and continued

population growth, and overconsumption, especially by the rich," say the scientists, who include Professor Paul Ehrlich, whose 1968 book "The Population Bomb" is a seminal and still controversial work.

(466) The Amazon rainforest is in grave danger due to extreme deforestation

- Up to 80% of the Amazon rainforest will be deforested by the end of the century. The Amazon rainforest generates half of its own rainfall, but deforestation (cutting down more trees and vegetation than what is re-planted) threatens to disrupt this cycle over the course of this century. The end result will be a shifting of large parts of the ancient Amazon forest to a dry, savannah habitat. If this tipping point is reached and exceeded, it will have a disastrous knock-on effect for climate and weather patterns around the globe. A quick and decisive transition to zero deforestation is the only way to avert catastrophic change to the Amazon, according to climate and environmental scientists. But conservationists fear the Brazilian government lacks the political will to take anything resembling this required action, as it continues to remove existing protections, and ignore the pleas from local indigenous groups to take control of the stewardship of the forest, and protect it from further excessive logging and clearing.

- Hi-tech intelligent buildings will revolutionize cities. The future of buildings will be smart. With the cost of interconnected sensors and cloud computing continuously falling, technology that intelligently monitors and controls the operations of buildings, will become more and more widespread. It is estimated that there will be as much as 10 billion devices installed in individual buildings in the coming decades, making it one of the fastest-growing industries worldwide. As a result, the smart building market is expected to grow from a size of $8.5 billion in 2016, to around $58 billion globally in 2022. What are the actual implications for the real estate industry when it comes to the smart building revolution? Here are some specific trends that will change the way buildings will be run in the coming years, helping organizations to save energy and other costs, deliver better occupant experiences, and achieve higher property values:

   a) Air Quality Monitoring – Buildings will increasingly be equipped with wireless sensors that will monitor carbon di oxide levels and harmful small particles, sending out warnings, and adjusting the ventilation if required.

   b) Smart Lighting – Smart LED lighting will automatically adjust to the preference of occupants. This is called human-centric lighting. These smart lights will be able to mimic the natural light progression of daylight to follow our circadian rhythm, or change their intensity according to the needs of different occupants.

   c) Building Security – Smart sensors, including more sophisticated cameras will be installed throughout buildings to make them even safer.

   d) The Importance of Cybersecurity – Building operators will stay ahead of potential online virtual threats by improving authorization controls, and implementing stronger data encryption.

   e) Occupant Control – smart buildings will give more control to its occupants. People will be able to more closely interact

with a building by directly adjusting temperatures, booking
meeting rooms, and changing lighting, all from one central
place, according to their preferences.

f) Intelligent Parking – Cameras and sensors will detect what
    parking spots are free, and send this information directly to
    commuters, reducing extra driving time and unnecessary
    fuel consumption.

g) Predictive Maintenance – Sensors placed around building
    machinery like pumps or heaters will be programmed to
    detect critical levels of noise, vibration, or heat. Above a
    certain threshold, a warning will be sent and the error
    fixed, before it escalates.

• Genetically engineered "designer babies." Embryos produced
  by IVF will be genetically screened. This will involve parts or all
  of the embryonic DNA being read to determine which gene
  variants they carry. The prospective parents will then be able to
  choose which embryos to implant in the hope of achieving a
  pregnancy with a desired outcome. In the future, human eggs
  will become much more abundant and will be readily available
  for quick, convenient, and inexpensive genome screening. By

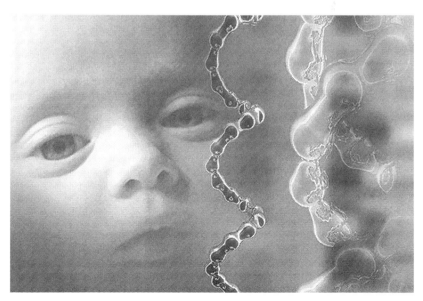

(467) Genetically engineered designer babies – will they become a reality?

the middle of this century, many scientists believe we will start seeing the use of gene editing and reproductive technologies for embryo enhancement; such as blond hair and blue eyes, for example; or improved athletic abilities; or enhanced reading and numeracy skills, and so on.

- Global average temperatures will rise by 4 degrees Celsius by the end of the century if no significant action is taken by the global community to reduce carbon and other greenhouse gas emissions. This is the nearly unanimous prediction and consensus of climate scientists around the world. These scientists have predicted the following devastating consequences of an average 4-degrees Celsius increase:

  a) The inundation of coastal cities due to rising sea levels. This will set off massive waves of permanently displaced homeless refugees, never seen before in the history of humanity. Major conflicts around the world are likely to be triggered by this humanitarian disaster.

  b) Increasing risks for global food production and security potentially leading to higher malnutrition rates. Food shortages will become a daily fact of life for billions of people, and will become more extreme.

  c) Many dry regions will become dryer, and wet regions will become wetter. This is going to destroy much of the planet's fertile agricultural land.

  d) Unprecedented heat waves in many regions, especially in the tropics.

  e) Extreme water scarcity in many regions of the globe. Future conflicts are likely to be fought over securing water resources, not non-renewable resources like oil.

  f) Increased frequency of high-intensity tropical cyclones.

  g) An irreversible loss of biodiversity, including coral reef systems. The Great Barrier Reef off the eastern coast of northern Australia is already in grave danger.

The scientific evidence is unequivocal about the fact that humans are the cause of global warming, and that major changes are already being observed. The global average temperature is now 0.8 degreesCelsius above pre-industrial levels; oceans have warmed by 0.09 Degrees Celsius since the 1950s, and are acidifying. Sea levels rose by an average of 20 centimetres since pre-industrial times, and are now rising at 3.2 centimetres per decade. An exceptional number of extreme heat waves occurred in the last decade; and major food cropgrowing areas around the globe are being increasingly affected by drought.

• Handheld MRI scanners will be common in the future. When it comes to brain scans for assessing head trauma, detecting brain cancer, and performing numerous other diagnostic tests, magnetic resonance imaging (MRI) is the best option; but MRI scanners are costly, require special infrastructure, and are immobile. A research team led by investigators at Massachusetts General Hospital has developed a low-cost, compact, portable, and low-power "head only" MRI scanner that can be mounted in an ambulance, wheeled into a patient's room, or put in small clinics or doctors' offices around the world. In the coming

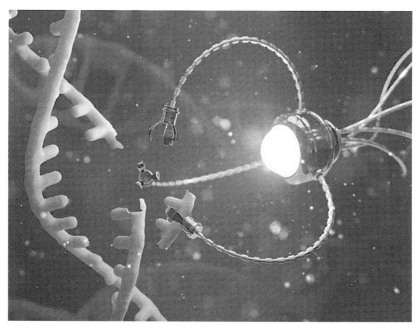

(468) Nanotechnology will dominate the future in science, medicine and industry

decades, these portable MRI scanners will be widely used by consumers in their homes, and will eventually be able to diagnostically scan the entire human body, not just the head and brain, for evidence of diseases and other heath or medical issues.

- Nanofabricators will become a mainstream consumer product. The ultimate manufacturing technology will be in the form of a nanofabricator, which will provide atom-by-atom control of the manufacturing process for complex objects, both large and small. The British science historian James Burke and a number of futurists have predicted that nanofabricators will become portable consumer devices by around 2040. Burke predicts that the year 2040 will see the beginning of the worldwide distribution of kits to make "nanofabricators" able to literally take "basic raw materials, air and water, and a bit of cheap, carbon-rich acetylene gas," manipulate the composite atoms and molecules of these materials and gases, and "produce anything you want, virtually free." Since a nanofabricator will be able to make a copy of itself, Burke predicts everybody will have one by 2042. This is Burke's prediction of a future with nanofabricators: "Sixty years later, we will have adapted to the new abundance and will be living in small, no-pollution, autonomous communities, anywhere. Energy from spray-on photovoltaics (producing electric currents) will make any object (like a house) its own power source. So, here you are in your nanofabricated dwelling, filled with Mona Lisa's if that's your wish, with holographic reality transforming any room into anywhere (like: beach, sun, wind ruffling hair). So nobody travels any more. Want to see a pal, have dinner with your mother, join a discussion group? No problem: they'll be there with you as 3D holograms, and you won't know Stork from butter, unless you try to make physical contact. The entire global environment will also be covered with quintillions of dust-sized nano-computers called motes. So your life will constantly be curated by an intelligent network of ubiquitous cyber-servants. The "motes" will know you need more food, or that it's a bit chilly today, or that you're supposed to call Charlie. And they'll take the relevant action. Your shirt (motes in the fabric) will call Charlie. Either his avatar will appear, or you'll hear his voice. Not sound waves, but brainwaves. Brain-to-brain communication." Burke points out that nanofabricators will eventually eliminate the need for infrastructure and for government, and that the resulting abundance will eliminate the need for crime, and with it the need for privacy. He also believes disease will be eliminated. Without jobs to qualify for, he predicts formal education will be replaced by "learning-for-fun." Entertainment will be "all in-brain" with accompanying holograms, tailored to a person's most specific idiosyncratic creative imaginary wishes.

- Longevity treatments will halt aging. Currently, anti-aging therapies are in clinical trials. However, as of yet, none have achieved regulatory approval for mass, widespread use. To reverse the aging process in humans, we will need something safe, effective, and ideally not expensive. Some therapies such as blood plasma transfusions would cost up to $8,000 USD a month, if approved and implemented. Reversing the aging process is already possible for human cells and simple organisms, in scientific experiments. From yeast and worms, science has moved on to being able to extend the lifespans of rats, mice, and monkeys. Although clinical trials have begun for regenerative stem cell therapies in degenerative diseases, such as heart failure; in the future, many organs will be harvested and grown from the cells of a single organ donor, alleviating the organ shortage for transplants, and the long waiting lists for patients around the world. This stem cell technology is very exciting and rapidly advancing. Many of the stem cells used are embryonic stem cells, taken from unfertilized embryos sourced from IVF (in vitro fertilisation) clinics. Giving these cells a second purpose clearly has a positive impact on humanity, but it is still controversial for many people. While reversing aging at a cellular level is attractive, it is not yet known whether this technology could, in the long term, lead to cancer, a disease in which individual cells reproduce uncontrollably, and become effectively immortal. Science will need to alleviate this problem in the future if this therapy is to become widespread. In future, this technology may be packaged into anti-aging pills, which we will be able to take as a preventative medication to prolong longevity, but it will be decades before we fully understand the effect of these pills and related interventions. When a reverse aging process becomes a reality, it will drastically alter humanity's perception of illness and aging, and ultimately, our perception of time itself.

- The first generation of antimatter-powered spacecraft will emerge. Most starships depicted in science fiction use antimatter as a source of energy fuel. Antimatter is sometimes referred to as the mirror image of normal matter because while it looks just like ordinary matter, some properties are reversed. For example, normal electrons, the familiar particles that carry electric current in everything from cell phones to plasma TVs, have a negative electric charge. Anti-electrons have a positive

(469) Antimatter powered spacecraft will travel into interstellar space

charge, so scientists dubbed them "positrons." When antimatter meets matter, both annihilate in a flash of energy. This complete conversion to energy is what makes antimatter so powerful. Even the nuclear reactions that power atomic bombs come in a distant second, with only about three percent of their mass converted to energy. NASA is developing an antimatter spacecraft that will cut fuel costs to a fraction of what they are today. In October 2020, NASA scientists announced early designs for an antimatter engine that could generate enormous thrust with only small amounts of antimatter fuel. Matter-antimatter propulsion will be the most efficient propulsion ever developed, because 100 percent of the mass of the matter and antimatter is converted into energy. When matter and antimatter collide, the energy released by their annihilation releases about 10 billion times more energy than the chemical energy such as hydrogen and oxygen combustion, used in conventional spacecraft such as the space shuttle. Matter-antimatter reactions are 1,000 times more powerful than the nuclear fission produced in nuclear power plants, and 300 times more powerful than nuclear fusion energy. As a result, matter-antimatter engines are going to take humanity further into space, with less fuel. Approximately 10 grams of antiprotons will be enough fuel to send a manned spacecraft to Mars. Today, it takes around 9 months for an unmanned spacecraft to reach Mars. With antimatter power the same trip will take around 3 months. As this form of energy is further developed over the coming decades, deep space missions to the outer solar system, and interstellar space travel will become viable. Most of the antimatter required will be man-made, and generated in particle physics accelerators at laboratories like Fermilab and CERN. With propulsion velocities at up to 40% of the speed of light, such technology will eventually cut travel time to the nearby Alpha Centauri star system to less than a decade; and more immediately, it would allow NASA to send a New Horizons-type space-probe to our outer solar system in less than a year.

- Space elevators will become operational. For more than half a century, rockets have been the only way to go to space. But in the not too distant future, we will have another option for sending up people and payloads. It will be a colossal elevator extending from the Earth's surface up to an altitude of 35,000 kilometres, where satellites orbit. NASA says the basic concept of a space elevator is sound, and researchers around the world are optimistic that one can be built. The Obayashi Corporation, a global construction firm based in Tokyo, has said it will build one by 2050, and China is planning to build one by 2045. "The space elevator is the Holy Grail of space exploration," says Michio Kaku, a professor of physics at City College of New York and a noted futurist. "Imagine pushing the 'up' button of an elevator and taking a ride into the heavens. It could open up space to the average person." Kaku isn't exaggerating. A space elevator would be the single largest engineering project ever undertaken and could cost close to $10 billion to build. But it could reduce the cost of putting payloads into orbit from around $3,500 per pound to as little as $25 per pound, says Peter Swan, president of International Space Consortium, based in Santa Ana, California. The idea of a space elevator was first conceived in 1895 by Konstantin Tsiolkovsky, a Russian scientist who did pioneering work in rocketry. As commonly conceived today, a space elevator would consist of motorized elevator pods that are powered up and down a ground-to-space tether. The tether would stretch from a spaceport at the equator to a space station in orbit overhead. Centrifugal forces caused by the Earth's rotation would hold the tether aloft. The ISS experiment, dubbed Space Tethered Autonomous Robotic Satellite-Mini elevator, or STARS-Me, was devised by physicists from Japan's Shizuoka University. It will simulate on a small scale the conditions that the components of such a system would encounter. Cameras will examine the movement of a pair of tiny "cubesats" along a 10 metre tether in a weightless environment. It's going to be the world's first experiment to test elevator movement in space.

(470) A Space Elevator will become an efficient way of getting into space

- Accurate simulations of viruses. For the first time in 2016, scientists created an accurate simulation of a virus invading a cell, which will lead to anti-viral therapies that will be much more effective than the ones we rely on today. The experiment was designed to examine how the protein shell of a virus, known as the capsid, changes as it prepares to inject genetic material into a healthy cell; changes that haven't been fully observed in previous research. Into this artificial cell surface, they inserted protein molecules from human cell receptors, to allow for outside signals into the 'cell' – something that has never been done before. An imaging technique called cryo-electron microscopy, through which beams of electrons can identify protein structures in minute detail, was then used to monitor the reaction between the virus capsids and the artificial cell receptor membranes. What they saw suggests that the researchers' hypothesis – that the virus's shape changes should only occur at the points where the cell receptors bind to the virus – was indeed correct. The researchers hope that by understanding more about how viruses get into cells, we can stop them more effectively. New drugs in the future will trigger viral mutations that will prevent them from gaining access to healthy cells.

(471) A Space Hotel orbiting the earth – will there be a space tourism boom?

- There will be widespread space tourism to the Moon and Mars. In humanity's first half-century as a spacefaring species, government-run space programs put people on the Moon and began to master Earth orbit. The next 50 years are going to bring a sea change, with commercial companies taking over near-Earth operations and freeing NASA and other national space agencies to send astronauts to the asteroids, planetary moons and Mars. As a result, by 2061, millions of people will have gone to space, and thousands will be living there. We will see permanently manned outposts on the Moon and Mars. The seeds of this transformation are being sown now, as private companies ramp up their space flight capabilities and start finding ways to make money in Earth orbit and beyond. Multiple companies are developing their own spaceships and their own plans for commercialising space. Virgin Galactic, for example, will start taking tourists on suborbital joyrides in the near future. Orbital tourist trips are not far behind. Various companies including Space-X are developing crewed vehicles to take paying customers to the International Space Station. By 2061, tourism facilities are likely to be established in orbit around the Earth, on the Moon, and on Mars; and space tourism will become a mainstream, popular and affordable industry.

The 22nd century is likely to be marked by the widespread emergence of post-scarcity and resource-based economies; the rapid growth of transhumanism; and major developments in space travel. By this stage, all of the world's energy may come from either nuclear fusion or renewable sources. Having begun to merge with human intelligence in the previous century, artificial intelligence will now surpass human intelligence, and will reach whole new levels of cognitive and intellectual capability. Though lacking the raw emotions and more-subtle traits of organic human minds, the sheer speed and power of AI will begin to profoundly alter the course of history. Almost every high level decision by government and transnational corporations will come directly from these sentient machines, which will oversee vast numbers of virtual employees, robots, and other automated systems. Developments in space during the 22nd century will include numerous permanent, manned settlements on the Moon and Mars; regular crewed journeys to the gas giant planets; hugely commercially lucrative mining operations in the Asteroid Belt; and various interstellar exploration probes. Space tourism will boom during this century, with trips to the Moon and Mars, and Earth orbiting space hotels becoming relatively commonplace. The speed and magnitude of progress now occurring, both on Earth and throughout the Solar System, will likely create what earlier futurists would have referred to as a "technological singularity." Will the achievement of technological singularity become another threshold of increasing complexity, in the Big History journey? It is quite possible that many of the scientific discoveries of the 22nd century will exceed the comprehension levels of "natural" technologically-unaided humans lacking the required brain upgrades and enhancements. The most notable scientific breakthroughs may relate to the cutting-edge field of quantum physics; in addition to whole new fields of science which did not exist in the 21st century.

Some of the developments in the 22nd century may include:

- The commencement of the terraforming of Mars. Could we make Mars into an Earth-like planet? For many years, Mars has existed as a hopeful "Planet B" – a secondary option if Earth can no longer support us as a species. From science fiction stories to scientific investigations, humans have considered the possibilities of living on Mars for a long time. A main staple of many Mars colonization concepts is terraforming – a process of changing the conditions on a planet to make it habitable for life that exists on Earth, including humans, without a need for life-support systems. To successfully make Mars Earth-like, we would need to raise temperatures, have water stably remain in liquid form, and thicken the atmosphere. By using the greenhouse gases already present on Mars, we could theoretically raise temperatures and change the atmosphere enough to make the planet Earth-like. Future technologies are likely to be developed in order to make the terraforming of Mars a reality.

(472) A Floating City of the future will be fully sustainable

- Nomadic floating cities will roam the oceans. As our cities grapple with overcrowding and undesirable living conditions, the ocean remains a frontier for sophisticated water-based communities. The United Nations has expressed support for further research into floating cities in response to rising sea levels, and to accommodate climate refugees. Two main types of very large floating structures (VLFS) technology can be used to carry the weight of a floating settlement. The first, pontoon structures, are flat slabs suitable for floating in sheltered waters close to the shore. The second, semi-submersible structures (such as oil rigs), comprise platforms that are elevated on columns off the water surface. These can be located in deep waters. Potentially, oil rigs could be repurposed for such floating cities in international waters. Technology is not a barrier to floating cities in international waters. Advances in technology will enable us to create structures for habitation in deep sea waters. Ultimately, there will be politically independent countries that float on the ocean in seasteads. Seasteads are floating islands of self-governing communities which hope to facilitate innovative business ideas in a low-regulation environment. How would seasteads be protected against piracy? Should they have their own armies? What about supplying food, drinking water, and electricity? These are some of the many questions which would need to be resolved for floating cities to become secure and sustainable.
- Human Intelligence will be vastly amplified by Artificial Intelligence. By the 22nd century, brain-computer interfaces will augment the visual cortex of the brain. This is going to significantly boost our spatial visualization and manipulation capabilities. For example, humans will be able to easily imagine and clearly visualise a complex engineering blueprint with high reliability and detail, or to learn completely new blueprints easily. There will also be highly advanced augmentations that will focus on the other portions of the sensory brain cortex, such as the tactile cortex and auditory cortex. Once

this happens, the next step will be the complete augmentation of the pre-frontal cortex with advanced artificial intelligence. This will result in human beings performing apparently impossible intellectual feats such as controlling the minds of other people, communicating their thoughts, and designing inventions that change the world overnight.

- Machine learning will enable us to predict earthquakes. Five years ago in 2014, Paul Johnson wouldn't have thought predicting earthquakes and tsunamis would ever be possible. Now he isn't so certain. "I can't say we will, but I'm much more hopeful we're going to make a lot of progress within decades," the Los Alamos National Laboratory seismologist says. "I'm more hopeful now than I've ever been." The main reason for that new hope is a technology Johnson started looking into about four years ago – machine learning. Many of the sounds and small movements along plate tectonic fault lines where earthquakes occur have long been thought to be meaningless. But machine learning, which involves training computer algorithms to analyse large amounts of data to look for patterns or signals, suggests that some of the small seismic signals might matter after all. These computer models are expected to advance to such a level, that by the 22nd century they will become the key to unlocking the ability to predict earthquakes. This machine learning will not only lead to accurately predicting the timing of an earthquake, but it will also ultimately extrapolate how big an earthquake is going to become, where its epicentre is, and what's going to be affected, including predicting tsunamis; all from the analysis of data from small seismic signals.

- Mind uploading will enter mainstream society. Mind uploading is the process by which the mind; a collection of memories, personality, and the attributes of a specific individual, is transferred from its original biological brain to an artificial computational substrate or platform. Once it is possible to move a mind from one substrate to another, it is then called

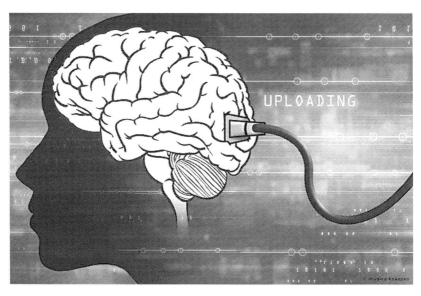

(473) Will mind uploading lead to immortality in humans?

a substrate-independent mind (SIM). The concept of SIM is inspired by the idea of designing software that can run on multiple computers with different hardware without needing to be rewritten. For example, Java's design principle "write once, run everywhere" makes it a platform independent system. In this context, substrate is a term referring to a generalized concept of any computational platform that is capable of universal computation. If we take a purely physiological position, the human mind is solely generated by the brain, and is a function of neural states. These neural states are computational processes and devices capable of universal computing sufficient to generate the same kind of computational processes found in the brain. This technology is going to allow a human mind to literally live for extremely long periods of time, or even achieve immortality, while it is housed within an artificially intelligent substrate, possibly even a humanoid robotic body. In 2011, the Russian billionaire Dmitry Itskov founded the 2045 Initiative, an organisation that aims to help humanity achieve immortality by 2045. "Within the next 30 years, I am going to make sure that we can all live forever," claims Itskov.

"The ultimate goal of my plan is to transfer somcone's person-ality into a completely new body." The 2045 Initiative has laid out its plan in three stages. The first stage involves building a humanoid robot called the Avatar, and a cutting-edge brain-computer interface system. The second stage consists of building a life support system for the human brain, and linking it with the Avatar. The third and final stage involves creating an artificial brain that will hold the original individual consciousness. The whole idea is a minefield of ethical and moral issues. Should humanity pursue this technology? Let's say that we've successfully uploaded a human mind onto a computer; does that mean a personal identity has also been transferred along with memories, and that this person is still the same? Or is it a new person with a different identity who just happens to share the same memories? What rights would this digital person have? And if you could create one copy of yourself, why stop there? Why wouldn't you create multiple copies? In that case, which one of those copies would be the 'real' you? And since you wouldn't have a physical body anymore and would essentially be reduced to a stream of data, who would that data belong to? Who would own you? How could you prevent major corporations from misusing your data?

- There will be large-scale civilian settlement of the Moon and Mars. What would it be like to build a full scale city on the Moon? The acclaimed architectural and engineering firm, Skidmore, Owings & Merrill, in partnership with the European Space Agency (ESA) and the Massachusetts Institute of Technology (MIT), have presented a conceptual design for a future "Moon Village." The Moon Village is imagined on the edge rim of the Shackleton Crater near the South Pole because this area receives continuous daylight throughout the whole lunar year. The Moon Village would sustain its energy from direct sunlight, and set up food generation and life sustaining elements through in-situ resource utilization by tapping into the Moon's natural resources. Water extracted from the depressions near the South Pole would create breathable air and rocket propellants to support the burgeoning industry in the settlement. As for housing the people, there will be individual pressurized modules which are inflatable, giving residents the flexibility to increase their living space as required. Most buildings will be three to four storey structures that will serve as a combined workspace, living quarters, and have the necessary environmental and life support systems integrated into one. The Moon Village design was created for the ESA's planning for future exploration beyond 2050 in partnership with NASA's strategic plan to "extend human presence deeper into space and to the Moon for sustainable long term exploration and utilization." In 2017, an MIT team developed a design for a settlement on Mars. It proposes to create domes or tree habitats that will house up to 50 people each. The dome will provide residents with open public spaces containing vegetation and water, which will be harvested from deep within the Martian northern plains. The tree habitats will be connected on top of a network of tunnels, or roots, providing transportation and access to both public and private spaces between other inhabitants of this proposed 10,000-person strong community. Advanced technology such as artificial light inside these pods, will strongly mimic the sight and feel of natural sunlight. Valentina Sumini from MIT describes the project's design fundamentals, and elaborates on the project's poetic forest metaphor: "On Mars, our city will physically and functionally mimic a forest, using local Martian resources such as ice and water, regolith (or soil), and sun to support life. Designing a forest also symbolizes the potential for outward growth as nature spreads across the Martian landscape. Each tree habitat incorporates a branching structural system and an inflated membrane enclosure, anchored by tunnelling roots. With regards to the entirety of the proposed Martian city, referred to as "Redwood Forest", System Design Management

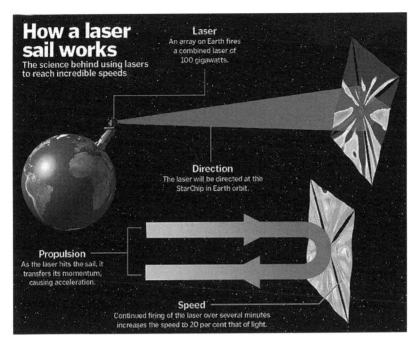

(474) The Breakthrough Starshot Laser Sail initiative

Fellow, George Lordos, describes it as follows: "Every tree habitat in Redwood Forest will collect energy from the Sun and use it to process and transport the water throughout the tree, and every tree is designed as a water-rich environment. Water fills the soft cells inside the dome providing protection from radiation, helps manage heat loads, and supplies hydroponic farms for growing fish and greens. Solar panels produce energy to split the stored water for the production of rocket fuel, oxygen, and for charging hydrogen fuel cells, which are necessary to power long-range vehicles as well as provide backup energy storage in case of dust storms."

- Interstellar exploration will become common. When humans first start to explore other star systems, the logical first choice will be our closest stellar neighbour, the triple star system of Alpha Centauri, 4.37 light-years away. In 2016, astronomers at the European Southern Observatory discovered an Earth-sized planet orbiting Alpha Centauri's red dwarf star, Proxima Centauri. The planet, named Proxima b, is at least 1.3 times the mass of the Earth, but has a very tight orbit around Proxima Centauri, taking just 11 Earth days to complete its 'year.' What has astronomers particularly excited, is that this planet is in the right temperature range for liquid water, which is a useful guide for the possibility of life evolving on this planet. How would we get there? Even at the fastest speeds of our current technology, it would take 18,000 years. But exciting interstellar engineering technology is being developed to speed up such a trip. One project is called Breakthrough Starshot. This is a $100 million initiative privately funded by Russian billionaires Yuri and Julia Milner. The primary focus is on propelling a tiny unmanned probe by hitting its extremely lightweight sail with a powerful Earth-based laser. The idea is that if the spacecraft is small enough – and we're talking barely one gram – and the sail is light enough, the impact of the laser will be enough to gradually accelerate the craft to around 20% of the speed of light, taking it to Alpha Centauri in around 20 years. The Milners are counting on nano (miniaturization) technologies to enable this tiny craft to carry a camera, thrusters, a power supply, communication, and navigation equipment, so it can report on what it sees as it flashes past Proxima b. A mission such as this, if successful, will then lay the foundation for the next and more challenging phase of interstellar travel: including human crews. A manned mission to Alpha Centauri is likely to be propelled by antimatter. An antimatter spacecraft would have a far higher energy density than any other known form of propulsion. If energy resources and efficient production methods are found on Earth to produce antimatter in the quantities required, and store it safely, it will be possible to reach speeds of at least 20% of the speed of light. Interstellar and intergalactic travel at the speed of light may be achievable by the 22nd century by developing very advanced technologies such as artificial black holes or wormholes. A theoretical idea for enabling interstellar travel is the propelling of a starship by creating an artificial black hole, and using a parabolic reflector to reflect its Hawking radiation. The black hole could act as a power source if the Hawking radiation it emits could be converted into energy and thrust. One potential method would involve placing the artificial black hole at the focal point of a parabolic reflector attached to the starship, creating enough energy for forward thrust. Wormholes on the other hand, are distortions in space-time that astrophysicists postulate could connect two arbitrary points in the Universe, across an Einstein-Rosen Bridge. Einstein's equation of general relativity allows for the existence of wormholes. Some astrophysicists argue that wormholes may have been created in the early Universe, stabilized by cosmic strings. The 100 Year Starship (100YSS) is the name of the overall effort that will, over the next 100 years, work towards achieving interstellar travel. Harold White from NASA's Johnson Space Center is a member of Icarus Interstellar, the non-profit foundation whose mission is to realize interstellar space travel before the year 2100. White has used a laser to try to warp space-time by 1 part in 10 million, with the aim of making interstellar travel possible. Some of the past and current interstellar starship and probe designs include:

a) Project Orion – human crewed interstellar ship design (1958-1968).

b) Project Daedalus – non-crewed interstellar probe design (1973-1978).

c) Starwisp – non-crewed interstellar probe design (1985).

d) Project Longshot – non-crewed interstellar probe design (1987-1988).

e) Starseed Launcher – a fleet of non-crewed interstellar probe designs (1996).

f) Project Valkyrie – human crewed interstellar ship design (2009).

g) Project Icarus – non-crewed interstellar probe design (2009-2014).

h) Project Dragonfly – small laser propelled interstellar probe
     design (2013-2015).

i)   Breakthrough Starshot – a fleet of non-crewed interstellar
     probes, announced on April 12th, 2016.

• The world's first humans living over 200 years. Futurist
  Doctor Ian Pearson, amongst many other futurists and scientists, argues that by using advanced biotechnology combined
  with AI, humanity will be in a position to merge our minds
  with machines, making our bodies obsolete. You could end up
  attending your own "body funeral." Pearson paints the picture
  of this future, stating: "One day, your body dies, and with it, your
  brain stops, but no big problem, because 99% of your mind is
  still fine, running happily on IT, in the cloud. Assuming you
  saved enough and prepared well, you connect to an android
  to use as your body from now on, attend your funeral, and
  then carry on as before, still you, just with a younger, highly
  upgraded body." And this is just the beginning. In the 22nd
  century, if not earlier, there could be many different ways you
  may be able to preserve your mind and consciousness. Humans
  might just switch to different humanoid bodies after a certain
  period, the same way you might buy a new car with new
  features. Or thanks to projects like Elon Musk's Neuralink, your
  mind may just be a few simple clicks away from downloading
  yourself into a computer, or robotic body. It is possible that in
  the more distant future humanity may decide it is better for
  humans to live in massive megastructures that literally generate
  our own reality. "The mind will be in the cloud, and be able
  to use any android that you feel like, to inhabit the real world,"
  says Pearson. It could get to a point in which you hire an
  android body for the day. Rather than travel to Italy, you just
  upload your brain to an android stationed in Italy. Or maybe
  there is a great concert that you want to see, but the band is in
  another city, thousands of kilometres away. You might be able
  to simply upload yourself to experience the show. The result
  of this is that humans may never need a flesh and blood body

again. Advances in biotechnology and genetic engineering
will eventually prevent the aging of cells, or completely
reverse it altogether. Lose a finger? Simply 3D-print a new
one. Need a new arm? Call up a doctor and have them reinstall
one. Cryogenic freezing has its fair share of sceptics, but over
the years, the scientific community has slowly embraced the
idea. This could come in handy in the future, during trips to
distant planets in other star systems hundreds of light-years
away. Also, by the time of the 22nd century, living in a virtual
world will become an alternative to living in the current one.
Humans will be able to upload their minds into a cloud-based
advanced virtual reality system; a place where they will be
able to spend all of eternity living in peace with an avatar
of their choice. Eventually, humanity will become so technologically advanced that it will be able to simulate reality on a
universal scale, using what is known as a matrioshka brain. A
matrioshka brain is a hypothetical megastructure proposed by
Robert J. Bradbury (1956-2011) based on the Dyson Sphere,
with immense computational capacity. The idea was proposed
when imagining the most advanced civilizations that may exist
in the Universe. This technology would be in the form of an
impressive Class B stellar engine, employing the entire energy
output of a star to drive computer systems. Our entire species
could be uploaded on this computer system, which would be
able to simulate reality and remake the Universe as we know
it. Think, the Matrix.
- Humanity will become a Type 1 civilisation on the Kardashev
  scale. The Kardashev scale was originally designed in 1964
  by the Russian astrophysicist, Nikolai Kardashev, who was
  looking for signs of extra-terrestrial life within cosmic
  radio-wave signals. It has 3 base classes, and 2 subsequently
  added more advanced classes, each with an energy disposal
  level: Type I (1016th W), Type II (1026th W), Type III (1036th
  W), Type IV (1046th W), and Type V. The energy available to
  a Type V civilization would equal all of the energy available

in not just our Universe, but in all universes and in all time-lines. It is important to note that currently in the early 21st century, the human race is not even on this scale yet! Since we still sustain our energy needs from dead plants and animals, here on Earth, we are a lowly Type 0 civilization. It is likely we will reach Type I in the next 100-200 years. What do each of these categories actually mean in literal terms? A Type I designation is given to species who have been able to harness most or all of the energy that is available from a neighbouring star, gathering and storing it to meet the energy demands of a growing population. This means that we would need to boost our current energy production to over 100,000 times to reach this status. However, being able to harness all of Earth's energy would also mean that we could have control over all natural forces. Human beings at this Type I level would be able to control volcanoes, the weather, and even earthquakes. As astounding as this level of technology will be; it is still basic and primitive compared to the capabilities of civilizations with higher rankings. The next step up – a Type II civilization – can harness the power of their entire star; not merely transforming starlight into energy, but controlling the star as well. Several methods to achieve this have been proposed. The most popular of which is the hypothetical 'Dyson Sphere.' This device would encompass every single centimetre of the star, gathering all of its energy output and transferring it to a planet for later use. Alternatively, if fusion power (the mechanism that powers stars) has been mastered by the intelligent beings, a reactor on a truly immense scale would be used to satisfy their energy needs. Nearby gas giant planets could be utilized for their hydrogen, slowly drained by an orbiting fusion reactor. What would this much energy mean for a species? It would mean that nothing known to science could wipe out a Type II civilization. For example, if humans survived long enough to reach this status, and a moon-sized asteroid entered our solar system on a collision course with our Earth, we would have the ability to vaporize it out of existence. Type III, is where a species become inter-galactic travellers with the knowledge of everything having to do with energy; making them into a master race. Type IV civilizations would almost be able to harness the energy content of the entire Universe and with that, they would be able to traverse the accelerating expansion of space. Furthermore, advanced races of these species may live inside supermassive black holes. A Type IV civilization would need to tap into energy sources unknown to us, using strange, or currently unknown, laws of physics. In Type V civilizations, beings would be like gods, having the knowledge to manipulate the Universe as they please.

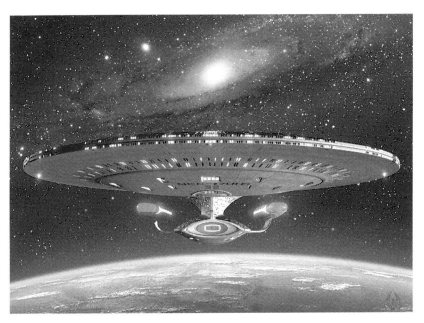

(475) Will the Star Trek USS Enterprise become a reality?

One of the most insightful futurists I have followed is Ray Kurzweil. He is a well-known and often controversial advocate for the role of machines in the future of humanity. He is a computer scientist and author of the 2005 futuristic book, "The Singularity is Near." As a scientist, his areas of interest are health, artificial intelligence, futurism, and transhumanism. He strongly argues that humanity can evolve beyond our current biological status through technology. In his book, Kurzweil defines singularity as a time period in which technological growth is at such a pace that it blurs the distinct lines between natural biology and technology. In the book, this singularity is presented in wide-ranging ideas depicting progressive technological leaps in the future, to the point where technology merges with, and overtakes biology and the human brain. He states that our intelligence will become super powerful, and will evolve into a non-biological entity with the extraordinary processing power of machines. This merger is the core of singularity.

(476) Time magazine featuring 2045   *(477) Ray Kurzweil*

Kurzweil formulates and predicts six major epochs in his futuristic journey to singularity. Each epoch progressively evolves into the next, with information processing technology ultimately merging with known science. It is a progression of information integration into all of the facets that make up existence today. Let us briefly examine each of these six epochs:

- Epoch 1: Physics and Chemistry – This discusses the storage of information in the basic structures scientists have identified as the building blocks of the Universe; including all matter and energy. He believes there is more to the building blocks of humanity beyond our mere existence.

- Epoch 2: Biology and DNA – Information is stored in DNA, which is known as the building block of genetic information in all living organisms. Kurzweil foresees the ability of the advancement of human creativity to inject new and upgraded information into DNA.

- Epoch 3: Brains – Evolving from Epoch 2, DNA and the other basic building blocks of life have enabled organisms to be able to detect information, using the sensory organs, and store it as data in their super-evolved organs, called brains.

- Epoch 4: Technology – With the brain capabilities developed during Epoch 3, human intelligence progresses to the creativity level of replicating its capabilities into machines. Human-created technology has become integral to the life we live today. The rate of evolution in humans is linear, but the rate of evolution in technology is and will continue to be, exponential.

- Epoch 5: Merging with Technology – Here we see a period where human intelligence is merged into technology, and the singularity is achieved. The light at the end of the tunnel for Kurzweil's futuristic theorems begins to be realized in this epoch.

- Epoch 6: The Universe Wakes Up – This is the pinnacle of Kurzweil's vision. The Universe waking up is a culmination of, or the aftermath of, the singularity of the Universe. In this epoch, intelligence will become the combined derivative from human biological brains and progressive human ingenuity in its developing technologies. The marriage of these two evolutionary paths will saturate the building blocks of humanity; birthing an optimal level of artificial intelligence. In this state, Kurzweil foresees the evolution of a new civilization where existence as we know it today will transcend our biological limitations by the sheer dynamism of our human creativity.

The merger of humans and machines has already started; we almost feel we have lost our limbs when we cannot locate our cell phones, for example. Elon Musk has already unveiled a working demonstration of Neuralink, which will eventually make a direct link between the human brain and the machine, known as brain-machine interface technology. At Neuralink, they are trying to create wires that will allow faster and better communication between humans and machines. These wires, called "neural laces" will be 10 times thinner than human hair, but very flexible and powerful. This is a historical moment because it is likely to be the start of a new era of superhuman intelligence. Musk believes that artificial intelligence is coming fast and it represents a danger to humanity. So, he says we might as well join artificial intelligence to avoid being replaced by it. This is captured in the mission statement of Neuralink which says: "If you can't beat it, join it." Elon Musk believes humanity must become like cyborgs in order to survive the upcoming artificial intelligence revolution. The AI is already here. Our slow-moving human brains will shortly become obsolete, as machines and robots are rapidly coming to replace us. So, the only way out for us human beings is to merge ourselves with machines. Kurzweil states: "Can the pace of technological progress continue to speed up indefinitely? Isn't there a point at which humans are unable to think fast enough to keep up? For unenhanced humans, clearly so. But what would 1,000 scientists, each 1,000 times more intelligent than human scientists today, and each operating 1,000 times faster than contemporary humans (because the information processing in their primarily non-biological brains is faster) accomplish? One chronological year would be like a millennium for them. What would they come up with?"

Here are some of Ray Kurzweil's predictions, as described in his 2005 book: "The Singularity is Near: When Humans Transcend Biology."

- He predicts the date in which singularity will first occur will be 2045. Kurzweil claims singularity will represent a profound transformation in humanity. Non-biological intelligence in 2045 will be 1 billion times more powerful than all human intelligence in existence today.

- The law of accelerating returns predicts an exponential increase in the technologies of genetics, nanotechnology, robotics, and artificial intelligence. Exponential growth is already being experienced in many more established technologies such as transistors, microprocessors, DNA sequencing, and nanotechnology patents.

- The speed of evolution will increase every year. Consider how many revolutionary technologies we have seen in the past few years: electric cars, self-driving cars, drones, reusable rockets, 3D printing, big data, the internet of things, machine learning and AI, virtual reality, augmented reality, robotics, quantum computing, brain-machine interfaces, and more.

- DNA-powered computers will give us 1000 times increases in processing power. At a price of around $1000, people of the future will be able to afford a computer that is smarter than all of humanity combined.

- The intelligence of humans will become increasingly non-biological and trillions of times more powerful than it is today. This is the dawning of a new civilization that will enable humans to transcend their biological limitations and amplify their creativity.

- In this new world, there will be no clear distinction between humans and machines; and no clear distinction between real reality and virtual reality.

- In this future, humans will be able to assume different bodies and take on a range of personae at their will. Our bodies will be transformed by nanotechnology to overcome the limits of biology.

- Cell technology will become so advanced that the aging process will be reversed.

- Animal muscle tissue will be created without creating the actual animal. This will give humanity a limitless supply of food with minimal environmental impact.

- Evolution will move towards greater complexity, greater elegance, greater intelligence, greater beauty, greater creativity, and greater love. Kurzweil says that these attributes are generally used to describe God; that means evolution is moving towards a conception of God; and the transition away from our biological roots is in fact a spiritual awakening and undertaking.

- When singularity occurs, it will not be long after that human aging and illness will be reversed; pollution stopped; and world hunger and poverty eliminated.

- In the human-machine civilization, human experience will shift from real reality to simulated virtual reality.

- We will be able to download the content of our brains to an external storage device. It is incredible to think how telepathy will transform the Internet as we know it.

- We will be able to create any physical product, just from the information about its chemical composition. This will result in radical wealth creation. The great majority of the world's wealth (GDP) will be created by non-human intelligence. Non-biological intelligence will be able to emulate the richness, subtlety, and depth, of human thinking.

(478) Human created intelligence – will it ultimately span the entire universe?

- Nanobots will soon replace doctors and repair the body from the inside. Nanobots will travel in our bloodstream and deliver medicine to specific parts of our bodies. They will regulate insulin and implement nutritional sustenance. They will also eliminate toxins or viruses from the body, or repair your genes when they are damaged.

- Nanobots will eliminate DNA transcription errors and turn off unwanted replication. This will literally mean the elimination of cancer.

- With nanobot technology, we will place sensitive and high-resolution sensors at billions of locations in the brain, and witness living cells in action.

- Nanobots will also intercept or supress inputs coming from our senses, and replace them with signals from a virtual reality or simulated reality environment.

- None of these technologies will be free of risk though. For example, if nanobots start multiplying uncontrollably outside of the human body, it would potentially be a disaster for humanity. Infinitely replicating nanobots would suck carbon from biological life including trees, animals, and even humans. It would take 130 iterations (or repetitions) until all life on Earth would be eliminated.

- When singularity is reached, machine intelligence will become more powerful than all human intelligence combined. Afterward, intelligence will radiate outward from the planet until it saturates the Universe.

We are fast approaching the singularity. At the rapid rate of technological progress which we are experiencing currently in the areas of nanotechnology, biotechnology, genetic engineering, robotics, and artificial intelligence; it is expected that at some time around the middle of the twenty first century, 2040-2050 CE, the point will be reached when artificial intelligence will exceed human intelligence. I believe that this singularity point will be the ninth threshold in the journey of Big History.

IMAGE CREDITS

The images in this book have been sourced from the following websites and organisations.

1.   En.wikipedia.org
2.   Egyptiangeographic.com
3.   Commons.wikimedia.org
4.   En.wikipedia.org
5.   En.wikipedia.org
6.   Commons.wikimedia.org
7.   En.wikipedia.org
8.   En.wikipedia.org
9.   Parabola.org
10. En.wikipedia.org
11. En.wikipedia.org
12. En.wikipedia.org
13. En.wikipedia.org
14. En.wikipedia.org
15. Crystalinks.com
16. Medium.com
17. Sciencephotos.com
18. Bighistoryproject.com
19. Sciencemag.org
20. En.wikipedia.org
21. Literariness.org
22. Math.wiki.org
23. Math.wiki.org
24. Wikiquote.org
25. Biography.com
26. En.wikipedia.org
27. Biography.com
28. Atomicarchive.com
29. Idmb.com
30. En.wikipedia.org
31. Bigbangcentral.com
32. Britannica.com
33. En.wikipedia.org
34. En.wikipedia.org
35. New York Times
36. Howitworksdaily.com
37. En.wikipedia.org
38. Imdb.com
39. Schoolobservatory.org
40. Universetoday.com
41. Lifeng.lamost.org
42. En.wikipedia.org
43. Commons.wikimedia.org
44. Commons.wikimedia.org
45. En.wikipedia.org
46. Earthsci.org
47. Eventhorizontelescope.org
48. Nature.com
49. En.wikipedia.org
50. Learnthesky.com
51. En.wikipedia.org
52. Imdb.com
53. NASA
54. Smithsonian National Museum
55. European Space Agency
56. Space.com

57. Space.com
58. European Space Agency
59. NASA
60. Space.com
61. NASA
62. Humanmars.net
63. NASA
64. NASA
65. Planetaryscience.com
66. NASA
67. NASA
68. NASA
69. NASA
70. En.wikipedia.org
71. University of Arkansas
72. European Space Agency
73. History.com
74. Interestingengineering.com
75. En.wikipedia.org
76. Universe Today
77. Astrobites.com
78. Geologyin.com
79. Khsappliedgeography.com
80. Cosmos Magazine
81. Britannica.com
82. The Atlantic
83. NASA
84. Thoughtco.com
85. Link.springer.com
86. Commons.wikimedia.org
87. Commons.wikimedia.org
88. Britannica.com
89. Fine Art America
90. Britannica.com
91. Thoughtco.com
92. Sciencephoto.com
93. Thoughtco.com
94. Bioninja.com.au
95. Evolution.berkely.edu
96. Commons.wikimedia.org
97. Commons.wikimedia.org
98. Slideshare.net
99. Sciencedirect.com
100. Humanorigins.si.net
101. Britannica.com
102. Commons.wikimedia.org
103. Britannica.com
104. Dian Fossey Gorilla Fund
105. University of Michigan
106. The New Yorker
107. Wordpress.com
108. National History Museum
109. History.com
110. History.com
111. History.com
112. Britannica.com
113. Genengnews.com
114. Ancient-origins.net
115. Usgs.gov
116. Humanorigins.si.net
117. Wordpress.com
118. Science Photo Library
119. Hierarchystructure.com
120. Stanford University
121. Ancient History Encyclopaedia
122. Earthchangesmedia.com
123. Ancienthistorylists.com
124. NASA
125. Commons.wikimedia.org

126. Commons.wikimedia.org
127. Commons.wikimedia.org
128. Kevin Jones Associates
129. Politicalscience.yale.edu
130. Commons.wikimedia.org
131. En.wikipedia.org
132. Ancient-origins.net
133. Ancient-origins.net
134. Ancient-origins.net
135. En.wikipedia.org
136. En.wikipedia.org
137. Ancient-origins.net
138. Ancient-origins.net
139. En.wikipedia.org
140. Researchgate.net
141. Thinglink.com
142. NASA
143. En.wikipedia.org
144. En.wikipedia.org
145. En.wikipedia.org
146. Commons.wikimedia.org
147. Commons.wikimedia.org
148. Commons.wikimedia.org
149. Commons.wikimedia.org
150. Commons.wikimedia.org
151. Livescience.com
152. En.wikipedia.org
153. En.wikipedia.org
154. En.wikipedia.org
155. Ancient-origins.net
156. En.wikipedia.org
157. Imdb.com
158. Thoughtco.com
159. Imdb.com
160. En.wikipedia.org
161. En.wikipedia.org
162. En.wikipedia.org
163. Rarenewspapers.com
164. En.wikipedia.org
165. En.wikipedia.org
166. En.wikipedia.org
167. En.wikipedia.org
168. Commons.wikimedia.org
169. Commons.wikimedia.org
170. Commons.wikimedia.org
171. Commons.wikimedia.org
172. En.wikipedia.org
173. En.wikipedia.org
174. En.wikipedia.org
175. Commons.wikimedia.org
176. Commons.wikimedia.org
177. Realhistoryww.com
178. Fine Art America
179. Ancient-origins.net
180. Ancient-origins.net
181. Commons.wikimedia.org
182. Britannica.com
183. Wikimedia.commons.org
184. Weaponsandwarfare.com
185. Livescience.com
186. Sciencesource.com
187. En.wikipedia.org
188. Britannica.com
189. En.wikipedia.org
190. En.wikipedia.org
191. En.wikipedia.org
192. History.com
193. Ancient-literature.com
194. En.wikipedia.org
195. En.wikipedia.org

196. About-history.com
197. En.wikipedia.org
198. History.com
199. Commons.wikimedia.org
200. Commons.wikimedia.org
201. Commons.wikimedia.org
202. History.com
203. Commons.wikimedia.org
204. History.com
205. En.wikipedia.org
206. Commons.wikimedia.org
207. Commons.wikimedia.org
208. Commons.wikimedia.org
209. Commons.wikimedia.org
210. En.wikipedia.org
211. En.wikipedia.org
212. En.wikipedia.org
213. Ancient-origins.net
214. En.wikipedia.org
215. En.wikipedia.org
216. En.wikipedia.org
217. Britannica.com
218. En.wikipedia.org
219. En.wikipedia.org
220. En.wikipedia.org
221. En.wikipedia.org
222. Hbo-rome.fandom.com
223. Ancientworldmagazine.com
224. Hbo-rome.fandom.com
225. Commons.wikimedia.org
226. Commons.wikimedia.org
227. Sciencephotolibrary.com
228. Ancient-origins.net
229. En.wikipedia.org
230. En.wikipedia.org
231. Commons.wikimedia.org
232. Commons.wikimedia.org
233. En.wikipedia.org
234. Commons.wikimedia.org
235. Commons.wikimedia.org
236. Epochtimes.com
237. Commons.wikimedia.org
238. Biography.com
239. En.wikipedia.org
240. Sciencealert.com
241. Ancient-origins.net
242. Nationsonline.org
243. En.wikipedia.org
244. En.wikipedia.org
245. En.wikipedia.org
246. Sci-news.com
247. En.wikipedia.org
248. En.wikipedia.org
249. En.wikipedia.org
250. En.wikipedia.org
251. Commons.wikimedia.org
252. En.wikipedia.org
253. En.wikipedia.org
254. En.wikipedia.org
255. Commons.wikimedia.org
256. Commons.wikimedia.org
257. En.wikipedia.org
258. En.wikipedia.org
259. En.wikipedia.org
260. En.wikipedia.org
261. En.wikipedia.org
262. En.wikipedia.org
263. En.wikipedia.org
264. En.wikipedia.org
265. En.wikipedia.org

266. En.wikipedia.org
267. Edu.glogster.com
268. En.wikipedia.org
269. En.wikipedia.org
270. En.wikipedia.org
271. En.wikipedia.org
272. En.wikipedia.org
273. En.wikipedia.org
274. History.com
275. En.wikipedia.org
276. Commons.wikimedia.org
277. Epicworldhistory.com
278. NASA
279. Info-base Publishing
280. Commons.wikimedia.org
281. Commons.wikimedia.org
282. En.wikipedia.org
283. En.wikipedia.org
284. En.wikipedia.org
285. Henry Hamlin
286. En.wikipedia.org
287. En.wikipedia.org
288. Sciencephoto.com
289. History.com
290. Commons.wikimedia.org
291. Commons.wikimedia.org
292. Australianstogether.org.au
293. Britannica.com
294. En.wikipedia.org
295. Teasa-govt.nz
296. Commons.wikimedia.org
297. Commons.wikimedia.org
298. Commons.wikimedia.org
299. Commons.wikimedia.org
300. Commons.wikimedia.org

301. Commons.wikimedia.org
302. Commons.wikimedia.org
303. Commons.wikimedia.org
304. Commons.wikimedia.org
305. Commons.wikimedia.org
306. Commons.wikimedia.org
307. Commons.wikimedia.org
308. Commons.wikimedia.org
309. Commons.wikimedia.org
310. Commons.wikimedia.org
311. Commons.wikimedia.org
312. Commons.wikimedia.org
313. Commons.wikimedia.org
314. History.com
315. Commons.wikimedia.org
316. Commons.wikimedia.org
317. Britannica.com
318. En.wikipedia.org
319. En.wikipedia.org
320. En.wikipedia.org
321. En.wikipedia.org
322. En.wikipedia.org
323. Smithsonianmag.com
324. Commons.wikimedia.org
325. En.wikipedia.org
326. En.wikipedia.org
327. Commons.wikimedia.org
328. En.wikipedia.org
329. En.wikipedia.org
330. Whbailey.weebly.com
331. Commons.wikimedia.org
332. Commons.wikimedia.org
333. Commons.wikimedia.org
334. En.wikipedia.org
335. En.wikipedia.org

336. En.wikipedia.org
337. En.wikipedia.org
338. En.wikipedia.org
339. En.wikipedia.org
340. En.wikipedia.org
341. En.wikipedia.org
342. En.wikipedia.org
343. En.wikipedia.org
344. En.wikipedia.org
345. En.wikipedia.org
346. En.wikipedia.org
347. En.wikipedia.org
348. En.wikipedia.org
349. En.wikipedia.org
350. En.wikipedia.org
351. En.wikipedia.org
352. En.wikipedia.org
353. Commons.wikimedia.org
354. Commons.wikimedia.org
355. Britannica.com
356. History.com
357. En.wikipedia.org
358. En.wikipedia.org
359. En.wikipedia.org
360. History.com
361. En.wikipedia.org
362. En.wikipedia.org
363. En.wikipedia.org
364. Commons.wikimedia.org
365. Britannica.com
366. En.wikipedia.org
367. En.wikipedia.org
368. En.wikipedia.org
369. En.wikipedia.org
370. En.wikipedia.org
371. En.wikipedia.org
372. En.wikipedia.org
373. En.wikipedia.org
374. En.wikipedia.org
375. En.wikipedia.org
376. En.wikipedia.org
377. En.wikipedia.org
378. En.wikipedia.org
379. En.wikipedia.org
380. En.wikipedia.org
381. En.wikipedia.org
382. En.wikipedia.org
383. Commons.wikimedia.org
384. Commons.wikimedia.org
385. Commons.wikimedia.org
386. Commons.wikimedia.org
387. En.wikipedia.org
388. En.wikipedia.org
389. Britannica.com
390. En.wikipedia.org
391. En.wikipedia.org
392. En.wikipedia.org
393. En.wikipedia.org
394. En.wikipedia.org
395. En.wikipedia.org
396. Imdb.com
397. Imdb.com
398. Commons.wikimedia.org
399. En.wikipedia.org
400. En.wikipdia.org
401. En.wikipedia.org
402. History.com
403. En.wikipedia.org
404. En.wikipedia.org
405. En.wikipedia.org

406. En.wikipedia.org
407. En.wikipedia.org
408. En.wikipedia.org
409. En.wikipedia.org
410. En.wikipedai.org
411. En.wikipedia.org
412. En.wikipedia.org
413. En.wikipedia.org
414. En.wikipedia.org
415. Imdb.com
416. En.wikipedia.org
417. Britannica.com
418. En.wikipedia.org
419. En.wikipedia.org
420. En.wikipedia.org
421. En.wikipedia.org
422. En.wikipedia.org
423. En.wikipedia.org
424. En.wikipedia.org
425. Commons.wikimedia.org
426. Commons.wikimedia.org
427. Commons.wikimedia.org
428. Commons.wikimedia.org
429. Commons.wikimedia.org
430. Commons.wikimedia.org
431. Commons.wikimedia.org
432. Commons.wikimedia.org
433. Commons.wikimedia.org
434. Commons.wikimedia.org
435. Commons.wikimedia.org
436. En.wikipedia.org
437. En.wikipedia.org
438. En.wikipedia.org
439. En.wikipedia.org
440. Commons.wikimedia.org
441. En.wikipedia.org
442. En.wikipedia.org
443. En.wikipedia.org
444. En.wikipedia.org
445. En.wikipedia.org
446. En.wikipedia.org
447. En.wikipedia.org
448. En.wikipedia.org
449. En.wikipedia.org
450. En.wikipedia.org
451. En.wikipedia.org
452. En.wikipedia.org
453. En.wikipedia.org
454. En.wikipedia.org
455. En.wikipedia.org
456. En.wikipedia.org
457. En.wikipedia.org
458. En.wikipedia.org
459. En.wikipedia.org
460. Imdb.com
461. En.wikipedia.org
462. En.wikipedia.org
463. Imdb.com
464. UNICEF
465. En.wikipedia.org
466. En.wikipedia.org
467. En.wikipedia.org
468. En.wikipedia.org
469. En.wikipedia.org
470. En.wikipedia.org
471. En.wikipedia.org
472. En.wikipedia.org
473. En.wikipedia.org
474. Howitworks.com
475. Commons.wikimedia.org

476. En.wikipedia.org
477. En.wikipedia.org
478. En.wikipedia.org

BIBLIOGRAPHY

Abbot, J. "Climate Change – The Facts", Stockade Books, (2015).

Adkins, L. and Adkins, R. "Handbook to Life in Ancient Rome", Oxford University Press, (1998).

Albert, L. "Greek Mythology – The Gods, Goddesses, and Heroes Handbook", Adams Media, (2021).

Alden-Mason, J. "The Ancient Civilizations of Peru", Penguin Books, (1988).

Alexander, C. "The Bounty – The True Story of the Mutiny on the Bounty", Penguin Books, (2004).

Alexander, J. "Catherine the Great – Life and Legend", Oxford University Press, (1988).

Ancient-Greece.org. (2021).

Appel, D. "5 Mythological Stories About How the World Was Created", (2018).

Arendt, H. "The Origins of Totalitarianism", Franklin Classics Trade Press, (2018).

Asbridge, T. "The Crusades – The Authoritative History of the War for the Holy Land", Ecco, (2011).

Ashby, L. "With Amusement for All – A History of American Popular Culture Since 1830", The University Press of Kentucky, (2006).

Attenborough, D. "Life on Earth", William Collins, (2020).

Aurelius, M. "Meditations", Random House Publishing Group, (2003).

Avari, B. "India: The Ancient Past – A History of the Indian Subcontinent", Routledge, (2016).

Backman, C. "The Worlds of Medieval Europe", Oxford University Press, (2014).

Baird, F. and Heimbeck, R. "Philosophic Classics: Asian Philosophy", Routledge, (2010).

Barker, G. "The Agricultural Revolution in Prehistory – Why did Foragers become Farmers?", Oxford University Press, (2009).

Barrett, O. "10 Greatest Scientific Discoveries and Inventions of the 21st Century", (2018).

Bartusiak, M. "The Day We Found the Universe", Vintage, (2010).

Baudin, L. "Daily Life of the Incas", Dover Publications, (2011).

Bauer, S. "The History of the Ancient World", W.W. Norton & Company, (2007).

Beard, M. "SPQR – A History of Ancient Rome", Liveright, (2016).

Benn, C. "Daily Life in Traditional China: The Tang Dynasty", Greenwood, (2001).

Bennett, J. "The Top Ten Scientific Discoveries of the Decade", (2019).

Bergreen, L. "Columbus – The Four Voyages", Penguin Books, (2012).

BiblioCommons. "Early Astronomers: Ptolemy, Aristotle, Copernicus, and Galileo", (2020).

Brandon, S. "Creation Legends of the Ancient Near East", London, Hodder & Stoughton, (1963).

Broad, W. "The Oracle – Ancient Delphi and the Science Behind Its Lost Secrets", Penguin Books, (2007).

Brown, M. "Copernicus' Revolution and Galileo's Vision: Our Changing View of the Universe in Pictures", from The Conversation (2016).

Brownworth, L. "The Sea Wolves – A History of the Vikings", Crux Publishing Ltd, (2014).

Bulfinch, T. "Bulfinch's Mythology", Vintage Books, (2009).

Bunson, M. "Encyclopedia of Ancient Egypt", Gramercy, (1999).

Campbell, J. "The Hero with a Thousand Faces", New World Library, (2008).

Carter, H. "The Discovery of the Tomb of Tutankhamen", Dover Publications, (1977).

Cartwright, M. "Aztec Civilization", Ancient History Encyclopaedia, (2020).

Cartwright, M. "Inca Civilization", Ancient History Encyclopaedia, (2020).

Cartwright, M. "Mycenae", Ancient History Encyclopaedia, (2009).

Cashman, S. "America in the Gilded Age", NYU Press, (1993).

Chernow, R. "Washington – A Life", Penguin Books, (2011).

Chomsky, N. "On Language", New Press, (1998).

Christensen, B. "10 Southeast Asian Kingdoms You Need to Know About", from Real Clear History, (2018).

Christian, D. etal. "Big History: Between Nothing and Everything", McGraw-Hill Education, (2013).

Christian, D. "Maps of Time: An Introduction to Big History", University of California Press, (2011).

Clements, J. "Confucius – A Biography", Albert Bridge Books, (2017).

Cline, E. "The Oxford Handbook of the Bronze Age Aegean", Oxford University Press, (2012).

Coe, M. "Angkor and the Khmer Civilization", Thames & Hudson, (2005).

Coe, M. "The Maya – Ancient Peoples and Places", Thames & Hudson, (2015).

Cohn, N. "Cosmos, Chaos, and the World to Come", Yale University Press, (2001).

Contera, S. "Nano Comes to Life – How Nanotechnology is Transforming Medicine and the Future of Biology", Princeton University Press, (2019).

Coogan, S. and Smith, S. "Stories from Ancient Canaan", John Knox Press, (2012).

Cook, J. "The Things People Ask About the Scientific Consensus on Climate Change", from The Conversation (2016).

Cooney, K. "When Women Ruled the World – Six Queens of Egypt", National Geographic, (2020).

Cortes, H. "Five Letters of Cortes to the Emperor: 1519-1526", Conquistador Books, (2017).

Cowling, M. Bendemra, H. and Zobel, J. "It's Back to the Future Day Today – So What Are the Next Future Predictions?", from The Conversation (2015).

Dalley, S. "Myths from Mesopotamia: Creation, The Flood, Gilgamesh, and Others", Oxford University Press, (2012).

Dalton, D. "James Dean – The Mutant King: A Biography", Chicago Review Press, (2001).

D'Altroy, T. "The Incas", Wiley-Blackwell, (2014).

David, R. "Religion and Magic in Ancient Egypt", Penguin Books, (2003).

Davis, P. "100 Decisive Battles", Oxford University Press, (2001).

Deal, W. "Handbook to Life in Medieval and Early Modern Japan", Oxford University Press, (2007).

DeGrasse Tyson, N. "Death by Black Hole – And Other Cosmic Quandaries", W.W. Norton & Company, (2014).

De Landa, D. "Yucatan Before and After the Conquest", Dover Publications, (2012).

Desmond, A. "Darwin – The Life of a Tormented Evolutionist", W.W. Norton & Company, (1994).

De Souza, P. "The Greek and Persian Wars 499-386 BC", Osprey Publishing, (2003).

Diamandis, P. "The Future Is Faster Than You Think", Simon & Schuster, (2020).

Diamond, J. "Guns, Germs, and Steel – The Fates of Human Societies", W.W. Norton & Co. (1999).

Dierenfield, B. "The Civil Rights Movement", Routledge, (2008).

Dio, C. "Dio's Roman History", Forgotten Books, (2019).

Dolan, B. "Wedgewood – The First Tycoon", Viking Adult, (2004).

Donald, D. "Lincoln", Simon & Schuster, (1996).

Donovan, J. "20 Memorable Moments of the 21st Century So Far", (2019).

Duignan-Cabrera, A. and Chao, T. "The Top 10 Intelligent Designs (or Creation Myths), (2004).

Durant, W. "Caesar and Christ", Simon & Schuster, (1993).

Durant, W. "Our Oriental Heritage", Simon & Schuster, (1954).

Durant, W. "The Life of Greece", Simon & Schuster, (2011).

DW.com. "What is China's World order for the 21st Century?", (2020).

DW.com. "No End in Sight to the US-China Confrontation", (2020).

DW.com. "The Chinese Dream and Xi Jinping's Power Politics", (2020).

DW.com. "China Unveils Plans to Step Up Military Power", (2020).

DW.com. "China's Ambitious Silk Road Strategy", (2020).

DW.com. "China's Ambitious Bid for Southeast Asia Hegemony", (2020).

DW.com. "Opinion: China is looking to Challenge the US", (2020).

Ebrey, P. "The Cambridge Illustrated History of China", Cambridge University Press, (2010).

Einhard and Notker the Stammer. "Two Lives of Charlemagne", Penguin Classics, (2008).

Encyclopaedia Britannica. "Henrietta Swan Leavitt", Encyclopaedia Britannica, Inc. (2020).

Eraly, A. "The Age of Wrath – A History of the Delhi Sultanate", Penguin, (2015).

Everitt, A. "Cicero – The Life and Times of Rome's Greatest Politician", Random House, (2002).

Farrokh, K. "Shadows in the Desert", Osprey Publishing, (2007).

Frankfort, H. "Kingship and the Gods", University of Chicago Press, (1978).

Frankopan, P. "The Silk Roads – A New History of the World", Vintage, (2017).

Freeman, M. "Ancient Angkor", River Books, (2007).

Freeman, P. "Alexander the Great", Simon & Schuster, (2011).

FutureTimeline.net. (2021).

Gaddis, J. "The Cold War – A New History", Penguin Books, (2006).

Gammage, B. "The Biggest Estate on Earth – How Aborigines made Australia", Allen & Unwin, (2011).

Gates, W. "Bill Gates Article Written for Time Magazine – 7th February 2019."

Geach, J. "Stephen Hawking's PhD Thesis Crashed its Host Website – Here's What It Says in Simple Terms", from The Conversation (2017).

George, M. "The Autobiography of Henry VIII – With Notes by His Fool, Will Somers: A Novel", St. Martin's Griffin, (1998).

Gibbon, E. "The History of the Decline and Fall of the Roman Empire", Penguin, (2000).

Gilbert, M. "A History of the Twentieth Century – The Concise Edition of the Acclaimed World History", William Morrow Paperbacks, (2002).

Gleick, J. "Isaac Newton", Vintage, (2004).

Glenny, M. "The Fall of Yugoslavia – The Third Balkan War", Penguin Books, (1996).

Goldhagen, D. "Hitler's Willing Executioners", Vintage, (1997).

Goldsworthy, A. "Caesar – Life of a Colossus", Yale University Press, (2008).

Goodall, J. "In the Shadow of Man", Mariner Books, (2010).

Goodman, R. and Soni, J. "Rome's Last Citizen – The Life and Legacy of Cato, Mortal Enemy of Caesar", Griffin, (2014).

Gordin, M. "A Well-ordered Thing – Dimitri Mendeleev and the Shadow of the Periodic Table", Basic Books, (2004).

Graham, A. "Black Holes are even Stranger than You Can Imagine", from The Conversation (2017).

Grant, M. "History of Rome", Scribner's, (1978).

Grant, M. "Readings in the Classical Historians", Scribner's, (1993).

Grant, M. "The Climax of Rome", Weidenfeld, (1993).

Graves, R. "The Greek Myths", Penguin, (1993).

Grey, G. "Polynesian Mythology", Whitcombe and Tombs Ltd, (1956).

Halberstam, D. "The Coldest Winter – America and the Korean War", Hachette Books, (2008).

Hamilton, E. "Mythology", Scribner's Publishing, (1998).

Harari, Y. "Sapiens – A Brief History of Mankind", Harper Perennial, (2018).

Harari, Y. "Homo Deus – A Brief History of Tomorrow", Harper Perennial, (2018).

Harari. Y. "21 Lessons for the 21st Century", Random House Publishing Group, (2019).

Harvey, B. "Daily Life in Ancient Rome", Focus. (2016).

Hastings, M. "Vietnam – An Epic Tragedy, 1945-1975", Harper, (2018).

Hawking, S. "A Brief History of Time", Bantam, (1998).

Heidal, A. "The Babylonian Genesis", University of Chicago Press, (1952).

Herwig, W. "History of the Goths", University of California Press, (1988).

Hibbert, C. "The House of Medici – Its Rise and Fall", William Morrow Paperbacks, (1999).

Higgins, R. "Minoan and Mycenaean Art", Thames & Hudson, (1997).

High Energy Astrophysics Science Archive Research Centre – Cosmology (2020).

Hill, D. "Ancient Rome: From The Republic to The Empire", Parragon Books, (2009).

Hill, T. and Lewis, G. "Einstein's Theory of Gravity Tested by a Star Speeding Past a Supermassive Black Hole", from The Conversation (2018).

History.com. "Native American Cultures", (2020).

History.com. "Mycenae", (2020).

Hobbs, B. "General Relativity: How Einstein's Theory Explains the Universe, and More", (2015).

Holmes, G. "The Oxford History of Medieval Europe", Oxford University Press, (2012).

Holst, S. "Phoenician Secrets – Exploring the Ancient Mediterranean", Santorini Books, (2011).

Homer. "The Illiad", translated by Fagles, R. Penguin Classics, (1998).

Hough, R. "Captain James Cook – A Biography", W. W. Norton & Company, (1997).

Hughes, B. "The Hemlock Cup – Socrates, Athens and the Search for the Good Life", Vintage, (2012).

Hughes, D. "Ecology in Ancient Civilizations", University of New Mexico Press, (1975).

Ikram, S. "Death and Burial in Ancient Egypt", The American University in Cairo Press, (2015).

Inalcik, H. "The Ottoman Empire – The Classical Age 1300-1600", Phoenix, (2001).

Isaacson, W. "Einstein – His Life and Universe", Simon & Schuster, (2008).

Isaacson, W. "Leonardo da Vinci", Simon & Schuster, (2018).

Ives, E. "The Life and Death of Anne Boleyn", Wiley-Blackwell, (2005).

Jacobsen, T. "The Treasures of Darkness", Yale University Press, (1978).

James, T. "Ramses II", Friedman, (2002).

Kagan, D. "Pericles of Athens and the Birth of Democracy", Free Press, (1998).

Kaku, M. "The Future of Humanity – Terraforming Mars, Interstellar Travel, Immortality, and Our Destiny Beyond Earth", Doubleday, (2018).

Karakas, F. "Why You Should Read 'The Singularity is Near: When Humans Transcend Biology' by Raymond Kurzweil, (2010).

Kaye, H. "Thomas Paine and the Promise of Ame", Hill & Wang, (2006).

Keay, J. "India: A History", Grove Press, (2010).

Keay, J. "China: A History", Basic Books, (2011).

Keay, J. "The Honourable Company", Harper Collins, (2010).

Kelly, C. "The Roman Empire", Oxford University Press, (2006).

Khan Academy. "Edwin Hubble: Evidence for an Expanding Universe", (2020).

Khan Academy. "Claudius Ptolemy: An Earth-Centred View of the Universe", (2020).

Khan Academy. "Nicolaus Copernicus: A Sun-Centred View of the Universe", (2020).

Khan Academy. "Galileo Galilei: Father of Modern Observational Astronomy", (2020).

Koller, J. "Asian Philosophies", Prentice Hall, (2007).

Kramer, S. "The Sumerians: Their History, Culture and Character", University of Chicago Press, (1971).

Krauss, L. "A Universe from Nothing – Why There is Something Rather Than Nothing", Atria Books, (2013).

Kriwaczek, P. "Babylon: Mesopotamia and the Birth of Civilization", St. Martin's Griffin, (2012).

Kulke, H. and Rothermund, D. "A History of India", Barnes & Noble Books, (1995).

Kurzweil, R. "The Singularity Is Near – When Humans Transcend Biology", Penguin Books, (2006).

Larsen, K. "Stephen Hawking – A Biography", Prometheus, (2007).

Larson, E. "Summer for the Gods – The Scopes Trial and America's Continuing Debate over Science and Religion", Basic Books, (1997).

Lasky, P. "Second Detection Heralds the Era of Gravitational Wave Astronomy", from The Conversation (2016).

Leach, M. and Fried, J. "Funk & Wagnall's Standard Dictionary of Folklore, Mythology, and Legend", Harper & Row Publishers, (1984).

Leakey, M. "The Sediments of Time – My Lifelong Search for the Past", Houghton Mifflin Harcourt (2020).

Leakey, R. "Origins Reconsidered – In Search of What Makes Us Human", Doubleday, (1992).

Leick, G. "Mesopotamia – The Invention of the City", Penguin Books, (2003).

Lewis, G. "Melbourne Researchers Rewrite Big Bang Theory – Or Not", from The Conversation (2012).

Lewis, G. "Timeline: The History of Gravity", from The Conversation (2016).

Lewis, J. "The Mammoth Book of Eyewitness Ancient Rome", Running Press, (2003).

Lightbody, D. "The Great Pyramid – 2950 BC Onwards", Haynes Publishing UK. (2019).

Lindow, J. "Norse Mythology", Oxford University Press, (2002).

Lindsay, L. "Captives as Commodities – The Transatlantic Slave Trade", Pearson, (2007).

Livy, T. "The History of Rome", Penguin Classics, (2002).

Long, J. "Historical Dictionary of Hinduism", Rowman & Littlefield Publishers, (2010).

Mackie, G. "Expand into 2013 by Toasting 100 years of Modern Cosmology", from The Conversation (2012).

MacQuarrie, K. "The Last Days of the Incas", Simon & Schuster, (2008).

Mahbubani, K. "Has China Won? – The Chinese Challenge to American Primacy", Public Affairs, (2020).

Man, J. "Marco Polo – The Journey that Changed the World", William Morrow Paperbacks, (2014).

Man, J. "The Mongol Empire – Genghis Khan, His Heirs and the Founding of Modern China", Transworld Publishers, (2016).

Mann, C. "1491", Vintage Press, (2006).

Mann, W. "The Contender – The Story of Marlon Brando", Harper, (2019).

Mark, J. "Ancient China", Ancient History Encyclopaedia, (2012).

Mark, J. "Ancient Greece", Ancient History Encyclopaedia, (2013).

Mark, J. "Ancient Rome", Ancient History Encyclopaedia, (2009).

Mark, J. "Angkor Wat", Ancient History Encyclopaedia, (2020).

Mark, J. "Indus Valley Civilization", Ancient History Encyclopaedia, (2020).

Mark, J. "Maya Civilization", Ancient History Encyclopaedia, (2012).

Mark, J. "Mythology", Ancient History Encyclopaedia, (2018).

Mark, J. "Roman Empire", Ancient History Encyclopaedia, (2018).

Martin, T. "An Overview of Classical Greek History from Mycenae to Alexander", Yale University Press, (2013).

Massie, R. "Peter the Great – His Life and World", Random House Trade Paperbacks, (1981).

Matloff, G. "Stellar Engineering", Curtis Press, (2019).

McCoy, R. "Ending in Ice – The Revolutionary Idea and Tragic Expedition of Alfred Wegener", Oxford University Press, (2006).

McCullough, C. "Antony and Cleopatra – A Novel", Simon & Schuster, (2008).

McEwan, C. "Moctezuma", British Museum Press, (2009).

Mellor, R. "The Historians of Ancient Rome", Routledge, (2012).

Meyerson, D. "In the Valley of the Kings – Howard Carter and the Mystery of King Tutankhamun's Tomb", Brecourt Academic, (2009).

Miles, R. "Carthage Must Be Destroyed", Penguin Books, (2008).

Mitchell, S. "Bhagavad Gita", Harmony, (2002).

Montgomery, S. "Walking with the Great Apes – Jane Goodall, Dian Fossey, Birute Galdikas", Chelsea Green Publishing, (2009).

Mooney, J. "Myths of the Cherokee", Nineteenth Annual Report of the Bureau of American Ethnology for 1897-1898, (1900).

Moseley, M. "The Incas and Their Ancestors", Thames & Hudson, (2001).

Mowat, F. "Woman in the Mists – The Story of Dian Fossey and the Mountain Gorillas of Africa", Grand Central Publishing, (1988).

Mundle, A. "The First Fleet", ABC Books, (2014).

Nagle, B. "The Ancient World", Pearson, (2009).

Narayan, R. "The Mahabharata", University of Chicago Press, (2016).

O'Brien, J. and Major, W. "In the Beginning: Creation Myths from Ancient Mesopotamia, Israel, and Greece", Scholars Press, (1982).

Olmstead, A. "History of the Persian Empire", Chicago University Press, (2009).

O'Neill, G. "The High Frontier – Human Colonies in Space", Space Studies Institute Inc., (2013).

Opler, M. "Myths and Tales of the Jicarilla Apache Indians: Memoirs of the American Folklore Society", (1938).

O'Reilly, B. "The United States of Trump – How the President Really Sees America", Henry Holt and Co. (2019).

O'Reilly, D. "Early Civilizations of Southeast Asia", Alta Mira Press, (2006).

Ovid. "Metamorphoses", translated by Rolfe Humphries. Indiana University Press, (1955).

Pernoud, R. "Joan of Arc – By Herself and Her Witnesses", Scarborough House, (1990).

Philbrick, N. "The Last Stand – Custer, Sitting Bull, and the Battle of the Little Bighorn", Penguin Books, (2011).

Phillips, C. "Aztec and Maya – An Illustrated History", Lorenz Books, (2019).

Pimbblet, K. "The Fate of the Universe: Heat Death, Big Rip or Cosmic Consciousness?", from The Conversation (2015).

Pinch, G. "Egyptian Mythology: A Guide to the Gods, Goddesses, and Traditions of Ancient Egypt", Oxford University Press, (2004).

Plokhy, S. "The Last Empire – The Final Days of the Soviet Union", Basic Books, (2015).

Plubins, R. "Khmer Empire", Ancient History Encyclopaedia, (2013).

Polybius. "The Rise of the Roman Empire", Penguin Classics, (1980).

Potter, L. "The Life of William Shakespeare – A Critical Biography", Wiley-Blackwell, (2012).

Price, M. "Life May Be a Guide to the Evolution of the Cosmos – Here's How", from The Conversation (2017).

Redd, N. "Einstein's Theory of General Relativity", (2017).

Reichley, L. "Our Solar System – An Exploration of Planets, Moons, Asteroids, and Other Mysteries of Space", Rockridge Press, (2020).

Roberts, A. "Evolution – The Human Story", DK Publishing, (2018).

Roberts, A. "Napoleon – A Life", Penguin Books, (2015).

Robinson, A. "The Indus – Lost Civilizations", Reaktion Books, (2016).

Rothschild, N. "Wu Zhao: China's Only Woman Emperor", Pearson-Longman, (2008).

Rouhiainen, L. "Artificial Intelligence – 101 Things You Must Know Today About Our Future", Createspace Independent Publishing Platform, (2018).

Rubin, P. "Future Presence – How Virtual Reality Is Changing Human Connection, Intimacy, and the Limits of Ordinary Life", Harper One, (2018).

Sagan, C. "Cosmos", Carl Sagan Productions, Inc. (2010).

Sanders, N. "The Epic of Gilgamesh", Penguin Classics, (1960).

Saniotis, A. and Henneberg, M. "Taking Over from Evolution: How Technology Could Enhance Humanity", from The Conversation (2012).

Scarre, C. and Fagan, B. "Ancient Civilizations", Pearson, (2011).

Scarre, C. "Chronicle of the Roman Emperors – The Reign by Reign Record of the Rulers of Imperial Rome", Thames & Hudson, (2012).

Scheuer, M. "Osama Bin Laden", Oxford University Press, (2012).

Schiff, S. "Cleopatra – A Life", Back Bay Books, (2011).

Scott, J. "Against the Grain – A Deep History of the Earliest States", Yale University Press, (2018).

Seddon, C. "Humans: from the beginning – From the first apes to the first cities", Glanville Publications, (2014).

Shaw, I. "The Oxford History of Ancient Egypt", Oxford University Press, (2006).

Shepard, J. "The Cambridge History of the Byzantine Empire c.500-1492", Cambridge University Press, (2019).

Silent Sky. "3 Great Reasons the World Needs to Know About Henrietta Leavitt), (2020).

Sima, Q. "Records of the Grand Historian", Columbia University Press, (1995).

Singh, S. "Big Bang – The Origin of the Universe", Harper Perennial, (2005).

Slangen, A. and Church, J. "Burning Fossil Fuels is Responsible for Most Sea-Level Rise Since 1970", from The Conversation (2016).

Slotkin, R. "Regeneration through Violence: The Mythology of the American Frontier, 1600-1860", Middletown, CT, Wesleyan University Press, (1973).

Smith, H. "Virgin Land: The American West as Symbol and Myth", Harvard University Press, (1973).

Smith, R. "Myths and Legends of the Australian Aborigines", Dover Publications, (2003).

Smith, W. "Smaller Classical Dictionary", New York, E.P. Dutton, (1958).

Smolin, L. "The Life of the Cosmos", Oxford University Press, (1997).

Southey, M. "Explainer: What is the Human Genome Project?", from The Conversation (2012).

Spathari, E. "Mycenae", Hesperos, Athens, (2001).

Sperber, J. "Karl Marx – A Nineteenth Century Life", Liveright, (2013).

Spitzer, M. "The Musical Human – A History of Life on Earth", Bloomsbury Publishing, (2021).

Spoto, D. "Marilyn Monroe – The Biography", Cooper Square Press, (2001).

Stearns, P. "The Industrial Revolution in World History", Routledge, (2012).

Steinhardt, P. "Endless Universe – Beyond the Big Bang", Doubleday, (2007).

Stokes-Brown, C. "Big History: From the Big Bang to the Present", The New Press, (2012).

Stokes-Brown, C. "Henrietta Leavitt", from Khan Academy, (2020).

Stone, R. "Art of the Andes", Thames & Hudson, (2012).

Strauss, B. "The Death of Caesar – The Story of History's Most Famous Assassination", Simon & Schuster, (2016).

Strnad, S. "The Story of How the World Began: An Anthropological Analysis of Creation Mythology", Senior Honors Theses, State University of New York, (2013).

Sturluson, S. "Odin and Ymir", (1987).

Stutley, M. and Stutley, J. "Harper's Dictionary of Hinduism", Harper & Row, (1990).

Summers, M. "Exoplanets – Diamond Worlds, Super Earths, Pulsar Planets, and the New Search for Life Beyond Our Solar System", Smithsonian Books, (2018).

Swinburne University of Technology. "The Big Bang", (2020).

Sykes, R. "Kindred – Neanderthal Life, Love, Death and Art", Bloomsbury Sigma, (2020).

Tacitus. "Annals of Tacitus", Cambridge University Press, (2001).

Tanner, H. "China: A History, From Neolithic Cultures through the Great Qing Empire", Hackett Publishing Company, Inc., (2010).

Taylor, A. "Plato – The Man and His Work", Dover Publications, (2011).

Taylor, J. "Journey Through the Afterlife – Ancient Egyptian Book of the Dead", Harvard University Press, (2013).

The Daily Galaxy. "Life in the Cosmos – The Mystery that Contradicts Big Bang Theory", (2020).

Thompson, C. "Sea People – The Puzzle of Polynesia", Harper, (2019).

Thorndike, L. "The History of Medieval Europe", Perennial Press, (2018).

Thucydides. "History of the Peloponnesian War", Penguin Classics, (1972).

Thunberg, G. "No One Is Too Small to Make a Difference", Penguin Books, (2019).

Tooze, A. "Crashed – How a Decade of Financial Crises Changed the World", Penguin Books, (2019).

Treistman, J. "The Prehistory of China: An Archaeological Exploration", Doubleday & Company, (1972).

Turok, N. "The Universe Within – From Quantum to Cosmos", House of Anansi Press, (2012).

Tyldesley, J. "Cleopatra – Last Queen of Egypt", Basic Books, (2010).

Tzu, S. "The Art of War", Special Edition Books, (2009).

Van De Mieroop, M. "A History of the Ancient Near East ca. 3000-323 BC", Wiley-Blackwell, (2015).

Waines, D. "The Odyssey of Ibn Battuta – Uncommon Tales of a Medieval Adventurer" University of Chicago Press, (2010).

Waldrop, M. "Einstein's Relativity Explained in 4 Simple Steps", from National Geographic (2017).

Walls, J. and Walls, Y. "Classical Chinese Myths", Joint Publishing Company, (1984).

Walsh, B. "End Times – A Brief Guide to the End of the World", Hachette Books, (2019).

Walsh, T. Dowe, D. and Lea, G. "Your Questions Answered on Artificial Intelligence", from The Conversation (2015).

Waterfield, R. "Creators, Conquerors, and Citizens – A History of Ancient Greece", Oxford University Press, (2020).

Waterfield, R. "The First Philosophers", Oxford University Press, (2009).

Weidokal, M. "The Ten Most Important Geopolitical Events of the 21st Century", (2020).

Weir, A. "Henry VIII – King and Court", Ballantine Books, (2002).

Weir, A. "The Life of Elizabeth I", Ballantine Books, (1999).

Wikipedia. "History of Indigenous Australians", (2021).

Wikipedia. "History of the Pacific Islands", (2021).

Wilkinson, T. "The Rise and Fall of Ancient Egypt", Random House, (2013).

Wilson, D. "Charlemagne – A Biography", Doubleday, (2006).

Wolchover, N. "Physicists Debate Hawking's Idea That the Universe Had No Beginning", (2019).

Wolkstein, D. and Kramer, S. "Inanna: Queen of Heaven and Earth", Harper Perennial, (1983).

Woodard, C. "American Nations – A History of the Eleven Rival Regional Cultures of North America", Penguin Books, (2012).

Young, A. and Sauter, M. "3D Printing, E-Cigarettes Among the Most Important Inventions of the 21st Century", (2020).

Zimmerman, J. "Dictionary of Classical Mythology" Bantam Books, (1964).

INDEX

Made in the USA
Middletown, DE
21 July 2021

44556478R00457